Sweet Osmanthus Research
桂花研究

杨秀莲 王良桂 等 编著

东南大学出版社·南京

前　言

桂花是木犀科木犀属植物,也是我国传统名花之一,为我国重要的观赏树种,集绿化、美化、香化于一身,并兼有生态效应、社会效应和经济效应。它树姿典雅,四季常青,碧叶如云,金秋时节繁花满树,桂蕊飘香,沁人心脾。其花味浓香甜,可用于化工、食品工业,利用价值很高。

作者所在研究团队长期致力于桂花种质资源收集评价、新优品种选育、繁育技术以及分子生物学等方面的研究,取得了一定的成果,在国内外各类刊物上发表论文40余篇。今将所发表论文重新整理和补充,编纂成《桂花研究》。

全书根据研究内容分为分子生物学研究、生理学研究、引种育种、繁殖栽培技术和应用研究5个部分。其中分子生物学研究论文9篇,生理学研究论文18篇,引种育种论文5篇,繁殖栽培技术论文6篇,应用研究论文2篇。研究团队早期主要以生殖生理和繁殖栽培研究为主,近年来,侧重于分子水平的研究,重点进行了桂花花色、花香等功能基因的挖掘和验证。目前,课题组已完成桂花的全基因组测序,相关文章已经在国际园艺学科期刊公开发表。

《桂花研究》的出版,是研究团队对以往研究成果的总结,也是希望借此机会和同行进行学术交流和探讨,促进桂花研究的进一步深入。由于此文集是黑白印刷,原文彩图中有些信息无法清晰表达,请查阅原文发表期刊或 web of science 或 ScienceDirect 网站,带来不便,还望读者见谅。

感谢研究团队成员完成了桂花的相关研究和论文的发表!

感谢在编著过程中,参与论文收集、编纂和文字校对的岳远征副教授、施婷婷讲师以及团队的博士、硕士研究生!

感谢江苏高校品牌专业建设工程项目(PPZY2015A063)、江苏高校优势学科建设工程项目的资助。

杨秀莲　王良桂

2018年10月

目 录

第一部分 分子生物学

The Chromosome-Level Quality Genome Provides Insights into the Evolution of the Biosynthesis Genes for Aroma Compounds of *Osmanthus fragrans* ………………… 3

Cloning and Expression Analysis of Three Critical Triterpenoid Pathway Genes in *Osmanthus fragrans* ……………………………………………………………… 28

Transcriptomic Analysis of the Candidate Genes Related to Aroma Formation in *Osmanthus fragrans* …………………………………………………………… 45

Identification and Validation of Reference Genes for Gene Expression Studies in Sweet Osmanthus (*Osmanthus fragrans*) Based on Transcriptomic Sequence Data …………………………………………………………………… 67

Cloning and Expression Analysis of MEP Pathway Enzyme-Encoding Genes in *Osmanthus fragrans* ……………………………………………………………… 81

桂花(＋)-新薄荷醇脱氢酶基因 *OfMNR* 的克隆与表达分析 …………………… 103

桂花(*Osmanthus fragrans* Lour.)4-香豆酸辅酶 A 连接酶(4CL)基因克隆与 表达分析 ……………………………………………………………………… 110

两个桂花品种花色色素相关基因的差异表达 …………………………………… 118

Transcriptome Sequencing and Analysis of Sweet Osmanthus (*Osmanthus fragrans* Lour.) ………………………………………………… 125

第二部分 生理学研究

'波叶金桂'花香成分释放规律 …………………………………………………… 147

3 个四季桂品种花瓣挥发性成分的 GC-MS 分析 ……………………………… 161

11 个桂花品种花瓣与叶片中矿质元素含量的比较 …………………………… 170

'紫梗籽银'桂种子内源抑制物质分析 …………………………………………… 177

3 个桂花品种对 NaCl 胁迫的光合响应 ………………………………………… 184

'朱砂丹桂'扦插技术及生根过程中生理生化分析 ……………………………… 191

不同桂花品种香气成分的差异分析 ……………………………………………… 198

冷藏对 5 个桂花品种主要营养成分的影响 …………………………………… 206

桂花花朵香气成分的研究进展 ………………………………………………………………… 211
保存方法对桂花精油提取及香气成分的影响 ……………………………………………… 222
两种植物生长延缓剂对盆栽'日香桂'的矮化效应 ………………………………………… 230
'波叶金桂'扦插生根过程中营养物质和激素含量变化 …………………………………… 238
25 个桂花品种花瓣营养成分分析 …………………………………………………………… 245
桂花花瓣营养成分分析 ……………………………………………………………………… 254
桂花种子对赤霉素处理的生理生化响应 …………………………………………………… 260
淹水对 2 个桂花品种生理特性的影响 ……………………………………………………… 267
'紫梗籽银'桂种子休眠原因的初步探讨 …………………………………………………… 273
桂花品种切花瓶插衰老生理研究 …………………………………………………………… 278

第三部分　引种育种

NaCl 胁迫对 5 个桂花品种叶片超微结构的影响 …………………………………………… 287
^{60}Co-γ 辐射对桂花种子萌发及幼苗生长的影响 ………………………………………… 303
'晚籽银桂''多芽金桂'花芽的形态分化 …………………………………………………… 314
桂花花粉活力测定与'晚籽银桂'柱头可授性分析 ………………………………………… 321
'晚籽银桂'胚和胚乳发育的研究 …………………………………………………………… 327

第四部分　繁殖栽培技术

覆盖物对土壤性质及微生物碳氮的影响 …………………………………………………… 333
Effects of Mulching on Soil Properties and Growth of Tea Olive
　　(Osmanthus fragrans) ……………………………………………………………………… 343
桂花花梗与叶片的愈伤组织诱导研究 ……………………………………………………… 356
控根栽培下桂花根系的动态生长与垂直分布特征 ………………………………………… 368
桂花不同品种的种子形态比较 ……………………………………………………………… 374
桂花种子休眠和萌发的初步研究 …………………………………………………………… 381

第五部分　应用研究

桂花露酒浸提及营养成分研究 ……………………………………………………………… 389
桂花专类园植物景观综合评价 ……………………………………………………………… 401

第一部分
分子生物学

The Chromosome-Level Quality Genome Provides Insights into the Evolution of the Biosynthesis Genes for Aroma Compounds of *Osmanthus fragrans* [*]

Xiulian Yang[1,2,†], Yuanzheng Yue[1,2,†], Haiyan Li[1,2], Wenjie Ding[1,2], Gongwei Chen[1,2], Tingting Shi[1,2], Junhao Chen[3], Min S. Park[4], Fei Chen[3], Lianggui Wang[1,2]

1. Co-Innovation Center for Sustainable Forestry in Southern China, Nanjing Forestry University, Nanjing 430073, China
2. College of Landscape Architecture, Nanjing Forestry University, Nanjing 350002, China
3. State Key Laboratory of Ecological Pest Control for Fujian and Taiwan Crops, Fujian Agriculture and Forestry University, Fuzhou 350002, China
4. Nextomics Bioscience Institute, Wuhan 430073, China

† These authors contributed equally to this work and should be considered co-first authors.

Abstract: Sweet osmanthus (*Osmanthus fragrans*) is a very popular ornamental tree species throughout Southeast Asia and USA particularly for its extremely fragrant aroma. We constructed a chromosome-level reference genome of *O. fragrans* to assist in studies of the evolution, genetic diversity, and molecular mechanism of aroma development. A total of over 118 Gb of polished reads was produced from HiSeq (45.1 Gb) and PacBio Sequel (73.35 Gb), giving 100X depth coverage for long reads. The combination of Illumina-short reads, PacBio-long reads, and Hi-C data produced the final chromosome quality genome of *O. fragrans* with a genome size of 727 Mb and a heterozygosity of 1.45%. The genome was annotated using *de novo* and homology comparison and further refined with transcriptome data. The genome of *O. fragrans* was predicted to have 45,542 genes, of which 95.68% were functionally annotated. Genome annotation found 49.35% as the repetitive sequences, with long terminal repeats (LTR) being the richest (28.94%). Genome evolution analysis indicated the evidence of whole-genome duplication 15 million years ago, which contributed to the current content of 45,242 genes. Metabolic analysis revealed that linalool, a monoterpene is the main aroma compound. Based on the genome and transcriptome, we further demonstrated the direct connection between terpene synthases (TPSs) and the rich aromatic molecules in *O. fragrans*. We identified three new flower-specific *TPS* genes, of which the expression coincided with the production of linalool. Our results suggest that the high number of *TPS* genes and the flower tissue- and stage-specific *TPS* genes expressions might drive the strong unique aroma production of *O. fragrans*.

Key words: *Osmanthus fragrans*; aroma compounds; chromosome-level; genome; terpene synthases

1 Introduction

Sweet osmanthus (Dicotyledons, Lamiales, Oleaceae, *Osmanthus*) is one of the most

[*] 原文发表于 *Horticulture Research*, 2018, 5:72。

popular, evergreen ornamental tree species in China due to its unique sweet aroma[1-2]. More than 160 cultivars of O. fragrans have been classified based on phenotypes such as the leaf shape, flower color, aroma, season and frequency of flower blooming[3]. The association between phenotypes and genotypes of O. fragrans has been examined through aroma compounds[4-6], essential oils[7-10], and taxonomy using various molecular markers[11-16]. Transcriptome studies have determined the genes that might be responsible for the emission of flower scent in O. fragrans[17-18]. Gene expression has also been modulated at different flowering stages of O. fragrans[19]. Differential gene expression studies have identified genes in the mediated isopentenol production (MEP) pathway, as well as the terpenoid- and carotenoid-synthesis pathways. Transcriptomics studies allowed researchers to make connections between the major flower aroma compounds, and differentially expressed genes and encoded proteins. The flower aroma compounds, (R)- and (S)-linalool are produced by a terpene synthase(s) (TPS)[20]. Another key flower aroma compound, β-ionone, is produced through the oxidative cleavage of β-carotene by carotenoid cleavage enzymes (CCD)[21-23]. Transcriptome studies have shown that TPS(s) and CCD(s) are differentially expressed at different flowering stages in O. fragrans[24-25]. Additionally, we and others have recently reported a set of transcription factors (TFs) associated with the expression of color and the emission of fragrance in O. fragrans[21,26-27]. All of these gene expression studies provide valuable insights on how flower blooming and aroma production are interlinked[28]. However, a genome sequence is largely needed to reveal the full genetic background of aroma production in sweet osmanthus and the evolution of aroma in family Oleaceae.

In this study, we generated a reference genome for O. fragrans to provide a solid foundation for our future understanding of the genome structure and evolution of the Oleaceae family. Furthermore, we conducted a detailed analysis of the aroma compounds, tissue and flowering time-specific differential gene expression to investigate the molecular mechanisms of sweet fragrance development in O. fragrans.

2 Results

2.1 Sequencing Summary

We generated 100-fold PacBio single molecule long reads (a total of 73.4 Gb with an N50 length of 13.0 kb), 77-fold k-mer depth Illumina paired-end short reads (45.1 Gb) and Hi-C data that produced 23 unambiguous chromosome scaffolds for a high-quality assembly. For stepwise assembly, we first performed an initial PacBio-only assembly, resulting in an assembly size of 733.5 Mb and a contig N50 of 1.59 Mb, and the assembled genome had a highly complete BUSCOs (96.1%) (Tab. S1). Then, the initial contigs were subsequently polished with PacBio long reads and Illumina short reads. As the final step, Hi-C data was used to polish the scaffolds generated by the PacBio and Illumina reads.

2.2 Determination of Genome Size and Heterozygosity

The k-mer method[29] and KmerFreq_AR[30] were used to determine the genome size of *O. fragrans* using the quality-filtered reads of Illumina data. The genome size was estimated based on the formula: Genome size = Modified k-mer number/Average k-mer depth, where Modified k-mer = Total k-mer number − Error k-mer number and the Average k-mer depth obtained from the main peak of k-mer distribution curve (Fig. S1). To determine the heterozygosity, *Arabidopsis* genome data was used to simulate Illumina PE reads, which was carried out by using pIRS software[31]. Then, a fitting KmerFreq_AR[30] was developed using the k-mer distribution curve of *O. fragrans*. When the two curves k-mer curves were consistent, the heterozygosity of *Arabidopsis* was considered the reference for the heterozygosity of *O. fragrans*. The final analysis produced ~1.45% heterozygosity of the *O. fragrans* genome.

2.3 Genome Assembly and Quality Assessment

The integrated work-flow of genome assembly is shown (Fig. S2). The full PacBio long reads were converted to FASTA format. Then, all subreads of genome data were assembled using Falcon v0.3.0[32] with specific parameters (length-cutoff pr = 8 kb, length-cutoff pr = 9 kb). We used Arrow (https://github.com) to polish the draft genome (G1) to obtain the corrected genome (G2). Then, G2 was polished again by Pilon[33], which mapped the next-generation sequencing data to G2 with bwa to obtain the twice-corrected genome (G3). The *O. fragrans* genome had high heterozygosity, which led to a G3 size larger than the estimation. To acquire the nonredundant genome, heterozygous and redundant sequences were removed from the corrected genome using Redundans[34] with the following parameters: heterozygosity = 0.0145 and sequencing depth = 86. The nonredundant genome (G4) was approximately 741 Mb, with a contig N50 size of 1.595 Mb (Tab. 1). Finally, BUSCO v3.0 analysis[35] was performed to assess G4 using the embryophyta_odb10 database with default parameters.

Tab. 1　Quality assessment statistics of the assembled genome of *O. fragrans*

Stat Type	Contig length/bp	Contig number/bp
N50	1,595,720	145
Longest	8,253,028	1
Total	740,635,307	774
Length>5 kb	740,625,951	765

The clustering of contig by hierarchical clustering of the Hi-C data was performed. Through a comparative analysis, the only pair of reads around the DpnII digestion site was determined. Hi-C linkage was used as a criterion to measure the degree of tightness of the association between different contigs by standardizing the digestion sites of DpnII on the genome sketch. Agglomerative hierarchical clustering and LACHESIS produced chromosome assembly maps with a karyotype of $2n=46$ (Fig. 1). As a result, the total number of contigs of the *O. fragrans* genome map was 5,327, and the total length was 740,635,307 bp. The combined length of Hi-C contigs was 740,404,543 bp, accounted for 99.97% of the total length of the final assembled genome, indicating the high quality of Hi-C data (Tab. 2).

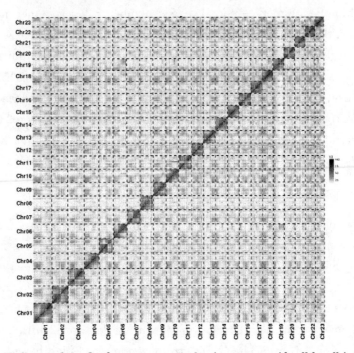

Fig. 1 Hi-C map of the *O. fragrans* genome showing genome-wide all-by-all interactions

The map shows a high resolution of individual chromosomes that are scaffolded and assembled independently.

Tab. 2 Summary statistics demonstrating the high quality of the Hi-C map of *O. fragrans*

Sample	*O. fragrans*
Draft contig total number	5,327
Draft contig total length	740,635,307
Final contig total number	5,305
Final contig total length	740,404,543
Contig coverage on full genome/%	99.97

2.4 Annotation of Repeat Sequences

The genome of *O. fragrans* had simply, moderately- and highly-repetitive sequences. Microsatellite was used to identify the repeat sequences in the genome of *O. fragrans* (MISA, RRID: SCR 010765). A total of 409,691 SSRs was obtained, including 305,868 mono-, 70,587 di-, 25,544 tri-, 3,934 tetra-, 2,081 penta-, and 1,991 hexa-nucleotide repeats, respectively (Tabs. S2 – S3). The tandem repeats finder (TRF, v4.07b)[36] identified over 400,000 tandem repeats, accounting for 0.076% of the *O. fragrans* genome.

We used homology-based and *de novo* approaches to identify transposable elements. RepeatMasker[37] was used to search against the Repbase (v. 22.11)[38] and Mips-REdat libraries[39]. Then, we used RepeatMasker v4.0.6 to search the *de novo* repeat library that we built using RepeatModeler v1.0.11 (RepeatModeler, RRID: SCR 015027). Finally, TEs were confirmed by searching the TE protein database using a RepeatProteinMask and WU-BLASTX. The repetitive sequence was 49.35%, of which LTR accounted for 28.49% of the assembled genome of *O. fragrans* (Tab. S4).

2.5 Annotation of Noncoding RNA (ncRNA)

We identified rRNA, miRNA and snRNA genes in the *O. fragrans* genome by searching the Rfam database (release 13.0)[40], using Blastn[41] (E-value$\leqslant 10^{-5}$). Software tRNAscan-SE (v1.3.1)[42] and RNAmmer v1.2[43] were used to predict tRNAs and rRNAs, resulting in an *O. frangrans* genome with 525 miRNAs, 847 tRNAs, 49 rRNAs, and 2,058 snRNAs (Tab. S5).

2.6 Gene Prediction

The protein-coding genes were identified using homology-based and *de novo* predictions-based approaches. The *O. fragrans* genome was mapped against the published sequences of *Arabidopsis thaliana*, *Olea europaea*, *Sesamum indicum*, *Solanum tuberosum*, and *Vitis vinifera*. To accurately identify spliced alignments, we used GeneWise v2.2.0[44] to filter all initially aligned coding sequences. For *de novo* prediction, the data from NGS and the full-length transcriptomes were analyzed with Hisat2-2.1.0 and PASApipeline-2.0.2 to predict the complete gene set. We randomly selected 1,000 genes to train the model parameters for Augustus v3.3[36], GeneID v1.4.4[45], GlimmerHMM[46] and SNAP[47]. The final consensus gene set was generated using EVidenceModeler (EVM) v1.1.1[48], which combined the genes predicted by the *de novo* and homology searches. The assembled genome had 45,542 genes with an average transcript length of 4,065 bp, an average CDS length of 1,142 bp, and a number of exons per gene of 5 (Tab. S6).

The functional validity of the predicted genes was further evaluated by searching the UniProt (release 2017_10), KEGG (release 84.0), and InterPro (5.21-60.0) databases using Blastall44, KAAS49, and InterProScan50. As a result, we were able to assign potential functions to 43,573 protein-coding genes out of the total of 45,542 genes in the *O. fragrans* genome (95.68%) (Tab. S7).

2.7 Genome Evolution

Gene family analysis. Although morphological investigation and a number of genes have placed *O. fragrans* in the Oleaceae family, there is still no whole genome-scale phylogenomic analysis of the evolutionary position of *O. fragrans*. Here, we compared the *O. fragrans* genome with the genome sequences of 11 other plants (*A. thaliana*, *Fraxinus excelsior*, *Glycine max*, *O. europaea*, *Oryza sativa*, *Petunia axillaris*, *Petunia inflata*, *Prunus mume*, *Rosa chinensis*, *Solanum lycopersicum*, and *V. vinifera*). We applied the OrthoMCL (v2.0.9) pipeline[51] (Blastp E-value $\leqslant 10^{-5}$) to identify the potential orthologous gene families between the genomes of these plants. Gene family clustering identified 17,513 gene families consisting of 38,808 genes in *O. fragrans*, of which, 1,086 gene families were unique to *O. fragrans*, *O. europaea*, and *F. excelsior* had the biggest number of shared gene families among these plants (Fig. 2).

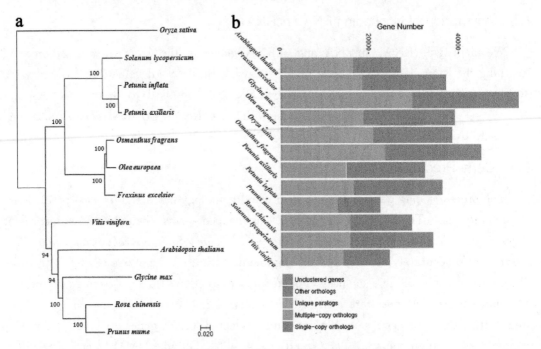

Fig. 2 Species tree and evolution of gene numbers

(a) The phylogenetic tree showing the close relationship between sweet osmanthus and the wild olive (*O. europaea*); (b) The number of genes in various plant species, showing the high gene number of *O. fragrans* compared to a model (*A. thaliana*) and other tree species. The number of multiple copy paralogs is high in *O. fragrans*.

Synteny analysis. We used the protein sequences of *O. fragrans* that were aligned against each other with Blastp (E-value$\leqslant 10^{-5}$) to achieve the conserved paralogs, Then, MCScanX (http://chibba.pgml.uga.edu/mcscab2) was used to find the collinearity block in the genome. Using the Circos tool (http://www.circos.ca), we mapped and gene density, GC content, Gypsy density, and Copia density, as well as the average expression value of genes expressed in flowers on individual chromosomes (Fig. 3).

Whole-genome duplication (WGD). To determine the source of the high number of genes (>45,000) in *O. fragrans*, the WGD events were analyzed by taking advantage of the high-quality genome of *O. fragrans*. We applied four-fold synonymous third-codon transversion (4DTv) and synonymous substitution rate (Ks) estimation to detect the WGD events. First, respective paralogous of *O. fragrans*, *G. max*, *O. europaea*, *V. vinifera*

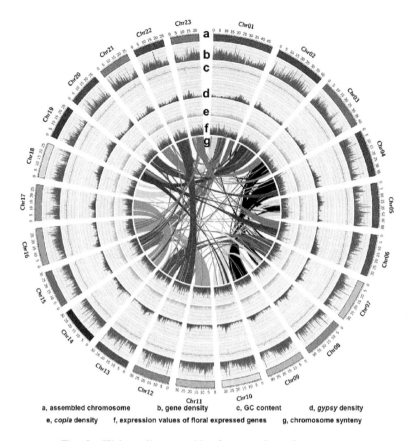

a, assembled chromosome   b, gene density   c, GC content   d, *gypsy* density
e, *copia* density   f, expression values of floral expressed genes   g, chromosome synteny

Fig. 3 High-quality assembly of twenty-three chromosomes

and *A. thaliana* were identified with OrthoMCL. Then, the protein sequences of these plants were aligned against each other with Blastp (E-value $\leqslant 10^{-5}$) to achieve the conserved paralogs of each plant. Finally, the WGD events of each plant were evaluated based on their 4DTv (Fig. 4a) or Ks (data not shown) distribution. The WGD analysis suggestted that *O. fragrans*, *G. max* and *O. europaea* experienced WGD events within less than 15 MYA, but *V. vinifer* and *A. thaliana* have not experienced WGD events recently (Fig. 4a). We also compared the number of duplicated genes (Fig. 4b), the chromosome-level duplications (Fig. 4c), and the number of a functional homologs of glycotransferase and bHLH-Myc transcription factor genes between *O. fragrans* and *V. vinifera* (Fig. 4d), further validating the WGD events.

2.8 Determination of Volatile Aroma Compounds

To make a direct connection between the biosynthetic genes and flower fragrance development, we determined the volatile aroma compounds. Headspace-SPME combined with GC-MS analysis identified over 40 volatile compounds, including linalool, dihydrojasmone lactone [2(3H)-furanone, 5-hexyldihydro-], 1-cyclohexene-1-propanol, 2,6,6-tetramethyl-, and β-ocimen as the major components. Linalool was present in the highest

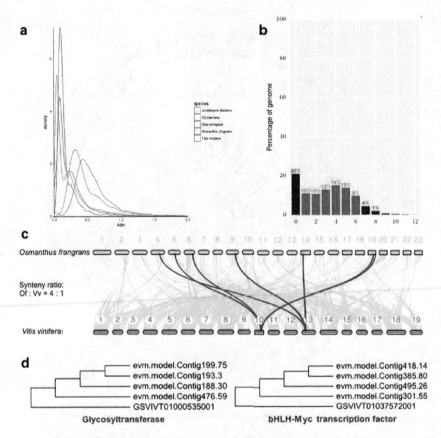

Fig. 4 Evidences for whole genome duplication events in *O. fragrans*

(a) 4DTv; (b) the most abundant genes (red bar, duplicated genes; green, nonduplicated; 0, tandem-duplicated or small-scale-duplicated genes; 0~12, number of duplicated copies) in a four-fold relationship when comparing *O. fragrans* and grapes; (c) a colinear relationship at the chromosome level; (d) four-fold expansion of the functional homologs of glycosyltransferase and bHLH-Myc genes in grape (GSVIVT) and *O. frgrans* (evm. model. Contig). The data suggest that *O. fragrans* experienced two WGD events.

amount at the early flowering stage (S1) and decreased afterwards (Tab. 3).

2.9 Expression Analysis

We also produced comprehensive transcriptome dataset using both HiSeq and the Iso-Seq pipeline. We focused our further analysis on identifying the specific genes responsible for floral development and the biosynthesis of volatile aroma compounds in *O. fragrans*. The members of MADs transcription factors that control plant development were highly expressed in all tissues tested. Among them, AG, AP3/PI, AP1, and SEP were predominantly expressed in the early flower stage (S1), whereas, the expression level of the *ANR*1 gene family was highly specific to the root tissue (Fig. 5b). Interestingly, the numbers of ABCE genes were higher than that of *Fraxinus chinensis*, a close relative of *O. fragrans* (Fig. 5a).

Tab. 3 Identity and quantity of volatile aroma compounds in the various flowering stages of *O. fragrans*

Fragrance molecules	S1(A)	P-value (A&B)	S2(B)	P-value (B&C)	S3(C)	P-value (A&C)
Linalool	6.67±1.63	0.001	2.91±2	0.007	0.22±0.15	0.000
2(3H)-Furanone, 5-hexyldihydro-	2.58±0.62	0.000	1.29±0.28	0.000	—	0.000
1-Cyclohexene-1-propanol, à,2,6,6-tetramethyl-	2.29±0.63	0.000	0.01±0.01	0.975	—	0.000
β-Ocimene	1.28±0.42	0.002	0.68±0.26	0.001	0.02±0.02	0.000
2-Furanmethanol, 5-ethenyltetrahydro-α,α,5-trimethyl-, *cis*-	1.25±0.58	0.213	0.82±0.21	0.010	1.79±0.77	0.123
trans-Linalool oxide (furanoid)	1.18±0.57	0.296	0.78±0.25	0.001	2.34±0.91	0.007
(3R,6S)-2,2,6-Trimethyl-6-vinyltetrahydro-2H-pyran-3-ol	0.7±0.26	0.050	0.41±0.11	0.003	0.6±0.29	0.162
α-Ionone	0.5±0.18	0.000	0.18±0.1	0.448	0.13±0.04	0.000
2-Butanone, 4-(2,2-dimethyl-6-methylenecyclohexyl)-	0.37±0.13	0.000	0.01±0	1.000	0.01±0	0.000
(3R,6R)-2,2,6-Trimethyl-6-vinyltetrahydro-2H-pyran-3-ol	0.35±0.13	0.038	0.2±0.05	0.000	0.51±0.15	0.033
2H-Pyran-3(4H)-one, 6-ethenyldihydro-2,2,6-trimethyl-	0.33±0.11	0.276	0.22±0.06	0.052	0.43±0.28	0.342
Butanoic acid, 3-hexenyl ester, (E)-	0.14±0.05	0.284	0.1±0.07	0.891	0.1±0.06	0.346
2,4,6-Octatriene, 2,6-dimethyl-, (E,E)-	0.1±0.03	0.055	0.06±0.02	0.253	0.08±0.04	0.387
Methyl salicylate	0.08±0.08	0.881	0.07±0.06	0.046	—	0.034
NA1	0.08±0.03	0.016	0.05±0.02	0.002	—	0.000
3-Buten-2-one, 4-(2,6,6-trimethyl-1-cyclohexen-1-yl)-	0.06±0.02	0.000	1.18±0.58	0.892	1.15±0.28	0.000
NA3	0.06±0.01	0.002	0.03±0.02	0.000	—	0.000
3-Hexen-1-ol, acetate, (Z)-	0.05±0.03	0.899	0.04±0.04	0.003	0.4±0.31	0.004
cis-3-Hexenyl isovalerate	0.05±0.02	0.224	0.03±0.02	0.030	0.06±0.02	0.278

Cont. Tab. 3

Fragrance molecules	S1(A)	P-value (A&B)	S2(B)	P-value (B&C)	S3(C)	P-value (A&C)
trans-β-Ocimene	0.05±0.01	0.002	0.03±0.01	0.000	—	0.000
Methyl isovalerate	0.04±0.03	0.376	0.03±0.01	0.898	0.03±0.02	0.314
(E)-4,8-Dimethylnona-1,3,7-triene	0.04±0.02	0.046	0.02±0	0.011	—	0.000
Hexanoic acid, methyl ester	0.03±0.02	1.000	0.03±0.01	0.011	—	0.011
cis-3-Hexenyl-α-methylbutyrate	0.03±0.01	0.091	0.02±0.01	0.091	0.03±0.01	1.000
3-Buten-2-ol, 4-(2,6,6-trimethyl-2-cyclohexen-1-yl)-, (3E)-	0.03±0.01	0.000	0.01±0.01	0.119	—	0.000
Octanoic acid, methyl ester	0.02±0.01	0.121	0.01±0	0.592	0.01±0.01	0.045
Tridecane	0.02±0.01	0.015	0.01±0	0.291	0.01±0.01	0.121
trans-β-Ionone	0.02±0.01	0.015	0.01±0	0.000	—	0.000
Megastigma-4,6(Z),8(E)-triene	0.02±0.01	0.000	—	1.000	—	0.000
2-Butanone, 4-(2,6,6-trimethyl-1-cyclohexen-1-yl)-	0.02±0.01	0.010	0.28±0.13	0.001	0.67±0.23	0.000
(3E,7E)-4,8,12-Trimethyltrideca-1,3,7,11-tetraene	0.02±0	0.173	0.03±0.01	0.000	—	0.000
Hexadecane	0.01±0.01	0.072	0.01±0	0.000	—	0.000
Butanoic acid, methyl ester	—	0.000	0.03±0.01	0.000	—	1.000
1-Butanol, 2-methyl-, acetate	—	1.000	—	0.000	0.06±0.03	0.000
NA2	—	1.000	—	0.000	0.01±0.01	0.000
Dodecane	—	1.000	—	0.000	0.02±0.01	0.000
NA4	—	1.000	—	0.000	0.02±0.01	0.000
2H-Pyran-3-ol, 6-ethenyltetrahydro-2,2,6-trimethyl-, acetate, trans-	—	0.000	—	0.000	0.02±0.01	0.000
1-Cyclohexene-1-ethanol, 2,6,6-trimethyl-	—	0.000	0.01±0	0.000	—	1.000

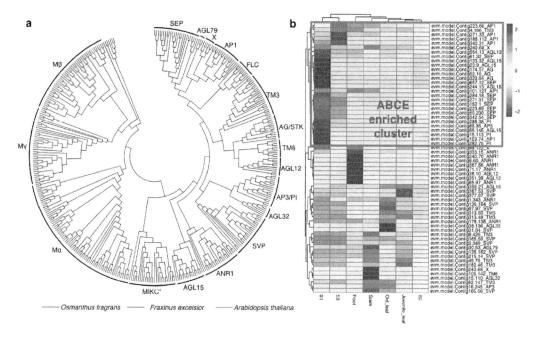

Fig. 5 MADS-box gene family in *O. fragrans*

(a) The evolution tree and expression values of the MADSs box genes; (b) Heatmap showing the tissue- and flowering stage- specific expressions of the members of ABCE genes.

The major component of the volatile compounds in the floral scent of *O. fragrans*, linalool (Fig. 6), is known to be synthesized by terpenoid synthetases (TPS). Therefore, we compared the expression profiles of *TPS* genes and identified over 40 genes that contain the functional motifs of TPS. Differential gene expression (DGE) analysis identified 7 *TPS* genes that are highly expressed in flowers, compared to roots, leaves, and stems (Fig. 7).

3 Discussion

Sweet osmanthus is one of the most beloved ornamental tree species in China and other parts of the world and has been cultivated for over two-thousand years in China due to its attractive traits of beautiful colors, unique aromas, a long flowering season, and medicinal efficacy. However, there is a limited number of studies that have investigated the genetic basis of the phenotypic diversity of sweet osmanthus. Recently, a set of genetic markers was identified[14-15], and an effort to construct a genetic linkage map was reported[16]. Additionally, several transcriptomics studies identified a large number of genes that are differentially expressed in some of the cultivars with attractive traits[17-18]. While these studies indirectly associate the diverse phenotypes with the genotypes of sweet osmanthus, there is no genome information that can directly link the specific genes to particular traits. Thus, we have sequenced, assembled and annotated the genome of sweet

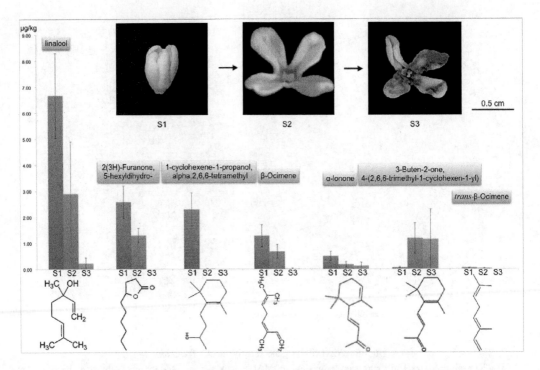

Fig. 6 The top 7 secondary metabolites produced by the osmanthus flower measured by GC-MS

Note that the top molecule is linalool, a monoterpene, which was produced most in the S1 stage of the flower, and then decreased its amount in the flower. S1, S2, and S3 stand for the early, middle, and late stages, respectively.

osmanthus. Furthermore, combining HiSeq- and IsoSeq-based transcriptome analyses, we gained deep insight into the genes that control aroma compounds synthesis in the flowers of *O. fragrans*.

3.1 The High-Quality Reference Genome Provides Deep Insights to the Evolution of *O. fragrans*

Currently, there are still no comprehensive analyses combining genomic, transcriptomic, and metabolic approaches to reveal the unique aroma of *O. fragrans*. Despite advances in second-generation sequencing, it is still very challenging to construct a high-quality plant genome due to the high complexity, large size, and high percentage of repeats and polyploidy. Therefore, we combined the second-generation short read to achieve high accuracy, the third-generation long reads for *de novo* assembly, and Hi-C to scaffold contigs into a chromosome-scale assembly. To guarantee a high-quality genome annotation, we combined *de novo*, homology-based, and experimental evidence obtained from the extensive transcriptomics data, including the full-length transcripts. We constructed a reference-quality genome that produced an unambiguous chromosome-scale assembly ($n=23$) and functionally annotated 43,573 genes out of the complete set of 45,542 genes of *O. fragrans* (95.68%).

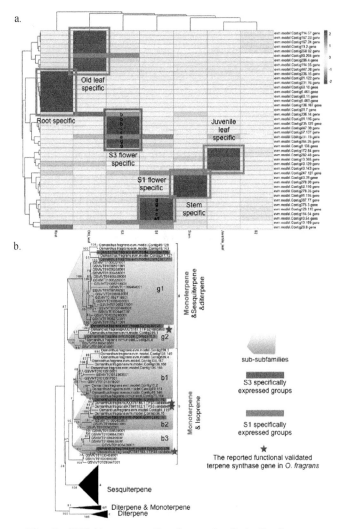

Fig. 7 TPS (terpene synthase) gene family in *O. fragrans*
(a) Heatmap showing the tissue- and flowering stage-specific expression of the members of *TPS* genes; (b) *TPS* gene subfamily members
Yellow: S1 early flowering stage-specific; Red: S3 late flowering stage-specific.

The number of genes, 45,542, is high and is more than the genes present in some of the plants that are related to *O. fragrans* (Fig. 2). This can be attributed to the repeated gene duplications which led to expansion of the gene families. The *O. fragrans* genome has higher number of multicopy genes compared to other plant species (Fig. 2). Furthermore, *O. fragrans* appears to have obtained and retained a large number of genes through the whole genome duplications (Fig. 4). The majority of plant species have experienced genome duplications in their evolutionary past[52-53]. The high gene number of *O. fragrans* might be a result of complex interactions among various factors such as the rate of evolution, number of duplication events, level of gene retention, expansion of gene family and selection pressure. The recent (~15 MYA) WGD and high retention might explain the large gene number. The number of genes involved in secondary metabolism is particularly high in O. fragrans (unpublished observation), and these genes might have been retained and/or expanded after the whole-genome duplications. This result may reflect the continuous interaction between O. fragrans and environmental factors, which imposes a constant pressure for adaptation[54].

The calculated level of heterozygosity (1.45%) is high in O. fragrans 'Rixiang Gui'. Considering that O. fragrans has been selectively bred for desirable traits for over 2,000 years in China, 160 cultivars with diverse phenotypes have been selected. The high heterozygosity (1.45%) in O. fragrans 'Rixiang Gui' might support an extensive breeding among cultivars throughout its history, although it is challenging to accurately determine the origin of the observed heterozygosity in the cultivar. Furthermore, as an androdioecious species[55], the coexistence of selfing and crossing poses an additional challenge to trace the origin of the high heterozygosity. Recently, the first genetic map of O. fragrans was created using the SLAF-seq method[16] to provide a framework for understanding the genome organization. This linkage map has helped us assemble the reference genome and can help to investigate the origin of the high heterozygosity and history of hybridization among the cultivars of O. fragrans.

The new genome can also be used as a reference for the whole genome resequencing of sweet osmanthus cultivars. Resequencing these whole genomes of various cultivars provides highly useful information on the potential drivers for the phenotype diversity, evolution, and population structure of a given species[56]. Our preliminary genome sequencing of 30 different cultivars of O. fragrans identified a large number of single nucleotide polymorphisms (SNP), copy number variation (CNV), insertion-sdeletion (InDel), structural variations (SV) and other mutation sites (unpublished results). Using the above mutation loci as new molecular genetic markers, researchers can study the history of cultivation, population dynamics and genetic diversity.

3.2 The Whole-Genome Duplication and the Tandem Duplication of the Biosynthetic Genes is Likely the Cause for the Strong Sweet Aroma of O. fragrans

Among *TPS*-family genes, the *TPS-b* and *TPS-g* subfamilies are known to synthesize monoterpenes[57]. Linalool, the major aroma compound identified in our study, is produced by the monoterpene synthesis pathway in O. fragrans. Using the high-quality genome and deep transcriptome information, we found a significant expansion of TPS as a whole, and of subfamilies *TPS-b* and *TPS-g* specifically, compared with the grape (*V. vinifera*), which did not have whole genome duplication (Fig. 5). In addition to TPS1, 2, 3, and 4, which have been previously functionally validated, we identified seven additional *TPS* genes that are specifically expressed in flower stages S1 and S3. Three *TPS* genes appear to be new genes that are flower specific, indicating that the production of fragrance is controlled by a complex network involving multiple *TPS* genes functioning in time- and flower specific manners. Our results suggest that the unique aromas of *O. fragrans* are some of the outcomes of the interrelationship between genome evolution, transcriptional regulation, and metabolic control. Our current work lays a solid foundation for further studies on the comparative genomics, molecular and biochemical mechanisms of aroma development in *O. fragrans*.

4 Conclusion

We constructed a high-quality reference genome of *O. fragrans* by combining Illumina, PacBio and Hi-C platforms. The genome of *O. fragrans* 'Rixiang Gui' is approximately 740 Mb and has a high heterozygosity of 1.45%. A large number of genes (45,542) was predicted by the gene models built with *de novo*, homology-based, and experimental data obtained from extensive transcription results. Our deep genome analysis indicates evidence of whole-genome duplication at ~15 MYA. Our new genome information should help the research community study the genome structure, genetic basis of genetic diversity, and regulation of the flowering process and scent development in *O. fragrans* and other related plant species.

5 Materials and Methods

5.1 Genome Sequencing

For genome sequencing, leaf samples were collected from a male tree (*O. fragrans* 'Rixiang Gui') on the campus of Nanjing Forestry University, Nanjing, China, and were processed for genomic DNA isolation and library construction. "Rixianggui" (Semperfloren) is a unique cultivar because it has a strong aroma and blooms continuously, except in hot summer months, while other cultivars, for example, Thunbergii, Latifolius, and Aurantiaeus, bloom only in autumn. Genomic DNA was extracted using the CTAB method, size fractionated with BluePipin (Sage Science, Inc., Beverly, MA, USA), used for library construction following the PacBio SMRT library construction protocol, and sequenced on the PacBio Sequel platform (Pacific Biosciences, Menlo Park, CA, USA). For Illumina library construction, the extracted DNA was fragmented and size-fractionated using g-tube and BluePipin, then subjected to paired-end library construction and sequenced on the HiSeq X ten platform (Illumina Inc., San Diego, CA, USA).

5.2 Hi-C Sequencing

To ensure the quality of the Hi-C library, leaf samples were initially examined for integrity of the nuclei by DAPI staining. Once confirmed for high quality nuclei, the samples were processed following the Hi-C procedure[58-60]. The Hi-C library was sequenced on the Illumina HiSeq X ten platform (Illumina Inc., San Diego, CA, USA), generating 740 million Hi-C read pairs, which were submitted to the Lachesis Hi-C scaffolding pipeline[58]. Hi-C libraries produce different molecular types, including invalid pairs of self-circles, dangling-ends, and dumped-pairs. According to the different molecular types that lead to the alignment of paired reads on the genome in different directions, the unique alignment of reads on the genome needs to be statistically analyzed. Once recognized as an effective interaction, the final data only retained effective interactions. According to the above rules, the position of the DpnII digestion site in the reference genome was used, because it can also provide useful information on the structural organization of individual chromosomes.

5.3 Transcriptome Sequencing

To obtain information that can assist in the empirical annotation of genes, full-length transcriptome sequencing was performed. The samples from flowers at three different blooming stages (S1: beginning, S2: middle, S3: late; see Fig. S3), leaves, stems, and roots were collected from the same tree described above and processed for library construction. The total RNAs were extracted according to the manufacturer's instructions of TRNzol Universal Reagent (Cat # DP424, TIANGEN Biotech Co. Ltd., Beijing, China). The quality and quantity of the RNA samples were evaluated using a NanoDrop™ One UV-Vis spectrophotometer (Thermo Fisher Scientific, Waltham, MA, USA), a Qubit© 3.0 Fluorometer (Thermo Fisher Scientific, Waltham, MA, USA) and an Agilent Bioanalyzer 2100 (Agilent Technologies, Santa Clara, CA, USA). All RNA samples with integrity values close to 10 were used for cDNA library construction and sequencing. The cDNA library was prepared using the TruSeq Sample Preparation (Illumina Inc., San Diego, CA, USA) and IsoSeq Library Construction kits (Pacific Biosciences, Menlo Park, CA, USA), and paired-end sequencing with 150 bp was conducted on a HiSeq X ten platform (Illumina Inc., San Diego, CA, USA).

5.4 Aroma Compound Analysis

Fresh flowers at three different stages (S1: beginning, S2: middle, S3: late), defined by the size of the flower (Fig. S3), were picked from the same tree at the time of sample collection for the transcriptome studies described above. Sampling was replicated five times, and the samples were quickly put into polyethylene bags impermeable to gases, kept frozen and stored at -20 ℃. Headspace solid phase microextraction (SPME) combined with gas chromatography-mass spectrometry (GC-MS) was used to determine the identity and quantity of the aroma volatiles. Flowers (0.3 g) were placed in a 4 mL solid-phase microextraction vial (Supelco Inc., Bellefonte, PA, USA), 1 μL of 1000X diluted ethyl caprate (Macklin Inc., Shanghai China) was added, and vials were capped with a 65 μm DB-5 ms extraction head (Supelco Inc., Bellefonte, PA, USA). Then, the vial was incubated for 40 min in a water bath at 45 ℃ to volatilize the aroma compounds and release them into the headspace. After the adsorption period, the fiber head was removed and introduced into the heated injector port of the GC for desorption at 250 ℃ for 3 min. The desorbed volatile compounds from DB-5 ms were analyzed on a Trace DSQ GC-MS (Thermo-Fisher Scientific, Waltham, MA, USA), equipped with a 30 m × 0.25 mm × 0.25 mm TR-5 MS capillary column (Supelco Inc., Bellefonte, PA, USA). The oven temperature was programmed at 60 ℃ for 2 min, increasing at 5 ℃/min to 150 ℃, then increasing at 10 ℃/min to reach 250 ℃, followed by maintaining the temperature of the transfer line at 250 ℃. Helium was taken as the carrier gas at a linear velocity of 1.0 mL/min. Mass detector conditions on MS were: source temperature: 250 ℃ and the electronic impact (EI) mode at 70 eV, with a speed of 4 scans/s over the mass range m/z 33~450 amu in a 1 s cycle. Volatile compounds were first auto-matched by mass spectra using the NIST98 database through ChemStation (Agilent Technologies, Santa Clara, CA, USA). A series of n-alkanes (C7~C30) (Sigma, St. Louis, MO, USA) was injected into the GC-MS set to obtain the linear retention indices of the volatile compounds, and they were analyzed under the same conditions. The data were also compared with published linear retention indices (NIST Chemistry WebBook, SRD 69). The normalization of peak-areas was used to calculate the quantities of the volatile aroma compounds.

Acknowledgements

This work was supported by research grants provided by the National Natural Science Foundation (31870695 and 31601785), the Project of Key Research and Development Plan (Modern Agriculture) in Jiangsu (BE2017375), the Selection and Breeding of Excellent Tree Species and Effective Cultivation Techniques (CX(16)1005), the Project of Osmanthus National Germplasm Bank, and the Top-notch Academic Programs Project of Jiangsu Higher Education Institutions.

Author Contributions

Lianggui Wang and Yuanzheng Yue designed and coordinated the whole project. Xiulian Yang, Lianggui Wang, Yuanzheng Yue, and Fei Chen, together lead and performed the whole project. Junhao Chen, Fei Chen, Tingting Shi, Haiyan Li, and Wenjie Ding performed the analyses of genome evolution, gene family analyses, and metabolic analyses, Min S. Park, Junhao Chen, Fei Chen, Yuanzheng Yue, and Gongwei Chen participated in manuscript writing and revision. All authors read and approved the final manuscript.

Conflict of Interest

None declared.

Supplementary Information

Tab. S1 Quality assessment of the assembled genome of *O. fragrans* using BUSCOs

Type	Number	Percent/%
Complete BUSCOs (C)	1,322	96.1
Complete and single-copy BUSCOs (S)	935	68.0
Complete and duplicated BUSCOs (D)	387	28.1
Fragmented BUSCOs (F)	35	2.5
Missing BUSCOs (M)	18	1.4
Total BUSCO groups searched	1,375	100

Tab. S2 Statistics of the annotated simple sequence repeats (SSR) in the *O. fragrans* genome

	Number / Size
Total examined sequences	768 / 740,633,826
Total identified SSR	409,691
Compound format	62,549
Sequences contain SSR	765
Sequences contain SSR (>1)	765

Tab. S3 Specific statistics of the annotated SSRs in the *O. fragrans* genome

Type	Unit size(repeat number)	Number
p1	1(≥10)	305,868
p2	2(≥6)	70,587
p3	3(≥5)	25,544
p4	4(≥5)	3,949
p5	5(≥5)	2,081
p6	7(≥5)	1,662

Tab. S4 Summary statistics of the annotated transposable elements (TE) in *O. fragrans* genome

Type	Repbase TEs		TE proteins		Repeat Modeler		Combined TEs	
	Length /Mb	% in genome	Length /Mb	% in genome	Length /Mb	% in genome	Length /Mb	% in genome
DNA	11,466,986	1.55	13,854,578	1.87	35,500,978	4.79	42,348,260	5.72
LINE	2,324,325	0.31	4,764,469	0.64	5,344,922	0.72	6,954,170	0.94
LTR	96,988,128	13.1	99,688,493	13.46	156,097,224	21.08	210,982,390	28.49
SINE	2,208	0	0	0	156,054	0.02	154,696	0.02
Other	1,333,525	0.18	1,208,971	0.17	2,980,292	0.4	29,308,813	3.95
Unknown	34,896	0	0	0	100,723,522	13.6	75,787,119	10.23
Total	112,150,068	15.14	119,516,511	16.14	300,802,992	40.61	365,535,448	49.35

Tab. S5 Summary statistic of the annotated noncoding RNA sequences in *O. fragrans* genome

Type	Copy number	Average length/bp	Total length/bp	Percentage/% of genome
rRNA	49	1,677.10	82,178	0.007,985
18S	13	1,779.54	23,134	0.002,248
28S	14	3,994.21	55,919	0.005,434
5.8S	15	154.80	2,322	0.000,226
5S	7	114.71	803	0.000,078
snRNA	2,058	108.23	222,746	0.021,644
CD-box	1,840	104.96	193,131	0.018,766
HACA-box	66	127.74	8,431	0.000,819
splicing	152	139.37	21,184	0.002,058
miRNA	525	116.98	61,417	0.005,968
tRNA	847	74.44	63,049	0.006,126

Tab. S6 Summary statistics of the functional genes of *O. fragrans*

Specie	Total number of gene	Average transcript length/bp	Average CDS length/bp	Average exon number per gene/bp	Average exon length/bp	Average intron length/bp
O. fragrans	45,542	4,065.24	1,142.72	5.24	2,217.92	688.66
A. thaliana	27,411	1,856.41	1,205.11	5.09	236.78	159.26
O. europaea	39,579	3,689.11	1,163.25	4.77	243.73	669.5
S. indicum	27,072	2,816.89	1,178.23	4.73	249.2	439.55
S. tuberosum	27,797	4,225.06	1,278.26	5.11	250.24	717.32
V. vinifera	24,821	5,697.52	1,343.47	5.22	257.57	1,032.77

Tab. S7 Summary statistic of annotated genes in *O. fragrans* genome. Genes were annotated by the combination of *de novo*, homology-based, and empirical data sets obtained from the full-length transcriptome analyses

	Gene set	Total number of gene	Average transcript length/bp	Average CDS length /bp	Average exons number per gene	Average exons length/bp	Average intro length/bp
De novo	Augustus	63,912	2,089.76	871.7	4.18	208.4	382.68
	GeneID	56,938	6,181.04	908.26	4.79	189.58	1,390.89
	Genscan	38,791	10,805.23	1,218.99	6.49	187.88	1,746.76
	SNAP	73,612	3,881.48	696.58	4.31	161.65	962.47
Homology	*A. thaliana*	48,327	2,357.62	926.58	3.43	269.89	588.2
	O. europaea	32,262	2,499.1	950.2	3.75	253.4	563.3
	S. indicum	44,207	2,129.39	864.78	2.99	289.36	636.0
	S. tuberosum	31,320	2,603.13	969.4	3.7	262.34	606.22
	V. vinifera	41,750	2,727.86	1,089.88	3.63	300.14	622.55
PASA	Transdecoder	59,536	4,242.07	1,235.38	5.3	232.89	689.49
Final set	EVM	45,542		1,142.72	5.24	217.92	688.66

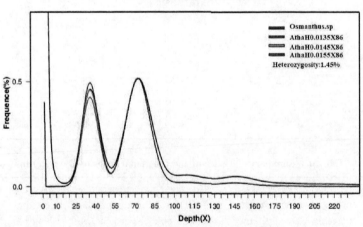

Fig. S1 The distribution of *O. fragrans* k-mers

The main peak is 73. Secondary peak is 37. The heterozygosity is consistent with the green line. AthaH0.0145X86 means that the depth of the simulated Illumina PE reads of *Arabidopsis thaliana* is 86 with the expected heterozygosity of 1.45%.

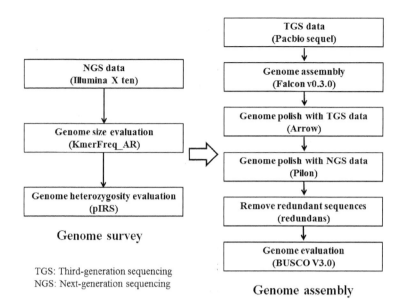

Fig. S2 Integrated work-flow for the assembly of the *O. fragrans* genome

Fig. S3 The flowers at different flowering stages
S1: early; S2: middle; S3: late.

References

[1] Shang, F. D., Yin, Y. J. & Xiang, Q. B. The culture of sweet osmanthus in China. *J. Henan Univ. Nat. Sci.* **43**, 136-139 (2003).

[2] Hao, R. M., Zang, D. K. & Xiang, Q. B. Investigation on natural resources of *Osmanthus fragrans* Lour. at Zhou luo cun in Hunan. *Acta Hortic. Sin.* **32**, 926-929 (2005).

[3] Zang, D. K., Xiang, Q. B., Liu, Y. L. & Hao, R. M. The studying history and the application to International Cultivar Registration Authority of sweet osmanthus (*Osmanthus fragrans* Lour.). *J. Plant Resour. Environ.* **12**, 49-53 (2003).

[4] Deng, C. H., Song, G. X. & Hu, Y. M. Application of HS-SPME and GC-MS to characterization of volatile compounds emitted from Osmanthus flowers. *Ann. Chim.* **94**, 921-927 (2004).

[5] Xin, H. P., Wu, B. H., Zhang, H. H., Wang, C. Y., Li, J. T., Yang, B. & Li, S. H. Characterization of volatile compounds in flowers from four groups of sweet osmanthus (*Osmanthus fragrans*) cultivars. *Can. J. Plant Sci.* **93**, 923-931 (2013).

[6] Cai, X., Mai, R. Z., Zou, J. J., Zhang, H. Y., Zeng, X. L., Zheng, R. R. & W, C. Y. Analysis of aroma-active compounds in three sweet osmanthus (*Osmanthus fragrans*) cultivars by gas-chromatography olfactometry and GC-mass spectrometry. *J. Zhejiang Univ. Sci. B.* **15**, 638-648 (2014).

[7] Hu, C. D., Liang, Y. Z., Li, X. R., Guo, F. Q., Zeng, M. M., Zhang, L. X. & Li, H. D. Essential oil composition of *Osmanthus fragrans* varieties by GC-MS and heuristic evolving latent projections. *Chromatographia* 70, 1163-1169 (2009).

[8] Wang, L. M., Li, M. T., Jin, W. W., Li, S., Zhang, S. Q. & Yu, L. J. Variations in the components of *Osmanthus fragrans* Lour. Essential oil at different stages of flowering. *Food Chem.* **114**, 233-236 (2009).

[9] Hu, B. F., Guo, X. L., Xiao, P. & Luo, L. P. Chemical composition comparison of the essential oil from four groups of *Osmanthus fragrans* Lour. flowers. *J. Essent Oil Plants.* **15**, 832-838 (2012).

[10] Lei, G. M., Mao, P. Z., He, M. Q., Wang, M. Q. Long, H. Liu, X. S. & Zhang, A. Y. Water-soluble essential oil components of fresh flowers of *Osmanthus fragrans* lour. *J. Essent Oil Res.* **28**, 177-184 (2016).

[11] Shang, F. D., Yin, Y. J. & Zhang, T. The RAPD analysis of 17 *Osmanthus fragrans* cultivars in Henan province. *Acta Hortic. Sin.* **31**, 685-687 (2004).

[12] Yuan, W. J., Han, Y. J., Dong, M. F. & Shang, F. D. Assessment of genetic diversity and relationships among *Osmanthus fragrans* cultivars using AFLP markers. *Electron. J. Biotechnol.* **14**, 2-3 (2011).

[13] Hu, W., Luo, Y., Yang, Y., Zhang, Z. Y. & Fan, D. M. Genetic diversity and population genetic structure of wild sweet osmanthus revealed by microsatellite markers. *Acta Hortic. Sin.* **41**, 1427-1435 (2014).

[14] Yuan, W. J., Li, Y., Ma, Y. F., Han, Y. J. & Shang, F. D. Isolation and characterization of microsatellite markers for *Osmanthus fragrans* (Oleaceae) using 454 sequencing technology. *Genet. Mol. Res.* **14**, 17154-17158 (2015).

[15] Han, Y. J., Chen, W. C., Yang, F. B., Wang, X. H., Dong, M. F., Zhou, P. & Shang, F. D. cDNA-AFLP analysis on 2 *Osmanthus fragrans* cultivars with different flower color and molecular characteristics of MYB1 gene. *Trees* 29, 931-940 (2015).

[16] He, Y. X., Yuan, W. J., Dong, M. F., Han, Y. J. & Shang, F. D. The First Genetic Map in Sweet osmanthus (*Osmanthus fragrans* Lour.) Using Specific Locus Amplified Fragment Sequencing. *Front. Plant Sci.* **8**, 1621 (2017).

[17] Zhang, X. S., Pei, J. J., Zhao, L. G., Tang, F. & Fang, X. Y. RNA-Seq analysis and comparison of the enzymes involved in ionone synthesis of three cultivars of Osmanthus, *J. Asian. Nat. Prod. Res.* **9**, 1-13 (2018).

[18] Yang, X. L., Li, H. Y., Yue, Y. J., Ding, W. J., Xu, C., Shi, T. T., Chen, G. W. & Wang, L. G. Transcriptomic Analysis of the Candidate Genes Related to Aroma Formation in *Osmanthus fragrans*, *Molecules* **23**, 1604 (2018).

[19] Xu, C., Li, H. G., Yang, X. L, Gu, C. S., Mu, H. N., Yue, Y. Z. & Wang, L. G. Cloning and Expression Analysis of MEP Pathway Enzyme-encoding Genes in *Osmanthus fragrans*. *Genes* **7**, 78 (2016).

[20] Zeng X. L., Liu C., Zheng R. R., Cai X., Luo J., Zou J. J. & Wang, C. Y. Emission and

Accumulation of Monoterpene and the Key Terpene Synthase (TPS) Associated with Monoterpene Biosynthesis in *Osmanthus fragrans* Lour. *Front. Plant Sci.* **6**, 1232 (2015).

[21] Baldermann, S., Kato, M., Kurosawa, M., Kurobayashi, Y., Fujita, A., Fleischmann, P. & Watanabe, N. Functional characterization of a carotenoid cleavage dioxygenase 1 and its relation to the carotenoid accumulation and volatile emission during the floral development of *Osmanthus fragrans* Lour. *J. Exp. Bot.* **61**, 2967-2977 (2010).

[22] Baldermann, S., Kato, M., Fleischmann, P. & Watanabe, N. Biosynthesis of α- and β-ionone, prominent scent compounds, in flowers of *Osmanthus fragrans*. *Acta. Biochim. Pol.* **59**, 79-81 (2012).

[23] Han, Y. J., Liu, L. X., Dong, M. F., Yuan, W. J. & Shang, F. D. cDNA cloning of the phytoene synthase (PSY) and expression analysis of PSY and carotenoid cleavage dioxygenase genes in *Osmanthus fragrans*. *Biologia* **68**, 258-263 (2013).

[24] Han, Y. J., Wang, X. H., Chen, W. C., Dong, M. F., Yuan, W. J., Liu, X. & Shang, F. D. Differential expression of carotenoid-related genes determines diversified carotenoid coloration in flower petal of *Osmanthus fragrans*. *Tree. Genet. Genom.* **10**, 329-338 (2014).

[25] Zhang, C., Wang, Y. G., Fu, J. X., Bao, Z. Y. & Zhao, H. B. Transcriptomic analysis and carotenogenic gene expression related to petal coloration in *Osmanthus fragrans* 'Yanhong Gui'. *Trees* **30**, 1207-1223 (2016).

[26] Mu, H. N., Li, H. G., Yang, X. L., Sun, T. Z., Xu, C. & Wang, L. G. Transcriptome sequencing and analysis of sweet osmanthus (*Osmanthus fragrans* Lour.). *Genes. Genom.* **36**, 777-788 (2014).

[27] Han, Y. J., Wu, M., Cao, L., Yuan, W. J., Dong, M. F., Wang, X., Chen, W. & Shang, F. D. Characterization of *OfWRKY*3, a transcription factor that positively regulates the carotenoid cleavage dioxygenase gene *OfCCD*4 in *Osmanthus fragrans*, *Plant Mol. Biol.* **91**, 485-496 (2016).

[28] Wang, L., Tan, N. N., Hu, J. Y., Wang, H., Duan, D. Z., Ma, L., Xiao J. & Wang, X. L. Analysis of the main active ingredients and bioactivities of essential oil from *Osmanthus fragrans* Var. thunbergii using a complex network approach. *BMC Syst. Biol.* **11**, 144 (2017).

[29] Guillaume, M. & Carl, K. A Fast, Lock-Free Approach for Efficient Parallel Counting of Occurrences of K-Mers. *Bioinformatics* **27**, 764-770 (2011).

[30] Luo, R., Liu, B., Xie, Y., Li, Z., Huang, W. H., Yuan, J., He, G., Chen, Y., Pan, Q., Liu, Y. et al. SOAP denovo2: an empirically improved memory-efficient short-read de novo assembler. *GigaScience* **27**, 18 (2012).

[31] Hu, X., Yuan, J., Shi, Y., Lu, J., Liu, B., Li, Z., Chen, Y., Mu, D., Zhang, H., Li, N., Yue, Z., Bai, F., Li H., & Fan W. pIRS: Profile-based Illumina pair-end reads simulator. *Bioinformatics* **28**, 1533-1535 (2012).

[32] Chin, C. S., Peluso, P., Sedlazeck, F. J., Nattestad, M., Concepcion, G. T., Clum, A., Dunn, C., O'Malley, R., Figueroa-Balderas, R., Morales-Cruz, A., Cramer, G. R., Delledonne, M., Luo, C., Ecker, J. R., Cantu, D., Rank, D. R., & Schatz, M. C. Phased diploid genome assembly with single-molecule real-time sequencing. *Nat. Methods.* **13**, 1050-1054 (2016).

[33] Walker, B. J., Abeel, T., Shea, T., Priest, M., Abouelliel, A., Sakthikumar, S., Cuomo, C. A., Zeng, Q., Wortman, J., Young, S. K. & Earl, A. M. Pilon: An Integrated Tool for Comprehensive Microbial Variant Detection and Genome Assembly Improvement. *PLoS ONE.* **9**, e112963 (2014).

[34] Pryszcz, L. P. & Gabaldón, T. Redundans: an assembly pipeline for highly heterozygous genomes.

Nucleic Acids. Res. **44**, e113 (2016).

[35] Simão, F. A., Waterhouse, R. M., Ioannidis. P., Kriventseva, E. V., Zdobnov, E. M. BUSCO: assessing genome assembly and annotation completeness with Single-Copy Orthologs. *Bioinformatics* **31**, 3210-3212 (2015).

[36] Stanke, M., Steinkamp, R., Waack, S. & Morgenstern, B. AUGUSTUS: a web server for gene finding in eukaryotes. *Nucleic Acids Res.* **32**, W309-312 (2004).

[37] Chen, N. Using RepeatMasker to identify repetitive elements in genomic sequences. *Curr. Protoc. Bioinformatics* **4**, 10 (2004).

[38] Bao, W., Kojima, K. K. & Kohany, O. Repbase Update, a database of repetitive elements in eukaryotic genomes. *Mob. DNA.* **6**, 11 (2015).

[39] Nussbaumer, T., Martis, M. M., Roessner, S. K., Pfeifer, M., Bader K. C., Sharma, S., Gundlach, H. & Spannagl, M. MIPS PlantsDB: a database framework for comparative plant genome research. *Nucleic Acids Res.* **41**, D1144-1151 (2013).

[40] Kalvari, I., Argasinska, J., Quinones-Olvera, N., Nawrocki, E. P., Rivas, E., Eddy, S. R. Bateman, A., Finn, R. D. & Petrov A. I. Rfam 13.0: shifting to a genome-centric resource for non-coding RNA families. *Nucleic Acids Res.* **46**, D335-342 (2017).

[41] Camacho, C., Coulouris, G., Avagyan V., Ma, N., Papadopoulos, J., Bealer, K. & Madden, T. L. BLAST+: architecture and applications. *BMC Bioinformatics.* **10**, 421 (2009).

[42] Lowe, T. M. & Eddy, S. R. tRNAscan-SE: a program for improved detection of transfer RNA genes in genomic sequence. *Nucleic Acids Res.* **25**, 955-964 (1997).

[43] Lagesen, K., Hallin, P., Rødland, E. A., Staerfeldt, H. H. & Rognes, T. RNAmmer: consistent and rapid annotation of ribosomal RNA genes. *Nucleic Acids Res.* **35**, 3100-3108 (2007).

[44] Birney, E. & Durbin, R. Using GeneWise in the Drosophila annotation experiment. *Genome Res.* **10**, 547-548 (2000).

[45] Blanco, E., Parra, G. & Guigó, R. Using geneid to identify genes. *Curr. Protoc. Bioinformatics.* **4**, 1-28 (2007).

[46] Majoros, W. H., Pertea, M. & Salzberg, S. L. TigrScan and GlimmerHMM: two open source ab initio eukaryotic gene-finders. *Bioinformatics.* **20**, 2878-2879 (2004).

[47] Bromberg, Y. & Rost, B. SNAP: predict effect of non-synonymous polymorphisms on function. *Nucleic Acids Res.* **35**, 3823-3835 (2007).

[48] Haas, B. J., Salzberg, S. L., Zhu, W., Pertea, M., Allen, J. E., Orvis, J., White, O., Buell, R. C. & Wortman R. J. Automated eukaryotic gene structure annotation using EVidenceModeler and the Program to Assemble Spliced Alignments. *Genome Biol.* **9**, R7 (2008).

[49] Moriya, Y., Itoh, M., Okuda, S., Yoshizawa, A. C. & Kanehisa, M. KAAS: an automatic genome annotation and pathway reconstruction server. *Nucleic Acids Res.* **35**, W182-W185 (2007).

[50] Quevillon, E., Silventoinen, V., Pillai, S., Harte, N., Mulder, N., Apweiler, R. & Lopez, R. InterProScan: protein domains identifier. *Nucleic Acids Res.* **33**, 116-120 (2005).

[51] De Bodt, S., Maere, S. & Van de Peer, Y. Genome duplication and the origin of angiosperms. *Trends Ecol. Evol.* **20**, 591-597 (2005).

[52] Li, L., Stoeckert, C. J. Jr. & Roos, D. S. OrthoMCL: identification of ortholog groups for eukaryotic genomes. *Genome Res.* **13**, 2178-2189 (2003).

[53] Ciu, L. Y., Wall, P. K., Leebens-Mack, J. H., Lindsay, B. G., Soltis, D. E., Doyle, J. J., Soltis, P. S., Carlson, J. E., Arumuganathan, K., Barakat, A., Abdelali B., Albert, V. A.,

Ma, H. & dePamphilis, C. W. Widespread genome duplications throughout the history of flowering plants. *Genome Res.* **16**, 738-749 (2006).

[54] Casneuf, T., De Bodt, S., Raes, J., Maere, S. & Van de Peer, Y. Nonrandom divergence of gene expression following gene and genome duplications in the flowering plants*Arabidopsis thaliana*. *Genome Biol.* **7**, R13 (2006).

[55] Xu, Y. C., Zhou, L. H., Hu, S. Q., Hao, R. M., Huang, C. J., & Zhao, H. B. The differentiation and development of pistils of hermaphrodites and pistillodes of males in androdioecious *Osmanthus fragrans* L. and implications for the evolution to androdioecy. *Plant Syst. Evol.* **300**, 843-849 (2014).

[56] Huang, X., Feng, Q., Qian, Q., Zhao, Q., Wang, L., Wang, A., Guan, J., Fan, D., Weng, Q., Huang, T., Dong, G., Sang, T. & Han, B. High-throughput genotyping by whole-genome resequencing. *Genome Res.* **19**, 1068-1076 (2009).

[57] Tholl, D. Terpene synthases and the regulation, diversity and biological roles of terpene metabolism. *Curr. Opin. Plant Biol.* **9**, 297-304 (2006).

[58] Kaplan, N. & Dekker, J. High-throughput genome scaffolding from in vivo DNA interaction frequency. *Nat. Biotechnol.* **31**, 1143-1147 (2013).

[59] Marie-Nelly, H. Marbouty, M. Cournac, A. Flot, J. Liti, G. Parodi, D. P. Syan, S. Guillén, N. Margeot, A. Zimmer, C. & Koszul, R. High-quality genome (re) assembly using chromosomal contact data. *Nat. Commun.* **5**, 5695 (2014).

[60] Jibran, R. Dzierzon, H. Bassil, H. Bushakra, J. M. Edger, P. P. Sullivan, S. Finn, C. E. Dossett, M. Vining, K. J. VanBuren, R. Mockler, T. C. Liachko, I. Davies, K. M. Foster, T. M. & Chagné, D. Chromosome-scale scaffolding of the black raspberry (*Rubus occidentalis* L.) genome based on chromatin interaction data. *Hortic. Res.* **5**, 18008 (2018).

Cloning and Expression Analysis of Three Critical Triterpenoid Pathway Genes in *Osmanthus fragrans*

Xiulian Yang[1,2,†], Wenjie Ding[1,2†], Yuanzheng Yue[1,2], Chen Xu[1,2], Xi Wang[1,2], Lianggui Wang[1,2]

1. College of Landscape Architecture, Nanjing Forestry University, No. 159 Longpan Road, Nanjing 210037, Jiangsu, China
2. Key Laboratory of Landscape Architecture of Jiangsu Province, No. 159 Longpan Road, Nanjing 210037, Jiangsu, China

† These authors contributed equally to this work.

Abstract: *Osmanthus fragrans* is an important ornamental tree and has been widely planted in China due to its pleasant aroma, which is mainly composed of terpenes. The monoterpenoids and sesquiterpenoids metabolic pathway of sweet osmanthus have been well studied. However, these studies were mainly focused on volatile small molecule compounds. The molecular regulation mechanism of large molecule compound (triterpenoids) synthesis is still confusing. Squalene synthase (*SQS*), squalene epoxidase (*SQE*), and beta-amyrin synthase (*BETA-AS*) are three critical enzymes of triterpenoids biosynthesis pathway. In this study, the full-length cDNA and gDNA sequences of *OfSQS*, *OfSQE*, and *OfBETA-AS* were isolated from sweet osmanthus. Phylogenetic analysis suggested *OfSQS* and *OfSQE* had the closest relationship with *Sesamum indicum*, and *OfBETA-AS* sequence shared the highest similarity of 99% with *Olea europaea*. The qRT-PCR analysis revealed the three genes were all highly expressed in flowers, especially *OfSQE* and *OfBETA-AS*, which were predominantly expressed in the flowers of both 'Boye Jingui' and 'Rixiang Gui' cultivars, suggesting they would play important roles in the accumulation of triterpenoids in flowers of *O. fragrans*. Furthermore, the expression of *OfBETA-AS* in two cultivars were significantly different during all the five flowering stages, suggesting that *OfBETA-AS* may be the critical gene for the differences in the accumulation of triterpenoids. The evidences indicate that *OfBETA-AS* could be the key gene in triterpenoids synthesis pathway, and it could also be used as a critical gene resource in the synthesis of essential oils through engineering bacterium.

Key words: *Osmanthus fragrans*, triterpenoid biosynthesis, flower-specific expression

1 Introduction

Osmanthus fragrans, commonly known as sweet osmanthus, is an evergreen shrub or small tree belonging to Oleaceae [1]. It is distributed in China and other Asian countries, such as South Korea, India, and Thailand [2-4]. Nowadays, it has been widely cultivated as an urban ornamental tree in China [2]. Its fresh flowers have a pleasantly fruity and sweet aroma. Also, the extracts from *O. fragrans* flowers are important sources in the perfume and cosmetic industries [6-7], which can be used to reduce inflammation, resist oxidation, and prevent aging [5].

As a group of secondary metabolites, triterpenoids are synthesized by the mevalonate pathway (MVA pathway) and 2-C-methyl-D-erythritol-4-phosphate pathway (MEP pathway) [8]. Although these pathways have been studied in some plants, the downstream genes of MEP pathway are still unclear, especially the synthesis processes of triterpenoids. Squalene synthase (SQS, EC 2.5.1.21), which catalyzes the first enzymatic step committing carbon pool away from the central isoprenoid pathway towards the biosynthesis of terpenes, could determine the yield of subsequent products, such as triterpenes, sterols, and cholesterol [9]. It has been reported that *SQS* genes could play an important role in regulating the biosynthetic pathway of triterpenes, and the overexpression of *SQS* genes in *Panax ginseng* could result in the enhanced accumulation of triterpenes [10]. This regulation could increase the mRNA accumulation of downstream genes such as squalene epoxidase (SQE, EC 1.14.99.7), and increase the production of phytosterols and triterpenes [11]. Squalene epoxidase, a downstream gene of *SQS*, catalyzes the formation of 2,3-oxidosqualene [12]. The precursor 2,3-oxidosqualene is catalyzed by beta-amyrin synthase (BETA-AS, EC 5.4.99.39) and cycloartenol synthase (*CAS*) to produce beta-amyrin and cycloartenol, which are further modified to form triterpenoid and phytosterol, respectively (Fig. 1) [13]. *BETA-AS* was defined as an important branch point along the metabolisms pathway of triterpenoids, and it played a regulatory role in the biosynthesis of triterpenoids [14].

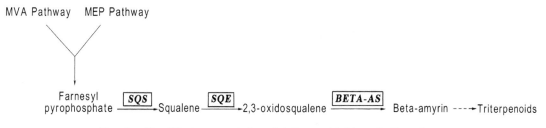

Fig. 1 Simplified representation of triterpenoids biosynthetic pathway

MVA pathway: mevalonate pathway; MEP pathway: 2 - C-methyl-D-erythritol-4 - phosphate pathway; *SQS*: squalene synthase gene; *SQE*: squalene epoxidase gene; *BETA-AS*: β-amyrin synthase gene. The three genes in the box are the genes involved in this study.

Triterpenoids isolated from *Osmanthus fragrans* exhibited hypolipidemic and antioxidation activity [15]. However, the biosynthesis of these compounds is confusing, and the molecular mechanism of triterpenoids in sweet osmanthus is still unclear. In this study, the full-length cDNA and gDNA sequences of three genes, *OfSQS*, *OfSQE*, and *OfBETA-AS*, were successfully isolated and further identified. Then, the expression patterns of the three genes in the different flower development stages of two *Osmanthus* cultivars were analyzed using qRT-PCR.

2 Materials and Methods

2.1 Plant Materials

Two cultivars of *O. fragrans*, 'Boye Jingui' (with dense aroma) and 'Rixiang Gui' (with light aroma) have been grown under a natural state since 2005 on the campus of Nanjing Forestry University, Jiangsu Province, China (112.32′E, 156.32′W). The flowers of 'Rixiang Gui' at full blooming stage (S4) collected in 2015 were used for gene cloning. Different organs, including root, stem, leaf, and flower at full blooming stage were collected in October 2016 for tissue-specific expression study. From September to October 2016, flowers were harvested during five flowering stages including bud-pedicel stage (S1), bud-eye stage (S2), primary blooming stage (S3), full blooming stage (S4), and flower fading stage (S5), respectively. Samples were immediately frozen in liquid nitrogen and stored at -80 ℃ for subsequent use.

2.2 RNA Extraction and First-strand cDNA Synthesis

Total RNA was extracted using RNAprep pure Kit (Tiangen, China) following the manufacturer's instructions. The quality of the extracted RNA was measured by NanoDrop2000 Spectrophotometer (Thermo Scientific, USA) and RNA integrity was evaluated by agarose gel electrophoresis. Then, the first strand cDNA was synthesized with Revert Aid First Strand cDNA Synthesis Kit (Thermo Scientific, USA) according to the manufacturer's protocol.

2.3 Cloning of Full-length cDNAs

According to the EST (Expressed sequence tags) sequences of the *O. fragrans* transcriptomic databases, specific primers (Tab. S1) were designed to obtain the core sequences of *OfSQS*, *OfSQE*, and *OfBETA-AS*. Next, using the 3′-Full RACE Core Set with PrimeScript™ RTase (Takara Biotechnology) and SMARTer™ RACE cDNA Amplification Kit (Clontech, USA), full-length sequences of these three genes were obtained. The specific primers for 3′ RACE and 5′ RACE (Tab. S2) were designed by Oligo 6.0 software. At last, the PCR products of these three genes were purified and cloned into pEASY®-T1 vector (Transgen Biotech, China) to confirm by sequencings.

2.4 Isolation of Genomic DNA Sequences of *OfSQS*, *OfSQE*, and *OfBETA-AS* Genes

The genomic DNA was extracted using Plant Genomic DNA Kit (Tiangen, China) following the manufacturer's instruction. The genomic DNA extracted from 'Rixiang Gui' petals was used as a template to get the genomic DNA sequences of *OfSQS*, *OfSQE*, and *OfBETA-AS* genes with the specific primers (Tab. S3). After PCR amplifications, the purified products were directly sequenced by Genescrip Inc. (Genescrip, China).

2.5 Bioinformatics Analysis

Open reading frame and protein prediction were made by NCBI ORF Finder (http://www.ncbi.nlm.nih.gov/gorf/gorf.html). The identification of nucleotide sequences was performed by the NCBI Blast program (http://www.ncbi.nlm.nih.gov/BLAST). Physical and chemical parameters of proteins were determined by ProtParam tool (http://web.expasy.org/protparam). The conservative domain was predicted by PFAM tool (http://smart.embl-heidelberg.de/). The signal peptide was predicted by SignalP 3.0 Server (http://www.cbs.dtu.dk/services/SignalP/). Transmembrane topology prediction was made by TMHMM Serverversion 2.0 Server (http://www.cbs.dtu.dk/services/TMHMM/). The structure of genomic organization was established by Gene Structure Display Server (http://gsds.cbi.pku.edu.cn/). The homology analysis was conducted by BLAST of GenBank (http://www.ncbi.nlm.nih.gov/BLAST/). Finally, the phylogenetic trees were constructed by ClustalX 2.1 and MEGA 5.0 with 1,000 bootstrap.

2.6 Gene Expression Analysis

Quantitative real-time RT-PCR (qRT-PCR) was performed using ABI StepOnePlus Systems (Applied Biosystems, USA) with SYBR Premix Ex Taq (Takara Biotechnology). The RNA samples were quantified by NanoDrop2000 Spectrophotometer (Thermo Scientific, USA). The cDNA was synthesized from 5 μg total RNA and diluted ten-fold for the gene expression experiment. The *OfRAN* and *OfRPB2* were taken as the reference genes for different organs and different flowering stages, respectively [16]. The specific primers used in the experiment to detect the genes expression levels were designed by Primer Premier 5.0 software (Tab. S4). The thermal cycle conditions were used as follows: 95 ℃ for 30 s, followed by 40 cycles of 95 ℃ for 5 s, 60 ℃ for 30 s, 95 ℃ for 15 s, 60 ℃ for 1 min, and 95 ℃ for 15 s [17]. The relative expression levels were calculated by the $2^{-\triangle\triangle CT}$ method. Data were presented as means with error bars indicating Standard Error. Different letters mean significant differences at the 0.01 level according to the Tukey's test.

2.7 Subcellular Localization Assay

The coding region (without stop codon) of *OfBETA-AS* was amplified from cDNA temple of 'Rixiang Gui' petals with gene-specific primers (Tab. S5). Vector (*35s::GFP-OfBETA-AS*) constructed by subcloning the coding region of *OfBETA-AS* into the Super1300::GFP vector (supplied by Dr. Yuanzheng Yue) was used for subcellular localization assay. The *35s::GFP-OfBETA-AS* and Super1300::GFP control vector were electroporated into *Agrobacterium tumefaciens* strain EHA105 using Eppendorf Eporator® 4309 (Eppendorf, Germany). Then, the tobacco (*Nicotiana benthamiana*) leaves were infiltrated by the vector-containing *Agrobacterium*, respectively. After 2 days of incubation at 21 ℃ and 14 h photoperiod, these infiltrated plants were used to observe the GFP fluorescence signal on a Laser Scanning Confocal Microscope Zeiss LSM 710 (Zeiss, Germany).

3 Results

3.1 Cloning and Sequence Analysis of cDNAs

In order to obtain the full-length cDNAs of *OfSQS*, *OfSQE*, and *OfBETA-AS*, the RACE technology was used to get the 3' region and 5' region of these genes. The full-length sequence of *SQS* cDNA was 1,672 bp and contained an open reading frame (ORF) of 1,245 bp encoding a protein with 414 amino acids, a 5'-untranslated region (5'-UTR) of 90 bp, and a 3'-UTR of 337 bp. The cDNA of *SQE* was 1,841 bp with an ORF of 957 bp encoding a protein with 318 amino acids, a 5'-UTR of 494 bp and a 3'-UTR of 390 bp. The cDNA of *BETA-AS* was 2,737 bp with an ORF of 2,289 bp encoding a protein with 762 amino acids, a 5'-UTR of 185 bp and a 3'-UTR of 263 bp. The sequences of *OfSQS*, *OfSQE*, and *OfBETA-AS* have been submitted to GenBank with the accession numbers of KY992860, KY992861, and KY992862.

The theoretical isoelectric points (pI) of OfSQS, OfSQE, and OfBETA-AS were 7.57, 9.01, and 5.97 and the putative molecular weights of these genes were 47.67, 34.51, and 87.21 kDa, respectively. The instability index showed that OfSQS, OfSQE, and OfBETA-AS belonged to unstable proteins. In amino acid sequences of the three genes, no signal peptide sequence was identified. Transmembrane region analysis of the amino acid sequence indicated that strong transmembrane helixes were found from 281 to 303 aa and 387 to 409 aa in OfSQS, 251 to 270 aa and 277 to 299 aa in OfSQE, and 609 to 631 aa in OfBETA-AS. The conserved domains of OfSQS and OfSQE proteins were from 44 to 316 aa and 4 to 277 aa, respectively, and for OfBETA-AS protein, the conserved domains were from 100 to 406 and 415 to 754 aa. These conserved domains evidenced that these genes could be categorized as a member of the SQS family, SQE family, SQHop cyclase C and SQHop cyclase N superfamily (Fig. 2a), respectively. Alignment result revealed that 6 conservative domains, I (58-76), II (77-90), III (166-188), IV (206-224), V (310-324), and VI (391-417) were found in OfSQS. The putative flavin adenine dinucleotide (FAD) binding domain was found in the amino acid sequence of OfSQE. Motif analysis revealed that OfBETA-AS contained the four QW motifs (111-117, 161-167, 603-609, 652-658), the DCTAE motif (485-489), and the MWCYCR (256-262) motif among the high conserved regions.

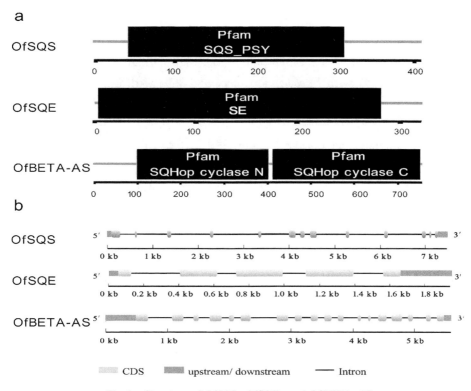

Fig. 2 Structure of OfSQS, OfSQE, and OfBETA-AS
(a) Conserved domains of OfSQS, OfSQE, and OfBETA-AS. The conserved domain analysis of the three proteins were performed using the Pfam Conserved Domain Database. The result indicated that OfSQS, OfSQE, and OfBETA-AS were categorized as a member of SQS family, SQE family, SQHop cyclase C and SQHop cyclase N superfamily, respectively. (b) Genomic organization of the OfSQS, OfSQE, and OfBETA-AS genes. Exons are represented as yellow lines, introns as fine lines, and UTRs as blue lines. Structure of genomic organization was performed using Gene Structure Display Server.

3.2 Cloning and Sequence Analysis of Genomic DNA

In order to analyze the exons/introns of the three genes, genomic sequences of *OfSQS*, *OfSQE*, and *OfBETA-AS* were obtained by PCR amplification with genomic DNA. Sequence analysis indicated that the sizes of *OfSQS*, *OfSQE*, and *OfBETA-AS* genomic DNA were 7,496 bp, 1,950 bp, and 5,606 bp, respectively. By homologous alignment of the cDNA and genomic DNA sequences, the genomic organizations of the three genes were elucidated and positions of introns/exons were determined. Our results showed that *OfSQS*, *OfSQE* and *OfBETA-AS* consisted of 12, 4 and 17 introns, respectively (Fig. 2b). All of these identified introns were began with the sequence GT and ended with AG, conforming to the GT/AG rules. Although the three genes contained multiple introns, no alternative splicing phenomenon was found in different organs and different flower development stages (Fig. S1).

3.3 Phylogenetic Analysis of *OfSQS*, *OfSQE*, and *OfBETA-AS*

BLAST tool was used to compare the obtained sequences with known sequences in the

GenBank database. The results of the BLASTp analysis of the three amino acid sequences demonstrated that the proteins shared high similarity with other plants. For OfSQS, the protein sequence similarity with other plants was higher than 85%, such as 90% with *Sesamum indicum*, 89% with *Bacopa monnieri*, and 86% with *Camellia oleifera*, *Salvia miltiorrhiza* and *Euphorbia tirucalli*. For OfSQE, the sequence similarity to other plants was 87% with *Ziziphus jujuba*, 86% with *Cucumis melo*, *Cajanus cajan* and *Prunus mume*, and 85% with *Cucumis sativus* and *Morus alba*. For OfBETA-AS, the sequence similarity was 99% with *O. europaea*, 91% with *S. indicum*, and 88% with *Dorcoceras hygrometricum*.

Cluster analysis of OfSQS with 20 SQS proteins from other plants showed that OfSQS was closest to the SQS of *S. indicum* and *B. monnieri*, and was also quite distant from SQS1 and SQS2 of *Arabidopsis thaliana* (Fig. 3a). For OfSQE, phylogenetic analysis of deduced amino acid sequences revealed a closer evolutionary relationship with SiSQE and AtSQE1(Fig. 3b). Moreover, OfBETA-AS shared the closest relationship with *O. europaea* (Fig. 3c).

3.4 Expression Analysis in Different Organs

To clarify the tissue-specific expression patterns of *OfSQS*, *OfSQE*, and *OfBETA-AS* genes in two cultivars of *O. fragrans*, 'Boye Jingui' and 'Rixiang Gui', qRT-PCR experiment were carried out with different organs including roots, stems, leaves, and flowers (Fig. 4). The tissue-specific expression patterns of *OfSQS*, *OfSQE*, and *OfBETA-AS* among the examined tissues were predominant in the flowers and then in the leaves and stems, while the roots had the lowest expression level. The *OfSQS* transcript level in the opening flower was 8-fold higher than that in the 'Rixiang Gui' root, while only 2.4-fold higher in 'Boye Jingui'. In 'Rixiang Gui' and 'Boye Jingui', the transcript levels of *OfSQE* in the opening flowers were 328-fold and 346-fold higher than in the root, respectively. For the *OfBETA-AS* gene, the transcript level of 'Rixiang Gui' and 'Boye Jingui' in the opening flower was 397-fold and 303-fold higher than that in the root, respectively.

3.5 Expression Analysis during Flower Development

To analyse the expression patterns of *OfSQS*, *OfSQE*, and *OfBETA-AS* genes during flower development, qRT-PCR was executed from five flowering stages: bud-pedicel stage (S1), bud-eye stage (S2), primary blooming stage (S3), full blooming stage (S4) and flower fading stage (S5) (Fig. 5). In 'Rixiang Gui', the expression level of the *OfSQS* gene did not show a significant change at the five stages, and the transcript level of *OfSQE* remained at a stable level from S1 to S3, and declined from S3 to S5. For *OfBETA-AS*, the expression showed a regularly down-regulated trend at five flowering stages. However, *OfSQS*, *OfSQE*, and *OfBETA-AS* showed the same expression trend in 'Boye Jingui', which increased from S1 to S2 and decreased from S3 to S5.

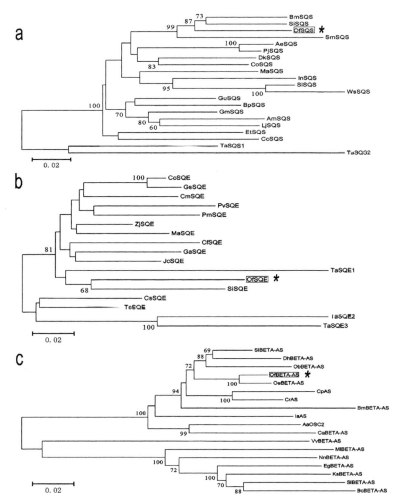

Fig. 3 Phylogenetic trees of *OfSQS*, *OfSQE*, and *OfBETA-AS* from *O. fragrans* and other plants using deduced amino acids. Phylogenetic trees were constructed by ClustalX 2. 1 and MEGA 5. 0 with 1000 bootstrap method

(a) Phylogenetic tree of *OfSQS* with other plants, such as *S. indicum* (XP_011092839. 1), *B. monnieri* (ADX01171. 1), *C. oleifera* (AGB05603. 1), *S. miltiorrhiza* (ACR57219. 1), *Diospyros kaki* (ACN69082. 1), *E. tirucalli* (BAH23428. 1), *Solanum lycopersicum* (NP_001234716. 2), *Panax japonicus* (ALB38664. 1), *Glycine max* (NP_001236365. 1), *Aralia elata* (ADC32654. 1), *Lotus japonicus* (BAC56854. 1), *Astragalus membranaceus* (ADW27427. 1), *Ipomoea nil* (XP_019169346. 1), *Betula platyphylla* (AKR76253. 1), *Glycyrrhiza uralensis* (ACS66750. 1), *Morus alba* (AOV62782. 1), *Withania somnifera* (ADW78251. 1), *Corchorus capsularis* (OMO74384. 1), *A. thaliana* (AtSQS1-AT4G34640. 1; AtSQS2-AT4G34650. 1); (b) Phylogenetic tree of *OfSQE* with other plants, such as *Z. jujuba* (XP_015885090. 1), *C. melo* (XP_016901349. 1), *C. cajan* (KYP34787. 1), *C. sativus* (XP_004141303. 1), *S. indicum* (XP_011092466. 1), *M. alba* (AOV62783. 1), *Glycine soja* (KHN17192. 1), *P. mume* (XP_008220536. 1), *Citrus sinensis* (XP_006466420. 1), *Panax vietnamensis* var. *fuscidiscus* (AIK23028. 1), *Cephalotus follicularis* (GAV57995. 1), *Theobroma cacao* (XP_007047610. 2), *Gossypium arboreum* (KHG06672. 1), *Citrus sinensis* (KDO78855. 1), *Jatropha curcas* (NP_001295656. 1), *A. thaliana* (AtSQE2-AT2G22830. 1; AtSQE3-AT4G37760. 1; AtSQE1-AT1G58440. 1); (c) Phylogenetic tree of *OfBETA-AS* with other plants, such as *O. europaea* (BAF63702. 1), *S. indicum* (XP_011096562. 1), *Coffea canephora* (CDP12624. 1), *D. hygrometricum* (KZV51042. 1), *Catharanthus roseus* (AEX99665. 1), *Ocimum basilicum* (AFH53506. 1), *Calotropis procera* (AND78515. 1), *Artemisia annua* (AHF22084. 1), *Ilex asprella* var. *asprella* (AIS39793. 1), *Bacopa monnieri* (ADM86392. 1), *Centella asiatica* (AAS01523. 1), *Erythranthe guttata* (XP_012842229. 1), *Daucus carota* subsp. Sativus (XP_017230789. 1), *Nelumbo nucifera* (XP_010245756. 1), *Maesa lanceolata* (AHF49822. 1), *Vitis vinifera* (XP_002270051. 2), *Solanum lycopersicum* (NP_001234604. 1), *Bupleurum chinense* (ABY90140. 2), *Eucalyptus grandis* (XP_010063692. 1), *Kalopanax septemlobus* (ALO23119. 1).

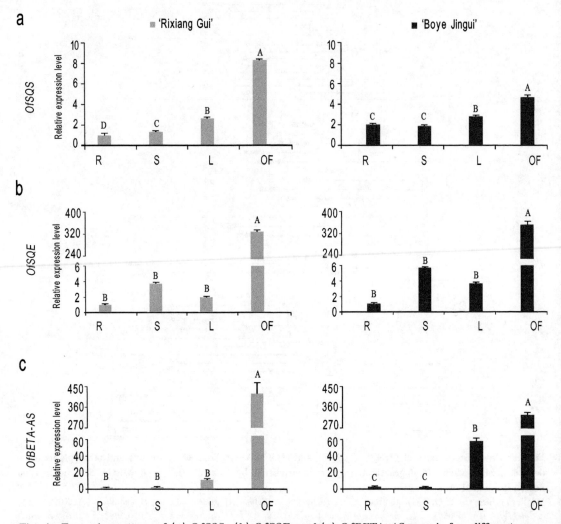

Fig. 4 Expression patterns of (a) *OfSQS*, (b) *OfSQE*, and (c) *OfBETA-AS* genes in four different organs of *O. fragrans*

R: root; S: stem; L: leaf and OF: opening flower. Data were presented as means with error bars indicating Standard Error. Different letters denote significant differences at the 0.01 level according to Tukey's test.

3.6 Subcellular Localization

To confirm the subcellular localization of OfBETA-AS protein, the $35s::GFP-OfBETA-AS$ vector was constructed and infiltrated into the tobacco leaves. Significant fluorescence signal can be detected in the cell nucleus and cell membrane (Fig. 6) indicated that *OfBETA-AS* was a nucleus and membrane located gene.

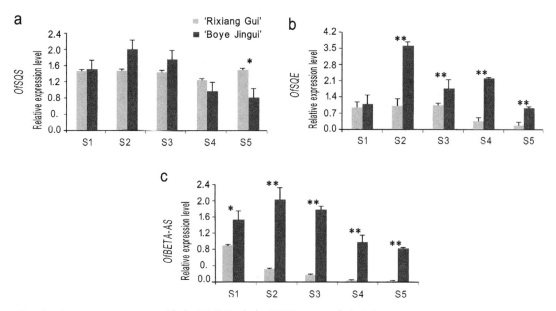

Fig. 5 Expression patterns of (a) *OfSQS*, (b) *OfSQE*, and (c) *OfBETA-AS* genes at five different flowering stages of *O. fragrans*

These cDNA templates were isolated from bud-pedicel stage (S1), bud-eye stage (S2), primary blooming stage (S3), full blooming stage (S4) and flower fading stage (S5). Data were presented with error bars indicating Standard Error of three independent biological samples with three technical repeats. Asterisk (*) indicate a significant difference at $P < 0.05$ and asterisks (* *) indicate a significant difference at $P < 0.01$ according to Tukey's test.

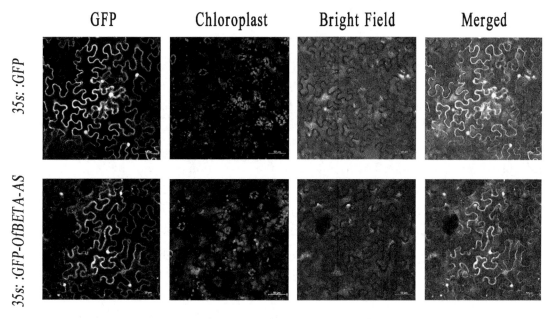

Fig. 6 Subcellular localization of *OfBETA-AS* in leaves of tobacco. *OfBETA-AS* fused with the GFP and GFP positive control were infiltrated into the leaves of tobacco through *Agrobacterium tumefaciens* strain EHA105. After incubated at 21 ℃ with 14 h photoperiod for 2 days, GFP fluorescence signals were observable by a Laser Scanning Confocal Microscope. Scale bar 50 μm

4 Discussion

Triterpenes are natural compounds that are mainly extracted from higher plants[18]. Genes involved in the biosynthesis of triterpenes have been identified and studied in numerous plants including, *A. thaliana* [19], *Panax ginseng* [21], *Glycyrrhiza uralensis* [21], and *Rleutherococcus senticosus* [22]. However, little research has been carried out on sweet osmanthus.

Analysis of amino acid sequence alignment revealed that among the 6 conserved domains of OfSQS, domain III, IV, and V were more conservative [23]. Moreover, large amounts of aspartic acid (DXXDD) existed in domain II and IV, which would influence the combination of FPP and Mg^{2+} [20,24]. Domain V was the binding domain of NADPH controlling the transition to format squalene [24]. Domain VI was considered as the anchoring signal of biological membrane [23]. Amino acid sequence (135-165) of OfSQE indicated to be the putative flavin adenine dinucleotide (FAD) binding domains, which played an important role in the biosynthesis of triterpenoids [25-26]. Sequence analysis revealed that OfBETA-AS contained four QW motifs, the DCTAE motif, and the MWCYCR motif among the high conserved regions, and it was reported that the QW and DCTAE motifs were present in all β-amyrin synthase and OSC superfamily [14,28]. The DCTAE motif could make squalene epoxide ring protonated, and it was thought to be in charge of initiating the cyclization reaction. The MWCYCR motif would be related with the specific formation of β-amyrin [27].

Homology analysis showed that the deduced protein sequences shared highly similar identity with other plants. OfSQS, OfSQE, and OfBETA-AS showed 91%, 86%, and 91% identity with known proteins from *S. indicum*, respectively. This suggests a relationship of evolutionary conservatism between *O. fragrans* and *S. indicum*, both of which belong to Tubiflorae. Moreover, OfBETA-AS shared the highest similarity of 99% with *O. europaea*, which also belongs to the Olea family, and the functional expressions of ObBETA-AS (*Ocimum basilicum*) led to the production of β-amyrin [28,29].

Analysis of the three genes revealed that the length and position of introns/exons in *OfSQS* and *OfBETA-AS* were consistent with previous report of *AtSQS*1 and *VvBETA-AS* (*Vitis vinifera*), respectively [30-31]. However, there were seven introns in *AtSQE*6 [32]. The introns numbers could vary dependent on species. No alternative splicing phenomenon was identified in different organs during flower developments, indicating that the functions of these three genes could be stable.

Triterpenoids are the downstream products of the MVA and MEP pathway. Up-regulation of *SQS* can significantly improve the expression level of *SQE* and *BETA-AS*, suggesting that *SQS* plays a pivotal role regulating the biosynthetic of triterpenoids [33]. Highest expression levels of *SQS* were reported in the vegetative organs of *Withania somnifera* and *Stevia rebaudiana* [34-35]. In addition, some similar observations found that higher *SQS* expression was in the root of other plants, such as *P. ginseng* and *Medicago truncatula* [10,18,20]. However, the transcript results of this study showed that *OfSQS* was constitutively expressed in the root, stem, leaf, and flower tissues, with the highest expression in the flowers in both 'Rixiang Gui' and 'Boye Jingui', which was different with results from *Aralia elata* flowers which showed lower level of expression [36]. Therefore, the high expression might result in the accumulation of specific triterpenoids in flowers of *O. fragrans*.

It was reported that the content of essential oil of *O. fragrans* was the highest at the initial flowering stage [7]. The three genes shared the same expression patterns in 'Boye', showing considerable variations with a clear peak in S2 stage, suggesting that the three genes could be involved in the synthesis of essential oil in sweet osmanthus. Previous studies showed that the aroma of 'Boye Jingui' was denser than 'Rixiang Gui', and in this study, the expression level of three genes was higher in 'Boye Jingui' than that of 'Rixiang Gui'. This indicates that further study of the relationship between these genes and the fragrance of sweet osmanthus is needed. Besides, the expression level of *OfSQS* showed no significant difference among the five stages, which was consistent with the result in *Withania somnifera* [34]. However, *OfSQE* and *OfBETA-AS* showed a typical trend of change at five stages. Therefore, *OfSQE* and *OfBETA-AS* would play an important role in the formation of triterpenes in sweet osmanthus. In addition, the role of *OfBETA-AS* in the formation of triterpenes could be more critical, for the expression difference of *OfSQE* in stage S1 was not significant in 'Boye Jingui' and 'Rixiang Gui', while *OfBETA-AS* showed a significant difference patterns during the five stages. OfBETA-AS: GFP fusion protein was detected in nucleus and membrane, which accorded with the localization characteristics of structural proteins.

5 Conclusions

In this study, three triterpenoids pathway genes, *OfSQS*, *OfSQE*, and *OfBETA-AS* from *O. fragrans* were cloned, and the exons/introns structure of the three genes were further analyzed. The phylogenetic analysis revealed an evolutionarily conserved relationship between genes of *O. fragrans* and other plants. The functions of the three genes could be stable, and no alternative splicing phenomenon was found. The qRT-PCR results showed that *OfSQS*, *OfSQE*, and *OfBETA-AS* genes had a clear flower-specific expression pattern, supporting the hypothesis that the enormous expression might result in the specific accumulation of triterpenes in the flower of *O. fragrans*. The expression level of the *OfSQS* gene did not show a significant change at the five flowering stages. However, the significantly different expression pattern of *OfSQE* gene in cultivars 'Boye Jingui' and 'Rixiang Gui' was detected from S2 to S5 stages. In addition, the expression patterns of the *OfBETA-AS* gene in cultivars 'Boye Jingui' and 'Rixiang Gui' were significantly different during the whole flowering stages, showing that the *OfBETA-AS* gene had an important influence on the production and accumulation of triterpenoids in two *Osmanthus* cultivars from the very beginning. Therefore, *OfBETA-AS* could play a more critical role in the synthesis of triterpenoids. The gene functional research will further be conducted in our following work.

Supplementary Materials

Tab. S1 Primers used for partial cDNA cloning

	Primer sequence (5'-3')	PCR condition
OfSQS-F	TTTGCTCTCGTCATTCAA	94 ℃ 3 min, 30 cycles of (94 ℃ 30 s;
OfSQS-R	TTGTTTCTAGCCTCTCCTTTG	52 ℃ 30 s; 72 ℃ 2 min), 72 ℃ 10 min
OfSQE-F	GGCGTTAGCTTATACACTTGG	94 ℃ 3 min, 30 cycles of (94 ℃ 30 s;
OfSQE-R	ACAAGATTCAAAGGGTGCGGA	53 ℃ 30 s; 72 ℃ 2 min), 72 ℃ 10 min
OfBETA-AS-F	CCTATGAAACAGCCACGACT	94 ℃ 3 min, 30 cycles of (94 ℃ 30 s;
OfBETA-AS-R	CACCAGACATAAGACCTAGCA	51 ℃ 30 s; 72 ℃ 2 min), 72 ℃ 10 min

Tab. S2 Primers used for 3 and 5 RACE reaction

Usage	Gene name	Primer sequence (5'-3')	PCR condition
3' RACE Outer PCR	OfSQS-OF	CAGCTAAACTTATGGACCGAACC	94 ℃ 3 min, 25 cycles of (94 ℃ 30 s;55/56/58 ℃ 30 s;72 ℃ 2 min),72 ℃ 10 min
	OfSQE-OF	TGCCAAATAGAAGTATGCCAGC	
	OfBETA-AS-OF	GGCATTCAAGCGTCTACATCCAG	
3' RACE Inner PCR	OfSQS-3'IF	ATTTCTCTCGCATGCTGAAGTCTAAGGTGG	94 ℃ 3 min, 35 cycles of (94 ℃ 30 s;67/65/65 ℃ 30 s; 72 ℃ 1.5 min),72 ℃ 10 min
	OfSQE-3'IF	ATTTAAATGATGCGGCTACCCTGTCCAAGT	
	OfBETA-AS-3'IF	TACAAACATCGTGGGCCATGCTAGGTCTTAT	
5' RACE Outer PCR	OfSQS-5'OR	ATTTCTCTCGCATGCTGAAGTCTAAGGTGG	94 ℃ 3 min, 25 cycles of (94 ℃ 30 s;68/69/68 ℃ 30 s; 72 ℃ 2 min),72 ℃ 10 min
	OfSQE-5'OR	CCACCTTCGGAGTACAAAGAGAGCGCCTCA	
	OfBETA-AS-5'OR	GCAATCGCACCATTACCATCGTCAGACCCT	
5' RACE Inner PCR	OfSQS-5'IR	CAGCTAAACTTATGGACCGAACC	94 ℃ 3 min, 35 cycles of (94 ℃ 30 s;54/55/55 ℃ 30 s; 72 ℃ 1.5 min), 72 ℃ 10 min
	OfSQE-5'IR	AAGCAAGGATGTTACCGTTC	
	OfBETA-AS-5'IR	TGCTCCACTGATATACAAGAC	

Tab. S3 Primers used for gDNA cloning

Gene name	Primer sequence (5'-3')	PCR condition
OfSQS-F	CTCATTTCGCACTAGTCGTT	94 ℃ 3 min, 35 cycles of (94 ℃ 30 s;
OfSQS-R	GTGCATATTTACTTGGATGCTT	54 ℃ 30 s; 72 ℃ 1 min), 72 ℃ 10 min
OfSQE-F	ACAATCAAAGGAGTGCAA	94 ℃ 3 min, 35 cycles of (94 ℃ 30 s;
OfSQE-R	ATTATCAAATGCCCATAGCG	52 ℃ 30 s; 72 ℃ 1 min), 72 ℃ 10 min
OfBETA-AS-F	CAGTAATTCTACGTGCAGCTTC	94 ℃ 3 min, 35 cycles of (94 ℃ 30 s;
OfBETA-AS-R	CTCTTTTCTTTCTCATGGACGG	56 ℃ 30 s; 72 ℃ 2 min), 72 ℃ 10 min

Tab. S4 Primers used for qRT-PCR

Gene name	Primer sequence (5'-3')	Amplification length
OfSQS-qF	AGTCACCGAGAAAGGCAGA	139 bp
OfSQS-qR	GGAAGCGACGCAAGAAAA	
OfSQE-qF	GGGCATTTGATGTAGCATA	122 bp
OfSQE-qR	ATCCTTCCCTTACCAATCT	
OfBETA-AS-qF	CCTCAAATTCCAAGGTGGGT	227 bp
OfBETA-AS-qR	GGATTCTCCGCAACTATGTC	
RAN-qF	AGAACCGACAGGTGAAGGCAA	117 bp
RAN-qR	TGGCAAGGTACAGAAAGGGCT	
RPB2-qF	CACCAAGCAAAGGACCAGCAAG	216 bp
RPB2-qR	TCACCAGGGAGAAGAGGATCAAGTA	

Tab. S5 Primers used for construction of vector

Gene name	Primer sequence (5'-3')	PCR condition
OfBETA-AS-F	aagcttctgcaggggcccgggATGTGGAAGCTTAAGATTGCTGAA	94 ℃ 3 min, 35 cycles of (94 ℃ 30 s; 54 ℃ 30 s; 72 ℃ 2 min), 72 ℃ 10 min
OfBETA-AS-R	gcccttgctcaccatggtaccCAGGCTTTGAGATGGCCACA	

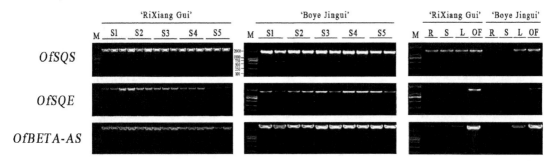

Fig. S1 Expression of *OfSQS*, *OfSQE*, and *OfBETA-AS* in different flowering stages (left) and different tissues (right) in 'Rixiang Gui' and 'Boye Jingui'
M (DNA Maker): DL 2000 (2,000 bp, Takara Biotechnology); S1: bud-pedicel stage; S2: bud-eye stage; S3: primary blooming stage; S4: full blooming stage; S5: flower fading stage; R: root; S: stem; L: leaf and OF: opening flower.

Acknowledgements

This work was funded by the Application Demonstration of the Key Technology and Innovation in Breeding the Six Precious Colored Tree Species of Jiangsu Province (No. BE2017375), the top-notch Academic Programs Project of Jiangsu Higher Education Institutions (TAPP, PPZY2015A063), and the Innovative Plan of Academic Degree Graduate Students in Jiangsu Province.

Conflict of interest

The authors declared that there no conflicts of interest concerning the publication of this paper.

References

[1] Huang B, Chen HQ, Shao LQ. The ethanol extract of *Osmanthus fragrans* attenuates *Porphyromonas gingivalis* lipopolysaccharide-stimulated inflammatory effect through the nuclear factor erythroid 2-related factor-mediated antioxidant signaling pathway. Archives of Oral Biology 2015;60:1030-8. https://doi.org/10.1016/j.archoralbio.2015.02.026.

[2] Zhang C, Wang YG, Fu JX, Bao ZY, Zhao HB. Transcriptomic analysis and carotenogenic gene expression related to petal coloration in *Osmanthus fragrans* 'Yanhong Gui'. Trees 2016;30:1207-23. https://doi.org/10.1007/s00468-016-1359-8.

[3] Baldermann S, Kato M, Kurosawa M, Kurobayashi Y, Fujita A, Fleischmann P, *et al*. Functional characterization of a carotenoid cleavage dioxygenase 1 and its relation to the carotenoid accumulation and volatile emission during the floral development of *Osmanthus fragrans* Lour. Journal of Experimental Botany 2010;61:2967-77. https://doi.org/10.1093/jxb/erq123.

[4] Xu C, Li HG, Yang X, Gu C, Mu HN, Yue YZ, *et al*. Cloning and expression analysis of MEP pathway enzyme-encoding genes in *Osmanthus fragrans*. Genes 2016;7:78. https://doi.org/10.3390/genes7100078.

[5] Wang LM, Li MT, Jin WW, Li S, Zhang SQ, Yu LJ. Variations in the components of *Osmanthus fragrans* Lour. essential oil at different stages of flowering. Food Chemistry 2009;114:233-6. https://doi.org/10.1016/j.foodchem.2008.09.044.

[6] Xiong LN, Mao SQ, Lu BY, Yang JJ, Zhou F, Hu YZ, *et al*. *Osmanthus fragrans* Flower Extract and acteoside protect against d-galactose-induced aging in an ICR mouse model. Journal of Medicinal Food 2016;19:54-61. https://doi.org/10.1089/jmf.2015.3462.

[7] Yao WR, Zhang YZ, Chen Y, Yang ZP. Aroma enhancement and characterization of the absolute *Osmanthus fragrans* Lour. Journal of Essential Oil Research 2010;22:97-102. https://doi.org/10.1080/10412905.2010.9700272.

[8] Tang Q, Ma XJ, Mo CM, Song C, Zhao H, Yang YF, *et al*. An efficient approach to finding *Siraitia grosvenorii* triterpene biosynthetic genes by RNA-seq and digital gene expression analysis. BMC Genomics 2011;12:343. https://doi.org/10.1186/1471-2164-12-343.

[9] Abe I, Rohmer M, Prestwich GD. Enzymatic cyclization of squalene and oxidosqualene to sterols and triterpenes. Cheminform 1993;93:2189-206. https://doi.org/10.1021/cr00022a009.

[10] Lee MH, Jeong JH, Seo JW, Shin CG, Kim YS, In JG, *et al*. Enhanced triterpene and phytosterol biosynthesis in *Panax ginseng* overexpressing squalene synthase gene. Plant Cell Physiol. 2004;45:976-984. https://doi.org/10.1093/pcp/pch126.

[11] Kim YS, Cho JH, Park S, Han JY, Back K, Choi YE. Gene regulation patterns in triterpene biosynthetic pathway driven by overexpression of squalene synthase and methyl jasmonate elicitation in *Bupleurum falcatum*. Planta 2011;233:343-55. https://doi.org/10.1007/s00425-010-1292-9.

[12] Ferrer M, Chernikova TN, Timmis KN, Golyshin PN. Expression of a temperature-sensitive esterase in a novel chaperone-based Escherichia coli Strain. Appl Environ Microbiol 2004;70:4499-504. https://doi.org/10.1128/AEM.70.8.4499-4504.2004.

[13] Gao K, Wu SR, Wang L, Xu YH, Wei JH, Sui C. Cloning and analysis of β-amyrin synthase gene in *Bupleurum chinense*. Genes & Genomics 2015;37:767-74. https://doi.org/10.1007/s13258-015-0307-0.

[14] Basyuni M, Oku H, Tsujimoto E, Kinjo K, Baba S, Takara K. Triterpene synthases from the *Okinawan mangrove* tribe, Rhizophoraceae. Febs Journal 2007;274:5028-42. https://doi.org/10.1111/j.1742-4658.2007.06025.x.

[15] Yue YM, Wang JM, Kang WY. *Osmanthus fragrans* triterpenoids and hypolipidemic effect. Chinese Journal of Experimental Traditional Medical Formulae 2013;19:126-128. https://doi.org/10.11653/syfj201324000.

[16] Zhang C, Fu JX, Wang YG, Bao ZY, Zhao HB. Identification of suitable reference genes for gene expression normalization in the quantitative Real-Time PCR analysis of Sweet Osmanthus (*Osmanthus fragrans* Lour.). PLoS ONE 2015;10:e0136355. https://doi.org/10.1371/journal.pone.0136355.

[17] Yue YZ, Yin CQ, Rui G, Peng H, Yang ZN, Liu GF, et al. An anther-specific gene PhGRP is regulated by PhMYC2 and causes male sterility when overexpressed in petunia anthers. Plant Cell Reports 2017;36:1401-15. https://doi.org/10.1007/s00299-017-2163-7.

[18] Iturbe-Ormaetxe I, Haralampidis K, Papadopoulou K, Osbourn AE. Molecular cloning and characterization of triterpene synthases from *Medicago truncatula* and *Lotus japonicus*. Plant Molecular Biology 2003;51:731-43. https://doi.org/10.1023/A:1022519709298.

[19] Mirjalili MH, Moyano E, Bonfill M, Cusido RM, Palazón J. Overexpression of the *Arabidopsis thaliana* squalene synthase gene in *Withania coagulans* hairy root cultures. Biologia Plantarum 2011;55:357-60. https://doi.org/10.1007/s10535-011-0054-2.

[20] Choi DW, Jung J, Ha YI, Park HW, In DS, Chung HJ, et al. Analysis of transcripts in methyl jasmonate-treated ginseng hairy roots to identify genes involved in the biosynthesis of ginsenosides and other secondary metabolites. Plant Cell Reports 2005;23:557-66. https://doi.org/10.1007/s00299-004-0845-4.

[21] Liu Y, Zhang N, Chen HH, Gao Y, Wen H, Liu CS, et al. Cloning and characterization of two cDNA sequences coding squalene synthase involved in glycyrrhizic acid biosynthesis in *Glycyrrhiza uralensis*. Lecture Notes in Electrical Engineering 2014;269:329-42. https://doi.org/10.1007/978-94-007-7618-0_32.

[22] Seo JW, Jeong JH, Shin CG, Lo SC, Han SS, Yu KW, et al. Overexpression of squalene synthase in *Eleutherococcus senticosus* increases phytosterol and triterpene accumulation. Phytochemistry 2005;66:869-77. https://doi.org/10.1016/j.phytochem.2005.02.016.

[23] Robinson GW, Tsay YH, Kienzle BK, Smithmonroy CA, Bishop RW. Conservation between human and fungal squalene synthetases: similarities in structure, function, and regulation. Molecular & Cellular Biology 1993;13:2706-17. https://doi.org/10.1128/MCB.13.5.2706.

[24] Tansey TR, Shechter I. Squalene synthase: structure and regulation. Progress in Nucleic Acid Research & Molecular Biology 2000;65:157-95. https://doi.org/10.1016/S0079-6603(00)65005-5.

[25] Han JY, In JG, Kwon YS, Choi YE. Regulation of ginsenoside and phytosterol biosynthesis by RNA interferences of squalene epoxidase gene in *Panax ginseng*. Phytochemistry 20107;1:36-46. https://doi.org/10.1016/j.phytochem.2009.09.031.

[26] Abe I, Prestwich GD. Identification of the active site of vertebrate oxidosqualene cyclase. Lipids 1995;30:231-4. https://doi.org/10.1007/BF02537826.

[27] Jin ML, Lee DY, Um Y, Lee JH, Park CG, Jetter R, et al. Isolation and characterization of an oxidosqualene cyclase gene encoding a β-amyrin synthase involved in *Polygala tenuifolia* Willd. saponin biosynthesis. Plant Cell Reports 2014;33:511-9. https://doi.org/10.1007/s00299-013-1554-7.

[28] Saimaru H, Orihara Y, Tansakul P, Kang YH, Shibuya M, Ebizuka Y. Production of triterpene acids by cell suspension cultures of *Olea europaea*. Chemical & Pharmaceutical Bulletin 2007;55:784-8. https://doi.org/10.1248/cpb.55.784.

[29] Misra RC, Maiti P, Chanotiya CS, Shanker K, Ghosh S. Methyl jasmonate-elicited transcriptional responses and pentacyclic triterpene biosynthesis in sweet basil. Plant Physiology 2014;164:1028-44. https://doi.org/10.1104/pp.113.232884.

[30] Kribii R, Arró M, Del AA, González V, Balcells L, Delourme D, et al. Cloning and characterization of the *Arabidopsis thaliana* SQS1 gene encoding squalene synthase-involvement of the C-terminal region of the enzyme in the channeling of squalene through the sterol pathway. European Journal of Biochemistry 1997;249:61-9. https://doi.org/10.1111/j.1432-1033.1997.00061.x.

[31] Jaillon O, Aury JM, Noel B, Policriti A, Clepet C, Casagrande A, et al. The grapevine genome sequence suggests ancestral hexaploidization in major angiosperm phyla. Nature 2007;449:463-7. https://doi.org/10.1038/nature06148.

[32] Tabata S, Kaneko T, Nakamura Y, Kotani H, Kato T, Asamizu E, et al. Sequence and analysis of chromosome 5 of the plant *Arabidopsis thaliana*. Nature 2000;408:823-6. https://doi.org/10.1038/35048507.

[33] Braga MV, Urbina JA, Souza WD. Effects of squalene synthase inhibitors on the growth and ultrastructure of *Trypanosoma cruzi*. International Journal of Antimicrobial Agents 2004;24:72-8. https://doi.org/10.1016/j.ijantimicag.2003.12.009.

[34] Bhat WW, Lattoo SK, Razdan S, Dhar N, Rana S, Dhar RS, et al. Molecular cloning, bacterial expression and promoter analysis of squalene synthase from *Withania somnifera* (L.) Dunal. Gene 2012;499:25-36. https://doi.org/10.1016/j.gene.2012.03.004.

[35] Kumar H, Kaul K, Bajpai-Gupta S, Kaul VK, Kumar S. A comprehensive analysis of fifteen genes of steviol glycosides biosynthesis pathway in *Stevia rebaudiana* (Bertoni). Gene 2012;492:276-84. https://doi.org/10.1016/j.gene.2011.10.015.

[36] Wu Y, Zou HD, Cheng H, Zhao CY, Sun LF, Su SZ, et al. Cloning and characterization of a β-amyrin synthase gene from the medicinal tree *Aralia elata* (Araliaceae). Genetics & Molecular Research 2012;11:2301-14. https://doi.org/10.4238/2012.August.13.4.

Transcriptomic Analysis of the Candidate Genes Related to Aroma Formation in *Osmanthus fragrans* *

Xiulian Yang †, Haiyan Li †, Yuanzheng Yue, Wenjie Ding,
Chen Xu, Tingting Shi, Gongwei Chen, Lianggui Wang

College of Landscape Architecture, Nanjing Forestry University, Nanjing 210037, China.
† These authors contributed equally to this work and should be considered co-first authors.

Abstract: *Osmanthus fragrans* 'Rixiang Gui' is an ornamental woody evergreen plant that is cultivated widely, because it blooms recurrently and emits strong fragrance. Recently, the germplasm resources, classification and aroma compositions of *O. fragrans* had been investigated. However, the molecular mechanisms of floral scent formation and regulation remained largely unknown. In order to obtain a global perspective on the molecular mechanism of aroma formation during blooming, nine RNA-Seq libraries were constructed from three flowering stages: the initial, full, and final flowering stage. In a word, a total of 523,961,310 high-quality clean reads were assembled into 136,611 unigenes, with an average sequence length of 792 bp. About 47.43% of the unigenes (64,795) could be annotated in the NCBI non-redundant protein database. A number of candidate genes were identified in the terpenoid metabolic pathways, and 1,327 transcription factors (TFs) which showed differential expression patterns among the floral scent formation stages were also identified, especially *OfMYB*1, *OfMYB*6, *OfWRKY*1, *OfWRKY*3, which could play critical roles in the floral scent formation. These results indicated that the floral scent formation of *O. fragrans* was a very complex process which involves a large number of TFs. This study would provide reliable resources for further studies of *O. fragrans* floral scent formation.

Key words: *Osmanthus fragrans*; transcriptome; flora scent; transcription factor

1 Introduction

Sweet osmanthus, a species of the Oleaceae family, is known for its fragrant flowers, which are rich in aromatic flavor when blooming. As one of the top ten traditional Chinese flowers, it had been cultivated for over 2,500 years. Through a long evolution, *O. fragrans* cultivars were classified into Thunbergii, Latifolius, Aurantiaeus and Semperfloren groups, the flowers of the first three cultivar groups bloomed in autumn, while the Semperfloren Group bloomed almost every month [1]. 'Rixiang Gui', which is a kind of cultivar in the Semperfloren Group with high economic value, was cultivated widely due to

---

* 原文发表于 *Molecules*, 2018, 23(7): 1604。

its important ornamental traits of floral scent and recurrent blooming.

Floral scent is one of the most important traits of ornamental plants and more than 1,700 floral constituents have been identified from 991 species of flowering plants and a few gymnosperms[2]. The components of floral scent are terpenoids, benzenes/phenylpropanes and fatty acid derivatives [3]. With the rapid development of science and technology, the separation method, analysis and identification technology of floral constitutes was promoted by using techniques such as headspace-solid phase, gas chromatography-olfactory, and gas chromatography-mass. According to the previous research, the main volatile constituents of sweet osmanthus were terpenoids [5]. The amount of concrete production of O. fragrans increased along with the flower opening, and was in the highest level at the full flowering stage, then decreased at the final flowering stage [4-6].

Terpenoids, the largest class of aromatic compounds, were derived from isopentenyl diphosphate and its homologous isomer dimethylallyl pyrophosphate. They were synthesized by different terpene synthases (TPSs) via the mevalonate (MVA) or methyl erythritol-4-phosphate (MEP) pathway [7]. The MEP pathway in plant cell plastids was a key approach for the synthesis of monoterpenes and their derivatives [8]. Xu et al. [9] isolated and identified 10 MEP pathway genes in sweet osmanthus. The release regularities of aroma were consistent with the circadian rhythms of the *OfDXS2* and *OfDHR1* expression patterns. Floral scent released from flowers or leaves was closely related to the expression level of TPSs [10, 12]. The major floral volatiles, linalool and its derivatives produced in *O. fragrans* flowers were encoded by *OfTPS*1, *OfTPS*2, and *OfTPS*3 [13]. Moreover, the dioxygenase cleavage step itself could generate volatile products. For example, carotenes were cleaved into α-ionone and β-ionone by *OfCCD*1 in *vitro* assays [14]. Except these, there was no further information available concerning the molecular mechanism of floral scent formation in *O. fragrans*.

Transcription factors (TFs) were proteins combined with *cis*-elements to regulate gene expression spatially and temporally, and also involved in the whole process of plant growth and development. Increasing reports indicated that terpene synthesis was governed by the transcriptional regulatory networks. In cotton, *GaWRKY*1 binded to the promoter of sesquiterpene cyclase positively regulated the gossypol biosynthesis [15]. WRKY3 and WRKY6 were found to be related to volatile terpene production for defense against pests in tobacco [16]. MYC2, a basic helix-loop-helix TF, interacted with DELLA protein to activate expression of sesquiterpene synthase genes TPS21 and TPS11 in *Arabidopsis* inflorescence [17]. MYB14 participated in the regulation of the volatile terpenoid biosynthesis preferentially via the MVA pathway in a conifer tree species [18]. The bZIP transcription factor HY5 interacted with the promoter of the monoterpene synthase gene QH6 in modulating its rhythmic expression [19]. *Arabidopsis* PAP1, a MYB TF, resulted in increasing phenylpropanoid-derived color and terpenoids in PAP1-transgenic rose, but the regulatory mechanism remained unclear[20]. *Solanum lycopersicum* WRKY73 activated the expression of three monoterpene synthase genes, suggesting that a single WRKY gene could regulate multiple distinct biosynthetic pathways [21]. The discoveries of these TFs provided an attractive strategy for modifying floral substances.

In the model plants Arabidopsis, Petunia and other herb plants like Snapdragon, Lavender and Nicotiana, there had been many studies on molecular mechanisms about formation of aromatic compounds. However, *O. fragrans* was a woody aromatic plant whose aromatic constitutes were terpenoids. The available gene resources were limited either. RNA sequencing (RNA-Seq) technology, an effective tool in genomic exploration and gene discovery in plants when no reference genome was available [22], could provide new insights into the biosynthesis and regulatory mechanisms of floral scents in *O. fragrans*. In this study, nine cDNA libraries of three different flowering stages including three biological replicates were constructed in *O. fragrans* for Illumina RNA-Seq. Analysis of differentially expressed genes (DEGs) and quantitative Real-Time PCR (qRT-PCR) were performed, aiming to excavate interesting genes and TFs associated with aromatic metabolism.

2 Results

2.1 Sequencing and *de novo* Assembly

The flowering process could be divided into four periods, i.e., the xiangyan stage, the initial flowering stage, the full flowering stage and the final flowering stage [23]. The amount of volatiles produced by *O. fragrans* petals gradually increased along with the flower opening, and reached the highest level at the full flowering stage, then declined at the final flowering stage [6]. To obtain genomics resource, narrow down the special pathways and get candidate genes of interest in *O. fragrans*, the petals from three individual plants were collected at the initial flowering (S1), full flowering (S2) and final flowering (S3) stages (Fig. 1). Totally, nine RNA-Seq libraries were constructed using above RNA samples and sequenced by Illumina HiSeq™ 4000.

Approximately 532,121,476 raw reads were obtained and 523,961,310 high-quality clean reads remained after filtering. The sequencing result was assembled into a transcriptome using Trinity to act as a reference sequence for subsequent analysis (Additional File 1). All transcriptomic data had been deposited in the NCBI Sequence Reads Archive (SRA) under the accession number SRP143423. The longest transcript of each gene was considered to be a unigene. Totally, there were 136,611 unigenes obtained with a maximum length of 16,876 bp, minimum length of 201 bp, average length of 792 bp, and N50 length of 1,424 bp (Tab. 1). The distribution of unigene length was shown as follows (Additional File 2).

Fig. 1　Photograps of flowers at different flowering stages(S1: initial flowering stage, S2: full flowering stage, S3: final flowering stage).

Tab. 1　Sumamary of the *O. fragrans* transcriptome

Total assembled bases	108,311,010
Total number of genes	136,611
Max length of unigenes (bp)	16,876
Min length of unigenes (bp)	201
Average length of unigenes (bp)	792
N50 (bp)	1,424
GC percentage (%)	38.921,5

2.2　Annotation and Functional Classification

To determine the putative functions of unigenes, the transcriptome was annotated using protein function, pathway, KOG function and GO annotations. The unigenes were aligned using the BLASTx program with an E-value threshold of 10^{-5} to the Nr, Swiss-Prot protein, KEGG and KOG databases, for which the percentages of annotated unigenes were 42.86, 34.62, 28.58, and 17.44%, respectively (Tab. 2). Generally speaking, approximately 47.43% of unigenes were aligned. According to the best alignment results to the species of homologous sequences, the number of homologous sequences of the top 10 species was shown as follows (Additional File 3).

Tab. 2　Statistics of annotation on unigenes against public databases

Database	Number of annotated unigene	Percentage of annotated unigenes (%)
Nr	58,556	42.86
Swiss-prot	47,294	34.62
KOG	39,043	28.58
KO	23,825	17.44
Total	64,795	47.43

The GO offered a strictly defined concept to comprehensively describe properties of genes and their products in any organism. The GO functional annotations of the 'Rixiang

Gui' unigenes were classified into three categories (Molecular Function, Cellular Component and Biological Process) and 50 subcategories (Fig. 2). There were 77,416 (56.67%) unigenes assigned to 20 subcategories in the Biological Process for which "metabolic process", "cellular process", and "single-organism process" were dominant subcategories. Molecular function categories with the most abundant members were "catalytic activity" and "binding"; in the Cellar category, "cell", "cell part" and "organelle" were abundant classes.

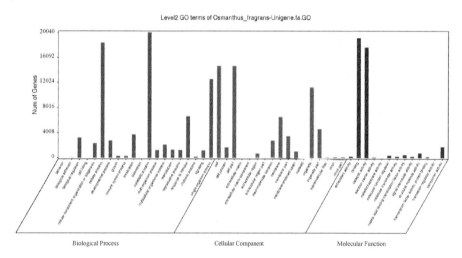

Fig. 2　GO functional classification of differentially expressed genes (DEGs) among different samples

In total, 136,611 of the 39,043 unigenes showing Nr hits were assigned to COG classifications and grouped into 25 categories (Additional File 4). The cluster of "General function prediction" was the largest group, representing 35.72%, followed by "Post-translational modification, protein turnover, chaperones" and "Signal transduction mechanisms". There were 1,537 unigenes in "Secondary metabolites biosynthesis, transport and catabolism".

A total of 23,825 unigenes were annotated and 23,166 sequences were assigned to 132 KEGG pathways (Additional File 5). Metabolism (~61.21%) had the largest number of sequences in the primary pathway hierarchy, and unigenes of Genetic Information Processing followed (~27.15%). Flower scents and colors were the main ornamental traits of O. fragrans, so we focused on secondary metabolic pathways, especially those of phenylpropanoids, carotenoids, flavonoids, flavones and flavonol, anthocyanins and terpenoids. These data provided a valuable resource and narrowed down the special pathways and candidate genes of interest in O. fragrans.

2.3　Analysis of DEGs

During the flowering process, the petals of O. fragrans released many aromatic compounds. To characterize gene expressions at S1, S2 and S3 flowering stages, a global expression analysis was performed for each stage using three biological replicates. DEG libraries were constructed, and then the unigenes were mapped to the *O. fragrans* reference transcriptome after sequencing by the Illumina platform. The expression of each gene was calculated in RPKM. We stratified the relationship of the samples based on the expression level of the whole gene. A dendrogram was plotted and all biological replicates were clustered together (Fig. 3). In addition, the results of two parallel experiments to evaluate the reliability and operational stability of the experiments (Additional File 6) showed that the correlation index between the two experiments for the same sample was close to be 1, indicating a high reproducibility. These revealed that all biological replicants were well correlated, and similarly for S2 and S3.

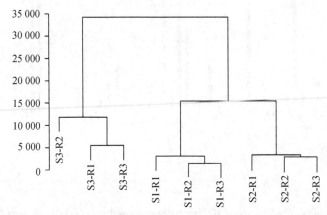

Fig. 3 Global analysis of transcriptome datasets of biological replicates and samples was shown
The dengrogram depicted global relationships of the samples. The bar plot described the number of expressed transcripts after filtering.

For better understanding of the gene expression profiles, Short Time-series Expression Miner (STEM) software was used to cluster gene expression patterns. Eight gene clusters with distinctive expression patterns were identified (Fig. 4). Trend blocks with color indicated a significant enrichment trend, clusters 7 and 8 containing 5,976 and 4,821 transcripts, respectively, showed opposite patterns (Fig. 4). Cluster 7 showed a gradual increase from S1 to S3, while cluster 8 showed a gradual decrease, indicating that some genes had opposite roles during flowering process. Cluster 4 (6,128 genes) remained constant, then gradually increased, but cluster 6 gradually increased firstly and then kept constant. The gene expression profiles revealed that the clusters were involved in the biological events during flowering process.

To further investigate the DEGs among groups, transcriptomes generated from different stages were compared. DEGs were filtered by using a false discovery rate (FDR) < 0.05 and $|\log_2 FC| > 1$. Comparing S1 with S2 resulted in 16,087 DEGs: 5,455 up-regulated and 10,632 down-regulated. When comparing S1 with S3, there were 29,631 DEGs: 16,386 were up-regulated and 13,245 were down-regulated. The comparison of S2 with S3 showed 16,072 DEGs: 10,404 were up-regulated and 5,668 were down-regulated (Fig. 5).

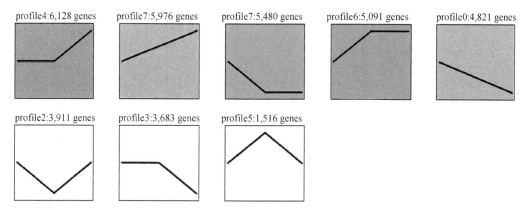

Fig. 4 Cluster analysis of differentially expressed genes was as above

Profiles in color indicated statistically significance ($P < 0.01$). The number on the top was the profile ID number and assigned genes. The black lines represented the model expression profiles of different genes.

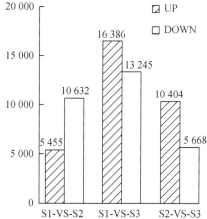

Fig. 5 Analysis of the differentially expressed unigenes (DEGs) during flowering in *O. fragrans* were listed with different comparisons, including S1 vs. S2, S1 vs. S3, S2 vs. S3

The red indicated up-regulated unigenes, while the green indicated down-regulated unigenes. The x-axis indicated the absolute expression levels (Log Conc). The y-axis indicated the log-fold changes between the two samples.

2.4 Identification of Unigenes Involved in Scent Metabolism

In previous studies, the *O. fragrans* floral constituents were found to be volatile terpeniods. The terpenoid metabolic pathway genes and many of the biosynthetic enzymes involved have been identified and functionally characterized. The key genes involved in the terpenoid pathway were identified according to the KEGG annotation and local BLASTx search (Tab. 3). A total of 9 unigenes involved in the terpenoid biosynthesis were selected to validate the sequencing data. The qRT-PCR was carried out with independent samples collected from the different flowering stages to analyze their differential expression pattern (Fig. 6A). The primer sequences were listed (Additional File 7).

Tab. 3 KEGG enrichment of DEGs among three flowering stages

KEGG Pathway	All genes with pathway annotation	DGEs genes with pathway annotation	Pathway ID
Phenylpropanoid	319	178	Ko00940
Monoterpenoid biosynthesis	17	8	Ko00902
Terpenoid backbone biosynthesis	184	78	Ko00900
Sesquiterpenoid and triterpenoid biosynthesis	69	45	Ko00909
Diterpenoid biosynthesis	45	25	Ko00904
Limonene and pinene degradation	15	10	Ko00903

2.5 Analysis of TFs During Blooming

The TFs are crucial in regulating gene expression involved in plant growth, development and other physiological functions. According to searches against the Pln TFDB, 2,269 TF genes were obtained and classified into 57 TF families. The top 10 families were C2H2 (234), ERF (193), bHLH (168), MYB-related (155), NAC (134), WRKY (134), MYB (96), bZIP (89), GRAS (82) and C3H (73). Number of the TFs identified in differential expression profiles were shown (Tab. 4). The MYB and WRKY families had been shown to be involved in regulating secondary metabolism in plants, especially floral volatiles and pigments. There were 93 MYB and 77 WRKY TFs with different expression patterns at different flowering stages, depicted with a heat map using MeV software (Additional File 8). Significant differential expression TFs were selected to perform qRT-PCR. A few of the TF family members had the highest expression levels at initial flowering stage, and the lowest expression levels at the final flowring stage (Fig. 6B). Interestingly, the expression level of *OfMYB*1, *OfMYB*6, *OfWRKY*1 and *OfWRKY*3 reached the highest at the full flowering stage, declined at the final flowering stage, which was consistent with the release of floral volatiles (Fig. 6B).

Tab. 4 Number of the transcription factors identified in the expression profiles

Transcription factors	Profile0	Profile1	Profile4	Profile6	Profile7	All Profiles
C2H2	3	2	1	0	2	234
ERF	56	88	50	21	50	193
bHLH	24	26	13	5	4	168
MYB-related	4	6	6	3	5	155
NAC	101	158	110	47	80	134
WRKY	3	4	41	7	24	112
MYB	18	13	24	9	19	96
bZIP	4	3	2	0	0	89
GRAS	2	0	8	11	15	82
C3H	0	1	0	0	2	73
FAR1	3	6	11	6	10	67
Dof	11	3	2	0	1	59
G2-Like	3	0	1	1	1	56
HD-ZIP	0	1	0	0	0	52
Trihelix	4	2	2	0	3	50
HSF	2	1	0	1	0	43

Cont. Tab. 4

Transcription factors	Profile0	Profile1	Profile4	Profile6	Profile7	All Profiles
TCP	6	11	2	0	7	41
B3	6	11	10	8	8	40
GATA	14	10	0	1	3	34
SBP	4	5	0	0	0	32
ARF	0	2	2	1	1	29
M-type	4	3	11	5	7	22
NF-YC	0	1	0	0	0	16
AP2	4	7	5	2	1	14
CPP	0	1	0	0	0	12
WOX	1	1	0	0	0	11
LSD	0	0	1	0	0	9
YABBY	1	1	0	0	0	8
BES1	2	2	0	0	0	8
SRS	1	0	1	2	1	6

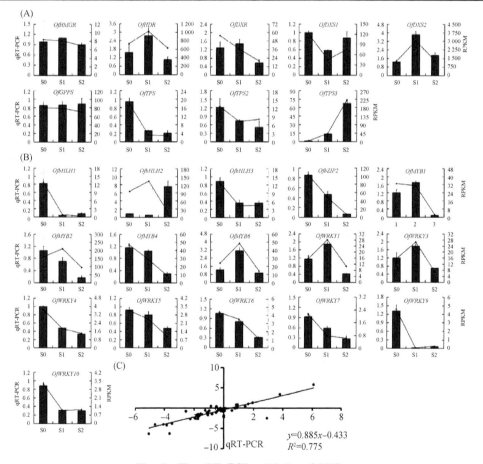

Fig. 6 The qRT-PCR validation of DEGs

(A) Expression patterns of genes encoding enzymes possibly involved in the terpenoid biosynthesis.
(B) Expression analysis of TFs in different flowering stages. Relative transcription level of flowers was set to be 1 (100%). Error bars indicated the calculated maximum and minimum expression quantity of three replicates. Different lowercase letters labeled on bars indicate statistically significant differences at the level of $P < 0.05$. S1, initial flowering stage; S2, full flowering stage; S3, final flowering stage.
(C) Correlation analysis of the gene expression ratios between qRT-PCR and RNA-Seq.

2.6 Validation of RNA-Seq Results by qRT-PCR

Nine key genes were selected in the terpenoid metabolic pathway and 16 TFs for qRT-PCR to verify the sequencing data. The primers of TFs were designed using Primer Primer 5.0 software, and were shown as follows (Additional File 9). The expression levels of these selected TFs in different flowering stages were depicted (Fig. 6B). The results showed that the expression patterns of these candidate genes obtained by qRT-PCR were largely consistent with RNA-Seq. Linear regression analysis revealed that the fold change values of qRT-PCR and RNA-Seq showed a strong positive correlation ($R^2 = 0.775$) at the level of $P \leqslant 0.01$ (Fig. 6C). These results indicated the RNA-Seq data were reliable.

3 Disscussion

Osmanthus fragrans 'Rixiang Gui', a traditional horticultural plant, known for its floral scent, is widely planted in Asia, especially in southern and central China [24]. Owing to their appealing fragrance and potentially exploitable chemical substances, the flowers were extensively used to produce tea, food, perfume and medicine [6]. However, little information is available to explain the transcriptional regulation of floral scent genes. As far as we know, there had been only two previous transcriptome reports based on RNA-Seq technology about the Oleaceae family. Using an orange-red-flowered cultivar 'Yanhong Gui' of *O. fragrans*, petal color, carotenoid content, transcriptome dynamics of flower buds, and expression of key genes for carotenoids were analyzed [25]. In our previous study, 197 and 237 candidate genes involved in fragrance and pigment biosynthesis were functionally annotated, respectively [26]. Here, nine libraries were constructed to perform RNA-Seq, and 136,611 unigenes were generated with the maximum length of 16,876 bp, the minimum length of 201 bp, the average length of 792 bp, and the N50 length of 1,424 bp. For the lack of available genetic information, only 47.43% of unigenes searched against the Nr database were matched, suggesting that a large portion of genes acting during blooming might involve some unique processes and pathways. The annotated unigenes of *O. fragrans* showed the highest homology to those of *Sesamum indicum* (Additional File 3), which may reflect a close evolutionary relationship between the two species.

Plants release an array of secondary metabolites to adapt to changing environments throughout their life cycles. Floral volatile organic compounds are a relatively large group which plays crucial roles in pollinator attraction, defense, communication and interaction with the surrounding environment. They were classified into three major groups: terpenoids, phenylpropanoids/benzenoids, fatty acid derivatives and aminoacid derivatives [27]. Terpenoids are the largest class of floral volatiles and encompass 556 scent compounds [28]. Flower scents vary between plant species on account of their different relative proportions of volatile compounds. Within a species the blend of emitted terpenoids differ quantitatively and qualitatively, with some compounds in common [29]. The β-ionone, cis-linalool oxide (furan), trans-linalooloxide (furan) and linalool were abundant in most *Osmanthus fragrans* cultivars, with other compounds typically present in smaller amounts [4].

The biosynthetic enzymes involved and emission of terpenes have been identified and functionally characterized in many plants, such as arabidopsis[30], snapdragon [31-32], *Clarkia breweri* [33-36], petunia [37], and rose [38-39]. Terpenoids, derived from prenyl diphosphate precursors, are synthesized by two independent pathways: the MVA pathway in the cytoplasm and the MEP pathway in plastids [40]. Based on sequence annotations and analysis of changes in gene expression during each flowering stage, the key floral scent-synthesizing genes DXS, DXR, HDR, TPS, GPPS were identified. The results indicated that expressions of DXS, DXR, HDR and monoterpene synthases were all positively correlated with monoterpene emission. DXS was the first gene identified and thought to be an important rate-controlling step of the MEP pathway [8]. Two *OfDXSs* showed differential expression patterns that the transcript level of *OfDXS1* was down-regulated at full flowering stage, while *OfDXS2* increased sharply. It seemed that *OfDXS2* transcript was correlated with monoterpene emission. These genes showed similar expression patterns in *Hedychium coronarium* [41]. DXR may also serve as a significant control factor of the metabolic flux through the MEP pathway since it catalyzed the first committed step of the pathway towards terpenoid biosynthesis [42-43]. HDR catalyzed the last step of the MEP pathway, and the highest expression was at full flowering stage. The TPSs were directly responsible for the synthesis of various terpenes, and *OfTPS2* and *OfTPS3* had almost opposite expression patterns at each flowering stage.

A total of 1,327 TF genes were differentially expressed in the *O. fragrans* transcriptome. The role of transcriptional regulation in reprogramming metabolic processes during plant development and in response to biotic and abiotic cues had been recognized in several plant systems. Among the regulatory networks governing secondary metabolism, there had been many studies on flavonoid biosynthesis, but very few was known for regulation at the molecular level of floral volatile production [44-47]. Only a few MYB and WRKY TF family members had been characterized and shown crucial role in regulating biosynthesis of floral volatiles. To date, of the R2R3-MYB TF family, *PhODO1*, *PhEOBI*, *PhEOBII* and *PhMYB4* had been characterized as regulators of volatile phenylpropanoids in petunia [48-51]. *AtPAP1* transcription factor enhanced production of phenylpropanoid and terpenoid scent compounds in rose flowers [20]. MYC2, a bHLH TF, directly binded to promoters of the sesquiterpene synthase genes TPS21 and TPS11 activated their expression in *Arabidopsis* flowers [17]. In the WRKY family, *GaWRKY1*, *NaWRKY3*, *NaWRKY6*, and *SlWRKY73* were found to be related to volatile terpenoid production [15-16, 21].

The qRT-PCR analysis revealed that bHLH, MYB and WRKY family members were differentially expressed at each flowering stage. The expression patterns of *OfMYB1*, *OfMYB6*, *OfWRKY1* and *OfWRKY6* were consistent with the release of floral volatiles, which may play critical roles in regulating the formation of floral scent. Further experiments, including yeast one-hybrid and trans-activation assays using promoters of key structural genes, would be carried out to determine their involvement in floral scent production.

4　Materials and Methods

4.1　Plant Materials and RNA Extraction

Plants of *O. fragrans* 'Rixiang Gui', cultivated over decades, were located in the Nanjing Forestry University campus, Jiangsu, China. Three plants (R1, R2, R3) grown under the same conditions were randomly selected as biological replicates. Flowers at initial flowering (S1), full flowering (S2) and final flowering (S3) stages were collected and immediately frozen in liquid nitrogen and stored at −80 ℃ until use (Fig. 1).

The RNA of samples were extracted using a total RNA extraction kit according to the manufacturer's protocol (Tiangen Biotech, Beijing, China). The RNA quantity was analyzed by NanoDrop2000, and the quality was determined by 1.2% agarose gel electrophoresis. For each developmental stage, equal amounts of high-quality RNA from the three plant sample were constructed into nine cDNA libraries for Illumina deep sequencing.

4.2　Libraries Construction and LIlumina Sequencing

After total RNA was extracted in the three stages of flowering containing three biological replicates for each stage, eukaryotic mRNA was enriched by Oligo (dT) beads. Then the enriched mRNA was fragmented into short fragments using fragmentation buffer and reverse transcribed into cDNA with random primers. Second-strand cDNA was synthesized by DNA polymerase I, RNase H, dNTP and buffer. The cDNA fragments were purified with QiaQuick PCR extraction kit, eluted by EB buffer, end repaired, poly (A) added and ligated to Illumina sequencing adapters. The ligation products were size-selected by agarose gel electrophoresis, and amplified by PCR. Finally all the libraries were sequenced using Illumina HiSeq™4000.

4.3　Assembly and Functional Annotation

To get high quality clean reads, raw reads were filtered by removing adapters, reads containing more than 10% of unknown nucleotides (N) and low quality reads containing more than 50% of low quality bases (Q-value ≤ 10). At the same time, Q20, Q30, GC-content and sequence duplication level of the clean data were calculated. De novo assembly of the transcriptome was carried out with Trinity software, which was used for subsequent analysis without a reference genome[52]. All the assembled unigenes were aligned to public databases, by BLASTX (http://www.ncbi.nlm.nih.gov/BLAST/) with an E-value threshold of 10^{-5}, including the NCBI non-redundant protein (Nr) (http://www.ncbi.nlm.nih.gov), Swiss-Prot protein (http://www.expasy.ch/sprot), Kyoto Encyclopedia of Genes and Genomes (KEGG) (http://www.genome.jp/kegg), and COG/KOG (http://www.ncbi.nlm.nih.gov/COG) databases. Classification of Gene Ontology (GO) was analyzed by Blast2GO software[53]. Functional classification of unigenes was performed using WEGO software[54].

4.4 Gene Expression Analysis

Every sample of the clean reads was mapped back onto the assembled transcriptiome by the bowtie[55]. Differential expression analysis of each two groups was performed using the DESeq R package. The results of aligned clean reads were counted and quantified to give the expression abundance of gene expression using the RSEM package (Reads Per Kilo bases per Million mapped Reads)[56]. Correlation analyses of biological replicates were performed. Differences in expression analysis between two samples were determined using the DESeq R package. The Q-value < 0.005 and $|\log_2(\text{foldchange})| > 1$ was set as the threshold to select the DEGs. Cluster of differentially expressed gene were analyzed using STEM software[57], profiles in color indicating statistically significant ($P < 0.01$). The GO enrichment analysis of the DEGs was implemented by the GOseq R packages based on the Wallenius non-central hyper-geometric distribution[58], which could adjust for gene length bias in DEGs. KOBAS software was used to test the statistical enrichment of DEGs in KEGG pathways[59].

4.5 qRT-PCR Validation

Nine DEGs involved in terpenoids were chosen for qRT-PCR validation. The first-strand cDNA was reverse transcribed with 5 μg of total RNA isolated from each of the three flowering stages and biological replicates using a RevertAid First Strand cDNA Synthesis Kit (Thermo Scientific). The primers designed using Primer Premier 5.0 software (Premier Biosoft International, Palo Alto, CA, USA) were listed (Additional File 7). The qRT-PCR was performed on 10 μL reactions systems containing cDNA template, forward and reverse primers, SYBR Premix Ex Taq (TaKaRa, Japan) and water, with an ABI StepOnePlus Systems (Applied Biosystems, Carlsbad, CA, USA). The PCR reaction conditions were as follows: 95 ℃ for 30 s, followed by 40 cycles of 95 ℃ for 5 s, 60 ℃ for 30 s, and 95 ℃ for 15 s. *OfRPB2* was used as the reference genes to calculate relative fold-differences based on comparative cycle threshold ($2^{-\Delta\Delta Ct}$) values[60]. Every reaction for each gene used three biological replicates, with three technical replicates.

5 Conclusion

The dynamic transcriptome sequencing analysis of 'Rixiang Gui' at different flowering stages were performed using Illumina RNA-Seq technology. Based on DEG analysis and previous research on the pattern of floral compounds releasing, several candidate genes related to monoterpenes, sesquiterpenes and phenylpropanoid biosynthesis were obtained. What's more, a series of TFs were identified, and qRT-PCR analysis indicated that $OfMYB1$, $OfMYB6$, $OfWRKY1$ and $OfWRKY6$ could participate in the terpenoid biosynthesis. These results would be helpful in exploring the molecular mechanism of floral scent formation in $O.\ fragrans$, and supply important gene resources for the floral scent breeding of ornamental plants.

Additional Files

Additional File 1 Summary of transcriptome sequencing data and transcriptome assembly

Sample	Raw reads	Clean reads	GC/%	Adaptor/%	Low quality/%	Q20/%	Q30/%
S1-R1	56,727,396	55,846,862	43.94	0.04	1.15	97.93	94.76
S1-R2	61,833,244	60,966,256	44.11	0.04	1.01	98.03	94.96
S1-R3	86,376,464	84,598,042	44.05	0.03	1.67	97.62	94.04
S2-R1	46,858,546	46,188,794	43.86	0.03	1.03	98.02	94.93
S2-R2	55,732,750	54,949,688	43.36	0.03	1.01	98.06	95.02
S2-R3	54,468,388	53,697,708	43.37	0.03	1.02	98.03	94.94
S3-R1	57,508,808	56,711,496	43.52	0.04	0.99	98.04	94.98
S3-R2	56,463,786	55,663,954	43.61	0.03	1.02	97.96	94.79
S3-R3	56,152,094	55,338,510	43.24	0.03	1.06	97.98	94.83
Total	532,121,476	523,961,310					

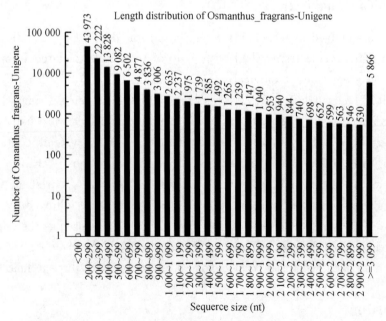

Additional File 2 Length distribution of assembled $O.\ fragrans$ unigenes
All the Illumina reads for each flowering stage were combined together with 136,611 transcripts obtained. The horizontal and vertical axes showed the size and the number of transcripts, respectively.

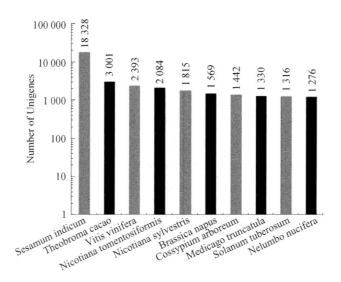

Additional File 3　Characteristics of homology search of unigenes against the Nr database

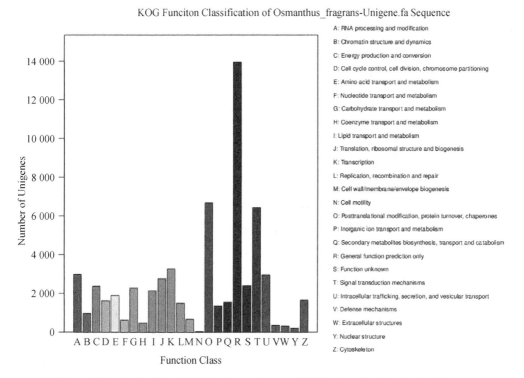

Additional File 4　Eu Karyotic Orthologous Groups (KOG) classifications in *O. fragrans*

A total of 136,611 sequences with KOG classifications within the 25 categories were shown.

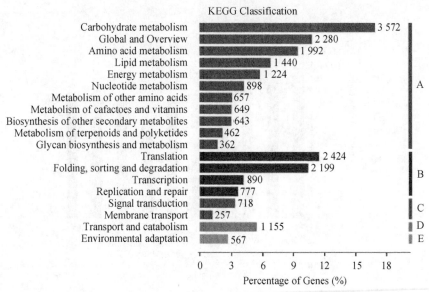

Additional File 5　Summary of KEGG pathways involved in the *O. fragrans* flower transcriptome

Additional File 6　A heat map plot of nine clusters displaying the relative expression levels of centroids

Additional File 7　Primers used for qRT-PCR

Gene	Description	Forward primer (5' to 3')	Reverse primer (5' to 3')
OfTPS2	Terpene synthase	GACCATCTCCTGTTGCGTGTAG	TCAATAATTCCCTAACATCCTTCAGT
OfTPS3	Terpene synthase	ATGGCACTTACCACCCTACTCC	TTCCCCACCTTCGTTCGG
OfDXS1	1-Deoxy-D-xylulose 5-phosphate synthase	AGTCACCGAGAAAGGCAGA	GGAAGCGACGCAAGAAAA
OfDXS2	1-Deoxy-D-xylulose 5-phosphate synthase	GGGCATTTGATGTAGCATA	ATCCTTCCCTTACCAATCT
OfDXR	1-Deoxy-D-xylulose 5-phosphate reductoisomerase	CACCTTCTTCTCCCTCGTCCT	CAACTATTACAGCCACCAATCTCC
OfHDR2	4-hydroxy-3-methylbut-2-enyl diphosphate reductase	ACAATCGGAAGGGGTTT	TTCTCGCCTCGTAAGCA
OfGPPS	geranyl pyrophosphate synthase	TCCGAGTTCGTTTCGTTTAGC	TTTCAGGATTACGGATTACCCA
OfHMGR	Hydroxymethylglutaryl-Co A reductase	GCTCCTCCCACGACGCTT	GGATTTCGCCCGAGACCA
OfFPPS	Farnesylpyrophosphate synthase	TGAAGACGCAGGCACATTTATT	TCCAGGTACATTGTAGTCCAACATC

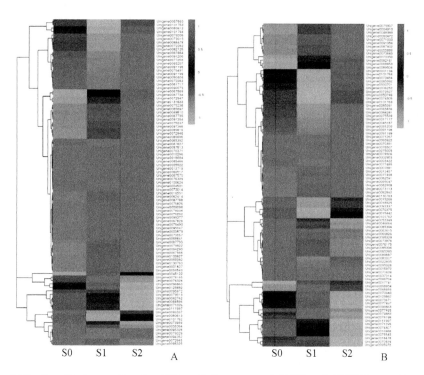

Additional File 8　Heat map for the differentially expressed MYB and WRKY genes by K-means method. Red indicated up-regulated genes and blue indicated down-regulated genes

(A) MYB unigenes. (B) WRKY unigenes.

Additional File 9 Primers of TFs used for qRT-PCR

Gene	Forward primer (5' to 3')	Reverse primer (5' to 3')
OfbHLH1	CGAAGTCATAGTGAATCGTGGC	GAAGTGTTCTTTCATTTGCTTTTGTA
OfbHLH2	GCACTTGGGTTCTGTTCTTCAC	TTTGGAGTCGAACGCTAATGG
OfbHLH3	CTTGAGTTGAGGGTATCCCACAT	GCAGTGATGTTGGCAGTAATGAT
OfbZIP2	TTTCCATAAAGGATTTTGAGGG	CTCATTCGGGATCGTTTCG
OfMYB1	CAGGTTACTGAGGCTTTGCG	CTTTGATAGCCAGTTGCGAGA
OfMYB2	CACTGGAATACCCGTATGCTCA	CGCTTGCGTTCAACTCCTG
OfMYB4	GCAGCGGAGGTTGAGGAG	ACTTGACGATGAACCAGAGCC
OfMYB6	CAATCCAGTCTCGGGCACC	CAACCTGGGATGGAATAGCC
OfWRKY1	AAAACTCCCGAGCTGGTAAGA	GGATACTGGCTGCCTTTGACT
OfWRKY3	GGCGGTGGAAATGATGGC	CGCTCTGGTGCTGGAGTTTG
OfWRKY4	GGTCCCTCCTCCTATTGATGC	CACAAGAAACCTGAGAAAACGAG
OfWRKY5	AGCCGATGATGCGGATAAA	TCCTCAAGACCCACAAAGAAAT
OfWRKY6	TGCTCCAACCCCATAACCA	TTCGGTTGTGGATAGAAATGGAGT
OfWRKY7	TCCCCGAGACGAACTTTGA	ACTCCCTCCACCTGTGCTACTA
OfWRKY9	GATTTGCGGTGCCCAGAC	CCCAACCTTGAGACCCCTTAT
OfWRKY10	TGAACAACCGAAACGCCATC	GCCACTATTACTACGCCGACAC

Acknowledgements

This work was supported by The Project of Key Research and Development Plan (Modern Agriculture) in Jiangsu(BE), the Project of Osmanthus National Germplasm Bank, and the Doctorate Fellowship Foundation of Nanjing Forestry University, Top-notch Academic Programs Project of Jiangsu Higher Education Institutions.

Author Contributions

Chen Xu, Lianggui Wang designed the experiments; Xiulian Yang, Haiyan Li, Wenjie Ding performed the experiment. Xiulian Yang, Haiyan Li, Yuanzheng Yue drafted the manuscript. Lianggui Wang, Yuanzheng Yue revised the manuscript, Wenjie Ding, Chen Xu, Tingting Shi, Gongwei Chen analyzed the data. All authors read and approved the final manuscript.

Conflicts of Interest

The authors declare no conflict of interest.

References

[1] Zang D K, Xiang Q B, Liu Y L, et al. The studying history and the application to international cultivar registration authority of sweet osmanthus (*Osmanthus fragrans* Lour)[J]. *J Plant Resour Environ*, 2003, 12: 49-53.

[2] Knudsen J T, Eriksson R, Gershenzon J, et al. Diversity and distribution of floral scent[J]. *Bot Rev*. 2006, 72, 1, 1-120.

[3] Gang D R, Evolution of flavors and scents[J]. *Annu Rev Plant Biol*. 2005, 56: 301-325.

[4] Xin H, Wu B, Zhang H, et al. Characterization of volatile compounds in flowers from four groups of sweet osmanthus (*Osmanthus fragrans*) cultivars[J]. *Can J Plant Sci*. 2013, 93, 5: 923-931.

[5] Cai X, Mai R Z, Zou J J, et al. Analysis of aroma-active compounds in three sweet osmanthus (*Osmanthus fragrans*) cultivars by GC-olfactometry and GC-MS[J]. *J Zhejiang Univ-Sci B (Biomed & Biotechnol)*. 2014, 15, 7: 638-648.

[6] Wang L M, Li M T, Jin W W, et al. Variations in the components of *Osmanthus fragrans* Lour. essential oil at different stages of flowering[J]. *Food Chem*. 2009, 114, 1: 233-236.

[7] Lichtenthaler H K. The 1-deoxy-D-xylulose-5-phosphate pathway of isoprenoid biosynthesis in plants[J]. *Annu Rev Plant Physiol Plant Mol Biol*, 1999, 50: 47-65.

[8] Cordoba E, Salmi M, Leon P. Unravelling the regulatory mechanisms that modulate the MEP pathway in higher plants[J]. *J Exp Bot*, 2009, 60, 10: 2933-2943.

[9] Xu C, Li H, Yang X, et al. Cloning and expression analysis of MEP pathway enzyme-encoding genes in *Osmanthus fragrans*[J]. *Genes*. 2016, 7, 10.

[10] Tholl D, Chen F, Petri J, et al. Two sesquiterpene synthases are responsible for the complex mixture of sesquiterpenes emitted from Arabidopsis flowers[J]. *Plant J*, 2005, 42, 5: 757-771.

[11] Roeder S, Hartmann A M, Effmert U, et al. Regulation of simultaneous synthesis of floral scent terpenoids by the 1, 8-cineole synthase of *Nicotiana suaveolens*[J]. *Plant Mol Biol*, 2007, 65: 1-2, 107-124.

[12] Nagegowda D A, Gutensohn M, Wilkerson C G, et al. Two nearly identical terpene synthases catalyze the formation of nerolidol and linalool in snapdragon flowers[J]. *Plant J*. 2008, 55, 2, 224-239.

[13] Zeng X, Liu C, Zheng R, et al. Emission and accumulation of monoterpene and the key terpene synthase (TPS) associated with monoterpene biosynthesis in *Osmanthus fragrans* Lour[J]. *Front in Plant Sci*, 2015, 6: 1232.

[14] Susanne B, Masaya K, Miwako K, et al. Functional characterization of a carotenoid cleavage dioxygenase1 and its relation to the carotenoid accumulation and volatile emission during the floral development of *Osmanthus fragrans* Lour[J]. *J Exp Bot*, 2010, 61, 11: 2967-2977.

[15] Xu Y H, Wang J W, Wang S, et al. Characterization of *GaWRKY1*, a cotton transcription factor that regulates the sesquiterpene synthase gene (+)-delta-cadinene synthase-A[J]. *Plant physiol*, 2004, 135, 1: 507-515.

[16] Skibbe M, Qu N, Galis I, et al. Induced plant defenses in the natural environment: *Nicotiana attenuata* WRKY3 and WRKY6 coordinate responses to herbivory[J]. *Plant Cell*, 2008, 20: 1984-2000.

[17] Hong G J, Xue X Y, Mao Y B, et al. Arabidopsis MYC2 interacts with DELLA proteins in regulating sesquiterpene synthase gene expression[J]. Plant cell, 2012, 24(6):2635-2648.

[18] Bedon F, Bomal C, Caron S, et al. Subgroup 4 R2R3-MYBs in conifer trees: gene family expansion and contribution to the isoprenoid- and flavonoid-oriented responses[J]. J Exp Bot, 2010, 61, 14, 3847-3864.

[19] Zhou F, Sun T H, Zhao L, et al. The bZIP transcription factor HY5 interacts with the promoter of the monoterpene synthase gene QH6 in modulating its rhythmic expression[J]. Front Plant Sci, 2015, 6:304.

[20] Zvi M M, Shklarman E, Masci T, et al. PAP1 transcription factor enhances production of phenylpropanoid and terpenoid scent compounds in rose flowers[J]. New phytol, 2012, 195, 2:335-345.

[21] Spyropoulou E A, Haring M A, Schuurink R C. RNA sequencing on *Solanum lycopersicum* trichomes identifies transcription factors that activate terpene synthase promoters [J]. BMC Genomics, 2014, 15, 402.

[22] Wang Z, Gerstein M, Snyder M. RNA-Seq: a revolutionary tool for transcriptomics[J]. Nature Rev Genet, 2009, 10, 1:57-63.

[23] Yang K M, Zhu W J. Osmanthus fragrans [M]. Shanghai Press of Science and Technology. 2000.

[24] Huang B, Chen H, Shao L. The ethanol extract of *Osmanthus fragrans* attenuates porphyromonas gingivalis lipopolysaccharide-stimulated inflammatory effect through the nuclear factor erythroid 2-related factor-mediated antioxidant signalling pathway[J]. Archives of Oral Biology, 2015, 60(7):1030-1038.

[25] Zhang C, Wang Y, Fu J, et al. Transcriptomic analysis and carotenogenic gene expression related to petal coloration in *Osmanthus fragrans* 'Yanhong Gui'[J]. Trees, 2016, 30(4):1207-1223.

[26] Mu H N, Li H G, Wang L G, et al. Transcriptome sequencing and analysis of sweet osmanthus (*Osmanthus fragrans* Lour.)[J]. Genes Genomics, 2014, 36:777-788.

[27] Dudareva N, Pichersky E. Biochemical and molecular genetic aspects of floral scents[J]. Plant Physiol, 2000, 122:627-633.

[28] Mc Garvey D J, Croteau R. Terpenoid metabolism[J]. Plant Cell, 1995, 7:1015-1026.

[29] Dudareva N, Negre F, Nagegowda D A, et al. Plant Volatiles: Recent advances and future perspectives[J]. Crit Rev Plant Sci, 2006, 25, 5, 417-440.

[30] Chen F, Tholl D, D'Auria J C, et al. Biosynthesis and emission of terpenoid volatiles from Arabidopsis flowers[J]. Plant Cell, 2003, 15(2): 481-494.

[31] Dudareva N. (E)-beta-Ocimene and myrcene synthase genes of floral scent biosynthesis in snapdragon: function and expression of three terpene synthase genes of a new terpene synthase subfamily[J]. Plant Cell, 2003, 15(5): 1227-1241.

[32] Muhlemann J K, Maeda H, Chang C Y, et al. Developmental changes in the metabolic network of snapdragon flowers[J]. PLOS ONE, 2012, 7, e40381.

[33] Dudareva N, Cseke L, Blanc V M, et al. Molecular characterization and cell type-specific expression of linalool synthase gene from Clarkia[J]. Plant Physiol, 1996, 111:815.

[34] Dudareva N, Cseke L, Blanc V M, et al. Evolution of floral scent in Clarkia: novel patterns of S-linalool synthase gene expression in the *C. breweri* flower[J]. Plant Cell, 1996, 8:1137-1148.

[35] Dudareva N, D'Auria J C, Nam K H, et al. Acetyl-Co A: benzylalcohol acetyltransferase-an enzyme involved in floral scent production in *Clarkia breweri*[J]. Plant J, 1998, 14:297-304.

[36] Tholl D, Kish C M, Orlova I, et al. Formation of monoterpenes in *Antirrhinum majus* and *Clarkia breweri* flowers involves heterodimeric geranyl diphosphate synthases[J]. *Plant Cell*, 2004, 16(4): 977-92.

[37] Boatright J, Negre F, Chen X, et al. Understanding in vivo benzenoid metabolism in petunia petal tissue[J]. *Plant physiol*, 2004, 135(4):1993-2011.

[38] Spiller M, Berger R G, Debener T. Genetic dissection of scent metabolic profiles in diploid rose populations[J]. Theor Appl Genet, 2010,120(7):1461-1471.

[39] Guterman I. Rose Scent: Genomics approach to discovering novel floral fragrance-related genes[J]. *Plant Cell*, 2002, 14(10): 2325-2338.

[40] Dudareva N, Klempien A, Muhlemann J K. et al. Biosynthesis, function and metabolic engineering of plant volatile organic compounds[J]. *New Phytol*, 2013, 198(1):16-32.

[41] Yue Y, Yu R, Fan Y. Transcriptome profiling provides new insights into the formation of floral scent in *Hedychium coronarium*[J]. *BMC Genomics*, 2015, 16:470.

[42] Mahmoud S S, Croteau R B. Metabolic engineering of essential oil yield and composition in mint by altering expression of deoxyxylulose phosphate reductoisomerase and menthofuran synthase[J]. *Proc Natl Acad Sci USA*, 2001, 98:8915-8920.

[43] Carretero-Paulet L, Ahumada I, Cunillera N, et al. Expression and molecular analysis of the Arabidopsis DXR gene encoding 1 - deoxy-D-xylulose 5 - phosphate reductoisomerase, the first committed enzyme of the 2-C-methyl-D-erythritol 4-phosphate pathway[J]. *Plant Physiol*, 2002, 129, 4:1581-1591.

[44] Koes R, Verweij W, Quattrocchio F. Flavonoids: a colorful model for the regulation and evolution of biochemical pathways[J]. *Trends Plant Sci*, 2005, 10(5):236-242.

[45] Schwinn K, Venail J, Shang Y J, et al. A small family of MYB-regulatory genes controls floral pigmentation intensity and patterning in the genus Antirrhinum[J]. *Plant Cell*, 2006, 18:831-851.

[46] Aharoni A, Galili G. Metabolic engineering of the plant primary-secondary metabolism interface[J]. *Curr Opin Biotechnol*, 2011, 22(2):239-244.

[47] Feller A, Machemer K, Braun E L, et al. Evolutionary and comparative analysis of MYB and bHLH plant transcription factors[J]. *Plant J*, 2011, 66(1):94-116.

[48] Verdonk J C, Haring M A, van Tunen A J, et al. ODORANT1 regulates fragrance biosynthesis in petunia flowers[J]. *Plant Cell*, 2005, 17(5):1612-1624.

[49] Spitzer-Rimon B, Marhevka E, Barkai O,et al. EOBII, a gene encoding a flower-specific regulator of phenylpropanoid volatiles' biosynthesis in petunia[J]. *Plant Cell*, 2010, 22(6):1961-1976.

[50] Spitzer-Rimon B, Farhi M, Albo B, et al. The R2R3 - MYB-like regulatory factor *EOBI*, acting downstream of *EOBII*, regulates scent production by activating ODO1 and structural scent-related genes in petunia[J]. *Plant Cell*, 2012, 24(12):5089-5105.

[51] Colquhoun T A, Kim J Y, Wedde A E, et al. PhMYB4 fine-tunes the floral volatile signature of Petunia x hybrida through *PhC4H*[J]. *J Exp Bot*, 2011, 62(3):1133-1143.

[52] Grabherr M G, Haas B J, Yassour M, et al. Full-length transcriptome assembly from RNA-Seq data without a reference genome[J]. *Nat Biotechnol*, 2011, 29(7): 644-652.

[53] Gotz S, Garcia-Gomez J M, Terol J, et al. High-throughput functional annotation and data mining with the Blast2GO suite[J]. *Nucleic Acids Res*, 2008, 36(10):3420-3435.

[54] Ye J, L Fang, Zheng H K, et al. "WEGO: a web tool for plotting GO annotations."[J] *Nucleic Acids Res*, 2006, W293-297.

[55] Langmead B, Trapnell C, Pop M, *et al*. Ultrafast and memory-efficient alignment of short DNA sequences to the human genome[J]. *Genome Biol*, 2009, 10(3):R25.

[56] Li B, Dewey C N. RSEM: accurate transcript quantification from RNA-Seq data with or without a reference genome[J]. *BMC Bioinformatics*, 2011, 12: 323.

[57] Ernst J, Bar-Joseph Z. STEM: a tool for the analysis of short time series gene expression data[J]. *BMC Bioinformatics*, 2006, 7: 191.

[58] Young M D, Wakefield M J, Smyth G K, *et al*. Gene ontology analysis for RNA-seq: accounting for selection bias[J]. *Genome Biol*, 2010, 11(2):R14.

[59] Mao X Z, Cai T, Olyarchuk J G, *et al*. Automated genome annotation and pathway identification using the KEGG Orthology (KO) as a controlled vocabulary[J]. *Bioinformatics*, 2005, 21(19): 3787-3793.

[60] Livak K J, Schmittgen T D. Analysis of relative gene expression data using real-time quantitative PCR and the $2^{-\Delta\Delta Ct}$ method[J]. *Methods*, 2001, 25:402-408.

Identification and Validation of Reference Genes for Gene Expression Studies in Sweet Osmanthus (*Osmanthus fragrans*) Based on Transcriptomic Sequence Data*

Hongna Mu[1,2], **Taoze Sun**[2], **Chen Xu**[1], **Lianggui Wang**[1], **Yuanzheng Yue**[1], **Huogen Li**[3] and **Xiulian Yang**[1]

1. College of Landscape Architecture, Nanjing Forestry University, Nanjing 210037, China
2. College of Horticulture and Gardening, Yangtze University, Jingzhou 434025, Hubei, China
3. The Key Laboratory of Forest Genetics and Gene Engineering, Nanjing Forestry University, Nanjing 210037, China

Abstract: Accurate normalized data is a primary requisite for quantifying gene expression using RT-qPCR technology. Despite this importance, however, suitable reference genes in *Osmanthus fragrans* are not available. In this study, seven potential candidate reference genes (*OfL25-1*, *OfL25-10*, *OfRP2*, *OfTUA*, *OfTUB3*, *OfUBQ* and *Of18S*) were evaluated to determine which one would be the most reliable reference genes. The expression levels of the candidate reference genes were analyzed by RT-qPCR in flower, leaf, pedicel, blossom bud tissues, as well as in floral organs at different developmental stages. GeNorm and NormFinder were used to statistically analyse transcript variation. Results indicated that *OfRP2* and *OfL25-10* were the optimal reference genes for use in RT-qPCR when analysing different stages of floral development; while *OfTUB3* and *OfL25-1* were optimal across tissues. The selected reference genes were used to examine *OfMYB1* expression. The results appeared to be useful for future gene expression analyses aiming to characterize developmental stages and tissues of *O. fragrans*.

Key words: reference gene selection; validation; *OfTUB*; *OfRP2*; *Osmanthus fragrans*.

1 Introduction

Osmanthus fragrans is a widely cultivated ornamental plant in China and a total of 189 cultivars have been reported to date (Zang *et al.* 2006, 2014; Hu *et al.* 2014; Xiang *et al.* 2014a, b). Fragrance (aroma) is regarded as the most important trait in *Osmanthus*. However, at present, the genes and metabolic pathways that regulate the biosynthesis of fragrance compounds in *Osmanthus* remain largely uncharacterized. In our previous research, transcriptome data from *Osmanthus* was utilized to analyse secondary metabolism at the transcript level (Mu *et al.* 2014).

Gene expression analysis has provided crucial information for understanding gene function. Among the different expression analysis techniques, real time quantitative

---

* 原文发表于 *Journal of Genetics*, 2017, 96(2): 273-281。

polymerase chain reaction (qRT-PCR) has been widely used due to its high accuracy, efficiency and low cost (Bustin 2002; Artico *et al.* 2010). The most common method used to ensure the accuracy of the obtained gene expression data utilizes endogenous genes as reference genes to normalize the expression level of other genes of interest (Vandesompele *et al.* 2009). The ideal reference gene should exhibit a stable expression level in different tissues and cell types under various developmental, environmental and experimental conditions (Nolan *et al.* 2006, Wan *et al.* 2010).

Recently, several studies on the validation of reference genes in different plant species have been reported, including *O. fragrans* (Zhang *et al.* 2015), *Polygonum convolvulus* (Demidenko *et al.* 2011), *Vitis vinifera* (Ferrandino and Lovisolo 2014), *Atropa belladonna* (Li *et al.* 2014), *Malus × domestica* (Bowen *et al.* 2014), *Petunia hybrida* (Mallona *et al.* 2010) and *Rosa hybrida* (Meng *et al.* 2013). Most of these studies used a list of reference genes from other plant species and tested them under their own experimental conditions. qRT-PCR data were subjected to analyses using software geNorm (Vandesompele *et al.* 2002), NormFinder (Andersen *et al.* 2004), or BestKeeper (Pfaffl *et al.* 2004) to statistically identify the most stable reference gene among a group of candidate genes in a defined set of biological samples. Typical genes examined as potential candidates include ubiquitin and tubulin genes. Ubiquitin directs the localization of various proteins to specific cellular compartments, including proteasomes that breakdown and recycle proteins (Callis *et al.* 1995). Tubulin proteins are composed of α-tubulin and β-tubulin, which play important roles in plant growth and development (Chuong *et al.* 2004). In this study, seven potential housekeeping genes, belonging to the *tubulin* and *ubiquitin* gene families, were selected from the *O. fragrans* transcriptome. The objective was to identify and validate optimal reference genes that could be used in RT-qPCR analysis in different vegetative tissues and floral tissues at different developmental stages in sweet osmanthus (*O. fragrans*).

2 Materials and Methods

2.1 Plant Material

Cultivated plants of *O. fragrans* ('Zao Yingui' and 'Chenghong Dangui') located at the campus of Nanjing Forestry University, aged-18 years were used for all the analyses. Stages of floral development are defined as described by Han *et al.* (2013) and Dong *et al.* (2014). All samples were collected in September 2013.

The collection of samples was designed to beable to identify potential reference genes that could be used in a variety of tissues and flowers at different developmental stages. Therefore, leaves, buds, flowers (initial flowering stage), pedicels and roots were collected. In addition, samples from various stages of blooms were also collected. These were classified as: stage 1 (the Dingke stage) with unfolded bud scales and protruded inflorescence (Fig. 1a); stage 2 (the Xiangyan stage), with petals which still remain folded (Fig. 1b); stage 3 (the Linggeng stage), with elongated pedicels and petals which remain unopened (Fig. 1c); stage 4, initial flowering stage (Fig. 1d); stage 5, peak flowering stage 1 (Fig. 1e); stage 6, peak flowering stage 2 (Fig. 1f); stage 7, peak flowering stage 3 (Fig. 1g) and stage 8 (the Mohua stage) with the initiation of floral organ senescence (Fig. 1h). For the RT-qPCR analyses, eight sets of petal samples (stages 1–8) were collected daily within a 1-h time period in the morning from 9 to 10 am. All samples were frozen in liquid nitrogen after being har-vested and were subsequently stored at −80 ℃ until total RNA extraction. All seven candidate reference genes were used in the RT-qPCR analysis for all the collected samples (see 'Reference gene detailed information' in electronic supplementary material at http://www.ias.ac.in/jgenet/).

Fig. 1 Classification of eight flowering stages in *O. fragrans* as described by Dong *et al*. (2014)

2.2 Extraction of Total RNA and cDNA Synthesis

Total RNA was extracted from the bud, petal, leaf, pedicel and root samples using a Tiangen Plus plant total RNA extraction kit (Tiangen, Beijing, China). The integrity of the RNA samples was assessed by gel electrophoresis on 1.5% agarose gel. Approximately 1 μg of total RNA was reverse-transcribed into first strand cDNA using a TransScript First-

Strand cDNA Synthesis SuperMix kit (Transgene, Beijing, China) according to the manufacturer's instructions.

2.3 Identification and Validation of Candidate Reference Genes

The expression stability of candidate reference genes was calculated using two different software packages (geNorm and NormFinder) that utilize two different methods to calculate stability. The first software package (geNorm) is a visual application tool for Microsoft Excel that operates on the assumption that the expression ratio of two ideal reference genes is constant in the different groups of templates. A gene expression stability value (M) was calculated for all the candidate reference genes that were examined. In a previous report, a M value <1.5 was recommended by Vandesompele et al. (2002). To identify suitable Osmanthus reference genes, unigenes of traditional housekeeping genes were selected from three O. fragrans transcriptome datasets; including one α-tubulin gene (OfTUA), one β-tubulin gene (OfTUB3), one ubiquitin gene (OfUBQ), two large subunit ribosomal protein genes (OfL25-1 and OfL25-10) and one RNA polymerase II subunit gene (OfRP2). However, the validity of the 18S gene as a reference gene was not assessed in a previous study of O. fragrans by Han et al. (2013). Therefore, Of18S was also assessed as one of the seven potential candidate reference genes in the current study. Each reference gene was used to normalize the expression profile of the OfMYB1 gene, which was cloned in our research lab from O. fragrans 'Chenghong Dangui' petals. The resulting RT-qPCR data was then further analysed to determine which candidate reference genes provided the best assessment of OfMYB1 expression.

2.4 Expression Profiles of Reference Genes

After verifying the size of the product amplified by the designed primer pairs using semiquantititative PCR (sqRT-PCR) and gel electrophoresis, the samples were analysed by RT-qPCR using an ABI 7,500 real-time PCR system (Applied Biosystems, Alameda, USA). The qRT-PCR reactions were carried out in a total volume of 20 μL liquid containing: 2 μL cDNA template (10-fold dilution), 0.4 μL (10 μmol) of each primer, 10 μL Fast-Start Universal SYBR Green Master (ROX) and 7.6 μL of sterile distilled water. The qRT-PCR programmer utilized was 95 ℃ for 10 min (denaturation), followed by 40 cycles of 95 ℃ for 30 s and 60 ℃ for 60 s, then 95 ℃ for 15 s, 60 ℃ for 60 s, 95 ℃ for 30 s and 60 ℃ for 15 s for constructing the melting curve. Each RT-qPCR analysis was performed in triplicate. Two different statistical packages, geNorm (Van-desompele et al. 2002) and Norm Finder (Andersen et al. 2004) were used to evaluate the expression stability of the candidate reference genes.

3 Results

Amplification Efficiency and Specificity of Primers for the Candidate Reference Genes

The sqRT-PCR data verified the accuracy and specificity of the primer pairs used to amplify the candidate reference genes (Fig. 1 in electronic supplementary material). Therefore, the same primers were used to evaluate the expression stability of the seven candidate reference genes by qRT-PCR. The PCR amplification efficiency of the seven reference genes ranged from 97.7 to 104.9% (Tab. 1). The Ct values of the seven genes in 10 different sample sets of floral organs were used to estimate the expression level the candidate reference genes. The highest Ct value (36.56) was obtained with the *UBQ2* gene in 'Chenghong Dangui' petal samples and the lowest Ct value (27.90) was also obtained with the *UBQ2* gene in 'Zaoyingui' petal samples (Fig. 2). Most of the Ct values among all of the candidate reference genes varied between 20 and 28. In the current analysis, the 18S gene that was previously used by Han *et al.* (2013) exhibited unstable expression patterns. Importantly, however, none of the seven candidate reference genes that were examined had an invariant, stable expression level. Ct values varied among tissue types and in the different stages of floral development. These results indicate that it is critical to select a reference gene(s) for *O. fragrans* that are stable for the tissue type, environmental, or experimental conditions in which gene expression analyses are to be conducted.

Tab. 1 Gene-specific primers utilized in RT-qPCR analyses

Name	Primer sequences (forward/reverse) 5'-3'	Ta (℃)	Product size	Amplification efficiency
TUA	AGGGAAGCAGTGATGGAAGACA	61.1	206	97.9
	TCTCTGTGGATTACGGAAAGAAGTC	61.2		
TUB	CCAGCACCAGACTGACCGAAGA	64.7	216	104.1
	TGAGCACGGCATAGACCCAACT	64.3		
L25-1	ACCAGGGCAAACATCTACTCCA	61.1	244	97.7
	GATTAAGGCGATCCTCAAGCA	59.7		
L25-10	ATACACGGTTTCAACGACTCCCA	63.3	224	100.5
	GATGCCATAACCAGGACCAAGAA	62.9		
RP2	CACCAAGCAAAGGACCAGCAAG	64.5	216	101.7
	TCACCAGGGAGAAGAGGATCAAGTA	63.4		
UBQ	GCTTCAATCCTCAACTTCGTGGTA	62.6	206	98.3
	CACAGATCGTCCAGGAGGCTAA	61.7		
18S	AGCCTGAGAAACGGCTACCAC	61.0	200	104.9
	ATACGCTATTGGAGCTGGAA	62.6		

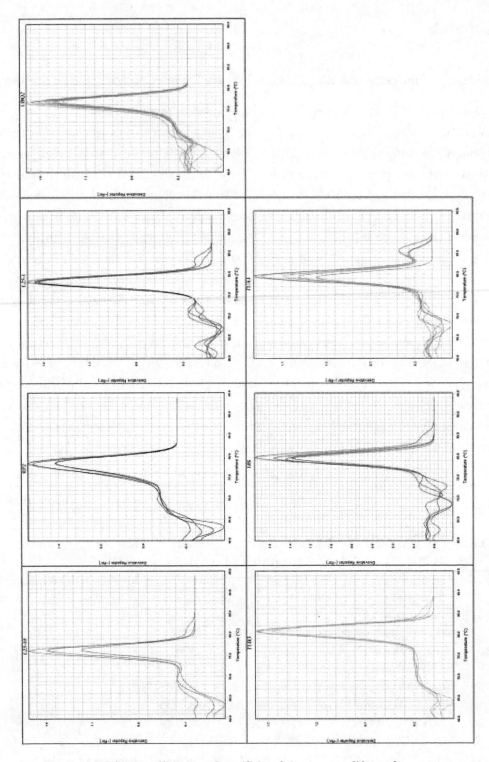

Fig. 2 Amplification efficiency and specificity of the seven candidate reference genes

Expression Stability of Candidate Reference Genes

In the present study, the M values of TUB and TUA were <1.5, which were the lowest values obtained in the gene expression analyses in the different tissue types (Fig. 3). In contrast, $OfL25-10$ and $OfRP2$ were the most stable genes (lowest M values) in samples representing different stages of floral development (Fig. 4). Data from the floral stage analyses revealed that cut-off values ranged from 0.133 to 1.02, and only the V3/4 value (0.133) with different stages was lower than the ideal value (0.15) (Fig. 3). These data indicate that the three reference genes ($OfL25-10$, $OfRP2$ and $OfL25-1$) would be suitable for normalizing gene expression in the different stages of $O.\ fragrans$ flower opening.

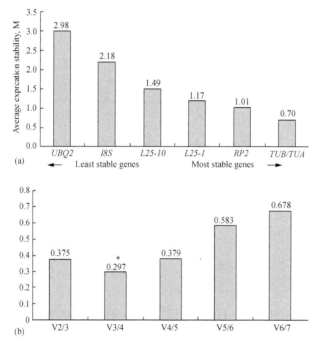

Fig. 3 Stability analysis of seven candidate reference genes in different tissues of $O.\ fragrans$ using geNorm software

* The optimal number of reference genes in this test.

Norm Finder software utilizes a different model-based algorithm to identify optimal reference genes. In this analysis, the candidate reference genes were grouped into five groups based on sample types (leaves, buds, flower, pedicels and roots), then Norm Finder calculated intragroup and intergroup variations. Genes which had the lowest variations were considered as the most stable. Norm Finder indicated that $OfRP2$ and $OfTUB$ were the most stable reference genes also for the different tissue types (leaves, buds, unopened flowers, pedicels and roots). $OfL25-10$ and $OfRP2$ were the most stable reference genes in samples representing the different stages of flower opening and senescence (Tab. 2; Fig. 2).

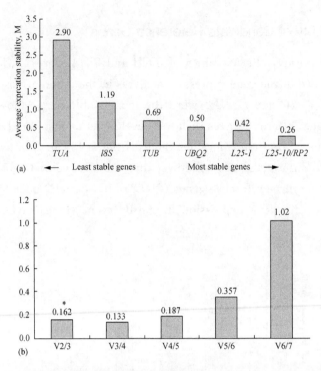

Fig. 4 Stability analysis of seven candidate reference genes in floral stages of *O. fragrans* using geNorm software

* The optimal number of reference genes in this experiment.

Tab. 2 Overall ranking of *O. fragrans* reference genes based on NormFinder analysis

	OfUBQ	OfTUA	OfL25-1	OfL25-10	OfTUB	OfRP2
Different tissues	1.627	0.257	0.219	0.768	0.122	0.311
Different stages	0.089	2.472	0.046	0.045	0.142	0.045

The utilization of reference genes, *TUA* and *TUB3* to normalize *MYB1* expression provided the most accurate expression profile. *L25-1* and *UBQ*, both overestimated the transcript level of *MYB1* in petals and pedicel samples. *RP2* also underestimated the expression level of *MYB1* and *L25-10* completely changed the expression profile of MYB1. The 18S reference gene also provided an inaccurate expression profile for MYB1 (Fig. 5).

4 Discussion

Candidate reference genes were identified for different tissues and flowering stages of *O. fragrans* from transcriptome datasets (Tab. 2). In the present study, we focused on the spatiotemporal expression stability of six house keeping genes and one reported 18S reference gene (Han *et al*. 2013) in two varieties of *O. fragrans*. Our data indicated that 18S was unsuitable as a reference gene, which was also indicated by Zhang *et al*. (2015).

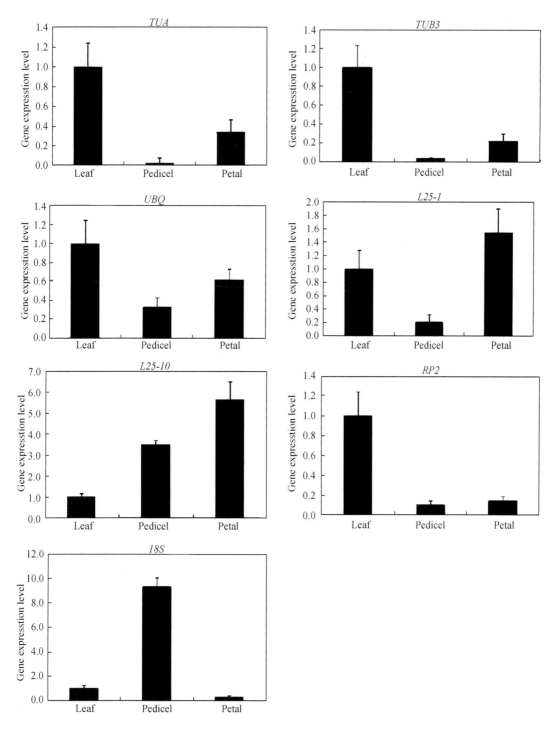

Fig. 5 Different results obtained on the qRT-PCR analysis of *OfMYB* expression in different tissues of sweet osmanthus (*O. fragrans*) based on the use of seven different reference genes for the normalization of gene expression

In contrast, 18S has been reported to be a suitable reference gene in other species, including *Panicum virgatum* (Jiang et al. 2014), *Nilaparvata lugens* (Wang et al. 2014), *Oryza sativa* (Kim et al. 2003), and *Cichorium intybus* (Maroufi et al. 2010). How-ever, in the present study, *Of18S* exhibited considerable variability in expression in different tissues and flower-ing stages. These results differ from the results obtained by Han et al. (2013) in the same species. Other recent studies have also shown that 18S is unstable in *Fragaria ×ananassa* (Galli et al. 2015), *Buglossoides arvensis* (Gadkar and Filion 2015), *Cymbidium kanran* (Luo et al. 2014), *Chlamydomonas* (Mou et al. 2014), *Setaria italica* (Kumar et al. 2013), *Nicotiana tabacum* (Cortleven et al. 2009), *Prunus persica* (Tong et al. 2009) and other plants.

OfTUA and *OfTUB* were the two most stable reference genes in the different tissue types that were examined. In all the three *O. fragrans* varieties ('Zaoyingui' 'Chenghong Dangui' and 'Boyejingui'), *OfTUB3* exhibited stable expression and thus appeared as a good reference for gene expression studies in this species. These results are similar to reported reference gene studies in *Citrus grandis* (Wang et al. 2013), *Nerviliao fordii* (Huang et al. 2013) and *Quercus suber* (Soler et al. 2008; Marum et al. 2012). *TUA* exhibited the least stability in gene expression among the different flowering stages of *O. fragrans*. This finding is also consistent with the results of experiments reported in *Cucurbita pepo* (Obrero et al. 2011) and *Cucumis sativus* (Migocka and Papierniak 2010). However, *TUA* has also been reported to be a suitable internal control gene in *Cucumis sativus* (Wan et al. 2010), *Linum usitatissimum* (Huis et al. 2010) and *Populus trichocarpa × P. deltoids* (Brunner et al. 2004). In this study, the *RP2* candidate reference gene exhibited the most stable expression in samples obtained from the different stages of flowering. A similar result was also reported in *Prunus persica* (Tong et al. 2009). *OfACT* has been previously reported as an optimal reference gene in sweet osmanthus for samples from different floral stages (Zhang et al. 2015). In contrast to Zhang et al. (2015), *OfACT* exhibited the least expression stability in our study of samples from different tissues and floral development stages. Actually, *OfACT* was eliminated as a potential candidate reference gene during the preliminary sqRT-PCR analysis (see Fig. 1). It is plausible that conflicting results were obtained by Zhang et al. (2015) and the present study as the two studies utilized different *ACT* gene family members for their analyses.

Unstable expression levels were observed for *UBQ* in samples from different tissues as well, when samples were taken from different floral development stages. These results are similar to the results reported for *Quercus suber*, where *UBQ* was ranked as the fourth most reliable reference gene (Marum et al. 2012). In contrast, *UBQ* exhibited the most stable expression in *Arabidopsis thaliana* (Czechowski et al. 2005). The *UBQ* gene family contains members which exhibit very stable patterns of expression (Smalle and Vierstra 2004). Therefore, further studies in *O. fragrans* may reveal additional members of the *UBQ* gene family that can serve as suitable reference genes.

A previous study identified the reference genes in four groups (Luteus, Albus, Aurantiacus and Asiaticus) within *O. fragrans*, representing 16 varieties (Zhang et al. 2015). Han et al. (2013, 2014) previously reported the expression of genes related to carotenoid metabolism using 18S as an internal control. Despite their importance when con-ducting RT-qPCR analyses, very little is known regarding suitable reference genes for *O. fragrans*. At present, our results and those obtained from previous studies, clearly indicate that a single universal reference gene for all tissues and conditions does not exist for *O. fragrans*. Instead, the specific tissues and experimental conditions need to be analysed before a suitable reference gene is selected for use in RT-qPCR analyses. However, this conclusion does not imply that future studies may be able to identify such a universal reference gene for use in *O. fragrans* under all experimental conditions.

In conclusion, in the present study, *OfTUA*, *OfTUB*, *OfL25-1*, *OfL25-10*, *OfRP2*, *OfUBQ* and *Of18S* genes were selected as candidate reference genes based on *O. fragrans* transcriptome data and previous reports in other species. The expression stability of these seven candidate reference genes were examined in different tissues and stages of floral opening. Results indicated that *OfTUB* and *OfTUA* were the most reliable reference genes in different tissues of *O. fragrans*, while *OfRP2* and *OfL25-10* in samples obtained from different flowering stages. Among all candidate reference genes, *OfTUB* and *OfTUA* provided the most accurate data on *OfMYB1* expression in different tissues. It is concluded that *OfTUB* and *OfTUA*, *OfRP2* and *OfL25-10* are the most suitable reference genes for the normalization of gene expression in different tissues and floral development stages, respectively.

Acknowledgements

This work was financially supported by the Agriculture-Forestry Plant Germplasm Resources Exploration, Innovation and Utilization Project of the National Twelve-Five Science and Technology Support Program (2013BAD01B06-4), and the Advantages Discipline Construction Engineering Project of Colleges and Universities in the Jiangsu Province (PAPD).

References

[1] Artico S, Nardeli S M, Brilhante O, et al. 2010 Identification and evaluation of new reference genes in *Gossypium hirsutum* for accurate normalization of real-time quantitative RT-PCR data[J]. *BMC Plant Biol.* 10, 49.

[2] Andersen C L, Jensen J L, Ørntoft T F, et al. 2004 Normalization of real-time quantitative reverse transcription-PCR data: a model-based variance estimation approach to identify genes suited for

normalization, applied to bladder and colon cancer data sets[J]. *Cancer Res*. 64:5245-5210.

[3] Bowen J, Ireland H S, Crowhurst R, et al. 2014 Selection of low-variance expressed *Malus × domestica* (apple) genes for use as quantitative PCR reference genes (housekeepers)[J]. *Tree Genet Genomes*. 10: 751-759.

[4] Brunner A M, Yakovlev I A, Strauss S H. 2004 Validating internal controls for quantitative plant gene expression studies[J]. *BMC Plant Biol*. 18:4-14.

[5] Bustin S A. 2002 Quantification of mRNA using real-time reverse transcription PCR (RT-PCR): trends and problems[J]. *J. Mol. Endocrinol*. 29:4021-4022.

[6] Chuong S D, Good A G, Taylor G J, et al. 2004 Large-scale identification of tubulin-binding proteins provides insight on subcellular traf-ficking, metabolic channeling, and signaling in plant cells[J]. *Mol. Cell. Proteomics*. 3:970-983.

[7] Callis J, Carpenter T, Sun C W, et al. 1995 Structure and evolution of genes encoding polyubiquitin and ubiquitin-like proteins in Arabidopsis thaliana ecotype Columbia[J]. *Genetics*, 139:921-939.

[8] Cortleven A, Remans T, Brenner W G, et al. 2009 Selection of plastid-and nuclear-encoded reference genes to study the effect of altered endogenous cytokinin content on photosynthesis genes in *Nicotiana tabacum*[J]. *Photosynth. Res*. 102:21-29.

[9] Czechowski T, Stitt M, Altmann T, et al. 2005 Genome-wide identification and testing of superior reference genes for transcript normalization in *Arabidopsis*[J]. *Plant Physiol*. 139:5-17.

[10] Demidenko N V, Logacheva M D, Penin A A. 2011 Selection and validation of reference genes for quantitative real-time PCR in buckwheat (*Fagopyrum esculentum*) based on transcriptome sequence data[J]. *PLoS One*. 6:e19434.

[11] Dong L G, Wang X R, Ding Y L. 2014 Study on the *Osmanthus fragrans* blooming season phenology [J]. *J. Nanjing For. Univ*. 38:51-56.

[12] Ferrandino A, Lovisolo C. 2014 Abiotic stress effects on grapevine (*Vitis vinifera* L.): focus on abscisic acid-mediated consequences on secondary metabolism and berry quality[J]. *Environ Exp Bot*. 103:138-147.

[13] Gadkar V J, Filion M. 2015 Validation of endogenous reference genes in *Buglossoides arvensis* for normalizing RT-qPCR-based gene expression data[J]. *SpringerPlus*. 4,1-12.

[14] Galli V, Borowski J M, Perin E C, et al. 2015 Validation of reference genes for accurate normalization of gene expression for real time-quantitative PCR in strawberry fruits using different cultivars and osmotic stresses[J]. *Gene*. 554:205-214.

[15] Han Y J, Li L, Dong M, et al. 2013 cDNA cloning of the phytoene synthase (PSY) and expression analysis of *PSY* and carotenoid cleavage dioxygenase genes in *Osmanthus fragrans*[J]. *Biologia*. 68:258-263.

[16] Han Y J, Wang X H, Chen W C, et al. 2014 Differential expression of carotenoid-related genes determines diversified carotenoid coloration in flower petal of *Osmanthus fragrans*[J]. *Tree Genet. Genomes* 10:329-338.

[17] Hu S Q, Xiang Q B, Liu Y L. 2014 Three new cultivars of *Osmanthus fragrans* in Zhejiang[J]. *J. Nanjing For. Univ*. 38:185.

[18] Huis R, Hawkins S, Neutelings G. 2010 Selection of reference genes for quantitative gene expression normalization in flax (*Linum usitatissimum* L.)[J]. *BMC Plant Biol*. 10: 1-14.

[19] Huang Q L, He R, Zhan R T, et al. 2013 Selection of reference genes for real-time fluorescence quantitative PCR in *Nerviliao fordii* folium[J]. *Chinese Tradi Herb Drugs* 43:1979-1983.

[20] Jiang X M, Zhang X Q, Yan H D, et al. 2014 Reference gene selection for real-time quantitative PCR normalization in switch grass (*Panicum virgatum* L.) root tissue[J]. *J Agr Biotech*. 22:55-63.

[21] Kim B R, Nam H Y, Kim S U, et al. 2003 Normalization of reverse transcription quantitative-PCR with housekeeping genes in rice[J]. *Biotechnol. Lett.* 25:1869-1872.

[22] Kumar K, Muthamilarasan M, Prasad M. 2013 Reference genes for quantitative real-time PCR analysis in the model plant foxtail millet (*Setaria italica* L.) subjected to abiotic stress conditions[J]. *Plant Cell Tiss. Org.* 115:13-22.

[23] Li J, Chen M, Qiu F, et al. 2014 Reference gene selection for gene expression studies using quantitative real-time PCR normalization in *Atropa belladonna*[J]. *Plant Mol. Biol. Rep.* 32:1002-1014.

[24] Luo H, Luo K, Luo L, et al. 2014 Evaluation of candidate reference genes for gene expression studies in *Cymbidium kanran*[J]. *Sci. Hortic.* 167:43-48.

[25] Mallona I, Lischewski S, Weiss J, et al. 2010 Validation of reference genes for quantitative real-time PCR during leaf and flower development in Petunia hybrida[J]. *BMC Plant Biol.* 10, 4. doi:10.1186/1471-2229-10-4.

[26] Maroufi A, Bockstaele E V, Loose M D. 2010 Validation of reference genes for gene expression analysis in chicory (*Cichorium intybus*) using quantitative real-time PCR[J]. *BMC Mol Biol*, 2010, 11:397.

[27] Marum L, Miguel A, Ricardo C P, et al. 2012 Reference gene selection for quantitative real-time PCR normalization in *Quercus suber*[J]. *PLoS One* 7:e35113.

[28] Meng Y, Li N, Tian J, et al. 2013 Identification and validation of reference genes for gene expression studies in postharvest rose flower (*Rosa hybrida*)[J]. *Sci. Hortic.* 158:16-21.

[29] Migocka M, Papierniak A. 2010 Identification of suitable reference genes for studying gene expression in cucumber plants subjected to abiotic stress and growth regulators[J]. *Mol. Breed* 28:343-357.

[30] Mou S, Zhang X, Miao J, et al. 2014 Reference genes for gene expression normalization in *Chlamydomonas* sp. ICE-L by quantitative real-time RT-PCR. J[J]. *Plant Biochem. Biotechnol.* 24:1-7.

[31] Mu H N, Li H G, Wang L G, et al. 2014 Transcriptome sequencing and analysis of sweet osmanthus (*Osmanthus fragrans* Lour.)[J]. *Gene Genome* 36:777-788.

[32] Nolan T, Hands R E, Bustin S A. 2006 Quantification of mRNA using real-time RT-PCR[J]. *Nat. Protoc.* 26:1559-1582.

[33] Obrero A, Die J V, Roman B, et al. 2011 Selection of reference genes for gene expression studies in zucchini (*Cucurbita pepo*) using qPCR[J]. *J. Agic. Food Chem.* 59:5402-5411.

[34] Pfaffl M W, Tichopad A, Prgomet C, et al. 2004 Determination of stable housekeeping genes, differentially regulated target genes and sample integrity: BestKeeper—Excel-based tool using pair-wise correlations[J]. *Biotechnol. Lett.* 26:509-515.

[35] Smalle J, Vierstra R D. 2004 The ubiquitin 26S proteasome proteolytic pathway[J]. *Annu Rev Plant Biol.* 55:555-590.

[36] Soler M, Serra O, Molinas M, et al. 2008 Seasonal variation in transcript abundance in cork tissue analyzed by real time RT-PCR[J]. *Tree Physiol.* 28:743-751.

[37] Tong Z, Gao Z, Wang F, et al. 2009 Selection of reliable reference genes for gene expression studies in peach using real-time PCR[J]. *BMC Mol. Biol.* doi:10.1186/1471-2199-10-71.

[38] Vandesompele J, De Preter K, Pattyn F, et al. 2002 Accurate normalization of real-time quantitative

RT-PCR data by geometric averaging of multiple internal control genes[J]. *Genome Biol*, 2002 (doi: 10.1186/gb-2002-3-7-research0034).

[39] Vandesompele J, Pfaffl M W, Kubista M. 2009 Reference gene validation software for improved normalization. In *Real time PCR: current technology and application* (ed. J. Logan, K. Edwards and N. Saunder), pp. 47-64. London: Caister Academic Press, 2009: 47-64.

[40] Wan H, Zhao Z, Qian C, et al. 2010 Selection of appropriate reference genes for gene expression studies by quantitative real-time polymerase chain reaction in cucumber[J]. *Analyt. Biochem.* 399: 257-261.

[41] Wang L H, Peng Y J, Yang L, et al. 2013 Validation of internal reference genes for qRT-PCR normalization in 'Guanxi Sweet Pummelo' (*Citrus grandis*)[J]. *J. Fruit Sci.* 30: 48-54.

[42] Wang W X, Lai F X, Li K L, et al. 2014 Selection of reference genes for gene expression analysis in *nilaparvata lugens* with different levels of virulence on rice by quantitative real-time PCR[J]. *Rice Sci.* 21: 305-311.

[43] Xiang M, Duan Y F, Xiang Q B. 2014a Establishment of a new group: *Osmanthus fragrans* color group[J]. *J. Nanjing For. Univ.* 38: 11-12.

[44] Xiang Q B, Wang X R, Liu Y L. 2014b Three new cultivars of *Osmanthus fragrans* from east China [J]. *J. Nanjing For. Univ.* 38: 181.

[45] Zang D K, Xiang X Q, Liu Y L. 2006 Notes on cultivar classification in osmanthus[J]. *Sci Silvae Sinicae* 42: 17-23.

[46] Zang D K, Sun M P, Dong B R, et al. 2014 Three new cultivars of *Osmanthus fragrans* from east China[J]. *J. Nanjing For. Univ.* 38: 181.

[47] Zhang C, Fu J X, Wang Y G, et al. 2015 Identification of suitable reference genes for gene expression normalization in the quantitative real-time PCR analysis of sweet osmanthus (*Osmanthus fragrans* Lour.)[J]. *PLoS One* 10: e0,136355.

Cloning and Expression Analysis of MEP Pathway Enzyme-Encoding Genes in *Osmanthus fragrans* *

Chen Xu[1], Huogen Li[2], Xiulian Yang[1], Chunsun Gu[3], Hongna Mu[4], Yuanzheng Yue[1], Lianggui Wang[1]

1. College of Landscape Architecture, Nanjing Forestry University, Nanjing 210037, China.
2. Key Laboratory of Forest Genetics & Gene Engineering of the Ministry of Education, Nanjing Forestry University, Nanjing 210037, China.
3. Institute of Botany, Jiangsu Province and Chinese Academy of Sciences, Nanjing 210014, China.
4. College of Horticulture and Gardening, Yangtze University, Jingzhou 434025, China.

Abstract: The 2-C-methyl-D-erythritol 4-phosphate (MEP) pathway is responsible for the biosynthesis of many crucial secondary metabolites, such as carotenoids, monoterpenes, plastoquinone, and tocopherols. In this study, we isolated and identified 10 MEP pathway genes in the important aromatic plant sweet osmanthus (*Osmanthus fragrans*). Multiple sequence alignments revealed that 10 MEP pathway genes shared high identities with other reported proteins. The genes showed distinctive expression profiles in various tissues, or at different flower stages and diel time points. The qRT-PCR results demonstrated that these genes were highly expressed in inflorescences, which suggested a tissue-specific transcript pattern. Our results also showed that *OfDXS1*, *OfDXS2*, and *OfHDR1* had a clear diurnal oscillation pattern. The isolation and expression analysis provides a strong foundation for further research on the MEP pathway involved in gene function and molecular evolution, and improves our understanding of the molecular mechanism underlying this pathway in plants.

Key words: *Osmanthus fragrans*; MEP pathway; tissue-specific; flower development; diel oscillations

1 Introduction

Osmanthus fragrans, also known as sweet osmanthus, sweet olive, and tea olive, is a traditional aromatic flowering tree that is native to China and has been cultivated for over 2,500 years. It is considered to be one of top 10 Chinese traditional flowers and is also cultivated as an urban ornamental tree[1]. Owing to its pleasant aroma and evergreen properties, it is now widely distributed in Asian countries, such as China, Japan, Thailand, and India[2]. Today, 166 registered cultivars of *O. fragrans* have been classified into five groups based on morphological characteristics and growth habit. These

---

* 原文发表于 *Genes*, 2016, 7 (10): 78。

are the Luteus group, the Albus group, the Aurantiacus group, the Asiaticus group, and the Colour group[3-4]. Generally, the cultivars in the Luteus group have golden-yellow flowers that only appear in the fall, whereas the Asiaticus group cultivars bloom all year round and have creamy-yellow flowers[5]. The fresh flowers are very abundant in aromatic compounds, including terpenoids, fatty acid derivatives, and phenylpropanoids/benzenoids[6-7]. Although the relative contents of the volatiles vary among different cultivars and developmental stages, the main aromatic components are the terpenoids, including monoterpenes ocimene and linalool[8-10]. Most of these substances are the primary components of perfumes and essential oils[11]. Because of the importance of these terpenoid compounds to the aesthetic value of *O. fragrans* plants, it has been of strong interest to understand their biosynthesis[12-14].

In plants, the biosynthesis of terpenoids is catalyzed by a family of enzymes collectively designated as terpene synthase (*TPSs*), which convert prenyl diphosphates to various subclasses of terpeneoids including monoterpenes[15]. Several TPS genes involved in the biosynthesis of volatile terpenoids from *O. fragrans* flowers have been isolated and characterized. The over-expressions of *OfTPS1*, *OfTPS2*, and *OfTPS3* in transgenic tobacco leaves results in the formation of the major monoterpenes, linalool and *trans*-β-ocimene[16]. In contrast to our knowledge about TPS genes, little is known about the biosynthesis of the substrates for TPSs, i. e., prenyl diphosphates. Generally, two biochemical pathways supply the prenyl diphosphates in plants: the mevalonate (MVA) pathway and the 2-C-methyl-D-erythritol 4-phosphate (MEP) pathway[17]. The MVA pathway functions in cytosol for the production of farnesyl diphosphate, which is the substrate for sesquiterpenes. In contrast, the MEP pathway is localized in plastids and produces geranyl diphosphate and geranylgeranyl diphosphate, which are substrates for monoterpenes and diterpenes, respectively[18]. Because the main terpenoids from *O. fragrans* flowers are monoterpenes, the MEP pathway is therefore of our interest for this study.

The MEP pathway consists of eight enzymatic catalysis stages, and each step is schematically represented (Fig. 1)[19-20]. This plastid-localized route begins with the production of 1-deoxy-D-xylulose 5-phosphate(DXP) by 1-deoxy-D-xylulose-5-phosphate synthase (DXS). The second step is catalyzed by 1-deoxy-D-xylulose-5-phosphate reductoisomerase(DXR), which transforms DXP to MEP[21]. Subsequently, MEP is converted to isopentenyl diphosphate(IPP) and dimethylallyl diphosphate(DMAPP) via 2-C-methyl-D-erythritol 4-phosphate cytidylyltransferase (MCT), 4-(cytidine 5r-diphospho)-2-C-methyl-D-erythritol kinase (CMK), 2-C-methyl-D-erythritol 2, 4-cyclodiphosphate synthase (MDS), 4-hydroxy-3-methylbut-2-enyl diphosphate synthase (HDS), and 4-hydroxy-3-methylbut-2-enyl diphosphate reductase (HDR)[22]. The transformation between IPP and DMAPP proceeds through isopentenyl-diphosphate isomerase (IDI), which is a reversible reaction with crosstalk[23]. Apart from HDS and HDR, other crystal structures of enzymes in

MEP route have been successfully represented[24].

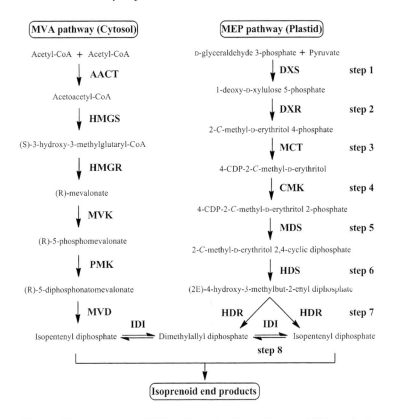

Fig. 1 The steps of the MEP pathway leading to Isoprenoid biosynthetic

Enzymes of MEP pathway are as follows: step 1, DXS; step 2, DXR; step 3, MCT; step 4, CMK; step 5, MDS; step 6, HDS; step 7, HDR; step 8, IDI.

The MEP pathway was originally detected in bacteria. However, further evidence has shown that it is widely found in phototrophic eukaryotes[25]. Various homologous genes have been isolated and cloned independently from many plant species, such as *Arabidopsis* (*Arabidopsis thaliana*)[26-27], periwinkle (*Catharanthus roseus*)[28-29], peppermint (*Mentha piperita*)[30-31], and tomato (*Lycopersicon esculentum*)[32-33]. Most enzymes in the MEP pathway are encoded by single genes in plants, including *Arabidopsis*, poplar (*Populus trichocarpa*), and rice (*Oryza sativa*)[24]. However, both *DXS* and *HDR* are reported to be encoded by a small gene family. For instance, there are three *DXS* genes encoding functional enzymes in maize (*Zea mays*), and two different genes encoding HDR have been identified in loblolly pine (*Pinus taeda*)[34-35]. Furthermore, previous studies have suggested that *DXS* and *DXR* have rate-limiting roles when controlling the metabolic flux through the MEP pathway[36]. Recently, the genetic transformation of *Artemisia annua* enhanced the biosynthesis of artemisinin by overexpressing *DXR* gene[37]. The metabolic engineering of plants is an effective way of improving desired characteristics, such as scent and color, which means research on MEP pathway enzyme-encoding genes is urgently needed because of their potential medical and industrial values[38].

In one of our recent studies, we analyzed the transcriptomes of *O. fragrans* by using Illumina technology. Many putative genes involved in floral scent biosynthesis were identified, including those of the MEP pathway[39]. The first objective of the present study is to isolate the full-length genes of the MEP pathway from *O. fragrans* and to compare them to the corresponding genes from other plant species. The second objective is to determine the expression patterns of the MEP pathway genes in order to understand their contribution to the biosynthesis of monoterpenes that are the major floral scent components of *O. fragrans*.

2 Materials and Methods

2.1 Plant Materials

Two cultivars of *O. fragrans* 'Boye Jingui' and *O. fragrans* 'Rixiang Gui' were grown in the campus of Nanjing Forestry University in Jiangsu, China. Florets of cymose inflorescences (FCI) with the same anthesis were harvested at bud-eye stage (S1), primary blooming stage (S2), full blooming stage (S3), and flower fading stage (S4) in September 2014 (Fig. 2). For tissue-specific gene expression studies, roots, stems and leaves, as well as florets of cymose inflorescences at full blooming stage were collected in September 2014. Materials used for diel analysis were collected every two hours (from 2:00 am to 24:00 pm) at full blooming stage (S3) on 11 October 2015. All these samples were immediately frozen in liquid nitrogen and stored at -80 ℃ for further use.

Fig. 2 Flowering stages of *O.* 'Boye Jingui' in (a) bud-eye stage (S1); (b) primary blooming stage (S2); (c) full blooming stage (S3); (d) flower fading stage (S4). Flowering stages of *O.* 'Rixiang Gui' in (e) bud-eye stage (S1); (f) primary blooming stage (S2); (g) full blooming stage (S3); (h) flower fading stage (S4)

2.2 Total RNA Extraction and Gene Cloning

The total RNA was obtained from the florets of cymose inflorescences of *O. fragrans* using RNAprep pure Kit (Tiangen Biotech, Beijing, China). The obtained RNA ratio of A260/280 was quantified by NanoDrop 2000 Spectrophotometer (Thermo Scientific, Waltham, MA, USA). The RNA integrity was evaluated by 1.5% agarose gel electrophoresis. Then the first strand cDNA reaction with 1 μg total RNA was performed using RevertAid First Strand cDNA Synthesis Kit (Thermo Scientific). According to the MEP pathway unigene sequences from the transcriptomic data of *O. fragrans*, the specific primers were designed to clone *OfDXS1*, *OfDXS2*, *OfDXR*, *OfMCT*, *OfCMK*, *OfMDS*, *OfHDS*, *OfHDR1*, *OfHDR2*, and *OfIDI* (Tab. S1). By using LA Taq (Takara Biotechnology, Dalian, China), purified DNA fragments of polymerase chain reaction (PCR) were ligated into pEASY©-T1 cloning vector (Transgen Biotech, Beijing, China) and transformed into *E. coli* chemically competent cells. Positive recombinant clones were identified and sequenced using the universal M13 primers.

2.3 Cloning of Full Length Genes by RACE

Rapid amplification of cDNA ends (RACE) was used to obtain the 3′ ends and 5′ ends of target genes according to the manufacture's procedure (Takara Biotechnology). The specific primers for 3′ RACE and 5′ RACE were designed using Oligo 6.0 software based on the obtained partial sequences. The primer sequences and PCR conditions were listed (Tab. S2). By sequential nested PCR, these unknown regions were amplified and sequenced. Then full-length genes were assembled together by the Lasergene 7.0 software (Dnastar, Madison, WI, USA). The coding regions were confirmed by PCR detection from start codon to stop codon.

2.4 Gene Expression Analysis

Following the MIQE guidelines, the primers of target genes for quantitative real-time PCR (qRT-PCR) were selected using primer premier 5.0 software (Premier biosoft, Palo Alto, CA, USA) (Tab. S3), and the absence of hairpin structure and primer dimer were predicted by Oligo 6.0 software (Molecular biology insights, Colorado Springs, CO, USA)[40]. Total RNA preparation was performed as described previously according to the manufacturer's instructions. Then first strand cDNA was synthesized from 1 μg total RNA and diluted five-fold for gene expression experiment. The qRT-PCR experiment was carried out by using an ABI StepOnePlus Systems (Applied Biosystems, Carlsbad, CA, USA) and SYBR Premix Ex Taq (Takara Biotechnology). The PCR conditions were as follows: 95 ℃ for 30 s, followed by 40 cycles of 95 ℃ for 5 s, and 58 ℃ for 30 s. The qRT-PCR for each sample was repeated three times. Previous validated genes *OfRAN*, *OfRPB2*, and *OfACT* were used as internal normalizations for different organs, different flowering stages and diel variations, respectively[12]. Each primer pair was validated the specificity by melt curve analysis, and the gene expression levels were calculated by the $2^{-\Delta\Delta CT}$ method. The qRT-PCR results were analyzed by using ABI StepOne software (Applied Biosystems).

3 Results

3.1 Cloning and Sequence Analysis of MEP Pathway Genes

To clone *OfDXS1*, *OfDXS2*, *OfDXR*, *OfMCT*, *OfCMK*, *OfMDS*, *OfHDS*, *OfHDR1*, *OfHDR2*, and *OfIDI*, degenerate primers were designed to obtain 2,232 bp, 1,085 bp, 1,453 bp, 596 bp, 1,306 bp, 628 bp, 2,366 bp, 359 bp, 604 bp, and 669 bp sized amplicons respectively. Based on the partial gene sequences, the 3' region and 5' region were amplified and sequenced by RACE. The size of full-length cDNA, open reading frame (ORF), amino acids, molecular weight, and isoelectric point (pI) were listed (Tab. 1). We submitted the 10 sweet osmanthus full-length cDNAs of MEP genes to NCBI GenBank with the accession number KX400841-KX400850.

Tab. 1 Sequence characteristics of 10 MEP pathway genes in *Osmanthus fragrans*

Gene	Accession No.	Full Length (bp)	ORF (bp)	Amino Acids(aa)	Molecular Weight(kDa)	PI
OfDXS1	KX400841	2,645	2,172	723	78.1	6.91
OfDXS2	KX400842	2,533	2,148	715	76.9	6.91
OfDXR	KX400843	1,687	1,425	474	51.3	6.04
OfMCT	KX400844	1,155	939	312	34.4	7.67
OfCMK	KX400845	1,543	1,206	401	44.4	5.75
OfMDS	KX400846	991	702	233	25.1	8.64
OfHDS	KX400847	2,551	2,229	742	82.5	5.78
OfHDR1	KX400848	1,642	1,386	461	52.0	5.51
OfHDR2	KX400849	1,593	1,380	459	51.8	5.73
OfIDI	KX400850	1,130	708	235	26.9	5.14

The amino acid sequences of 10 MEP genes were aligned with other plants to reveal identities and conserved domains in the NCBI database. Multiple alignments showed that the identities ranged from 73% to 92%, in which *OfDXS1*, *OfDXR*, *OfHDS*, and *OfIDI* shared the higher identities between 88% and 92%, *OfMCT*, *OfCMK*, and *OfMDS* shared the lower identities between 73% and 82%, and *OfDXS2* had the intermediate identity between 85% and 87%. Both *OfDXS1* and *OfDXS2* contained three conserved domains, a thiamine diphosphate-dependent domain at the N-terminus, a pyrimidine binding domain at the medial position, and a transketolase domain at the C-terminus. *OfDXR* also contained three reductoisomerase domains, which was located at the position 80-208 aa, 222-305 aa, and 337-459 aa. *OfMCT* had an IspD domain at the C-terminus, which is also known as YgbP domain. *OfCMK* contained two GHMP kinase domains at the position 176-234 aa and 287-360 aa. *OfMDS* showed a trimer YgbB domain at the position 76-230 aa. Conserved domain of GcpE was found in OfHDS at the position 89-731 aa. In *OfHDR1* and *OfHDR2*, LytB domain was shown at the C-terminus. *OfIDI* showed a NUDIX hydrolase domain at the position 53-205 aa. The amino acid sequences among MEP proteins were composed of multiple conserved residues, which were crucial to form distinct dimensional structures and specific biological functions (Tab. S4).

3.2 Expression Analysis of MEP Genes in Different Organs

To investigate the tissue-specific expressions of the MEP genes, the transcript levels of two fragrans cultivars, O. 'Boye Jingui' and O. 'Rixiang Gui', were detected in different organs including roots, stems, leaves, and inflorescences by using qRT-PCR. The transcript levels of *OfDXS1*, *OfDXS2*, *OfDXR*, *OfMCT*, *OfCMK*, *OfMDS*, *OfHDS*, *OfHDR1*, *OfHDR2*, and *OfIDI* were measured (Fig. 3).

The tissue-specific results suggested that *OfDXS1*, *OfDXS2*, *OfDXR*, *OfCMK*, and *OfHDR2* expressions were significantly abundant in the inflorescences, compared with other organs. In cultivar O. 'Boye Jingui', the *OfDXS1* and *OfCMK* transcript levels in the inflorescences were almost 8-fold and 22-fold higher than that in the roots respectively, while in cultivar O. 'Rixiang Gui' they were only 2.8-fold and 11-fold, respectively. In both cultivar O. 'Boye Jingui' and O. 'Rixiang Gui', *OfDXR* showed over 15-fold transcript levels in the inflorescences than in the roots. As for *OfDXS2*, the inflorescences were found to contain the highest transcript level. In cultivar O. 'Boye Jingui', the *OfDXS2* transcript level in the inflorescences was virtually 427-fold higher than that in the roots, whereas in cultivar O. 'Rixiang Gui' it was merely 224-fold. For cultivar O. 'Boye Jingui' and O. 'Rixiang Gui', *OfMCT*, *OfMDS*, *OfHDS*, and *OfHDR1* showed higher transcript levels in the leaves and the inflorescences than in the roots and the stems. In addition, the *OfMCT* transcript levels were paralleled in the leaves and the inflorescences. In cultivar O. 'Boye Jingui', *OfHDS* and *OfHDR1* showed higher transcript levels in the inflorescences than in the rest of the organs. Whereas in cultivar O. 'Rixiang Gui', *OfMDS* and *OfHDR1* showed higher transcript levels in the leaves. Furthermore, *OfIDI* showed slightly different transcript profiles among the four organs, which were consistent in the two cultivars. The *OfIDI* transcript level in the petals was 1.5-fold higher than that in the roots, and almost 3-fold higher than that in the stems or the leaves.

3.3 Expression Analysis of MEP Genes Over Flower Development

To determine the expression patterns during flower development, qRT-PCR were conducted to detect the transcript levels of MEP genes at four flowering stages, including

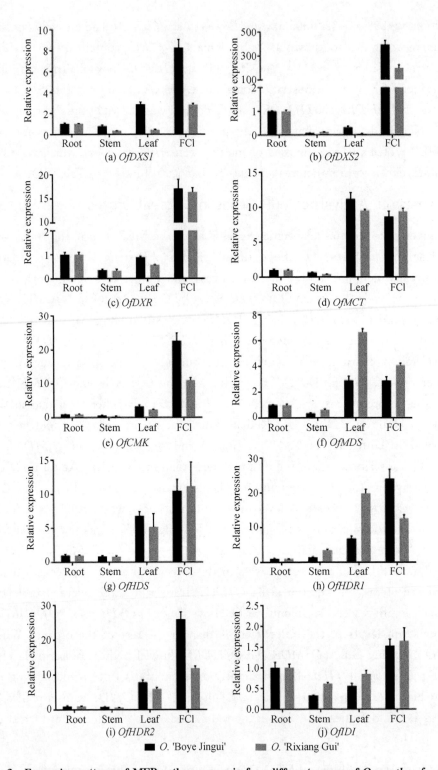

Fig. 3 Expression patterns of MEP pathway genes in four different organs of *Osmanthus fragrans*

FCI: florets of cymose inflorescences at full blooming stage. Data were presented as means with error bars indicating standard deviation.

bud-eye stage (S1), primary blooming stage (S2), full blooming stage (S3), and flower fading stage (S4) (Fig. 4).

The experimental results showed the MEP genes were all detected at four flowering stages, however their transcript patterns varied from each other. For $OfDXS1$, the transcripts in O. 'Boye Jingui' and O. 'Rixiang Gui' both displayed downregulated trends coincidently during the first three stages. The $OfDXS2$ transcript levels showed entirely opposite trends in the two cultivars: in O. 'Boye Jingui' high transcript level was maintained in the first three stages, whereas in O. 'Rixiang Gui', it first increased steadily from S1 to S3 stage, and then declined sharply at S4 stage. The $OfDXR$ transcript levels declined constantly in the two cultivars from S1 to S4 stage. For $OfMCT$ and $OfCMK$, their transcript levels in O. 'Boye Jingui' showed a regularly downregulated trend at the four stages. However, their transcripts in O. 'Rixiang Gui' remained at a high level from S1 to S3 stage, and then decreased considerably from S3 to S4 stage. For $OfMDS$ and $OfHDS$, their transcript levels in O. 'Boye Jingui' did not show a significant change at the four stages. However, their transcript levels in O. 'Rixiang Gui' showed a slight rise from S1 to S2 stage, and then declined from the S3 to S4 stage. For $OfHDR1$ and $OfHDR2$, their transcript levels in O. 'Rixiang Gui' showed a similar profile with that of $OfMDS$ and $OfHDS$, while their transcript levels in O. 'Boye Jingui' showed a reverse trend from S1 to S2 stage. For $OfIDI$, the transcripts in O. 'Boye Jingui' and O. 'Rixiang Gui' maintained almost identical levels during the first three stages, but ascended to 1.25-fold and 2.15-fold from S3 to S4 stage, respectively.

3.4 Expression analysis of MEP Genes during Diel Oscillations

To further study the expression patterns during diel oscillations from sweet osmanthus flowers over time, we chose 12 sampling time points with two-hour intervals in full blooming stage for daily analysis. Using qRT-PCR, the transcript levels of the 10 MEP genes were detected in the two cultivars (Fig. 5).

Each of the MEP genes showed a particular oscillating pattern during the daytime and night cycles. In cultivar O. 'Boye Jingui' and O. 'Rixiang Gui', the $OfDXS1$ transcript levels both exhibited a clear peak in the morning. The $OfDXS2$ transcript levels showed a typical diurnal oscillation both in cultivars O. 'Boye Jingui' and O. 'Rixiang Gui', which escalated in the morning, reached the peak at midday, then de-escalated in the afternoon. For $OfDXR$, the transcript level in O. 'Boye Jingui' showed a slight peak at 06:00, whereas peak transcript of O. 'Rixiang Gui' appeared to a later time point (10:00). The $OfMCT$ transcript in O. 'Boye Jingui' achieved the highest level at 06:00, while in O. 'Rixiang Gui' slight oscillations occurred during the whole day. Compared with other time points, the $OfCMK$ transcript in O. 'Boye Jingui' maintained higher level from 12:00 to 22:00. Yet in O. 'Rixiang Gui', the $OfCMK$ transcript level decreased after 14:00. For $OfMDS$, the transcript level in O. 'Boye Jingui' oscillated steadily during the whole

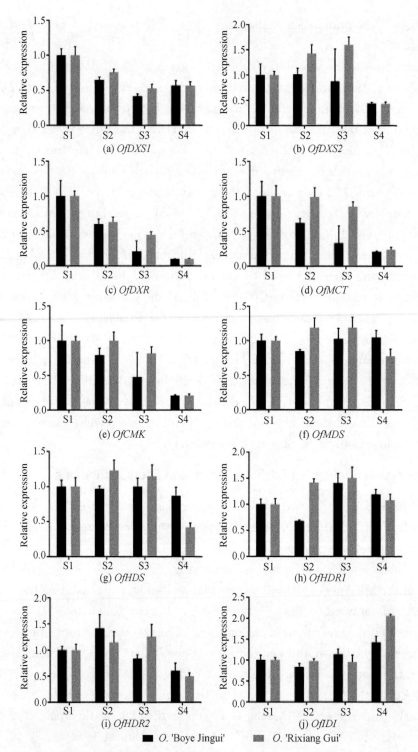

Fig. 4 Expression patterns of MEP pathway genes at four different flowering stages of *Osmanthus fragrans*

These cDNA templates were isolated from bud-eye stage (S1), primary blooming stage (S2), full blooming stage (S3), flower fading stage (S4). Data were presented as means with error bars indicating standard deviation.

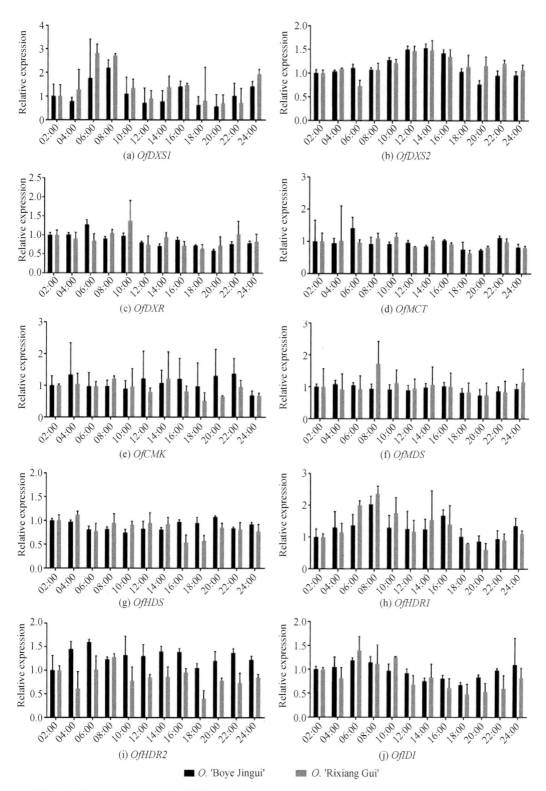

Fig. 5 Expression patterns of MEP pathway genes during diel oscillations of *Osmanthus fragrans*

These cDNA templates were isolated from 02:00 am to 24:00 pm with two-hour intervals in full blooming stage. Data were presented as means with error bars indicating standard deviation.

day, whereas the transcript level in O. 'Rixiang Gui' revealed a significant peak in the morning with 1.73-fold higher than the corresponding predawn level. In O. 'Boye Jingui', the *OfHDS* transcript peaked to 1.5-fold higher levels at 04:00, 14:00, and 20:00, while significant oscillation was undetected in O. 'Rixiang Gui'. For *OfHDR1*, the transcript levels in O. 'Boye Jingui' and O. 'Rixiang Gui' firstly crested at 08:00, thereafter reaching another peak at 14:00 and 16:00, respectively. While the *OfHDR2* transcript levels showed a slight peak between 06:00 and 08:00 in the two cultivars. As for *OfIDI*, the transcript levels showed a gradual decline in the daytime.

4 Discussion

The MEP pathway genes have been isolated and identified in a number of plant species, including Arabidopsis[26-27], peppermint[30-31], tomato[32-33], and rice[24]. However, this pathway has not yet been studied in sweet osmanthus. In this study, gene cloning allows the analysis of the MEP pathway genes sequences in sweet osmanthus and the results will facilitate further research on gene function and molecular evolution.

4.1 The MEP Pathway Genes of *O. fragrans* Are Highly Related to Those from Other Plants

The MEP pathway contains eight enzymatic steps and previous research has shown that terpenoids biosynthesis is regulated by a series of structural and functional genes.[24]. DXS, the first committed enzyme in the MEP pathway, influencing the accumulation of downstream isoprenoids, is encoded by a small multigene family in higher plants. In this study, we successfully isolated two *OfDXS* genes from sweet osmanthus. In the second enzymatic step of the MEP pathway, DXR also has a rate-limiting effect on the accumulation of MEP-derived isoprenoids[41-42]. Furthermore, the biosynthesis of MEP limits the production of downstream isoprenoids in Arabidopsis[43]. In the third step of the MEP pathway, AtMCT contain a plastid targeting sequence in Arabidopsis[44]. The CMK genes, which contain putative ATP binding sites and plastid target sequences, were also cloned from tomato and peppermint[45]. The GbMDS from ginkgo biloba is well conserved in the protein family and highly similar (over 70% identity) to other plants[46]. Although less is known about HDS and HDR than other genes in the MEP pathway, it has been shown that HDR is encoded by muticopy genes in plants, and we obtained two *OfHDR* genes from sweet osmanthus[35]. The last step in the MEP pathway is an isomerization reaction, which is catalyzed by IDI, and is also a rate-limiting step during isoprenoid synthesis[47].

In this study, through the efforts of transcriptome mining and RACE, we successfully obtained 10 full-length MEP pathway cDNAs from sweet osmanthus. These included *OfDXS1*, *OfDXS2*, *OfDXR*, *OfMCT*, *OfCMK*, *OfMDS*, *OfHDS*, *OfHDR1*, *OfHDR2*, and *OfIDI*. Their deduced protein sequences were all highly similar to those of other plants.

Interestingly, the sequence alignments of *OfDXS1*, *OfDXS2*, *OfDXR*, *OfMCT*, *OfCMK*, *OfHDS*, *OfHDR1*, and *OfHDR2* showed 92%, 87%, 91%, 80%, 82%, 92%, 86%, and 88% identity with reported corresponding proteins from *Sesamum indicum* respectively, which suggested that there was an evolutionary conserved relationship between sweet osmanthus and *S. indicum*. However, *OfMDS* shared a high similarity with *Salvia miltiorrhiza*. By bioinformatics analysis, we found that the MEP genes were conserved over their protein sequences, but their detailed evolutionary relationships need further investigation.

4.2 Expression Patterns of the MEP Pathway Genes Suggest that Enhanced Biosynthesis of Substrate Contributes to the Production of Monoterpenes in *O. fragrans* Flowers

The biosynthesis of terpenoids can be regulated at, at least, two levels: the level of terpene synthases and the level of the substrates of TPSs. While a previous study has shown the importance of the regulation of TPS genes[16], the present study indicates the importance of the regulation of the substrate biosynthesis.

The transcript results showed that the MEP genes were all highly expressed in the inflorescences compared with other organs. This suggested that there was a tissue-specific expression profile among these genes, which led to the biosynthesis of specific downstream isoprenoid-derived products[24]. Similarly, it has been reported that terpene synthase (TPS) genes involved in volatile terpenoid synthesis have been cloned and shown to be flower-specific in *Clarkia breweri*[48], *Antirrhinum majus*[49], and *A. thaliana*[50]. *OfDXS2*, which is involved in the first step of the MEP pathway, showed a clear flower-specific transcript profile and its transcript level was several hundred-fold higher than in the roots. It has been suggested that the floral and vegetative tissues are the main scent sources in many plants[51]. Therefore, the enormous expression might lead to the specific accumulation of monoterpenes and sesquiterpenes in flowers[52].

Previous studies have shown that the pigment and essential oil compositions vary in *O. fragrans* floral developmental process[9,53]. Furthermore, previous research has suggested that the aromatic compounds and relative contents differ among *O. fragrans* cultivar groups, including cultivars O. 'Boye Jingui' and O. 'Rixiang Gui'[54]. In this study, the expression profiles of these MEP pathway genes were investigated at different developmental stages by qRT-PCR. Notably, *OfDXS1* and *OfDXS2* remained continuous high expression during the anthesis, as well as *OfMDS*, *OfHDS*, and *OfHDR1*. These results were consistent with the expression level of *OfDXS* in O. 'Yanhong Gui', where there was a substantial accumulation of α- and β-carotene[1]. Moreover, carotenoids are considered to be the crucial substrates for the *OfCCD1* enzyme, which produces α- and β-ionone aroma compounds in flowers of *O. fragrans*[8]. In contrast, the *OfDXR* expression level was down-regulated dramatically from S1 to S4 stage. Interestingly, the *OfMCT* and *OfCMK* expression patterns in O. 'Boye Jingui' and O. 'Rixiang Gui' were not the same. These variations in expression might be caused by cultivar differences.

4.3 Expression Patterns of MEP Pathway Genes during Diel Oscillations

It has been observed that the MEP pathway genes fluctuated rhythmically over a daily light/dark cycle[55]. In *A. thaliana*, the expression of MEP pathway genes was reported to be controlled by light[56]. However, in snapdragon flowers, MEP gene expressions follows a diurnal rhythm, which is regulated by an endogenous circadian clock[57]. Recent study has shown that in many plants, scent emission can be regulated either by circadian clock or by light, mostly by the gene expression levels[58].

Previous research has suggested that the flower volatile emissions of sweet osmanthus follow a diurnal pattern, and sweet osmanthus flowers release the highest total amount of volatiles at 10:00 in the daytime[2]. In this study, we monitored the transcript levels of the MEP pathway genes every 2 h for 24 h at the *O. fragrans* full blooming stage. The *OfDXS2* expression level, not *OfDXR*, showed a diurnal oscillation profile that increased under light conditions with the highest accumulation occurring between 12:00 and 14:00. Similarly, the transcript levels of *OfDXS1* and *OfHDR1* showed considerable intraday variations with a clear peak in the morning. Apart from *OfDXS1*, *OfDXS2*, and *OfHDR1*, the rest of the MEP pathway genes lacked obvious diurnal oscillation patterns. Although previous mathematical modelling data suggested that flux through the MEP pathway is due to the photosynthesis-dependent supply of metabolic substrates, additional experimental work is needed to clarify the contribution of enzymatic substrate biosynthesis to diurnal patterns of volatile emission[59].

5 Conclusions

In this study, the genes in the MEP pathway in *O. fragrans* were cloned, compared with those of other plants, and analyzed for their expression patterns. The sequence alignment analysis revealed that the MEP pathway genes of *O. fragrans* had high sequence identities with other reported proteins, which suggests that an evolutionarily conserved relationship exists. The qRT-PCR results showed that many MEP pathway genes had a higher-level of expression in the inflorescences, supporting that the enhanced production of the prenyl diphosphates—the substrates of terpene synthases—contributes to the biosynthesis of monoterpene floral volatiles. In addition, the expressions of several genes, such as *OfDXS1*, *OfDXS2*, and *OfHDR1*, exhibited a diurnal oscillation pattern. Our results lay an important foundation for future research on functional and molecular evolutionary analysis of the terpene pathway genes involved in the production of terpene floral volatiles in *O. fragrans*.

Supplementary Materials

The following are available online at www.mdpi.com/2073-4425/7/10/78. Tab. S1:

Primers used for partial cDNA cloning; Tab. S2: Primers used for RACE reaction; Tab. S3: Primers used for qRT-PCR; Tab. S4: Conserved domain analysis for MEP genes.

Tab. S1 Primers used for partial cDNA cloning

Gene name	Primer sequence(5′-3′)	PCR condition
OfDXS1-F	GATGGCTCTCTTTACACTTGC	94 ℃ 3 min, 30 cycles of (94 ℃ 30 s;
OfDXS1-R	AGCCAGAGTCCATCCTAAGAC	61 ℃ 30 s; 72 ℃ 2 min), 72 ℃ 10 min
OfDXS2-F	CGGTCCTGATCCATATTGT	94 ℃ 3 min, 30 cycles of (94 ℃ 30 s;
OfDXS2-R	TAGAGCTTCTTTGGGCCTT	58 ℃ 30 s; 72 ℃ 2 min), 72 ℃ 10 min
OfDXR-F	TGACTGGTAGTAGAGTGACACAC	94 ℃ 3 min, 30 cycles of (94 ℃ 30 s;
OfDXR-R	CAACTCATACAAGAGCAGGAC	61 ℃ 30 s; 72 ℃ 2 min), 72 ℃ 10 min
OfMCT-F	GCCAAAGCAGTATCTTCCAC	94 ℃ 3 min, 30 cycles of (94 ℃ 30 s;
OfMCT-R	CAGCAAGTAACAAATCATCAGG	62 ℃ 30 s; 72 ℃ 1 min), 72 ℃ 7 min
OfCMK-F	AGTCAACTGTGTTCATCTTC	94 ℃ 3 min, 30 cycles of (94 ℃ 30 s;
OfCMK-R	GACAACAAAGTACAACAGTC	51 ℃ 30 s; 72 ℃ 2 min), 72 ℃ 10 min
OfMDS-F	ATGGCGACTTCAAACCAC	94 ℃ 3 min, 30 cycles of (94 ℃ 30 s;
OfMDS-R	CCTCGTAAGGAGAACCACT	58 ℃ 30 s; 72 ℃ 1 min), 72 ℃ 7 min
OfHDS-F	AACGGTCGGGATTTGGAGTTG	94 ℃ 3 min, 30 cycles of (94 ℃ 30 s;
OfHDS-R	GCCCTTCTGCCCATAAACTTA	65 ℃ 30 s; 72 ℃ 3 min), 72 ℃ 10 min
OfHDR1-F	TGTGGACTACTGTTGAAAAGC	94 ℃ 3 min, 30 cycles of (94 ℃ 30 s;
OfHDR1-R	ATTTTCCACTCCATACTTGCG	60 ℃ 30 s; 72 ℃ 1 min), 72 ℃ 7 min
OfHDR2-F	TTGGAGCTGCTGTGGATGA	94 ℃ 3 min, 30 cycles of (94 ℃ 30 s;
OfHDR2-R	CGAGCCTCTGCTATTTCTTGTA	57 ℃ 30 s; 72 ℃ 1 min), 72 ℃ 7 min
OfIDI-F	CGCCACTGATTCCGGTATG	94 ℃ 3 min, 30 cycles of (94 ℃ 30 s;
OfIDI-R	TCGGCTGCTTTGCTCAGAG	62 ℃ 30 s; 72 ℃ 1 min), 72 ℃ 7 min

Tab. S2 Primers used for RACE reaction

Usage	Gene name	Primer sequence(5′-3′)	PCR condition
3′ RACE Outer PCR	OfDXS1-3′OF	TTTGTGCAATCTACTCGTCCTTC	94 ℃ 3 min, 20 cycles of (94 ℃ 30 s; 55 ℃ 30 s; 72 ℃ 2 min), 72 ℃ 10 min
	OfDXS2-3′OF	TCTGTGCCATCTATTCGTCATTC	
	OfDXR-3′OF	AGGTCCATTTGTACTTCCTCTTG	
	OfMCT-3′OF	AATTTGCATTGCCCGGAAAGGAG	
	OfCMK-3′OF	TCCATTGGTTCTCATAAAGCCTC	
	OfMDS-3′OF	TTACACTTCCAGCAGAAACCTTC	
	OfHDS-3′OF	CGATGAATCCCAAGAAGAGTTTG	
	OfHDR1-3′OF	GGACTACTGTTGAAAAGCACAAG	
	OfHDR2-3′OF	ATGAAGAGACTGTAGCAACTGCT	
	OfIDI-3′OF	CTCTTGGATGAACTTGGTATTCC	
	OfDXS1-3′IF	AGTGATGGCTCCTTCTGATGAGGCTGAG	
	OfDXS2-3′IF	TGTAGCATACATGGCTTGTTTGCCCAAC	
	OfDXR-3′IF	GTGGTGCTTTTAGGGATTTGCCTGCTGA	

Cont. Tab. S2

Usage	Gene name	Primer sequence(5'-3')	PCR condition
3' RACE Inner PCR	OfMCT-3'IF	GTTCCTGCCAAAGCTACTATCAAAGAGG	94 ℃ 3 min, 30 cycles of (94 ℃ 30 s; 60 ℃ 30 s; 72 ℃ 1.5 min), 72 ℃ 10 min
	OfCMK-3'IF	ATTGCCGCAGGTCGAGGACAGTATGATG	
	OfMDS-3'IF	ATCGAGGCTGTGAAGCTCACTCAGATGG	
	OfHDS-3'IF	CACAGGGACGATTTAGTCATTGGTGCTG	
	OfHDR1-3'IF	GTGTGCGATTACATTTTGGGTGGTGAAC	
	OfHDR2-3'IF	TGTTGAAGGGAGAAACAGAGGAGATTGG	
	OfIDI-3'IF	GATTACCTGCTCTTCATTGTTCGGGATG	
5' RACE Outer PCR	OfDXS1-5'OF	GATATGACTGGTGCCCAACATCC	94 ℃ 3 min, 20 cycles of (94 ℃ 30 s; 55 ℃ 30 s; 72 ℃ 2 min), 72 ℃ 10 min
	OfDXS2-5'OF	GACGAATAGATGGCACAGAACGG	
	OfDXR-5'OF	TCCCGTAACTACTGTGACAGCGT	
	OfMCT-5'OF	ACTTCAAGACCTTCCCTGTTCAC	
	OfCMK-5'OF	ACATGAAAGAGCGACGCCAAATC	
	OfMDS-5'OF	GCCTCCTTGTGTGGGCTTACTTT	
	OfHDS-5'OF	CGCTTCTTTCTTTCCCTGCACTG	
	OfHDR1-5'OF	ACGTTGTGCCGGTATTGAGACCT	
	OfHDR2-5'OF	TCTCCTCTGTTTCTCCCTTCAAC	
	OfIDI-5'OF	GCCGACGGCATTTCTAAGTCAAC	
5' RACE Inner PCR	OfDXS1-5'IF	TTGATCGTGTCCAAAAGAGGAGTCGGAG	94 ℃ 3 min, 30 cycles of (94 ℃ 30 s; 60 ℃ 30 s; 72 ℃ 1.5 min), 72 ℃ 10 min
	OfDXS2-5'IF	GTTCCACCACCCATTGCGGCATGAATAG	
	OfDXR-5'IF	TCACCTGGTCGGCAAGAAGAGTCACATT	
	OfMCT-5'IF	CCCAATAAGCCAACCATCCTTTAGGACC	
	OfCMK-5'IF	TCATACGCTACCTCCACTTTTCTGCCAG	
	OfMDS-5'IF	TATATCTGGAAGCCCCAATGCCCCCAAT	
	OfHDS-5'IF	CTCACTACCAAGAGCCACATTTCCAACC	
	OfHDR1-5'IF	TTACAGGACCTTCTGGTAGCCAGTTCTC	
	OfHDR2-5'IF	GTTCACCACCCAAAATGTAATCGCACAC	
	OfIDI-5'IF	GCTTAATACCACCCTCACCAGCATCTGC	

Tab. S3 Primers used for qRT-PCR

Gene name	Primer sequence(5'-3')	Amplification length
OfDXS1-qF	AGTCACCGAGAAAGGCAGA	239 bp
OfDXS1-qR	GGAAGCGACGCAAGAAAA	
OfDXS2-qF	GGGCATTTGATGTAGCATA	208 bp
OfDXS2-qR	ATCCTTCCCTTACCAATCT	
OfDXR-qF	CCTCAAATTCCAAGGTGGGT	219 bp
OfDXR-qR	GGATTCTCCGCAACTATGTC	
OfMCT-qF	CCTAAAGGATGGTTGGCTTAT	214 bp

Cont. Tab. S3

Gene name	Primer sequence(5'-3')	Amplification length
OfMCT-qR	TGATACATCGTCGGTTACTTC	
OfCMK-qF	GTAGTAGTAATGCCGCCACA	198 bp
OfCMK-qR	ATGGAACGTCAAAGGGTAGA	
OfMDS-qF	GCTTGCTTTACACGGAACT	200 bp
OfMDS-qR	CCGATAATCAGCGGATAAC	
OfHDS-qF	ACGAAGCCATTACCCTACA	214 bp
OfHDS-qR	TTGCTGCATGAACTCTGCT	
OfHDR1-qF	GCTAGAAAACAATTCCCAA	159 bp
OfHDR1-qR	TACCACAACATCACCCTTA	
OfHDR2-qF	ACAATCGGAAGGGGTTT	195 bp
OfHDR2-qR	TTCTCGCCTCGTAAGCA	
RAN-qF	AGAACCGACAGGTGAAGGCAA	117 bp
RAN-qR	TGGCAAGGTACAGAAAGGGCT	
RPB2 qF	CACCAAGCAAAGGACCAGCAAG	216 bp
RPB2-qR	TCACCAGGGAGAAGAGGATCAAGTA	
ACT-qF	CCCAAGGCAAACAGAGAAAAAT	143 bp
ACT-qR	ACCCCATCACCAGAATCAAGAA	

Tab. S4 Conserved domain analysis for MEP genes

Gene	Size (ORF)	Characteristic domain
OfDXS1	2,172 bp	Pfam DXP_synthase_N; Transket_pyr; Pfam Transketolase_C
OfDXS2	2,148 bp	Pfam DXP_synthase_N; Transket_pyr; Pfam Transketolase_C
OfDXR	1,425 bp	Pfam DXP_reductoisom; Pfam DXPR_C
OfMCT	939 bp	Pfam IspD
OfCMK	1,206 bp	Pfam; Pfam
OfMDS	702 bp	Pfam YgbB
OfHDS	2,229 bp	Pfam GcpE
OfHDR1	1,386 bp	Pfam LytB

Cont. Tab. S4

Gene	Size (ORF)	Characteristic domain
OfHDR2	1,380 bp	Pfam LytB
OfIDI	708 bp	Pfam NUDIX

Acknowledgements

This work was supported by the study on germplasm resources exploration, innovation, and utilization of forest tree species (2013BAD01B06) and the Doctorate Fellowship Foundation of Nanjing Forestry University.

Author Contributions

Chen Xu, Huogeng Li and Lianggui Wang conceived and designed the experiments; Chen Xu and Hongna Mu performed the experiments; Chen Xu, Xiulian Yang and Chunsun Gu analyzed the data; Huogeng Li and Lianggui Wang contributed reagents/materials/analysis tools; Chen Xu and Yuanzheng Yue wrote the paper.

Conflicts of Interest

The authors declare no conflict of interest.

Abbreviations

The following abbreviations are used in this manuscript:
DXS   1-deoxy-D-xylulose-5-phosphate synthase
DXR   1-deoxy-D-xylulose-5-phosphate reductoisomerase
MCT   2-C-methyl-D-erythritol 4-phosphate cytidylyltransferase
CMK   4-(cytidine 5r-diphospho)-2-C-methyl-D-erythritol kinase
MDS   2-C-methyl-D-erythritol 2,4-cyclodiphosphate synthase
HDS   4-hydroxy-3-methylbut-2-enyl diphosphate synthase
HDR   4-hydroxy-3-methylbut-2-enyl diphosphate reductase
IDI   isopentenyl-diphosphate

References

[1] Zhang C, Wang Y, Fu J, et al. Transcriptomic analysis and carotenogenic gene expression related to petal coloration in *Osmanthus fragrans* 'Yanhong Gui'[J]. Trees 2016, 30(4):1-17.

[2] Baldermann S, Kato M, Kurosawa M, et al. Functional characterization of a carotenoid cleavage dioxygenase 1 and its relation to the carotenoid accumulation and volatile emission during the floral development of *Osmanthus fragrans* Lour[J]. J. Exp. Bot. 2010, 61:2967-2977.

[3] Xiang Q B, Liu Y L. Classification system of sweet osmanthus cultivars. In An Illustrated Monograph of the Sweet Osmanthus Cultivars in China [M]. Hangzhou: Zhejiang Science & Technology Press, 2008:80-91.

[4] Yuan W J, Li Y, Ma Y F, et al. Isolation and characterization of microsatellite markers for *Osmanthus fragrans* (Oleaceae) using 454 sequencing technology [J]. Genet. Mol. Res. 2015, 14: 4696-4702.

[5] Han Y J, Chen W C, Yang F B, et al. cDNA-AFLP analysis on 2 *Osmanthus fragrans* cultivars with different flower color and molecular characteristics of *OfMYB1* gene[J]. Trees 2015, 29:931-940.

[6] Deng C H, Song G X, Hu Y M. Application of HS-SPME and GC-MS to characterization of volatile compounds emitted from osmanthus flowers[J]. Ann. Chim. 2004, 94:921-927.

[7] Cai X, Mai R Z, Zou J J, et al. Analysis of aroma-active compounds in three sweet osmanthus (*Osmanthus fragrans*) cultivars by GC-olfactometry and GC-MS[J]. J. Zhejiang Univ. B 2014, 15: 638-648.

[8] Baldermann S, Kato M, Fleischmann P, et al. Biosynthesis of α- and β-ionone, prominent scent compounds, in flowers of *Osmanthus fragrans*[J]. Acta Biochim. Pol. 2012, 59:79-81.

[9] Wang L M, Li M T, Jin W W, et al. Variations in the components of *Osmanthus fragrans* Lour. essential oil at different stages of flowering[J]. Food Chem. 2009, 114:233-236.

[10] Xin H P, Wu B H, Zhang H H, et al. Characterization of volatile compounds in flowers from four groups of sweet osmanthus (*Osmanthus fragrans*) cultivars[J]. Can. J. Plant Sci. 2013, 93:923-931.

[11] Lei G M, Mao P Z, He M Q, et al. Water-soluble essential oil components of fresh flowers of *Osmanthus fragrans* Lour[J]. J. Essent. Oil Res. 2016, 28:177-184.

[12] Zhang C, Fu J, Wang Y, et al. Identification of suitable reference genes for gene expression normalization in the quantitative real-time PCR analysis of sweet osmanthus (*Osmanthus fragrans* Lour.)[J]. PLoS ONE 2015, 10: e0,136,355.

[13] Moronkola D O, Aiyelaagbe O O, Ekundayo O. Syntheses of eight fragrant terpenoids [ionone derivatives] via the aldol condensation of citral and eight ketones. J. Essent. Oil Bear[J]. Plants 2005, 8:87-98.

[14] Muhlemann J K, Klempien A, Dudareva N. Floral volatiles: From biosynthesis to function[J]. Plant Cell Environ. 2014, 37:1936-1949.

[15] Tholl D. Terpene synthases and the regulation, diversity and biological roles of terpene metabolism. Curr[J]. Opin. Plant Biol. 2006, 9:297-304.

[16] Zeng X L, Liu C, Zheng R R, et al. Emission and accumulation of monoterpene and the key terpene synthase (TPS) associated with monoterpene biosynthesis in *Osmanthus fragrans* Lour[J]. Front.

Plant Sci. 2016, 6:1-16.

[17] Lichtenthaler H K. The 1-deoxy-D-xylulose-5-phosphate pathway of isoprenoid biosynthesis in plants [J]. Annu. Rev. Plant Physiol. Plant Mol. Biol. 1999, 50:47-65.

[18] Gutensohn M, Orlova I, Nguyen T T H, et al. Cytosolic monoterpene biosynthesis is supported by plastid-generated geranyl diphosphate substrate in transgenic tomato fruits[J]. Plant J. 2013, 75: 351-363.

[19] Singh H, Gahlan P, Kumar S. Cloning and expression analysis of 10 genes associated with picrosides biosynthesis in *Picrorhiza kurrooa*[J]. Gene 2013, 515:320-328.

[20] Pulido P, Perello C, Rodríguez-Concepción M. New insights into plant isoprenoid metabolism[J]. Mol. Plant 2012, 5:964-967.

[21] Lichtenthaler H. Non-mevalonate isoprenoid biosynthesis: Enzymes, genes and inhibitors[J]. Biochem. Soc. Trans. 2000:785-789.

[22] Rohdich F, Kis K, Bacher A, et al. The non-mevalonate pathway of isoprenoids: Genes, enzymes and intermediates[J]. Curr. Opin. Chem. Biol. 2001, 5:535-540.

[23] Hunter W N. The non-mevalonate pathway of isoprenoid precursor biosynthesis. J. Biol[J]. Chem. 2007, 282: 21573-21577.

[24] Cordoba E, Salmi M, León P. Unravelling the regulatory mechanisms that modulate the MEP pathway in higher plants[J]. J. Exp. Bot. 2009, 60:2933-2943.

[25] Rohmer M. The discovery of a mevalonate-independent pathway for isoprenoid biosynthesis in bacteria, algae and higher plants[J]. Nat. Prod. Rep. 1999, 16:565-574.

[26] Estévez J M, Cantero A, Romero C, et al. Analysis of the expression of CLA1, a gene that encodes the 1-deoxyxylulose 5-phosphate synthase of the 2-C-methyl-D-erythritol-4-phosphate pathway in Arabidopsis[J]. Plant Physiol. 2000, 124:95-104.

[27] Carretero-paulet L, Ahumada I, Cunillera N, et al. Expression and molecular analysis of the Arabidopsis DXR Gene encoding 1-deoxy-D-xylulose 5-phosphate reductoisomerase, the first committed enzyme of the 2-C-methyl-D-erythritol 4-phosphate pathway[J]. Plant Physiol. 2002, 129:1581-1591.

[28] Chahed K, Oudin A, Guivarc H N, et al. 1-deoxy-D-xylulose 5-phosphate synthase from periwinkle: cDNA identification and induced gene expression in terpenoid indole alkaloid-producing cells[J]. Plant Physiol. Biochem. 2000, 38:559-566.

[29] Veau B, Courtois M, Oudin A, et al. Cloning and expression of cDNAs encoding two enzymes of the MEP pathway in *Catharanthus roseus*[J]. Biochim. Biophys. Acta 2000, 1,517:159-163.

[30] Lange B M, Wildung M R, McCaskill D, et al. A family of transketolases that directs isoprenoid biosynthesis via a mevalonate-independent pathway[J]. Proc. Natl. Acad. Sci. USA 1998, 95:2100-2104.

[31] Lange B M, Croteau R. Isoprenoid biosynthesis via a mevalonate-independent pathway in plants: Cloning and heterologous expression of 1-deoxy-D-xylulose-5-phosphate reductoisomerase from peppermint[J]. Arch. Biochem. Biophys. 1999, 365:170-174.

[32] Lois L M, Rodríguez-Concepción M, Gallego F, et al. Carotenoid biosynthesis during tomato fruit development: Regulatory role of 1-deoxy-D-xylulose 5-phosphate synthase[J]. Plant J. 2000, 22: 503-513.

[33] Rohdich F, Wungsintaweekul J, Lüttgen H, et al. Biosynthesis of terpenoids: 4-diphosphocytidyl-2-C-methyl-D-erythritol kinase from tomato[J]. Proc. Natl. Acad. Sci. 2000, 97:6451-6456.

[34] Cordoba E, Porta H, Arroyo A, et al. Functional characterization of the three genes encoding 1-deoxy-D-xylulose 5-phosphate synthase in maize[J]. J. Exp. Bot. 2011, 62:2023-2038.

[35] Kim S M, Kuzuyama T, Kobayashi A, et al. 1-hydroxy-2-methyl-2-(E)-butenyl 4-diphosphate reductase (IDS) is encoded by multicopy genes in gymnosperms Ginkgo biloba and Pinus taeda[J]. Planta 2008, 227:287-298.

[36] Tong Y R, Su P, Zhao Y J, et al. Molecular cloning and characterization of DXS and DXR genes in the terpenoid biosynthetic pathway of Tripterygium wilfordii[J]. Int. J. Mol. Sci. 2015, 16:25516-25535.

[37] Xiang L, Zeng L, Yuan Y, et al. Enhancement of artemisinin biosynthesis by overexpressing DXR, CYP71AV1 and CPR in the plants of Artemisia annua L[J]. Plant Omics 2012, 5:503-507.

[38] Roberts S C. Production and engineering of terpenoids in plant cell culture[J]. Nat. Chem. Biol. 2007, 3:387-395.

[39] Mu H N, Li H G, Wang L G, et al. Transcriptome sequencing and analysis of sweet osmanthus (Osmanthus fragrans Lour.)[J]. Genes Genomics 2014, 36:777-788.

[40] Bustin S A, Benes V, Garson J A, et al. The MIQE guidelines: Minimum information for publication of quantitative real-time PCR experiments[J]. Clin. Chem. 2009, 55:611-622.

[41] Yan X M, Zhang L, Wang J, et al. Molecular characterization and expression of 1-deoxy-D-xylulose 5-phosphate reductoisomerase (DXR) gene from Salvia miltiorrhiza[J]. Acta Physiol. Plant 2009, 31:1015-1022.

[42] Munos J W, Pu X, Mansoorabadi S O, et al. A secondary kinetic isotope effect study of the 1-deoxy-D-xylulose-5-phosphate reductoisomerase-catalyzed reaction: Evidence for a retroaldol-aldol rearrangement[J]. J. Am. Chem. Soc. 2009, 131:2048-2049.

[43] Carretero-Paulet L, Cairó A, Botella-Pavía P, et al. Enhanced flux through the methylerythritol 4-phosphate pathway in Arabidopsis plants overexpressing deoxyxylulose 5-phosphate reductoisomerase[J]. Plant Mol. Biol. 2006, 62:683-695.

[44] Rohdich F, Wungsintaweekul J, Eisenreich W, et al. Biosynthesis of terpenoids: 4-diphosphocytidyl-2C-methyl-D-erythritol synthase of Arabidopsis thaliana[J]. Proc. Natl. Acad. Sci. USA 2000, 97:6451-6456.

[45] Dubey V S, Bhalla R, Luthra R. An overview of the non-mevalonate pathway for terpenoid biosynthesis in plants[J]. J. Biosci. 2003, 28:637-646.

[46] Kim S M, Kuzuyama T, Chang Y J, et al. Cloning and characterization of 2-C-methyl-D-erythritol 2,4-cyclodiphosphate synthase (MECS) gene from Ginkgo biloba[J]. Plant Cell Rep. 2006, 25:829-835.

[47] Sedkova N, Tao L, Rouvière P E, et al. Diversity of carotenoid synthesis gene clusters from environmental enterobacteriaceae strains[J]. Appl. Environ. Microbiol. 2005, 71:8141-8146.

[48] Dudareva N, Cseke L, Blanc V M, et al. Evolution of floral scent in Clarkia: Novel patterns of S-linalool synthase gene expression in the C. breweri flower[J]. Plant Cell 1996, 8:1137-1148.

[49] Nagegowda D A, Gutensohn M, Wilkerson C G, et al. Two nearly identical terpene synthases catalyze the formation of nerolidol and linalool in snapdragon flowers[J]. Plant J. 2008, 55:224-239.

[50] Chen F, Tholl D, D'Auria J C, et al. Biosynthesis and emission of terpenoid volatiles from Arabidopsis flowers[J]. Plant Cell 2003, 15:481-494.

[51] Wang J H, Dudareva N, Bhakta S, et al. Floral scent production in Clarkia breweri[J]. Plant Physiol. 1998, 116:599-604.

[52] Xiang L, Zhao K G, Chen L Q. Molecular cloning and expression of *Chimonanthus praecox* farnesyl pyrophosphate synthase gene and its possible involvement in the biosynthesis of floral volatile sesquiterpenoids[J]. Plant Physiol. Biochem. 2010, 48:845-850.

[53] Han Y J, Wang X H, Chen W C, *et al*. Differential expression of carotenoid-related genes determines diversified carotenoid coloration in flower petal of *Osmanthus fragrans*[J]. Tree Genet. Genomes 2014, 10: 329-338.

[54] Yang X L, Shi T T, Wen A L, *et al*. Variance analysis of aromatic components from different varieties of *Osmanthus fragrans*[J]. Dongbei Linye Daxue Xuebao 2015, 43:83-87. (In Chinese)

[55] Nagegowda D A, Rhodes D, Dudareva N. The role of methyl erythritol-phosphate pathway in rhythmic emission of volatiles. In Biology of Floral Scent[M]. Boca Raton: CRC Press, 2006: 139-153.

[56] Mandel M A, Feldmann K A, Herrera-Estrella L, *et al*. *CLA1*, a novel gene required for chloroplast development, is highly conserved in evolution[J]. Plant J. 1996, 9:649-658.

[57] Dudareva N, Andersson S, Orlova I, *et al*. The nonmevalonate pathway supports both monoterpene and sesquiterpene formation in snapdragon flowers[J]. Proc. Natl. Acad. Sci. USA 2005, 102:933-938.

[58] Hendel-Rahmanim K, Masci T, Vainstein A, *et al*. Diurnal regulation of scent emission in rose flowers[J]. Planta 2007, 226:1491-1499.

[59] Pokhilko A, Bou-Torrent J, Pulido P, *et al*. Mathematical modelling of the diurnal regulation of the MEP pathway in *Arabidopsis*[J]. New Phytol. 2015, 206:1075-1085.

桂花(十)-新薄荷醇脱氢酶基因 OfMNR 的克隆与表达分析*

徐晨[1], 李火根[2], 杨秀莲[1], 母洪娜[1], 顾春笋[3], 王良桂[1]

(1. 南京林业大学 风景园林学院,江苏 南京 210037; 2. 南京林业大学 林木遗传和基因工程重点实验室,江苏 南京 210037; 3. 江苏省中国科学院植物研究所,江苏 南京 210014)

摘 要: 利用 cDNA 末端快速扩增技术,从桂花花瓣中克隆获得 1 个桂花新薄荷醇脱氢酶 MNR 基因的 cDNA 全长,命名为 OfMNR。OfMNR 基因序列长度为 1 092 bp,包含 936 bp 的开放阅读框,编码 312 个氨基酸。序列分析表明: OfMNR 编码的氨基酸序列与其他植物中预测的 MNR 具有较高的相似性。采用 qPCR 技术检测了 OfMNR 在桂花 2 个品种的不同组织和不同花期中的表达量,发现 OfMNR 在两个品种中花瓣的表达量均最高,并且在 4 个花期阶段中呈现不同的变化趋势。

关键词: 桂花花瓣;新薄荷醇基因;基因克隆;表达分析

桂花(Osmanthus fragrans)是我国传统名花,是香花中最具观赏和实用价值的树种。桂花栽培历史悠久,不仅具有很高的观赏价值,而且还是重要的香料植物,兼有食用和药用价值[1-2]。一直以来,人们往往侧重于研究植物花色等花朵的性状,而对于花香的研究则相对滞后[3]。桂花花香馥郁,清新高雅,其精油是一种高级天然香料,因此十分适合作为研究花香的植物材料[4-5]。花香是桂花最重要的观赏性状,花香挥发物主要有: 萜烯类化合物(terpenoids)、苯丙酸类化合物/苯环型化合物(benzenoids/phenylpropanoids)和脂肪酸衍生物(aliphatic compounds)三大类,其中萜类物质在桂花花香挥发物中占主导成分[6-8]。在桂花中一些与花色花香相关的基因已被成功克隆,如 OfPSY、OfLCY、OfCCD1 和 OfCCD4 等[9-14]。

新薄荷醇脱氢酶(MNR)属于氧化还原酶家族,这种酶也被称为单萜脱氢酶,参与单萜物质的合成。MNR 最早发现于薄荷(Mentha piperita)的叶片中,在该酶的作用下可将(一)-薄荷酮转化生成(十)-新薄荷醇,亦可将(十)-异薄荷酮转化生成(十)-异薄荷醇,这种脱氢酶类被认为是在植物发育过程中负责单萜酮的代谢[15-17]。此次以桂花为实验材料,克隆获得了 OfMNR 的基因全长 cDNA 序列,检测了它们在不同组织不同发育阶段的表达模式,并利用生物信息学软件对序列进行分析,为研究桂花单萜基因的功能奠定基础。

1 材料与方法

1.1 试材

植物材料包括 2 个桂花品种,分别是'波叶金桂'Osmanthus fragrans 'Boye Jingui'和

---

* 原文发表于《分子植物育种》,2016,14(6): 1389-1395。
基金项目: 国家科技部"十二五"科技支撑计划项目(2013BAD01B06-04)。

'日香桂'Osmanthus fragrans 'Rixiang Gui',均来自南京林业大学校园内,树龄分别为30a和10a,栽培管理条件一致。(1)花瓣的采集:东、南、西、北4个方位均进行采花,采集香眼期、初花期、盛花期、末花期4个发育时期的花瓣。(2)茎段的采集:取当年生枝条,用枝剪将其分段。(3)叶片的采集:取当年生叶片,用剪刀将叶片中脉去掉。(4)根的采集:将校园内'波叶金桂'和'日香桂'作为母本进行扦插,获得1年生扦插苗,取其根系。以上植物材料在采集后迅速装于离心管内,液氮速冻后,-80 ℃超低温冰箱保存备用[18]。

1.2 基因克隆

按照植物总RNA提取试剂盒(Tiangen),对桂花花瓣RNA进行提取,用紫外分光光度计检测RNA的浓度和OD值,并用1%的琼脂糖凝胶电泳检测。以总RNA为模板,按照RevertAid™第一链cDNA Synthesis试剂盒(Thermo Scientific)合成第一链cDNA。基因片段的PCR扩增,反应体系如下:cDNA模板50 ng,10×LA Taq Buffer(Mg^{2+} free) 5.0 μL,MgCl$_2$(25 mM)5 μL,dNTP Mixture(2.5 mM each)8 μL,正反引物(10 mM)各2.0 μL,TaKaRa LA Taq(5 U/μL)0.5 μL,灭菌超纯水补至50.0 μL。PCR扩增程序为:94 ℃ 3 min;94 ℃ 30 s,58 ℃ 30 s,72 ℃ 2 min,30个循环;72 ℃ 10 min。PCR产物经琼脂糖凝胶电泳检测后,切下目的条带,用胶回收试剂盒(TransGen Biotech)回收,产物与pEASY-T1克隆载体连接,转化,测序。根据测序获得目的基因片段设计RACE引物,进行目的基因3′RACE和5′RACE巢式PCR扩增,具体按照3′-Full RACE Core Set with PrimeScript™ RTase和5′-Full RACE Kit with TAP(Takara)试剂盒说明书进行。采用TransGen TransStart FastPfu高保真酶扩增目的基因的ORF(open reading frame)序列,PCR、回收、连接和转化方法同上。

表1 所用引物序列
Tab. 1 Primer sequences

引物名称 Primer_ID	引物序列(5′→3′) Primer sequence	用途 Usage
OfMNR-F	CCCAAGAACCTACCAAGAATACT	Gene specific amplification
OfMNR-R	AAGCCAATCTTACCGGACTTTCT	
OfMNR-3′Outer	TTATGGTGCGAAGAGAACTGCTG	3′RACE
OfMNR-3′Inner	AGGAAGAATCCCTGGAAACCAAAGGCTG	
OfMNR-5′Outer	CCGTGGTGAATTAGATAGCTGGAG	5′RACE
OfMNR-5′Inner	GCTCCACCAACTCCAGCATTATTCACCA	
OfMNR-OF	ATGAATTCAAGCATGGCCC	ORF amplification
OfMNR-OR	TTACTCCACGACAGACGA	
OfMNR-qF	CTATCTAATTCACCACGGAT	qPCR
OfMNR-qR	TGCTTTCGACAGTATGTATG	

1.3 基因序列分析及表达分析

利用ORF Finder(http://www.ncbi.nlm.nih.gov/projects/gorf/)预测拼接序列的开放阅读框和编码的蛋白质序列。使用在线软件ProtParam(http://web.expasy.org/protparam/)分析蛋白质分子质量、理论等电点、氨基酸组成。使用在线软件Pfam(http://

pfam. xfam. org/)和 SMART(http://smart. embl-heidelberg. de/)对蛋白质结构域进行预测。使用 TMHMM Server v. 2.0(http://www. cbs. dtu. dk/services/TMHMM/)分析蛋白质跨膜区。使用 SignalP 4.1 Server(http://www. cbs. dtu. dk/services/SignalP/)分析信号肽。蛋白质二级结构分析采用 SOPMA(https://npsa-prabi. ibcp. fr/cgi-bin/npsa_automat. pl?page=npsa_sopma. html)工具进行,利用 MEGA 5 构建系统进化树。

以不同的组织和不同花期阶段的材料为 cDNA 模板进行目的基因的实时定量 PCR 表达分析。使用 Oligo 6 软件设计扩增引物,RPB2 和 RAN 基因作为内参基因,引物见表 1[19]。采用 SYBR Premix Ex Taq(TaKaRa)在 ABI StepOnePlus Real-Time PCR Systems 上完成,每个样品 3 次重复。反应体系为:2×SYBR Premix Ex Taq(Tli RNaseH Plus)10 μL,50×ROX Reference Dye 0.4 μL,正反引物(10 mmol/L)各 0.4 μL,cDNA 2 μL,灭菌超纯水 6.8 μL。反应程序为:95 ℃ 30 s;95 ℃ 5 s,60 ℃ 30 s,40 个循环。

2 结果与分析

2.1 OfMNR 基因的克隆和序列分析

在桂花转录组数据库中确认 comp82987_c0 为目的基因模板[20]。通过目的基因片段的 PCR 扩增,获得长度为 852 bp 的 cDNA 序列,以此序列设计引物进行 3′RACE 和 5′RACE 巢式 PCR,利用 seqman 对获得的 3′端和 5′端序列进行拼接,获得总长为 1 092 bp 的 cDNA 全长序列,根据 ORF Finder 预测全长 cDNA 的 ORF 区。设计引物 ORF-MNR-F、ORF-MNR-R 进行 ORF 扩增,测序验证。测序结果表明:OfMNR 包含长度为 936 bp 的 ORF,编码 311 个氨基酸。

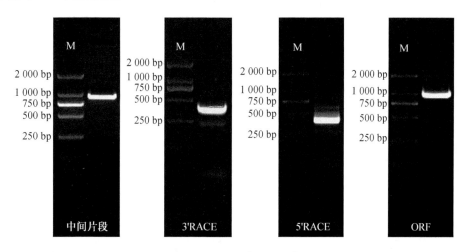

图 1 OfMNR 基因的中间片段、3′RACE、5′RACE 和 ORF 扩增
Fig. 1 Fragment, RACE and ORF amplication of OfMNR
注:M 为 marker(捷瑞 DL2501)。

使用在线预测软件预测 OfMNR 蛋白质分子质量为 33.79 kDa,理论等电点为 5.44,带负电荷的残基数(Asp + Glu)为 37 个,带正电荷的残基数(Arg + Lys)为 32 个,总亲水性值为-0.137,不稳定指数为29.05,该蛋白为亲水性稳定蛋白。该蛋白不具有跨膜区域,不存在信号肽酶切位点,属于非分泌型蛋白。

2.2 蛋白质结构分析与系统进化树构建

Pfam 和 Smart 数据库检索显示,*OfMNR* 蛋白具有 adh_short 结构域,位于 K17～K190 氨基酸之间。该结构域属于 SDR 短链脱氢酶家族,大多数脱氢酶拥有至少 2 个结构域,一个结合辅酶 I(NAD),另一个与底物结合。后者决定底物的特异性,并含有催化氨基酸的作用。SOPMA 分析表明,α螺旋(Alpha helix)是 *OfMNR* 蛋白二级结构的主要成分,含有 40.84% 的 α 螺旋、22.51% 的 β 折叠、12.86% 的 β 转角和 23.79% 的无规则卷曲。

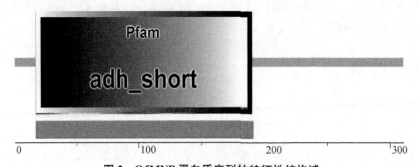

图 2 *OfMNR* 蛋白质序列的特征性结构域
Fig. 2 Characteristic domain of *OfMNR*

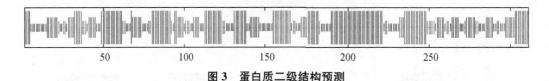

图 3 蛋白质二级结构预测
Fig. 3 Protein secondary structure prediction

通过 NCBI 上的 protein blast 进行同源性搜索,发现 *OfMNR* 与它们同源的蛋白质并不多,且相似性位于 60%～70%。*OfMNR* 与毛果杨(*Populus trichocarpa*)、野生大豆(*Glycine soja*)、南天竹(*Nandina domestica*)、拟南芥(*Arabidopsis thaliana*)、苹果(*Malus domestica*)相似性分别为:70%、68%、66%、65%、65%。

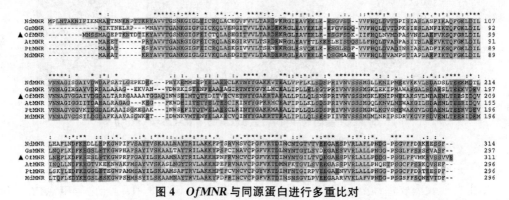

图 4 *OfMNR* 与同源蛋白进行多重比对
Fig. 4 Homologous protein multiple alignment of *OfMNR*

应用 Clustal X 软件将 *OfMNR* 蛋白与 NCBI 数据库下载的 5 个同源蛋白进行多重比对,再利用 MEGA 5.1 软件中的邻接法构建系统进化树。结果表明:*OfMNR* 与 5 个同源蛋白可以聚为三大类,第 I 类为 AtMNR,第 II 类包括 PtMNR 和 MdMNR,第 III 类可划为 2 个分支,*OfMNR* 和 NdMNR 同属第 1 分支,GsMNR 属于第 2 分支。

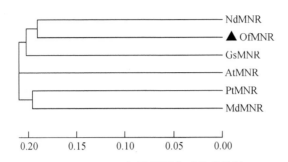

图 5 *OfMNR* 与同源蛋白系统进化树
Fig. 5 Phylogenetic tree of *OfMNR* and homologous protein

2.3 *OfMNR* 基因的表达

利用实时荧光定量 PCR(qPCR)技术对 *OfMNR* 基因的表达进行分析和研究。*OfMNR* 基因在'波叶金桂'和'日香桂'的 4 个不同组织器官根、茎、叶和花中均有不同程度表达,且 2 个品种的结果具有一致性,均在花中表达量最高,在根和叶片中次之,而在茎中表达量最低。*OfMNR* 基因在'波叶金桂'和'日香桂'的 4 个花期阶段:香眼期、初花期、盛花期和末花期同样均有表达,但 2 个品种的结果具有差异性,'波叶金桂'在香眼期有小幅下降,随后逐步升高,在末花期到达峰值。'日香桂'从香眼期表达量逐步升高,到盛花期达到峰值,随后在末花期降低。

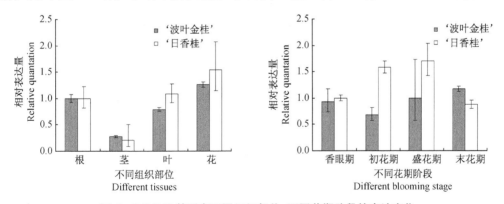

图 6 *OfMNR* 基因在不同组织部位、不同花期阶段的表达变化
Fig. 6 Expression change of *OfMNR* in different tissues and blooming stage

3 讨论

MNR 基因在植物单萜类的代谢过程中有重要的调控作用,目前对辣薄荷(*Mentha piperita*)中 *MNR* 基因研究较多,对其酶活性和代谢途径研究比较透彻。由于薄荷脑的含量影响薄荷精油的质量,因此薄荷酮还原酶 *MpMMR* 基因和 *MpMNR* 基因对产量和成分

有直接影响[21]。在其他一些植物中,虽已成功克隆出 MNR 基因,对其编码的蛋白质结构及表达有一定的了解,但并不清楚其具体的功能和调控机制。随着基因组测序和转录组测序技术的发展,与性状相关的基因已被大量克隆和鉴定,对该基因的功能研究将会更加深入。

本研究通过 EST 数据从桂花花瓣中成功克隆出 1 个 MNR 基因,通过 RACE 技术获得其完整 ORF,对其氨基酸保守结构域进行分析,发现保守结构域位于 N 端,属于 SDR 短链脱氢酶家族[22]。通过 BLAST 同源比对,发现与其他植物同源性的蛋白质相似性较高。通过 2 个桂花品种的 qPCR 检测,发现 OfMNR 基因在不同组织部位和不同花期阶段均有表达。试验结果表明本研究克隆的 OfMNR 基因在不同组织中表达量差异较大,其中在茎中表达量最低,而在花中表达量最高。2 个桂花品种在 4 个花期阶段中表达量趋势差异较大。在'波叶金桂'中 OfMNR 基因呈现先下降后逐步上升的趋势,而在'日香桂'中 OfMNR 基因呈现先逐步上升后下降的趋势。OfMNR 基因在不同组织均有表达,说明其在桂花次生代谢过程中处于不可或缺的地位。2 个桂花品种在 4 个时期(香眼期、初花期、盛花期、末花期)表达趋势不同,推测可能是由于品种之间的差异,不同的桂花品种花期变化和衰老的速率不同,导致变化趋势也不同。

后续可进行 OfMNR 基因功能的研究,构建正义表达载体并转化模式植物,研究该基因在植株内的表达情况和植株的表型变化,以揭示桂花单萜代谢途径的分子机理。

参考文献

[1] 向其柏,刘玉莲. 中国桂花品种图志[M]. 浙江科学技术出版社,2009

[2] 王丽梅. 桂花有效成分合成转化规律与药学研究[D]. 华中科技大学,2009.

[3] 樊荣辉,黄敏玲,钟淮钦,等. 花香的生物合成、调控及基因工程研究进展[J]. 中国细胞生物学学报,2011(09): 1028-1036.

[4] 杨秀莲,施婷婷,文爱林,等. 不同桂花品种香气成分的差异分析[J]. 东北林业大学学报. 2015(01): 83-87.

[5] 施婷婷,杨秀莲,赵林果,等. 保存方法对桂花精油提取及香气成分的影响[J]. 南京林业大学学报(自然科学版). 2014(S1): 105-110.

[6] 向林,陈龙清. 花香的基因工程研究进展[J]. 中国农业科学. 2009,42(6): 2076-2084.

[7] 赵印泉,周斯建,彭培好,等. 植物花香代谢调节与基因工程研究进展[J]. 热带亚热带植物学报,2011,19(4): 381-390.

[8] 孙宝军. 中国部分桂花品种芳香成分研究[D]. 河南大学,2011.

[9] Han Y, Li L, Dong M, et al. cDNA cloning of the phytoene synthase (PSY) and expression analysis of PSY and carotenoid cleavage dioxygenase genes in Osmanthus fragrans[J]. Biologia. 2013, 68(2):258-263.

[10] Han Y, Wang X, Chen W, et al. Differential expression of carotenoid-related genes determines diversified carotenoid coloration in flower petal of Osmanthus fragrans[J]. Tree Genetics & Genomes. 2014, 10(2): 329-338.

[11] Baldermann S, Kato M, Kurosawa M, et al. Functional characterization of a carotenoid cleavage dioxygenase 1 and its relation to the carotenoid accumulation and volatile emission during the floral development of Osmanthus fragrans Lour. [J]. Journal of Experimental Botany. 2010, 61(11): 2967-2977.

[12] Huang F C, Molnar P, Schwab W. Cloning and functional characterization of carotenoid cleavage dioxygenase 4 genes[J]. Journal of Experimental Botany. 2009, 60(11): 3011-3022.

[13] 郑玉娟,周文化,张党权,等.月桂 AACT 基因的 cDNA 全长克隆及生物信息学分析[J].中南林业科技大学学报,2015,35(06):87-92.
[14] 张威威,许锋,张芙蓉,等.桂花苯丙氨酸解氨酶基因的克隆与序列分析[J].华北农学报,2008,25(6):622-626.
[15] Croteau R B, Davis E M, Ringer K L, et al. (−)-Menthol biosynthesis and molecular genetics[J]. Naturwissenschaften, 2005, 92(12):562-577.
[16] Kjonaas R, Martinkus-Taylor C, Croteau R. Metabolism of monoterpenes: Conversion of l-Menthone to l-Menthol and d-Neomenthol by stereospecific dehydrogenases from peppermint (*Mentha piperita*) leaves[J]. Plant Physiology, 1982, 69(5):1013-1017.
[17] Davis E M. Monoterpene metabolism. cloning, expression, and characterization of menthone reductases from peppermint[J]. *Plant Physiology*, 2005, 137(3):873-881.
[18] 母洪娜,孙陶泽,杨秀莲,等.两个桂花品种花色色素相关基因的差异表达[J].南京林业大学学报(自然科学版),2015(3):183-186.
[19] Zhang C, Fu J, Wang Y, et al. Identification of suitable reference genes for gene expression normalization in the quantitative real-time PCR analysis of sweet Osmanthus (*Osmanthus fragrans* Lour.)[J]. PLOS ONE, 2015, 10(8):e136355.
[20] Mu H N, Li H G, Wang L G, et al. Transcriptome sequencing and analysis of sweet osmanthus (*Osmanthus fragrans* Lour.)[J]. Genes & Genomics, 2014, 36(6):777-788.
[21] 杨致荣,毛雪,李润植.植物次生代谢基因工程研究进展[J].植物生理与分子生物学学报,2005,31(1):11-18.
[22] Geissler R, Brandt W, Ziegler J. Molecular modeling and site-directed mutagenesis reveal the benzylisoquinoline binding site of the short-chain dehydrogenase/reductase salutaridine reductase[J]. Plant Physiology, 2007, 143(4):1493-1503.

Gene Cloning and Expression Analysis of *OfMNR* Gene Encoding *Osmanthus fragrans* (＋)-Neomenthol Dehydrogenase

XU Chen[1], Li Huogen[2], YANG Xiulian[1], MU Hongna[1], GU Chunsun[3], WANG Lianggui[1]

(1. College of Landscape Architecture, Nanjing Forestry University, Nanjing 210037, China; 2. Key Laboratory of Forest Genetics & Gene Engineering of the Ministry of Education, Nanjing Forestry University, Nanjing 210037, China; 3. Institute of Botany, Jiangsu Province and Chinese Academy of Sciences, Nanjing 210014, China)

Abstract: The full-length cDNA of *MNR* gene was cloned from *Osmanthus fragrans* petals by using the RACE strategy, named *OfMNR*. It is 1 092 bp, containing 936 bp ORF, encoding 312 amino acids. Sequence analysis showed *OfMNR* has high similarity with other plants in predicting amino acid sequence. QPCR technique was used to detect the expression of *OfMNR* in different tissues and different stages in two cultivars of *Osmanthus fragrans*. The result showed the expressions of *OfMNR* are both highest in the petals, and revealed different change trends in four blooming stages.

Key words: petals of *Osmanthus fragrans*; (＋)-neomenthol dehydrogenase; gene cloning; expression analysis

桂花(*Osmanthus fragrans* Lour.) 4-香豆酸辅酶 A 连接酶(4CL)基因克隆与表达分析[*]

母洪娜[1,2]，孙陶泽[2]，徐晨[1]，李火根[3]，杨秀莲[1]，王良桂[1]

(1. 南京林业大学 风景园林学院，江苏 南京 210037；2. 长江大学 园艺园林学院，湖北 荆州 434025；
3. 南京林业大学 林木遗传和基因工程重点实验室，江苏 南京 210037)

摘　要：运用 RT-PCR 和 RACE 方法，从'橙红丹桂'花瓣中克隆出了桂花 4-香豆酸辅酶 A 连接酶(4CL)基因。生物信息学分析和实时定量分析表明，所克隆的基因(全长 2 054 bp)是 *4CL* 基因家族的成员之一；时空表达分析发现，*4CL* 基因在花组织中特异性表达，营养器官根、叶中表达量均很低；*4CL* 基因在开花过程中的时间表达趋势是"低—高—低"，其表达水平变化趋势与花青素和类胡萝卜素含量变化趋势类似。推测 *4CL* 基因与桂花花色代谢相关。

关键词：桂花；*4CL*；基因克隆；时空表达

　　4-香豆酸辅酶 A 连接酶(4-coumarate：CoA ligase, 4CL)是连接苯丙烷代谢途径与木质素生物合成途径的关键酶。4CL 不仅是木质素合成途径上的限速酶(Wei *et al*., 2013)，而且其催化产物香豆酰辅酶 A 是类黄酮代谢途径上的查耳酮合酶的代谢底物(Cukovica *et al*., 2001)。4CL 蛋白的氨基酸序列存在保守基序，其中 N 端的 LP Y/F SSGTTGPKG 基序保守性较高，是催化反应中保守的 AMP 结合功能域；而 C 端的 GEICVRG 在所有 4CL 中完全保守(Stuible *et al*., 2000)。由于 *4CL* 基因处于植物代谢网络中的关键节点，发挥着重要的功能，据已有文献报道：拟南芥(Ehlting *et al*., 1999)、黑麦草(Heath *et al*., 2002)、欧洲山杨(Harding *et al*., 2002)、覆盆子(Kumar and Ellis, 2003)、毛白杨(杨向东等，2006)、亚麻(王进，2009)、青稞(杨晓云等，2014)、甜高粱(周龚等，2014)等多种植物中已成功克隆出了 *4CL* 基因。

　　桂花(*Osmanthus fragrans* Lour.)是木犀属中栽培应用最广泛的物种；桂花在我国栽培历史悠久，向其柏和刘玉莲(2008)把现有的桂花品种划分为金桂、银桂、四季桂、丹桂和彩桂五大品种群。自 20 世纪 80 年代至今，桂花的研究主要集中在地理分布与分类(臧德奎等，2004)、生殖生物学研究(杨秀莲和向其柏，2007)、花色提取方法研究(蔡璇等，2010；陈洪国，2006)、花香分析(金荷仙等，2006)等方面，而对桂花类黄酮合成途径上相关酶的研究较少。本实验通过桂花类黄酮合成关键基因 *4CL* 的克隆与时空表达谱的分析，为进一步研究 *4CL* 基因在整个桂花类黄酮代谢网络中的作用奠定基础。

---

[*] 原文发表于《分子植物育种》，2016,14(3):536-541。
基金项目：江苏省农业科技自主创新资金项目(CX(14)2031)；江苏省高校优势学科建设工程项目(PAPD)。

1 结果与分析

1.1 桂花 4CL 基因的克隆

利用全式金的反转录试剂盒得到桂花的第一链 cDNA。根据桂花转录组组装和注释的基因信息，设计特异性引物，首先克隆出中间片段(1 545 bp)；再根据中间片段运用 RACE 技术快速克隆出 5'-和 3'-端序列,3'-RACE 扩增出 525 bp 的序列，其中包含了终止密码子；序列拼接后得到了 2 054 bp 的基因序列，命名为 Of4CL。拼接后的序列经过 ORF Finder 在线分析，发现 Of4CL 的两个基因序列包含了完整的编码区，其 ORF 阅读框为 1 698 bp，编码 565 个氨基酸(图 1)。NCBI Blast p 分析桂花 4CL 的氨基酸序列，结果显示，本实验克隆的基因序列中包含了完整的 4CL 保守结构域(67-546，cd05904)及 AMP 结合位点(71-457，pfam00501)、AMP-binding C (465-540，pfa-m13193)、CoA 结合位点(57-546，PLN02246)(图 2)。

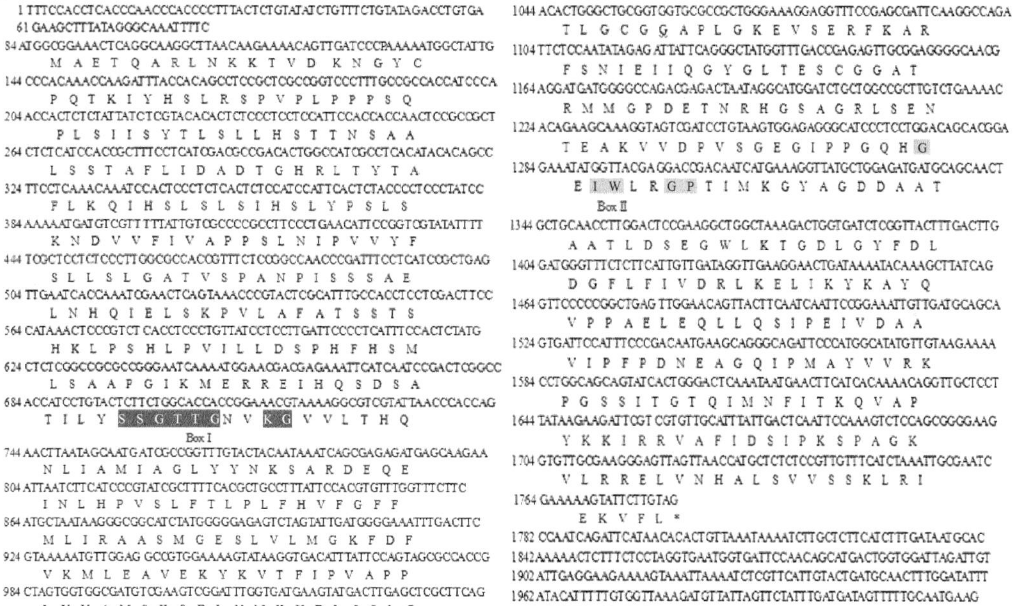

图 1 Of4CL 基因的全长序列及其编码的氨基酸

Fig. 1 Full-length sequence of cDNA and its coded amino acid for Of4CL gene

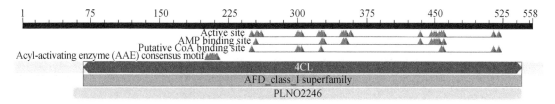

图 2 Of4CL 编码氨基酸保守性分析

Fig. 2 Amino acids conservation analysis for Of4CL

1.2 桂花 Of4CL 基因的生物信息学分析

把本试验所克隆的桂花 4CL 基因的氨基酸序列与其他 10 种植物的 4CL 的氨基酸序列进行多序列比对,发现:Of4CL 氨基酸序列与已知的其他 10 种植物的 4CL 氨基酸序列的一致性为 55.1%~74.6%,其中与水曲柳 4CL 氨基酸序列的一致性很高,达到 73.4%;与拟南芥 4CL 氨基酸序列的一致性最低,仅为 55.1%(表1)。Of4CL 基因编码的氨基酸序列与其他 10 种植物 4CL 基因编码的氨基酸序列存在较大的差异(表1)。多重序列比对发现:与其他植物相比较而言,桂花 4CL 氨基酸的 N 端保守区域是 SSGTTG NVKG,C 端保守区也有 2 个氨基酸存在差异(图1)。

表 1 桂花 4CL 基因和其他 10 种植物的 4CL 基因推导的氨基酸序列相似性比较
Tab. 1 Alignment of Of4CL and other 10 plants 4CL amino acid in GenBank

名称 Name	蛋白质登录号 Protein accessoion No.	氨基酸一致性/% Amino acid identity
水曲柳 Fraxinus mandshurica	W8Q9X0	73.4
葡萄 Vitis vinifera	F6GST0	67.1
蒺藜状苜蓿 Medicago truncatula	G7J515	63.5
亚洲棉 Gossypium arboreum	A0A0B0NWN1	61.9
香瓜 Cucumis melo subsp. melo	E5GBV5	62.3
毛果杨 Populus trichocarpa	B2Z6P2	61.9
川桑 Morus notabilis	W9SMF5	58.2
拟南芥 Arabidopsis thaliana	Q84P23	55.1
丹参 Salvia miltiorrhiza	U3N8B9	74.6
野大豆 Glycine soja	A0A0B2QU30	70.6

为进一步了解不同植物之间的 4CL 蛋白之间的进化关系,本研究运用 MEGA6.0 软件进行聚类分析,聚类结果:桂花 4CL 与水曲柳(Fraxinus mandshurica)、丹参(Salvia miltiorrhiza)、亚洲棉(Gossypium arboreum)先聚类,再与川桑(Morus notabilis)、葡萄(Vitis vinifera)、蒺藜状苜蓿(Medicago truncatula)、野大豆(Glycine soja)、毛果杨(Populus trichocarpa)、拟南芥(Arabidopsis thaliana)聚类(图3)。聚类分析表明:本研究所克隆的 Of4CL 基因与木犀科的水曲柳、唇形科的丹参关系更近,与川桑、拟南芥 4CL 基因的关系更远。

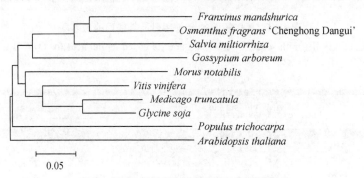

图 3 '橙红丹桂' 4CL 基因的系统发育树
Fig. 3 The phylogenetic tree of 4CL gene in 'Chenghong Dangui'

依据在线软件 ExPASy 预测出 $Of4CL$ 其编码蛋白质的分子量为 6.151 kD,理论等电点(PI) 8.64,属于碱性蛋白。运用跨膜分析软件 TMHMM 分析发现桂花 4CL 蛋白存在明显的跨膜结构,属于膜结合蛋白;亚细胞定位预测结果表明 4CL 蛋白定位到过氧化物酶体细胞器中。综上所述,本研究所克隆的序列是 $4CL$ 基因。

1.3 桂花 $4CL$ 基因的时空表达分析

$4CL$ 基因在'橙红丹桂'已革质化的当年生叶片中表达量最低,在当年生的根中表达量也很低,但根中的表达量略高于叶;花梗和花瓣中的表达量均很高,分别是根中表达量的 57 倍、95 倍(图 4)。由此可以看出,$4CL$ 基因在'橙红丹桂'不同组织中差异表达,并且 $4CL$ 基因在花(花梗,花冠)中表达量相对最高。

$4CL$ 基因在'橙红丹桂'顶壳期(Stage 1)到铃梗期(Stage 2)表达量逐渐升高,初花期(Stage 3)表达量达到峰值,然后表达量逐渐下降,其中盛花初期(Stage 4)、盛花中期(Stage 5)的表达量较为稳定,末花期(Stage 7)表达量下降(图 5)。由此可见,$4CL$ 基因在'橙红丹桂'花的不同发育阶段差异表达,并且其表达谱的曲线呈现"低—高—低"趋势,表达量最高的发育阶段是初花期。

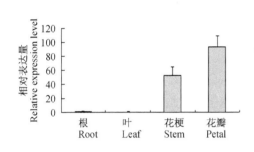

图 4 '橙红丹桂'不同组织中 $4CL$ 基因的表达谱
Fig. 4 Gene expression level among different tissues in 'Chenghong Dangui'

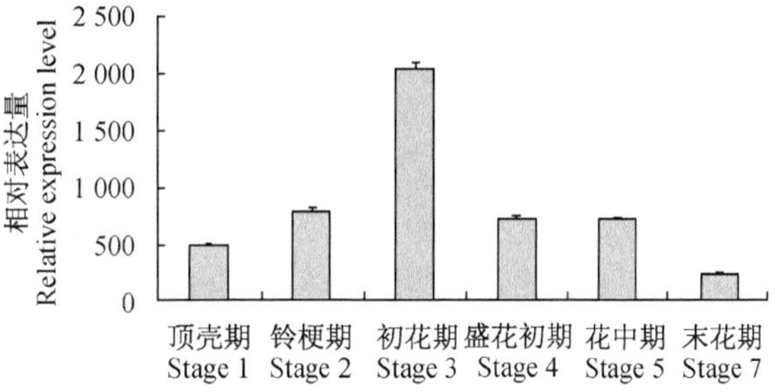

图 5 '橙红丹桂'不同发育阶段 $4CL$ 基因的表达谱
Fig. 5 Gene expression level during different flowering stages in 'Chenghong Dangui'

2 讨论

序列信息比对及其他生物信息学分析表明:本研究所克隆的基因序列是 $4CL$ 基因家族的成员。不同植物中的 $4CL$ 基因数量存在差异,拟南芥中已克隆出 4 个 $4CL$ 基因,黑麦草中有 3 个 $4CL$ 基因(Heath et al.,2002)。覆盆子 $4CL$ 的 N-端保守基序是 LP Y/F SS-GTTG PKG (Kumar and Ellis, 2003);本研究从桂花花瓣中克隆出 2 个 $4CL$ 基因(本研究只对其中的一个进行了时空表达分析)。桂花 N 端保守基序中(LYSSGT-TGNVKG)的天冬酰胺(Asn, N)和缬氨酸(Val, V)替代了 N-端保守基序中的脯氨酸(Pro, P)(图 1);C 端的保守基序由半光氨酸(Cys, C)、异亮氨酸(Ile, I)分别突变为桂花 $4CL$ 保守基序(GEICIRG)中的色氨酸(Trp, W)和亮氨酸(Leu, L);在覆盆子的 $Ri4CL3$ 基序中的异亮氨酸也被缬氨酸替代,发生了轻微的变异(GEICVRG)(Kumar and Ellis, 2003),这样的变异在黑麦草的 $Lp4CL3$ 中也有报道,保守基序中的氨基酸发生突变导致其相应的酶的催化活性低于野生型,表明这种高度保守的基序在酶的催化活动中发挥了十分重要的作用(Heath et al.,

2002)。

  4CL 基因在'橙红丹桂'不同组织中的表达量差异很大，其中在营养器官根和叶中表达量较低，在花梗中表达量较高，在花瓣中表达量最高(图 4)。试验结果表明本研究所克隆的 4CL 基因在花器官中特异性表达。4CL 基因从顶壳期表达量逐渐升高，到初花期到达峰值，随后表达量降低，盛花期虽然表达水平低于初花期，但是盛花期的表达水平较为稳定，末花期的表达量较低。该基因的时间表达变化趋势与桂花前 3 个生育期的花色表型变化一致。初花期以后，花瓣颜色基本上维持橙红色，花色素含量测定发现初花期以后类胡萝卜素、花青素含量逐渐下降，下降幅度均很小(母洪娜等,2015)。这表明初花期以后 4CL 基因的表达量与花色素含量的变化趋势相似。桂花盛花期以后 4CL 基因的表达水平明显下降，其可能的原因是：该基因在植物次生代谢网络中处于非常重要的节点，桂花开花过程中，通过合成类黄酮化合物来延缓花组织的逐渐衰老，该基因很可能通过影响类黄酮代谢路径碳流的方向和流速，进而调控花色的变化。因此，我们推测，桂花 4CL 基因与花的发育、花色变化存在密切的关系。

3 材料与方法

3.1 材料

  '橙红丹桂'(*Osmanthus fragrans* L. 'Chenghong Dangui')栽植于南京林业大学校园内，树龄约为 18～20 年生，喜光，在南京地区可以露地越冬。叶墨绿，花橙红色，有香味，从顶壳期到铃梗期和初花期花色变化明显，初花期以后花色一直保持橙红色。'橙红丹桂'不同时间表达的样品采集从顶壳期到末花期，其中样品的 Stage 1 代表顶壳期，Stage 2 代表铃梗期，Stage 3 代表初花期，Stage 4 代表盛花初期，Stage 5 代表盛花中期，Stage 7 末花期；空间表达的样品分别采集了根(当年生)、叶片(当年生)、花梗和花瓣 4 种不同的组织。

3.2 引物设计和桂花总 RNA 提取

  在桂花转录组中拼接的基因序列信息和同源序列比对基础上，使用 Primer 5.0 软件根据桂花 4CL 基因的中间保守区序列设计特异性引物，先克隆中间片段，再分别设计 3′-RACE 和 5′-RACE 特异性引物(表 2)，由上海捷瑞生物公司合成。利用天根 plus 植物总 RNA 提取试剂盒(DP437)提取桂花总 RNA。本试验按照北京全式金的反转录试剂盒 Thermo Scientific Revert Aid First Strand cDNA Synthesis Kit 说明书中 RNA 纯化以及合成 cDNA 第一条链的步骤进行操作。

表 2 桂花 *4CL* 基因克隆的引物序列
Tab. 2 Primer sequences of *4CL* gene

引物名称 Primer name	引物序列(5′-3′) Primer sequence(5′-3′)
4CL-F	TTTCCACCTCACCCAACCCA
4CL-R	ATGAGAATACATACGGAATCACTGC

续 表

引物名称 Primer name	引物序列(5′-3′) Primer sequence(5′-3′)
4CL3GSP1	CACCCAACCCACCCCTT
4CL3GSP2	CCACAAACCAAGATTTACCACAGCC
4CL5GSP1	ACAATGAAGAGAAACCCATCCAAGT
4CL5GSP2	GCATAACCTTTCATGATTGTCGGTCCTCG

3.3 桂花 4CL 基因的克隆

将 PCR 产物经 1.0% 琼脂糖凝胶电泳分析，按照全式金 Agarose Gel Extraction Kit 凝胶回收试剂盒(TRANSGENE，北京)说明书对 PCR 的扩增产物进行回收。回收后的 DNA 片段与 Peasy-T1 载体(TRANSGENE，北京)连接，25 ℃连接 8～20 min (10 min/1 kb，10-15 min/1-2 kb，15-20 min/2-3 kb)，转化入 DH5α 感受态细胞，37 ℃ 200 r 摇菌 1 h，离心，弃去部分上清，用无菌枪头吹匀菌液，涂布在含 100 mg/mL Amp$^+$ 的 LB 平板上，37 ℃培养 12 h 后，挑单菌落进行 PCR 检测，选取 6 个重组质粒委托南京金斯瑞公司进行测序。

根据克隆得到的花色基因序列，设计 3′-RACE 和 5′-RACE 特异引物。3′-和 5′-RACE 分别用 3′-Full RACE Core Set Ver. 2.0 D314 和 5′-Full RACE Kit D315 (Takara，日本)试剂盒；用 50 μL 反应体系，按照试剂盒说明书进行操作。

3.4 桂花 4CL 基因的生物信息学分析

对'橙红丹桂'花瓣中克隆的 4CL 基因利用 DNAMAN 软件进行序列拼接，随后用 ORF Finder 预测该基因的编码区。为确认所克隆的基因序列是不是 4CL 基因家族的成员，本研究从核苷酸和氨基酸同源性大小、结构域预测与分析(http://blast.ncbi.nlm.nih.gov/Blast.cgi/)，对编码蛋白质的理化性质(ExPASy 在线软件 http://expasy.org/tools/pi_tool.html；TMHMM 跨膜分析软件，亚细胞定位 http://linux1.softberry.com/berry.phtml)和系统发育树(MEGA 6.0，Neighbor-Joining 算法)几个方面进行分析。

3.5 桂花 4CL 基因的实时定量分析

荧光定量的试剂是罗氏的 FastStart Univeral SYBR Green Master (Rox)(货号：No. 04913850001)；反应体系和 PCR 程序参见罗氏 FastStart Univeral SYBR Green Master (Rox)说明书。本研究所设计的样品均设置了生物学重复，与此同时每个 qRT-PCR 反应设置 3 个技术重复以增加其可信度。

参考文献

[1] Cai X., Su F., Jin H. X., Yao C. H., and Wang C Y, 2010, Components and extraction methods for petal pigments of *Osmanthus fragrans* 'Siji Gui', Zhejiang Linxueyuan Xuebao (Journal of Zhejiang Forestry College), 27(4): 559-564.
(蔡璇，苏霏，金荷仙，姚崇怀，王彩云，2010，四季桂花瓣色素的初步鉴定与提取方法，浙江林学院学报，27(4): 559-564.)

[2] Chen H. G., 2006, Changes of pigment, soluble sugar and protein content of peta during florescence and senescence of *Osmanthus fragrans* Lour., Wuhan Zhiwuxue Yanjiu (Journal of Wuhan Botanical Research), 24(3): 231-234.

(陈洪国, 2006, 桂花开花进程中花瓣色素、可溶性糖和蛋白质含量的变化, 武汉植物学研究, 24(3): 231-234.)

[3] Cukovica D., Ehlting J., Van Ziffle J. A., and Douglas C. J., 2001, Structure and evolution of 4-coumarate: coenzyme a ligase (4CL) gene families, Biol. Chem., 382(4): 645-654.

[4] Ehlting J., Büttner D., Wang Q., Douglas C. J., Somssich I. E., and Kombrink E., 1999, Three 4-coumarate: coenzyme aligases in *Arabidopsis thaliana* represent two evolutionarily diver-gent classes in angiosperms, The Plant Journal, 19(1): 9-20.

[5] Harding S. A., Leshkevich J., Chiang V. L., and Tsai C. J., 2002, Differential substrate inhibition couples kinetically distinct 4-coumarate: coenzyme aligases with spatially distinct metabolic roles in quaking aspen, Plant Physiol., 128 (2): 428-438.

[6] Heath R., McInnes R., Lidgett A., Huxley H., Lynch D., Jones E., Mahoneya N., and Spangenberg G., 2002, Isolation and character risation of three 4-coumarate: CoA-ligase homologue cDNAs from perennial ryegrass (*Lolium perenne*), J. Plant Physiol., 159(7): 773-779.

[7] Jin H. X., Zheng H., Jin Y. J., Chen J. Y., and Wang Y., Research on major volatile components of 4 *Osmanthus fragrance* cultivars in Hangzhou Manlong Guiyu Park, Linye Kexue Yanjiu (Forest Research), 19(5): 612-615.

(金荷仙, 郑华, 金幼菊, 陈俊愉, 王雁, 2006, 杭州满陇桂雨公园4个桂花品种香气组分的研究, 林业科学研究, 19(5): 612-615.)

[8] Kumar A., and Ellis B. E., 2003, 4-Coumarate CoA ligase gene family in Rubus idaeus cDNA structures, evolution, and expression, Plant Molecular Biology, 51(3): 327-340.

[9] Mu H. N., Sun T. Z., Yang X. L., and Wang L. G., 2015, Differential expression of flower color related genes of *Osmanthus fragrans* Lour. 'Chenghong Dangui' and 'Zaoyingui', Nanjing LinyeDaxue Xuebao (Journal of Nanjing Forestry University (Natural Sciences Edition)), 39(3): 183-186.

(母洪娜, 孙陶泽, 杨秀莲, 王良桂, 2015, 两个桂花品种花色色素相关基因的差异表达, 南京林业大学学报(自然科学版), 39(3): 183-186.)

[10] Stuible H. D., Büttner D., Ehlting J., Hahlbrock K., and Kombrink E., 2000, Mutational analysis of 4-coumarate: CoA ligase identifies functionally important amino acids and verifies its close relationship to other adenylate-forming enzymes, Febs Letters, 467(1): 117-122.

[11] Wang J., 2009, Cloning and expression analysis of Flax (*Linum usitatissimum*) critical lignin metabolism genes, Thesis for M. S., Institute of Bast Fiber Crops, CAAS, Supervisor: Chen X. B., pp. 25-38.

(王进, 2009, 亚麻(*Linum usitatissimum*)木质素合成关键酶基因的克隆及表达分析, 硕士学位论文, 中国农业科学院麻类研究所, pp. 25-38.)

[12] Wei H. Y., Rao G. D., Wang Y. K., Zhang L., and Lu H., 2013, Cloning and analysis of a new 4CL-like gene in *Populus tomentosa*, Forest Science and Practice, 15(2): 98-104.

[13] Xiang Q. B., and Liu Y. L., eds., 2008, An illustrated monograph of the sweet osmanthus cultivars in China, Science and Technolgy Press, Hangzhou, China, pp. 3-24.

(向其柏, 刘玉莲, 2008, 中国桂花品种图志, 浙江科学技术出版社, 中国, 杭州, pp. 3-24.)

[14] Yang X. D., Wang H. Z., Hu Z. M., Liu G. H., and Shen J. X., 2006, The cloning, expression and

activity assays of 4 - CL cDNA from the *Populus tomentosa*, Huazhong Nongye Daxue Xuebao (Journal of Huazhong Agricultural University), 25(2): 101-105.

(杨向东,王汉中,胡赞民,刘贵华,沈金雄,2006,毛白杨 4-CL cDNA 的克隆、表达及活性分析,华中农业大学学报(自然科学版),25(2): 101-105.)

[15] Yang X. L., and Xiang Q. B., 2007, Pollen vitality and stigma receptivity of *Osmanthus fragrans*, Linye Keji Kaifa (China Forestry Science and Technology), 21(3): 22-25.

(杨秀莲,向其柏,2007,桂花花粉活力测定与'晚籽银'桂柱头可授性分析,林业科技开发,21(3): 22-25.)

[16] Yang X. Y., Yang Z. M., Luo X. J., Kong D. Y., Yuan J. E., Liu X. C., and Feng Z. Y., 2014, Cloning and expression analysis of 4-Ceoumarate: CoA ligase gene *4CL* in hulless barley, Mailei Zuowu Xuebao (Journal of Triticeae Crops), 34(12): 1603-1610.

(杨晓云,杨智敏,罗小娇,孔德媛,袁金娥,刘新春,冯宗云,2014,青稞 4 -香豆酸辅酶 A 连接酶基因 *4CL* 的克隆及表达分析,麦类作物学报,34(12): 1603-1610.)

[17] Zang D. K., Xiang Q. B., and Hao R. M., 2004, Study on distribution and utilization of plants in genus *Osmanthus*, Xinan Linxueyuan Xuebao (Journal of Southwest Forestry College), 24(1): 23-26.

(臧德奎,向其柏,郝日明,2004,木犀属植物的分布与开发利用,西南林学院学报,24(1): 23-26.)

[18] Zhou G., Wang Y. F., Lou L. Q., Zhou H., Chen S. L., Cai Q. S., and Teng S., 2014, Cloning, characterization and spatiotemporal expression analysis of 4-coumarate: CoA ligase genes (*4CL*) in sweet sorghum, Nanjing Nongye Daxue Xue bao (Journal of Nanjing Agricultural University), 37(3): 9-19.

(周龚,王玉锋,娄来清,周华,陈素丽,蔡庆生,滕胜,2014,甜高粱 4 -香豆酸辅酶 A 连接酶基因(*4CL*)的克隆与鉴定及时空表达分析,南京农业大学学报,37(3): 9-19.)

Gene Clone and Expression Analysis of 4-coumarate-CoA Ligase in Sweet Osmanthus (*Osmanthus fragrans* Lour.)

Mu Hongna[1, 2], Sun Taoze[2], Xu Chen[1], Yang Xiulian[1], Wang Lianggui[1]

(1. College of Landscape Architecture, Nanjing Forestry University, Nanjing 210037, China; 2. College of Horticulture and Gardening, Yangtze University, Jingzhou 434025, China; 3. The Key Laboratory of Forest Genetics and Gene Engineering, Nanjing Forestry University, Nanjing 210037, China)

Abstract: In this study, *4CL* gene had been cloned from petals of sweet osmanthus 'Chenghong Dangui' by using RT-PCR technique and RACE method. Furthermore, bioinformatics tools and real time fluorescence quantification technique were employed. The results indicated that cloned gene, full-length 2 054 bp, belong to the gene family of *4CL*. Spatiotemporal expression profile showed that *4CL* gene specifically expressed in flower tissues, but low expressed in leaves and roots. As for its temporal expression level, presented the overall variation trend, low - high - low, which variation trends was similar to flower color compounds. Therefore, we propose that the cloned *4CL* gene was likely to contact with flower pigments metabolism.

Key words: *Osmanthus fragrans*; *4CL*; flower color; anthocyanin

两个桂花品种花色色素相关基因的差异表达

母洪娜[1,2,3]，孙陶泽[3]，杨秀莲[1,2]，王良桂[1,2]

(1.南京林业大学 风景园林学院,江苏 南京 210037;2.南京林业大学 南方现代林业协同创新中心,江苏 南京 210037;3.长江大学 园艺园林学院,湖北 荆州 434025)

摘　要：结合桂花花青素、类胡萝卜素的含量与转录组测序检测到的花色代谢相关基因的差异表达,对'橙红丹桂'和'早银桂'两个桂花品种的花色变化过程进行了初步研究。结果发现:苯丙氨酸裂解酶、4-香豆酸 CoA 连接酶、乙酰辅酶 A 合酶、查尔酮合酶、R2R3-MYB1、细胞色素 P450、番茄红素 ε-环化酶基因的表达与花青素和类胡萝卜素含量变化相似,这两类色素代谢路径上的其他基因均在'早银桂'中表达量较高。此外,这种现象很可能是多酚氧化酶(PPO)、细胞色素 P450 基因在转录因子 R2R3-MYB 调控下差异表达所致,并且还发现桂花类黄酮的代谢路径是从柚皮素开始转向二氢槲皮素进而在 DFR、ANS 催化下形成矢车菊色素。

关键词：桂花;花青素;类胡萝卜素;花色基因差异表达

桂花(*Osmanthus fragrans* Lour.)是中国传统十大名花之一,原产于长江流域至华南、西南各地。其花色主要有乳白、浅黄、黄、金黄、橙、橙红,甚至还有浙江最新选育的桃红色的桂花。花色是桂花重要的观赏性状之一,与桂花花色相关的色素主要分为类胡萝卜素和类黄酮两大类,不同桂花品种的色素含量存在明显差异。植物类黄酮化合物代谢是苯丙烷类代谢途径的一个分支,由莽草酸代谢途径合成的苯丙氨酸在苯丙氨酸解氨酶(PAL)催化下转变成反式肉桂酸,在肉桂酸-4-羟化酶(C4H)作用下生成反式-4-香豆酸,4-香豆酰 CoA 和丙二酰 CoA 在查尔酮合酶(CHS)作用下形成查尔酮,在查尔酮异构酶(CHI)催化下形成柚皮素进入类黄酮的其他途径。类黄酮化合物代谢的 3 个阶段[1]:①黄酮和异黄酮合成,由黄烷酮为底物和黄酮合酶(FS)催化而形成;②黄酮醇合成,柚皮素在黄烷酮 3′-羟化酶(F3H)和类黄酮 3′,5′-羟化酶催化下生成二氢黄酮醇,二氢黄酮醇经黄酮醇合酶途径降解成槲皮素、山奈酚、杨梅素等黄酮醇;③花色素、原花色素合成,二氢黄酮醇在二氢黄酮醇还原酶(DFR)作用下,合成无色花色素,经过花色素合酶(ANS)催化形成花色素,之后在糖苷转移酶作用下发生糖基化形成花色素苷,在细胞内经过修饰的色素在谷胱甘肽转移酶(GST)和位于液泡膜上的特殊运载工具[2]共同作用下运输至液泡中。CHS 是类黄酮代谢途径的一个限速酶,截至目前,已经从金鱼草(*Antirrhinum majus*)[3]、百合(*Lilium hybrid*)[4]、苦荞(*Fagopyrum tataricum*)[5]、拟南芥(*Arabidopsis thaliana*)[6]等多个物种中克隆出 CHS 基因。黄烷酮 3′-羟化酶(flavanone 3-hydroxylase,F3H)基因是整个黄酮类化合物代谢途

---

* 原文发表于《南京林业大学学报(自然科学版)》,2015,39(3):183-186。
基金项目：国家林业公益性行业科研专项项目(201204607);江苏省普通高校研究生科研创新计划资助项目(CXZZ13-0553);江苏省高校优势学科建设工程资助项目(PAPD)。

径的中枢位点,调控着合成途径中的代谢流向。李鹏[7]在对粘毛黄芩 $F3H$ 基因转化烟草的研究中发现,转基因烟草植株总黄酮的平均含量明显高于反义植株及野生型烟草总黄酮类化合物含量;此外,$F3H$ 基因在红巴梨[8]、猕猴桃[9]等植物中的表达谱和转基因的初步研究已经开展。

目前的研究主要从生理学的视角去研究桂花的花色变化[10-11],由于缺少基因组和转录组的研究基础,从分子水平揭示其花色变异的研究很少。笔者研究'早银桂'('Zaoyingui')、'橙红丹桂'('Chenghong Dangui')两个桂花品种开花期间的花色素含量变化和其转录水平的花色素代谢路径上相关基因的差异表达,以期通过对生理指标及转录水平的分析,揭示可能导致桂花花色变异原因。

1 材料与方法

1.1 供试材料

供试的两个桂花品种'早银桂'和'橙红丹桂'均采自南京林业大学校园内,树龄分别为15 a 和 20 a,栽培管理条件一致,东、南、西、北 4 个方位均进行采花。自铃梗期开始,每天上午 9:00—10:00 采花带回实验室,其中'早银桂'花期 8 月下旬至 9 月上旬,花色浅黄,浓香;'橙红丹桂'花期 9 月下旬至 10 月上旬,花色橙红,香味较淡。二者均为长江流域普遍栽培的品种,观赏价值很高。于桂花的铃梗期、香眼期、盛花期、末花期分别采集'早银桂'和'橙红丹桂'的花装在自封袋中,去掉花梗后存于 -20 ℃冰箱中备用。

1.2 指标测定

花青苷提取采用 1‰盐酸法[12]。花青素含量(C)依据公式 $C = 1/958 \cdot V/100 \cdot 1/M \cdot D_{530} \cdot 100000$。式中:$V$ 为体积,M 为样品质量,D_{530} 为吸光度。类胡萝卜素提取参考颜少宾等[13]的方法,标准曲线绘制参考张岩等[14]的方法。β-胡萝卜素标准品(分析纯)溶于丙酮与乙醇混合液(以体积比配制),之后分别稀释成 1、2、4、8、16 μg/mL,用分光光度计测定吸光度,绘制标准曲线,得到 β-胡萝卜素的标准曲线方程 $y = 0.100\ 4x + 0.027\ 6$($R^2 = 0.997$)。桂花色素的吸收波长用 DU800 紫外分光光度计(Beckman)测定。

转录组样品准备和测序方法详见文献[15]。将过滤后的 clean reads 用 Trinity 软件拼接得到的转录组作为参考序列(ref),把每个样品的 clean reads 往 ref 上做 mapping,用 RSEM 软件统计 map 到 ref 上的 reads 数,即得到每个样品 map 到每个基因上的 reads 数,并对其进行 RPKM(每 1 百万个 map 上的 reads 中 map 到外显子的每 1 000 个碱基上的 reads 数)转换。由于桂花测序样品无生物学重复,需再采用 TMM 对 reads 数据进行标准化处理后(得到样品 1 和样品 2 的表达量数值即 value 1 和 value 2),再进行差异分析,差异分析的筛选阈值为 $q<0.005$ 且 $|\log_2(\text{value 1/value 2})|>1$,同时满足这两个筛选条件即判定为差异表达基因。$\log_2$(value 1/value 2)正值表示该基因在样品 1 中上调表达,负值表示在样品 1 中下调表达。q 为校正后的 P 值,q 值越小,表示基因表达差异越显著。文中样品 1(实验组)代表'橙红丹桂'花组织,样品 2(对照组)代表'早银桂'花组织。

2 结果与分析

2.1 桂花色素含量的变化分析

花青素是类黄酮中的重要成员,经观测,整个开花期间,花青素含量总体上呈下降趋势(图1)。'早银桂'开花的第1天到第2天花青素含量下降了2/3,2 d之后变化平缓,到5 d时含量略有增加;而开花第1天的'橙红丹桂'花青素含量高于'早银桂',并且在1~5 d内较'早银桂'花青素含量下降缓慢。'早银桂'的类胡萝卜素含量缓慢下降,从第1天时的1.76 mg·g^{-1}下降到第5天时1.16 mg·g^{-1},'橙红丹桂'的类胡萝卜素含量约是'早银桂'的10倍,并且在开花第2天达到峰值(32.05 mg/g)(图1)。

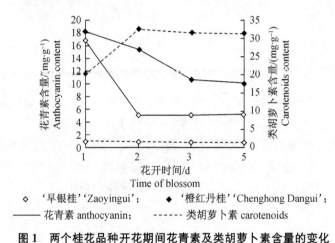

图1 两个桂花品种开花期间花青素及类胡萝卜素含量的变化
Fig. 1 The content variation of anthocyanin and carotenoids of *Osmanthus fragrans* L.

2.2 桂花初花期色素相关基因的差异表达分析

对'早银桂''橙红丹桂'两个品种的转录组数据进行了分析,研究发现,黄酮类色素代谢路径上的花色相关基因如苯丙氨酸裂解酶(PAL),4-香豆酸CoA连接酶(4CL)、乙酰辅酶A合酶(Acyl-CoA Synthetase)、查尔酮合酶(CHS)、R2R3-MYB1 |log$_2$(value 1/value 2)|>1,在'橙红丹桂'和'早银桂'花组织中差异表达,而且上述基因在'橙红丹桂'初花期花组织中上调表达(表1)。类胡萝卜素代谢路径上的差异表达基因 *CYP450*、*LCYe*、*ISPS*、*ZEP*、*GES*、*PPO*、*MAT*、*LOX* 在'橙红丹桂'花组织中上调表达,*CCD4* 在'橙红丹桂'花组织中下调表达(表1)。此外,桂花转录组数据中还有未表征的蛋白、未命名的蛋白质产品、预测性的蛋白和未被功能注释到的基因和转录因子(表中未列出),它们在'橙红丹桂'中上调表达,这些基因很可能直接或间接地参与桂花花色素的代谢。

表 1 两个桂花品种初花期与色素代谢相关的主要基因及其表达量
Tab. 1 The code and different expression level of genes related to flower pigments metabolism of *O. fragrans* L.

基因(酶) Gene/Enzyme name	表达量 Expression level		$\log_2$(value 1/ value 2)	P	q
	value 1	value 2			
VDE	9.24	18.71	−1.018 8	1.857×10^{-1}	9.969×10^{-1}
CCD4	1 793.96	10 786.12	−2.588 0	0	0
LCYb	61.15	80.07	−0.388 9	6.422×10^{-1}	9.969×10^{-1}
ZEP	26.28	38.25	2.175 3	2.669×10^{-9}	1.354×10^{-7}
PDS	2 485.23	3 735.82	−0.588 0	2.259×10^{-17}	2.367×10^{-15}
LCYe	416.54	191.80	1.118 9	1.043×10^{-30}	2.167×10^{-28}
ZDS	4 032.59	4 099.45	−0.023 7	2.522×10^{-15}	2.245×10^{-13}
CHS	2.04	1.0	1.019 2	4.467×10^{-1}	9.969×10^{-1}
PAL	83.88	28.32	1.566 5	2.640×10^{-10}	1.505×10^{-8}
ANS	2.04	5.78	−1.504 3	2.637×10^{-1}	9.969×10^{-1}
F3H	2.73	13.82	−2.338 3	1.203×10^{-2}	1.471×10^{-1}
CYP450	5 580.78	134.51	3.577 7	0	0
4CL	165.67	10.93	3.922 4	2.969×10^{-41}	9.004×10^{-39}
ACS	190.41	91.87	1.051 4	6.427×10^{-14}	5.031×10^{-12}
ISPS	1 446.3	151.38	3.256 1	0	0
MAT	101.93	278.56	1.864 7	5.428×10^{-12}	6.497×10^{-124}
LOX	1 053.5	3.17	1.940 9	2.543×10^{-19}	4.284×10^{-195}
PPO	250.97	40.69	2.624 7	2.249×10^{-36}	5.779×10^{-34}
GES	19 185.73	3 548.36	2.434 8	0	0
Predicted Protein	256.91	27.19	2.419 1	2.969×10^{-44}	9.784×10^{-42}
R2R3-MYB1	94.49	27.19	1.750 5	2.059×10^{-12}	1.419×10^{-10}

3 讨论

类黄酮和类胡萝卜素色素在使植物呈现固有颜色的同时,也与植物的生长发育密切相关[16]。桂花初花期,'橙红丹桂'花色素代谢路径上的基因大多没有表现出上调表达,而是比'早银桂'中的表达量还低。这有可能是还有其他的基因或转录因子调控着色素的代谢与合成,如 R2R3-MYBs 转录因子,在拟南芥[17]、杨梅[18]、苹果[19]、猕猴桃[20]中参与类黄酮和类胡萝卜素的代谢,这与此次研究中'橙红丹桂'初花期 R2R3-MYB1 上调表达结果一致。此外,研究发现橙红丹桂花器官的转录组数据的多酚氧化酶(PPO)、细胞色素 P445(CYP450)、F3H 基因差异表达,尤其是槲皮素葡萄糖苷上调表达揭示了桂花类黄酮代谢从柚皮素(Naringenin)转向槲皮素分支,二氢槲皮素在 DFR、ANS 催化下最后生成矢车菊色素(Cyanidin)。多酚氧化酶能进行去甲基化反应,黄色金鱼草中的查尔酮专一性 PPO 则参与橙酮的合成[21-22];CYP450 参与黄酮、类黄酮、类胡萝卜素等化合物的生物合成[23]。上述这些研究成果为此次研究中丹桂花瓣中类胡萝卜素和花青素含量较高提供了相对合理的佐证。

此外，S-腺苷甲硫氨酸合酶(S-adenosylmethionine synthetase，MAT)在甲基转移中是一个重要的甲基供体，影响基因表达，被称为控制基因表达的重要开关。该基因在'橙红丹桂'中的表达量高于'早银桂'，可能调控了某些基因不能表达。香叶醇合酶(geraniol synthase，GES)的高表达，可能为类胡萝卜素代谢间接提供了原料或底物。此外，酰基转移酶在两个桂花品种中均有表达，'橙红丹桂'中表达量略高(表中未列出)。综上所述，桂花的花色素代谢调控网络十分复杂，需要对差异表达的基因进行功能验证后才能确认上述基因在代谢网络中所发挥的功能。此外，桂花转录组数据中还有很多未表征的蛋白、未命名的蛋白质产品、预测性的蛋白和未被功能注释到的基因，这些蛋白编码的基因在'橙红丹桂'中均为上调表达，它们的功能在模式植物或其他植物中还未被验证，尚需进一步研究。

参考文献

[1] 乔小燕,马春雷,陈亮. 植物类黄酮生物合成途径及重要基因的调控[J]. 天然产物研究与开发,2009(21):354-360,207.
Qiao X Y, Ma C L, Chen L. Plant flavonoid biosynthesis pathway and regulation of its important genes[J]. Natural Product Research and Development, 2009(21): 354-360,207.

[2] Koes R, Verweij W, Quattrocchio F. Flavonoids: a colorful model for the regulation and evolution of biochemical pathways [J]. Trends in Plant Science, 2005,10(15):236-242.

[3] Sommer H, Saedler H. Structure of the chalcone synthase gene of *Antirrhinum majus* [J]. Molecular and General Genetics, 1986,202(3):429-434.

[4] Nakatsuka A, Izumi Y, Yamagishi M. Spatial and temporal expression of chalcone synthase and dihydroflavonol 4-reductase genes in the Asiatic hybrid lily [J]. Plant Science, 2003, 165(4): 759-767.

[5] 吴琦,李成磊,陈惠,等. 苦荞查尔酮合酶基因 *CHS* 的结构及花期不同组织表达量分析[J]. 中国生物化学与分子生物学报,2010,26(12):1151-1160.
Wu Q, Li C L, Chen H, *et al*. Molecular structure characterization of *Fagopyrum tataricum* chalcone synthase and its tissue-specific expression during florescence [J] Chinese Journal of Biochemistry and Molecular Biology,2010,26(12):1151-1160.

[6] Feinbaum R L, Ausubel F M. Transcriptional regulation of the *Arabidopsis thaliana* chalcone synthase gene [J]. Molecular and Cell Biology, 1988,8(5):1985-1992.

[7] 李鹏. 粘毛黄芩 *CHS* 基因植物表达载体的构建及其在烟草中的功能验证[D]. 重庆:西南大学,2010.
Li P. Constructing of plant expression vectors of chalcone synthase gene form *Scutellaria viseidula* Bunge and its function verification in tobacco [D]. Chongqing: Southwest University,2010.

[8] 张大生,崔丽洁,王景明,等. 红巴梨 *F3H* 基因的克隆及植物表达载体的构建[J]. 南京林业大学学报(自然科学版),2005,29(2):65-68.
Zhang D S, Cui L J, Wang J M, *et al*. Cloning *F3H* gene from red ballet fruit and its construction of plant express vector[J]. Journal of Nanjing Forestry University(Natural Sciences Edition),2005,29(2):65-68.

[9] 杨红丽,王彦昌,姜正旺,等. '红阳'猕猴桃 cDNA 文库构建及 *F3H* 基因的表达初探[J]. 遗传,2009, 31(12):1265-1272.
Yang H L, Wang Y C, Jiang Z W, *et al*. Construction of cDNA library of 'Hongyang' kiwifruit and analysis of *F3H* expression[J]. Hereditas, 2009,31(12):1265-1272.

[10] 蔡璇,苏蘩,金荷仙,等. 四季桂花瓣色素的初步鉴定与提取方法[J]. 浙江林学院学报,2010,27(4):

559-564.

Cai X, Su F, Jin H X, et al. Components and extraction methods for petal pigments of Osmanthus fragrans 'Siji Gui'[J]. Journal of Zhejiang Forestry College, 2010, 27(4):559-564.

[11] 陈洪国. 桂花开花进程中花瓣色素、可溶性糖和蛋白质含量的变化[J]. 武汉植物学研究, 2006, 24(3): 231-234.

Chen H G. Changes of pigment, soluble sugar and protein content of peta during florescence and senescence of Osmanthus fragrans Lour. [J]. Journal of Wuhan Botanical Research, 2006, 24(3): 231-234.

[12] 刘桂玲, 李海霞, 郭宾会, 等. 不同提取方法对甘薯花青素含量测定的影响[J]. 中国农学通报, 2007, 23(4):91-94.

Liu G L, Li H X, Guo B H, et al. Effects of different extraction methods on anthocyanin content detection in sweet potato[J]. Chinese Agricultural Science Bulletin, 2007, 23(4):91-94.

[13] 颜少宾, 张妤艳, 马瑞娟, 等. 杏果肉中类胡萝卜素的提取方法研究[J]. 江西农业学报, 2011, 23(9): 159-161.

Yan S B, Zhang Y Y, Ma R J, et al. Study on extraction method of carotenoid from apricot fruit [J]. Acta Agriculturae Jiangxi, 2011, 23(9): 159-161.

[14] 张岩, 时威, 吴燕燕, 等. 白莲花粉中β-胡萝卜素的提取工艺条件的优化[J]. 食品与发酵科技, 2011, 47(30):68-71.

Zhang Y, Shi W, Wu Y Y, et al. Optimization of extraction conditions of β-Carotene in white lotus pollen[J]. Food and Fermentation Technology, 2011, 47(30):68-71.

[15] Mu H N, Li H G, Wang L G, et al. Transcriptome sequencing and analysis of sweet osmanthus (Osmanthus fragrans Lour.) [J]. Genes and Genomics, 2014, 36:777-788.

[16] 江曼. 下调 GhDFR 和 GhF3H 基因对棉花生长发育的影响[D]. 重庆:西南大学, 2013.

Jiang M. Down-Regulation of GhDFR and GhF3H Influence the growth and development of cotton (Gossypium hirsutum L.) [D]. Chongqing: Southwest University, 2013.

[17] Stracke R, Werber M, Weisshaar B. The R2R3-MYB gene family in Arabidopsis thalialiana[J]. Current Opinion in Plant Biology, 2001, 4(5): 447-456.

[18] 牛姗姗. MYB 对杨梅果实花青素苷合成的调控及其机制[D]. 杭州:浙江大学, 2011.

Niu S S. Regulation of anthocyanin biosynthesis by MYB in Chinese bayberry fruit and the mechanisms [D]. Hangzhou: Zhejiang University, 2011.

[19] Sornkanok Vimolmangkang, Han Y P, Wei G C, et al. An apple MYB transcription factor, MdMYB3, is involved in regulation of anthocyanin biosynthesis and flower development [J]. BMC Plant Biology, 2013, 13(1):176.

[20] Lena G Fraser, Alan G Seal, Mirco Montefiori, et al. An R2R3 MYB transcription factor determines red petal colour in an Actinidia (kiwifruit) hybrid population [J]. BMC Genomics 2013, 14: 28.

[21] Nakayama T, Yonekur-askakibara K, Sato T, et al. Aureusidin synthase: A polyphenol oxidase homolog responsible for flower coloration [J]. Science, 2000, 290: 1163-1166.

[22] Strack D, Schliemann W. Bifunctional polyphenol oxidase: novel function in plant pigment biosynthesis[J]. Angewandte Chemie: International Edition, 2001, 40(20):3791-3794.

[23] 贺丽虹, 赵淑娟, 胡之壁. 植物细胞色素 P450 基因与功能研究进展[J]. 药物生物技术, 2008, 15(2): 142-147.

He L H, Zhao S J, Hu Z B. Gene and function reasearch progress of plant cytochrome P450[J]. Pharmaceutical Biotechnology, 2008, 15(2):142-147.

Differential Expression of Flower Color Related Genes of *Osmanthus fragrans* Lour. 'Chenghong Dangui' and 'Zaoyingui'

MU Hongna[1,2,3], SUN Taoze[3], YANG Xiulian[1,2], WANG Lianggui[1,2]

(1. College of Landscape Architecture, Nanjing Forestry University, Nanjing 210037, China; 2. Co-Innovation Center for the Sustainable Forestry in Southern China, Nanjing Forestry University, Nanjing 210037, China; 3. College of Horticulture and Gardening, Yangtze University, Jingzhou 434025, China)

Abstract: The analysis of anthocyanin, carotenoid content and flower color gene expression of *Osmanthus fragrans* transcriptome between 'Chenghong Dangui' and 'Zaoyingui' were carried out. Some genes such as phenylalanine ammonia lyase, 4-CoA ligasecoumaric acid, acetyl CoA synthase, chalcone synthase, *R2R3-MYB*1, cytochrome *P*450, lycopene epsilon cyclase did present up regulation in 'Chenghong Dangui' petals not 'Zaoyinggui', which was consistent with the anthocyanin and carotenoid content in 'Chenghong Dangui' and 'Zaoyinggui'. In contrast, the expression profile of other genes was up regulation in 'Zaoyingui'. In addition, polyphenol oxidase (PPO), cytochrome *P*450 gene and *R2R3-MYB* that probably regulated pigments accumulation of *O. fragrans* were analyzed in this article. Meanwhile the novel discovery that the flavonoids metabolic pathway direct to dihydroquercetin at metabolism web junction, that is naringenin, and then synthesized cyanidin with catalyzing of *DFR* and *ANS* was found out in this study.

Key words: *Osmanthus fragrans* Lour.; anthocyanin; carotenoids; differential expression of flower color genes

Transcriptome Sequencing and Analysis of Sweet Osmanthus (*Osmanthus fragrans* Lour.)*

Hongna Mu, Huogen Li, Lianggui Wang, Xiulian Yang, Taoze Sun, Chen Xu

College of Landscape Architecture, Nanjing Foresty university, Nanjing 210037, China

Abstract: *Osmanthus fragrans* is a woody, evergreen species of shrubs and small trees that is extensively planted in sub-tropical and temperate climates as an ornamental plant in gardens and for its health benefits. The flower color ranges from ivory to orange to pink among different varieties and even color difference during the whole blossom in the sweet osmanthus. Sweet osmanthus is widely cultivated throughout China and other countries due to its prominent fragrance, colorful flowers, and medicinal properties. However, the scanty genomic resources in the Oleaceae family have greatly hindered further exploration of its genetic mechanism on these economically important traits. In this study, transcriptome sequencing of *O. fragrans* was performed using the Illumina HighSeqTM2000 sequencing platform. Next generation sequencing (NGS) of the transcriptome of *O. fragrans* produced 31.7G of clean bases (211,266,818 clean reads) that were assembled into 256,774 transcripts and 117,595 unigenes. Of them, 197 and 237 candidate genes involved in fragrance and pigment biosynthesis respectively were identified based on function annotation. Meanwhile, 1 unnamed protein and 468 functional unknown genes were also identified. Furthermore, mRNA sequencing expression profiling of *O. fragrans* were compared to previous genes'. In summary, this comprehensive transcriptome dataset allows the identification of genes associated with several major metabolic pathways and provides a useful public information platform for further functional genomic studies in *O. fragrans* Lour.

Key words: sweet osmanthus; transcriptome sequencing; bioinformatic analysis; flower pigment

Abbreviations

GGPP	Geranylgeranyl pyrophosphate
PSY	Phytoene synthase
PDS	15-*Cis*-phytoene desaturase
ZDE	Zeta-carotene desaturase
PLYISO	Prolycopene isomerase
CRTISO	Carotenoids isomerase
β-LCY(LCYb)	β-lycopene cyclase
ε-LCY(LCYe)	ε-lycopene cyclase

---

* 原文发表于 *Genes Genom*, 2014, 36(6):777-788。

βOH	β-carotene 3-hydroxylase
ZE	Zeaxanthin epoxidase
VDE	Violaxanthin de-epoxidase
CCS	Capsanthin/capsorubin synthase
NS	Neoxanthin synthase
NCED	9-*Cis*-epoxycarotenoid dioxygenase
XDH	Xanthoxin dehydrogenase
AAO	Abscisic-aldehyde oxidase
AAH	(+)-Abscisic acid 8′-hydroxylase
CYP73A	*Trans*-cinnamate 4-monooxygenase
CHS	Chalcone synthase
CHI	Chalcone isomerase
CYP93A	Coumaroylquinate (coumaroyl shikimate) 3′-monooxygenase
FDO	Naringenin 3-dioxygenase
HID	2-Hydroxyisoflavanone dehydratase
FMO	Flavonoid 3′-monooxygenase
LAD	Leucoanthocyanidin dioxygenase
DFR	Bifunctional dihydroflavonol 4-reductase/flavanone 4-reductase
ANS	Anthocyanidin synthase
AGT	Anthocyanidin 3-*O*-glucosyltransferase
ANR	Anthocyanidin reductase
FLS	Flavonol synthase
FMT	Flavonol 3-*O*-methyltransferase
CMT	Caffeoyl-CoA O-methyltransferase
SHCT	Shikimate O-hydroxycinnamoyl transferase
CMO	Coumaroylquinate (coumaroylshikimate) 3′-monooxygenase

1 Introduction

Osmanthus fragrans, commonly known as sweet osmanthus, is a woody, ever green species of shrubs and small trees in the Oleaceae family. Sweet osmanthus has considerable economic value and cultural significance and is one of the top ten traditional flowers in China. Twenty-four of the thirty-five species in the *Osmanthus* genus are distributed in China, with the mostly highly representative species being *O. fragrans*. To date 166 variety of *O. fragrans* have been identified and divided into four groups: *O. fragrans* Asiaticus Group, *O. fragrans* Albus Group, *O. fragrans* Luteus Group, and *O. fragrans* Aurantiacus Group (Xiang and Liu 2007). The Albus, Luteus and Aurantiacus Group can be characterized as an autumn group, because they bloom in September and October. In contrast, the Asiaticus Group flowers throughout the year with the exception of the hot summer months. The flower color in the Aurantiacus Group ranges from orange to orange-red, while the flower color of other three groups range from ivory to yellow.

Sweet osmanthus is widely cultivated throughout China and other countries due to its prominent fragrance, colorful flowers, and medicinal properties. In China, it also has cultural significance and is cultivated extensively both as an ornamental and commercially for its flowers which have high economic value. In Asia, sweet osmanthus are ubiquitous landscape architecture trees (Shang *et al.* 2003). Traditionally, sweet osmanthus has been used as a tea, juice, wine and as a garnish on food. The medicinal value of sweet osmanthus was recognized in Chinese medicinal classics, such as the *Compendium of Materia Medica*.

Using modern methods of extraction and analytic chemistry, the major natural compounds contributing to the fragrance of sweet osmanthus have been identified to include ionone, linalool, and other terpene compounds (Sun 2012). The major flower pigment compounds are lavonoids and carotenoids (Cai *et al.* 2010). There is potential to commercially utilize the natural compounds found in sweet osmanthus since they are believed to have anticancer, antibacterial and antioxidant medical proper-ties, and also to prevent Alzheimer's disease (Lee *et al.* 2010, 2011; Wang *et al.* 2008).

Interestingly, flower color in sweet osmanthus is not related to fragrance, which varies among the four groups and even at different stages of blossom development. To begin to understand the genetic basis of differences in the main characteristics of sweet osmanthus, two representative varieties, 'Zaoyingui' (early silver) and 'Chenghong Dangui' were selected for use in this study. 'Zaoyingui' cultivated widely in Yangtze River basin, has light yellow flower color (RHS-CC 1D), blooms early, and is strongly aromatic. Therefore, 'Zaoyingui' can be considered to be representative of the Albus group. 'Chenghong Dangui', has red orange flowers, is lightly aromatic, and generally blooms late, although it bloomed only four days later than 'Zaoyingui' in 2013. Nevertheless, 'Chenghong Dangui' is representative of the Aurantiacus Group. The Albus Group is considered evolutionarily more primitive than both the Luteus and Aurantiacus Group, but the Asiaticus Group is the most primitive (Hao *et al.* 2005).

While the two varieties mentioned represent the Albus and Aurantiacus Groups, respectively, neither genome nor transcriptome sequence data is available for the species, *Osmanthus fragans*. In order to facilitate further research on secondary metabolite biosynthesis in sweet osmanthus, including the identification of key enzyme genes and transcription factors, we sequenced the transcriptome of the two varieties using an Illumina platform.

Next generation sequencing can provide deep coverage and RNA sequencing offers an attractive alternative to whole-genome sequencing since it allows the identification of specific genes associated with a specific developmental stage or physiological condition (Margulies et al. 2005; Huse et al. 2007; Novaes et al. 2008). Transcriptome profiling can provide information on relative gene expression, and can be used to identify the amino acid composition of the translated proteins. Transcriptome analysis is a powerful tool that can be used to identify functional elements of the genome, metabolic pathways, and differential gene expression in both model and non-model organisms (Wang et al. 2009). In general, the increased throughput of NGS technologies has shown great potential for expanding sequence databases of all type organisms (Meyer et al. 2009; O'Neil et al. 2010; Zhu et al. 2013).

In the present study, transcriptome sequencing of *O. fragrans* was performed using the Illumina High-SeqTM2000 sequencing platform. A total of 256,774 different transcripts and 117,595 unigenes were identified. This is the first reported study of the *O. fragrans* transcriptome. The dataset obtained in this study will serve a useful resource for future genetic and genomic studies in sweet osmanthus. The useful transcriptome datasets of sweet osmanthus was getting ready to upload public database.

2 Materials and Methods

2.1 Plant Materials and RNA Extraction

Two varieties of *O. fragrans*, 'Zaoyingui' and 'Chenghong Dangui', used in this study were located on the Nanjing Forestry University campus, Jiangsu Province, and were approximately 15-17 years old. Flower and leaf issues were collected from sweet osmanthus trees, immediately frozen in liquid nitrogen and stored at −80 ℃ until further analysis. Total RNAs were extracted from these materials using a Qiagen RNA Purification Kit (Qiagen, Valencia, CA, USA). The purity of total RNA was analyzed using a Neanodrop 2000 Spectrophotometer (IM-PLEN, CA, USA) and the RNA concentration of each sample was measured using a Qubit® RNA Assay Kit in a Qubit® 2.0 Fluorometer (Life Technologies, CA, USA). RNA integrity was assessed by agarose gel electrophoresis. For ensuring the representative of gene difference expression, high-quality RNA from different replicates of each sample material was pooled using equal quantities.

2.2 mRNA-seq Library Construction and LIumina Sequencing

The cDNA libraries were constructed using a mRNA-Seq Sample Preparation Kit (Cat # RS-930-1001, Illumina Inc, San Diego, CA) according to the manufacturer's instructions. Briefly, total RNA was isolated from flower and leaves of the two varieties of *O. fragrans* as described. Magnetic Oligo (dT) beads were subsequently utilized to isolate poly-(A) mRNA which was then fragmented by adding fragmentation buffer (Ambion, Austin, TX). Frag-mentation was carried out using divalent cations under elevated temperature in NEB Next First Strand Synthesis Reaction Buffer (5X). First strand cDNA was synthesized using random hexamer primer and M-MuLV Reverse Transcriptase (RNaseH-) by PCR using random hexamers. Second-strand cDNA was then synthesized by PCR employing a buffer consisting of dNTPs, DNA polymerase I (New England BioLabs, Ipswich, MA, USA) (NEB, USA) and Rnase H (Invitrogen). A mRNA-Seq Sample Preparation Kit (Cat # RS-930-1001, Illumina Inc, San Diego, CA, USA) was employed to construct a paired-end library using 300 bp fragments of sweet osmanthus cDNA according to the manufacturer's instructions (http:// mmjggl. caltech. Edu/sequencing/ mRNA-Seq_ Sample Prep_1004898_D. pdf). Double stranded cDNA was purified using AMPure XP beads (Beckman Coulter, Brea, CA, USA). Subsequently, end-repair and the addition of poly (A) was performed by PCR. PCR amplification of the mRNA-Seq library was carried out using a mRNA-Seq Sample Preparation Kit (Cat # RS 930 1001, Illumina Inc, San Diego, CA, USA). To quantify the DNA con-centration and to verify the overall quality of the constructed library, quality control analysis of each sample library was performed. mRNA-Seq libraries were constructed from flowers and leaves of each of the two varieties, 'Zaoyingui' and 'Chenghong Dangui'. The four mRNA-seq libraries were sequenced using an Illumina HiSeqTM2000 platform at the Novogene Bioinformatics Institute (Beijing, China). Data analysis and base calling were performed with Illumina instrument software.

2.3 Sequence Data Analysis and Assembly

After removal of adapter sequences, low-quality reads, and reads in which $N>10\%$ were removed. The remaining reads were considered to be clean reads. Reads less than 60 bp or more than 340 bp were then eliminated, based on the assumption that outlier reads might represent sequencing artifacts (Meyer *et al.* 2009). The trimmed, size-selected reads were assembled using Trinity (Grabherr *et al.* 2011; Tao *et al.* 2012) software. The overlap settings used for the assembly were 30 bp and 80% similarity, with all other parameters set to their default values.

2.4 Sequence Annotation

Sequence identification and annotation were carried out by comparing assembled sequences against the NCBI Nr and Nt database and Swiss-Prot database using BLASTn and BLASTx with an *E*-value of 10^{-5}. Gene names were assigned to each assembled sequence based on the best BLAST hit with an annotated gene. Each query was limited to the first 10 significant hits.

GO terms were employed to further annotate the assembled sequences in the categories biological processes, molecular functions and cellular components. GO annotations were assigned using Blast2Gov2. 5 (Conesa et al. 2005; Conesa and Götz 2008). Genes from the each of the transcriptome libraries that were assigned GO terms were further enriched using ANNEX (Moriya et al. 2007). Functional categories presented in the results represent a GO analysis of the sequence data at level 2.

The KOG database and KEGG Automatic Annotation Server online tools (http://www.genome.jp/kegg/akas/) were utilized to enable further prediction, classification and assignment of KEGG pathways for the functions of the assembled unigenes. The bidirectional best hit (BBH) method was subsequently applied to obtain a KEGG Orthology (KO) assignment (Moriya et al., 2007).

2.5 Differential Gene Expression Analysis

An analysis of differential gene expression was performed in the two varieties for which transcriptomic sequence data had been obtained. Read counts were standardized using TMM (weighted trimmed mean of the log expression ratios, trimmed mean of M values) (Robinson and Oshlack 2010), and then DEGseq R package (1.10.1) (Wang et al. 2010) was employed to conduct the differential expression analysis. A q-value of <0.005 and the absolute value of $\log_2$ fold change >1 (difference) or 2 (significant difference) was regarded as a standard for evaluating differentially expressed genes. In this study, the absolute value of $\log_2$ fold change >1 was set up for identifying differentially expressed genes. By controlling the selected thresh-old value, the false positive rate was minimized.

3 Results and Discussion

3.1 Sequence Analysis and Assembly

In order to begin understanding the genetic basis for differences in fragrance and flower color among different variety of *O. fragrans*, Illumina sequencing of the transcriptome of flowers and leaves of two varieties, 'Zaoyingui' and 'Chenghong Dangui', was performed. Sequencing of the transcriptome of 'Chenghong Dangui' flowers and leaves yielded 50,696,660 and 66,731,528 raw reads; resulting in 48,008,492 and 63,422,560 clean reads (16.72 G), respectively. After filtering, The Q20 reached $>98\%$ and the GC content was approximately 43% in each of the sequenced transcriptomes. In the cultivar 'Zaoyingui', a total of 105,117,296 raw reads were generated, resulting in 99,835,766 clean reads (14.98 G) after stringent quality assessment and data filtering. The Q20 and GC content were similar to that obtained in 'Chenghong Dangui' (Tab. 1).

Tab. 1 Quality parameters of Illumina transcriptome sequencing of *O. fragrans* Lour.

Sample	Read length	Raw reads	Clean reads	GC Content (%)	Q20 (%)
Fzy	150+150 bp	47,830,908	45,486,774	43.62	98.41
Lzy	150+150 bp	57,286,388	54,348,992	43.26	98.38
Fchd	150+150 bp	50,696,660	48,008,492	43.30	98.40
Lchd	150+150 bp	66,731,528	63,422,560	43.18	98.50

F and L indicate flowers and leaves respectively, zy present *Osmanthus fragrans* 'Zaoyingui', chd presents *Osmanthus fragrans* 'Chenghong Dangui'

The clean reads obtained from all four of the sequenced transcriptomes were combined and, using Trinity software, assembled into 256,774 transcripts and 117,595 unigenes, with an average length of 1,155 and 708 bp, respectively (Tab. 2). The N50 values of the transcripts and unigenes were 1,963 and 1,119 bp, respectively. The high number of unigenes obtained in this study is consistent with other studies using Illumina technology (Emrich et al., 2007). The length distribution pattern and mean length of contigs and unigenes was also similar to those reported in previous Illumina transcriptome studies (Hanson and Bentolila, 2004; Tanaka et al., 2004). Collectively, the data indicates that the transcriptome sequencing data from *O. fragrans* was assembled for the first time.

Tab. 2 Length distribution of transcripts and unigenes for the transcriptome assembly of *O. fragrans* Lour.

Category	Min length	Mean length	Max length	N50	N90	Total nucleotides
Transcript	201	1,155	17,007	1,963	481	296,680,960
Unigene	201	708	17,007	1,119	285	83,267,725

The sequences were assembled into 256,774 transcripts using the Trinity assembly program. The distribution of the lengths for both the transcripts and unigenes is shown in Figs. 1 and 2. There were 103,256 scaffolds coding for transcripts ≥1 kb and 47,327 scaffolds coding for transcripts ≥2 kb (Fig. 1). Approximately 117,595 unigenes were obtained after cluster and assembly analysis, in which 21,496 genes (18.28%) were ≥ 1 kb in length and more than 45,821 unigenes (∼38.97%) were ≥500 bp in length (Fig. 2). These results indicated that the Illumina sequencing platform was capable of capturing a great deal of transcriptomic information.

3.2 Sequence Annotation

Several complementary approaches were utilized to annotate the assembled sequences. Unigenes were identified and annotated by aligning them with sequences contained in the NCBI Nr database and Nt database; and the KO, SwissProt, PFAM, GO and KOG databases. A minimum E-value of $\leqslant 10^{-5}$ was used as a cutoff for sequence homology. The

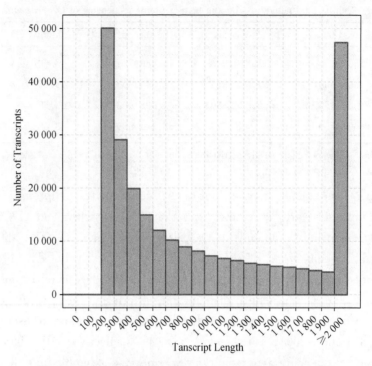

Fig. 1 Overview of the transcriptome assembly in *Osmanthus fragrans* Lour. Length distribution of transcripts

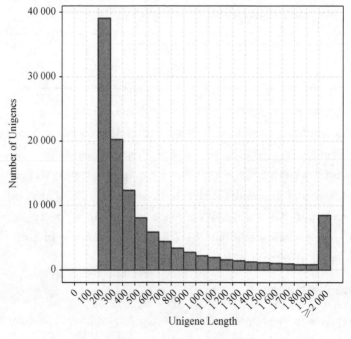

Fig. 2 Overview of the transcriptome assembly in *Osmanthus fragrans* Lour. Size distribution of unigenes

results of the functional annotation of sweet osmanthus are summarized in Tab. 3. In the process of identifying and annotating the obtained unigene sequences, a sequence similarity search against the NCBI Nr and Nt database, and SwissProt protein database was conducted using the Basic Local Alignment Search Tool (BLAST) algorithm specifying E-values of $\leqslant 10^{-5}$. The analysis indicated that out of a total of 117,595 unigenes, 36,094 unigenes (30.69%) had significant matches in the Nr database and 15,288(13.0%) had significant matches in the Nt database, while 6,540(5.56%), 24,802(21.09%), 25,750 (21.89%), 29,945(25.46%), and 11,795 (10.03%) unigenes had similarity to data in the KO, Swiss-Prot, PFAM, GO, and KOG databases, respectively. Out of the total of 117,595 unigenes, 2,401 (2.04%) were successfully annotated in all seven of the databases mentioned above. Collectively, the number of annotated genes obtained in this study was not as high as expected. One possible reason is the length of the reads produced by NGS. The length of the query sequence is a vital parameter when using BLAST to obtain matches with a low E-value to known genes and short reads would make this difficult (Novaes et al. 2008). Therefore, the low number of annotated unigenes obtained in the current study may be mainly attributed to the short reads, especially if they do not contain a conserved functional domain (Hou et al. 2011). Another possible reason is that the unmatched genes might represent genes specific to sweet osmanthus.

Tab. 3 Functional annotation of the unigenes obtained from the assembly of the Transcriptome of *O. fragrans* Lour.

Annotated Databases	Unigenes	Percentage/%
NR Annotation	36,094	36.09
NT Annotation	15,288	13
KO Annotation	6,540	5.56
SwissProt Annotation	24,802	21.09
PFAM Annotation	25,750	21.89
GO Annotation	29,945	25.46
KOG Annotation	11,795	10.03
Annotated in All Databases	2,401	2.04
Annotated in at Least One Databases	42,538	36.17
Total	117,595	100

3.3 GO Annotation

Gene Ontology (GO) analysis was carried out after Nr annotation, using the categories molecular function, cellular component, and biological process. As illustrated in Fig. 3, 117,595 unigenes were annotated, among which 46.33% in biological processes, 21.33% in molecular functions, and 32.44% in cellular components. In biological processes, genes involved in cellular processes (GO: 0009987), metabolic processes (GO:

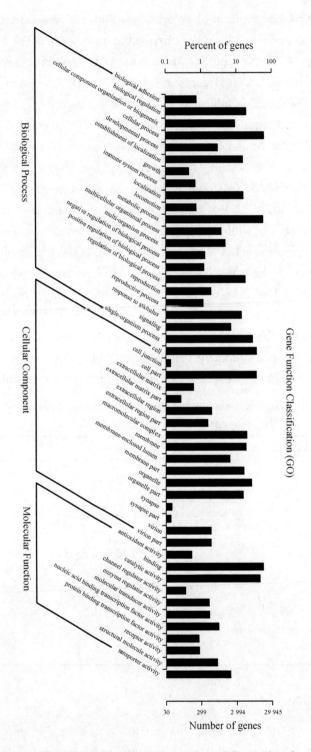

Fig. 3　Functional annotation of assembled sequences of *Osmanthus fragrans* based on gene ontology (GO) categorization

GO analysis was performed at the level 2 for three main categories (cellular component, molecular function, and biological process)

0008152), and single organism processes (GO: 0044699) were highly represented. In molecular functions, binding activity (GO: 0005488) was the most represented GO term, followed by catalytic activity (GO: 0003824). In cellular components, the most represented term was cells part (GO: 0044464).

3.4 KOG Classification

A total 117 595 unigene sequences were searched against the KOG database to obtain further functional prediction and classification (Fig. 4). Results indicated that, out of the 117 595 unigene sequences, 36,904 were clustered into 26 groups. The largest group was group R-the general function prediction cluster (2,317; 19.64%), followed by group O-post-translational modification, protein turnover and chaperones (1,576; 13.36%), group T-signal transduction mechanisms (19;9.74%), group J-translation (823; 6.98%), group

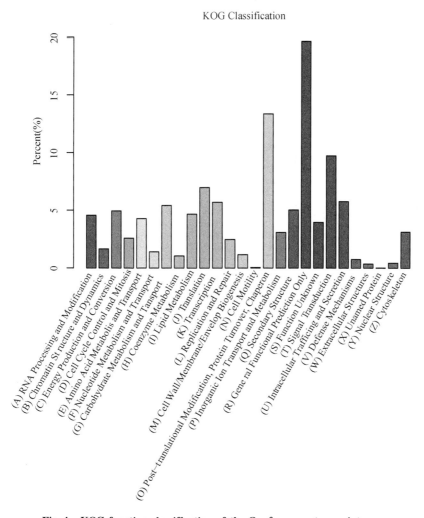

Fig. 4 KOG function classification of the *O. fragrans* transcriptome

In total, 36,094 of the 117,595 sequences with Nr hits were grouped into 26 KOG categories

U-intracellular trafficking and secretion (680; 5.77%), group K-transcription (672; 5.7%), group G-carbohydrate metabolism and transport (640; 5.43%), group Q-secondary structure (594; 5.04%), group C-energy production and conversion (584; 4.95%), group I-lipid metabolism (551; 4.67%), group A-RNA processing and modification (540; 4.58%), and group E-amino acid metabolism and transport (505; 4.28%). Relatively few unigenes were assigned to cell motility, extracellular structure, nuclear structure, defense mechanisms, coenzyme metabolism, cell wall/membrane/envelope biogenesis (5,42,51,88,122 and 136 unigenes, respectively). Additionally, 306 unigenes were allocated to cell cycle control and mitosis and 364 unigenes were assigned to inorganic ion transport and metabolism while 468 were classified as unknown function and 1 was classified as an unnamed protein (Fig. 4).

3.5 KEGG Classification

To explore the potential function of the identified unigenes in sweet osmanthus, the biochemical pathways and functions associated with the unigenes were assigned by using the KEGG Pathway Tools. As a result, 29,945 unigenes were identified as being associated with 247 biochemical. Path ways within branch 5 of the overall KEGG metabolite pathway organization (Fig. 5). Within branch five of the KEGG metabolite

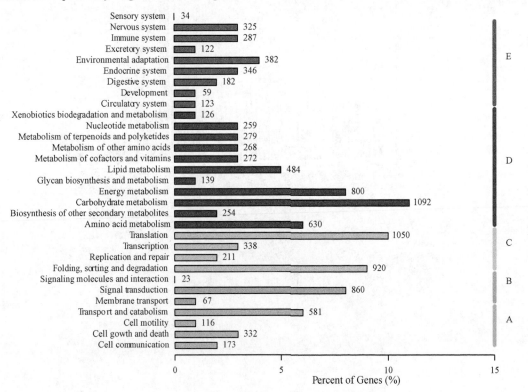

Fig. 5 KEGG classification of *Osmanthus fragrans*

The color scheme represents different metabolite pathway of the annotated genes, numbers in the right side of the columns indicate the number of genes involved in the metabolite pathway. Descriptive words to the left side of the bar indicate the name of corresponding metabolite pathways.

pathway classification system, pathways associated with secondary metabolite (such as carotenoids, flavonoids, anthocyanins, phenylpropanoids, mono-, di-, sesqui-, and triterpenoids, and terpenoid backbones) bio-synthesis are of special interest due to their involvement in flower color and aroma. Results indicated that genes coding for enzymes involved in the biosynthesis of floral pigments and those involved in fragrance were both identified in our unigene dataset. Genes related to carbohydrate, energy, and translation metabolism, as well as other metabolic pathways that play an important role in growth and development, were also present in our dataset but were not the focus of this study. In addition, a number of genes involved in forty-two different signaling pathways were also contained in our dataset, including the major signaling pathways associated with plant hormone signal transduction, PI3 K-Akt, HIF-1, MAPK, Jak-STA, Wnt, Phosphatidylinositol, and the mitogen-activated protein kinase (MAPK) signaling pathway. Since genomic and transcriptome data are lacking for the genus *Osmanthus*, the sequence information obtained in the present study will provide a foundation for further genetic, metabolite, and other molecular studies in sweet osmanthus.

3.6 Analysis of Differential Gene Expression

To explore the molecular basis for differences in floral color between 'Chenghong Dangui' (red-orange flowers) and 'Zaoyingui' (light yellow flowers), a comparison of gene expression in the two varieties was conducted using 'Zaoyingui' as a control. As illustrated in the Venn diagram (Fig. 6), 1,594 genes were differentially expressed in the flower organs of two varieties (Fchd VS Fzy) and 646 genes were differentially expressed in the leaves (Lchd VS Lzy). Gene expression also differed between different organs

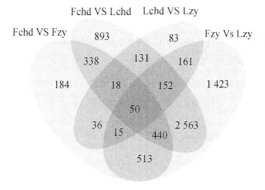

Fig. 6 Venn diagram of differential gene expression in leaves and flowers of two varieties of *Osmanthus fragrans* and between leaves and flowers within the same variety

The different colors represent four different subsets (Fchd VS Fzy, Lchd VS Lzy, Fchd VS Lchd, Fzy VS Lzy). The subset "Fchd VS Fzy" indicates the number of differentially expressed genes between flowers of the two varieties. "Lchd VS Lzy" indicates the number of differentially expressed genes between leaves of the two varieties. "Fchd VS Lchd" indicated the number of differentially expressed genes between the flower and leaves of 'Chenghong Dangui' while "Fzy VS Lzy" indicates the number of differentially expressed genes between the flower and leaves of 'Zaoyingui' (leaves VS flowers). There were 4,585 genes differentially expressed between the flower and leaves of 'Chenghong Dangui' (Fchd VS Lchd) and 5,317 genes differentially expressed between the flower and leaves of 'Zaoyingui' (Fzy VS Lzy). There were a total of 3,205 genes that were differentially expressed between the two varieties. A volcano plot was constructed for further analysis of the differential expression level in the two varieties (summarized in Tab. 4). When differential gene expression was compared in flowers of the two varieties, 652 and 942 genes were specific among the total of 1,594 differentially expressed genes. Among the differentially expressed genes in leaf tissues of the two varieties, 217 and 429 were specific among the total of 646 differentially expressed genes. In a comparison of differential gene expression in flowers VS leaves within a variety, 2,530 genes and 2,082 genes were specific in 'Chenghong Dangui', while 2,513 and 2,804 genes were specific in 'Zaoyingui'.

Tab. 4 Summary of differential gene expression in flowers and leaves of two *O. fragrans* varieties ('Chenghong Dangui' and 'Zaoyingui')

Differential Comparisons	Number of Differentially Expressed Genes	Number of Up-Regulated Genes	Number of Down-Regulated Genes
Fchd VS Fzy	1,594	652	942
Lchd VS Lzy	646	217	429
Fchd VS Lchd	4,585	2,503	2,082
Fzy VS Lzy	5,317	2,513	2,804

Relative to other traits, aroma and pigmentation are among the most important traits in sweet osmanthus. In this regard, special consideration was given to obtain the best stage (Xiangyan stage or initial flowering stage) for both aroma and petal color when conducting this study. As a result, genes associated with pigmentation and fragrance were both contained in our overall dataset and in the data obtained for differential gene expression stage (Xiangyan stage or initial flowering stage) for both aroma and petal color when conducting this study. As a result, genes associated with pigmentation and fragrance were both contained in our overall dataset and in the data obtained for differential gene expression.

3.7 Genes Involved in Fragrance Biosynthesis

Volatile organic compounds (VOCs), which contribute to fragrance, are represented by a large group of terpenoids (homo-, mono-, di-, sesquiterpenoids), phenylpropanoid aromatic compounds, and certain alkanes, alkenes, alcohols, esters, aldehydes, and ketones (Maffei 2010). These VOCs are synthesized through a series of enzymatic reactions. One-hundred and ninety-seven fragrance-related unigenes were identified among the 1,594 differentially expressed genes (Tab. 4). Seven KEGG pathways were putatively associated with fragrance biosynthesis, including styrene degradation, limonene and pinene degradation, diterpenoid biosynthesis, sesquiterpenoid and triterpenoid biosynthesis, monoterpenoid biosynthesis, terpenoid backbone biosynthesis, stilbenoid, diarylheptanoid and gingerol biosynthesis, and phenylpropanoid biosynthesis. Phenylpropanoid is the most common substrate or precursor both for the synthesis of VOCs, as well as pigments. Based on our observations (data not shown), the light yellow flowers of 'Zaoyingui' emit an intense aroma, while the red-orange flowers of 'Chenghong Dangui' emit little aroma. Therefore, it appears that the red-orange pigments in 'Chenghong Dangui' must affect aroma forming to some extent. However, the molecular regulation of aroma and the role of flower color on aroma, remains unclear and will require more focused studies in the future.

3.8 Genes Involved in Pigments Biosynthesis

Plant pigments such as carotenoids, flavonoids and anthocyanins, play a crucial role in determining the color of flowers, fruits, and seeds. Color is an important economic trait in many crops and ornamentals. This is especially true for sweet osmanthus. Flavonoids and carotenoids are the main floral pigments in flowers of the O. fragans cultivar 'Sijigui' (Cai et al. 2010). This was also verified in the flowers of 'Chenghong Dangui' and 'Zaoyingui' in the current study but the levels were different in the two varieties (data not shown). The data on differentially expressed genes and the KEGG metabolite pathway classification were combined to identify candidate genes involved in pigment biosynthesis in sweet osmanthus (Fig. 7). In this regard, 237 candidate genes were identified in our Illumina dataset, such as zinc finger protein CONSTANS-LIKE$_{14}$-like, isoprene synthase (ISPS), PAL, cinnamoyl-CoA reductase, Beta-glucoside enzyme, coumaroylquinate (coumaroylshikimate) 3'-monooxygenase 2, CYP98A, ABA8 hydroxylase, LCYe genes expression presented up regulation; genes of PSY, GGPS and DELLA protein present down regulation. These genes probably contribute to the biosynthesis of flavonoids (flavone, flavonol, anthocyanin) and carotenoids. However, the specific function of those genes need verification in future by integrating the target gene into tobacco and petunia. In this paper, the metabolic pathway of 36 genes (Fig. 7a) and 30 candidate genes (Fig. 7b) that involving in carotenoids and flavonoids metabolism were outlined based on KEGG analysis. The biosynthesis of flavonoids can be derived from the phenylpropanoid metabolite pathway which is part of the shikimic acid pathway. Carotenoid biosynthesis, however, is derived via the isoprenoid pathway. GGPP, the first precursor of carotenoid, is also a precursor for gibberellin, tocopherol, phytol and other natural compounds (Chappell 1995). A shortage of GGPP could limit the biosynthesis of catotenoids and other compounds biosynthesis, which could lead to the down-regulation of related genes or transcription factors for related common substrates.

In this study, GGPS differential expression was detected, with higher expression quantity in 'Zaoyingui' than in 'Chenghong Dangui'. Meanwhile, 1 unnamed protein and 468 functional unknown genes, which have not been verified in model plants, were identified based on KOG analysis.

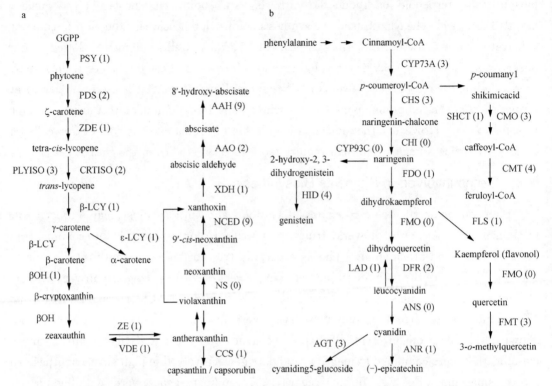

Fig. 7 Genes identified in *O. fragrans* that are involved in two pigment biosynthesis pathways

(a) *O. fragrans* genes involved in the carotenoid biosynthesis pathway. (b) *O. fragrans* genes involved in the flavonoid biosynthesis pathway. The two pigment biosynthesis pathways in *O. fragrans* were inferred from its transcriptome dataset. The numbers in the brackets following each gene name indicates the number of corresponding *O. fragrans* genes present in the transcriptome dataset.

Some transcription factors and protein complexes (such as MYB prollowing each gene name indicates the number of corresponding *O. fragrans* genes present in the transcriptome datasetotein family, bHLH, and WD repeat protein) have been linked to pigment biosynthesis (Huang et al. 2013; Gao et al. 2011). In the current study, genes coding for RING finger and CHY zinc finger domain-containing protein 1 (comp110024_c0), F-box protein family (comp107616_c0), WD repeat containing protein family (comp105673_c0), and the WRKY protein family (comp355212_c0) were identified in the transcriptome dataset. These genes may play a role in the regulation of pigment accumulation in cells. Transporters are also needed for moving pigments to the vacuole and other cellular locations. In our dataset, 56 ABC transporters (ko02010) were identified which would likely be involved in pigment transport, In which ABC transporters (comp105673_c0) was up regulated in 'Chenghong Dangui' (red-orange) and down regulated in 'Zaoyingui' (light yellow), which probably involve in sweet osmanthus pigment metabolism. Currently, a determination of what pigments are synthesized and allowed to accumulate in plants is still largely unclear despite the existence of model plant systems. Therefore, considerably more research will be needed to understand this aspect of plant metabolism in O. fragans.

3.9 Comparison of Gene Expression Profiling

PSY gene expression profiling of sweet osmanthus transcriptome revealed that its expressive abundance in 'Zaoyingui' flower was much higher than that in 'Chenghong Dangui' flower. However, Han et al. (2013) discovered that no significant difference of PSY gene expression and the carotenoid content of "Dangui" was clearly higher than that in"Yingui" petals of three flowering stage. Comparing the expression profiling of PSY between our transcriptome and Han's semiRT-PCR, the possible reason was that the sensitivity of semi RT-PCR was not as good as NEG. Zhong et al. (2012) cloned DELLA protein gene of Prunus avium L., PaGAI, its expression can be detected in flowers, fruit, leaf and phloem but expressive abundance was much higher in flowers and phloem than other tissues. The transcriptome expression profiling of DELLA protein gene of sweet osmanthus showed its expression presented in flower and leaves, there was significant expression between 'Zaoyingui' flower than 'Chenghong Dangui', the expression of DFR, CHS, F3H and ANS in sweet osmanthus transcriptome were similar to chrysanthemum semiRT-PCR (Han et al. 2012) except for CHI gene. Interestingly, CHI gene expression had not been detected in transcriptome sequencing.

4 Conclusions

This is the first transcriptome sequencing project for flowers and leaves of two varieties of O. fragrans. In total, 31.7 G bp of clean bases (Tab. 1) were generated and assembled into 117,595 unigenes. Numerous candidate genes potentially involved in plant secondary metabolic pathways were identified, some of which may potentially play critical role in the biosynthesis of compounds involved in floral fragrance and pigmentation in O. fragrans and are thus a valuable resource for further research on these processes. Differential gene expression analysis in flowers and leaves of 'Zaoyingui' and 'Chenghong Dangui', as well as between leaves and flowers within a cultivar, resulted in the identification of 237 candidate genes putatively involved in carotenoid and flavonoid biosynthesis, and 167 candidate genes putatively involved in floral aroma. Additionally, 18,327 SSRs were identified that can be used for genetic marker development (data not shown). The dataset obtained in the current study will serve as a resource for future studies on flower aroma and pigmentation, as well as other biochemical processes in sweet osmanthus.

Acknowledgements

We appreciate the technical support for Illumina sequencing and initial data analysis from Novogene Bioinformatics Institute Beijing, China. We are grateful to Zongming

Cheng from University of Tennessee, Knoxville, for comments on the manuscript. We thank the anonymous referees and the editor for their comments and suggestions that helped improve the manuscript. We also thank the assistance provided by Yali Wang, Chunjun Wang, Tingting Shi, Ping Wang and Min Ling in preparing sweet osmanthus flowers. This study was supported by the Agriculture-Forestry Plant Germplasm Resources Exploration, Innovation and Utilization Project of the National Twelve-Five Science and Technology Support Program (2013BA001B06), and the Advantages Discipline Construction Engineering Project of Colleges and Universities in the Jiangsu province.

References

[1] Cai X, Fan SU, Jin HX, Yao CH, Wang CY (2010) Components and extraction methods for petal pigments of *Osmanthus fragrans* 'Siji Gui'. J. Zhejiang For Coll 244:559-564. (in Chinese with English abstract)

[2] Chappell J (1995) Biochemistry and molecular biology of the isoprenoid biosynthetic pathway in plants. Ann Rev Plant Physiol Mol Bio 46:521-547.

[3] Conesa A, Götz S (2008) Blast2GO: A comprehensive suite for functional analysis in plant Genomics. Int J Plant Gen. doi:10.1155/2008/619832.

[4] Conesa A, Götz S, García-Gómez JM, Terol J, Tólon M, Robles M (2005) Blast2GO: a universal tool for annotation, visualization and analysis in functional genomics research. Bioinformatics 21:3674-3676.

[5] Emrich SJ, Barbazuk WB, Li L, Schnable PS (2007) Gene discovery and annotation using LCM-454 transcriptome sequencing. Gen Res 17:69-73.

[6] Gao ZM, Li CL, Peng ZH (2011) Generation and analysis of expressed sequence tags from a normalized cDNA library of young leaf from Ma bamboo (*Dendrocalamus latiflorus* Munro). Plant Cell Rep 30:2045-2057.

[7] Grabherr MG, Haas BJ, Yassour M, Levin JZ, Thompson DA, Amit I, Adiconis X, Fan L, Raychowdhury R, Zeng Q *et al.* (2011) Full-length transcriptome assembly from RNA-Seq data without a reference genome. Nature Biotech 29:644-652.

[8] Han KT, Zhao L, Tang XJ, Hu K, Dai SL (2012) The relationship between the expression of key genes in anthocyanin biosynthesis and the color of chrysanthemum. Acta Hortic Sin 39:516-524.

[9] Han YJ, Liu LX, Dong MF, Yuan WJ, Shang FD (2013) cDNA cloning of the phytoene synthase (PSY) and expressio nanalysis of PSY and carotenoid cleavage dioxygenase genes in *Osmanthus fragrans*. Biologia 68:258-263.

[10] Hanson MR, Bentolila S (2004) Interactions of mitochondrial and nuclear genes that affect male gametophyte development. Plant Cell 16:154-169.

[11] Hao RM, Zang DK, Xiang QB (2005) Investigation on natural resources of *Osmanthus fragrans* Lour. at Zhouluocun in Hunan. Acta Hortic Sin 32:926-929. (in Chinese with English abstract)

[12] Hou R, Bao Z, Wang S, Su H, Li Y, Du H, Hu J, Wang S, Hu X (2011) Transcriptome sequencing and de novo analysis for Yesso scallop (*Patinopecten yessoensis*) using 454 GS FLX. PLoS

One 6: e21560.

[13] Huang W, Sun W, Lv H, Luo M, Zeng S, Pattanaik S, Yuan L, Wang Y (2013) A R2R3-MYB transcription factor from *Epimedium sagittatum* regulates the flavonoid biosynthetic pathway. PLoS One 8: e70778.

[14] Huse SM, Huber JA, Morrison HG, Sogin ML, Welch DM (2007) Accuracy and quality of massively parallel DNA pyrosequencing. Genome Bio 8R:14.

[15] Lee DG, Choi JS, Yeon SW, Cui EJ, Park HJ, Yoo JS, Chung IS, Baek NI (2010) Secoiridoid glycoside from the flowers of *Osmanthus fragrans* var. *aurantiacus* Makino inhibited the activity of b-secretase. J Kor Soc Appl Bio Chem 53:371-374.

[16] Lee DG, Lee SM, Bang MH, Park HJ, Lee TH, Kim YH, Kim JY, Baek NI (2011) Lignans from the flowers of *Osmanthus fragrans* var. *aurantiacus* and their inhibition effect on no production. Arch Pharm Res 34:2029-2035.

[17] Maffei ME (2010) Sites of synthesis, biochemistry and functional role of plant volatiles. S Afr J Bot 76:612-631.

[18] Margulies M, Egholm M, Altman WE, Attiya S, Bader JS, Bemben LA, Berka J, Braverman MS, Chen YJ, Chen Z et al. (2005) Genome sequencing in micro fabricated high-density picolitre reactors. Nature 437:376-380.

[19] Meyer E, Aglyamova GV, Wang S, Buchanan-Carter J, Abrego D, Colbourne JK, Willis BL, Matz MV (2009) Sequencing and denovo analysis of a coral larval transcriptome using 454 GSFlx. BMC Genom 10:219.

[20] Moriya Y, Itoh M, Okuda S, Yoshizawa AC, Kanehisa M (2007) KAAS, an automatic genome annotation and pathway reconstruction server. Nucleic Acids Res 35: 182-185.

[21] Novaes E, Drost DR, Farmerie WG, Pappas GJ, Grattapaglia D, Sederoff RR, Kirst M (2008) High-through put gene and SNP discovery in *Eucalyptus grandis*, an uncharacterized genome. BMC Genom 9:312.

[22] O'Neil ST, Dzurisin JD, Carmichael RD, Lobo NF, Emrich SJ, Hellmann JJ (2010) Population level transcriptome sequencing of nonmodel organisms *Erynnis propertius* and *Papilio zelicaon*. BMC Genom 11:310.

[23] Robinson Mark D, Oshlack Alicia (2010) A scaling normalization method for differential expression analysis of RNA-seq data. Genome Biol 11: R25.

[24] Shang FD, Yin YJ, Xiang QB (2003) The culture of sweet osmanthus in China. J Henan Univ (Social Science) 43:136-139.

[25] Sun BJ (2012) HS-SPME-GC-MS analysis of different *Osmanthus fragrans* variety from Guilin Garden in Shanghai. J Fujian Coll Fores 32:39-42. (in Chinese with English abstract)

[26] Tanaka N, Fujita M, Handa H, Murayama S, Uemura M, Kawamura Y, Mitsui T, Mikami S, Tozawa Y, Yoshinaga T, Komatsu S (2004) Proteomics of the rice cell: systematic identification of the protein populations in subcellular compartments. Mol Genet Genomics 271:566-576.

[27] Tao X, Gu YH, Wang HY, Zheng W, Li X, Zhao CW, Zhang YZ (2012) Digital gene expression analysis based on integrated de novo transcriptome assembly of sweet potato (*Ipomoea batatas* (L.) Lam.). PLoS One 7: e36234.

[28] Wang LM, She LJ, Cui RM, Xiang J (2008) Flavonoids from *Osmanthus fragrans*: extraction and bacteriostatic activities. Natural Prod Res Dev 20:717-720. (in Chinese with English abstract)

[29] Wang Z, Gerstein M, Snyder M (2009) RNA-Seq: a revolutionary tool for transcriptomics. Nat Rev

Gene 10:57-63.

[30] Wang LK, Feng ZX, Wang X, Wang XW, Zhang XG (2010) DEGseq: an R package for identifying differentially expressed genes from RNA-seq data. Bioinformatics 26:136-138.

[31] Xiang QB, Liu YL (2007) An illustrated monograph of the sweet osmanthus variety in China. Hangzhou:Zhejiang Science & Technology Press.

[32] Zhong F, Shen XJ, Liu F, Yuan HZ, Liu LY, Li TH (2012) Cloning and expression analysis of PaGAI gene of DELLA protein from *Prunus avium*. Acta Hortic Sin 39:143-150. (in Chinese with English abstract)

[33] Zhu JY, Li YH, Yang S, Li QW (2013) De novo assembly and characterization of the global transcriptome for *Rhyacionia leptotubula* using illumina paired-end sequencing. PLoS One 8: e81096.

第二部分
生理学研究

'波叶金桂'花香成分释放规律

施婷婷,杨秀莲,王良桂

(南京林业大学 风景园林学院,江苏 南京 210037)

摘 要:为探索'波叶金桂'花香成分释放规律,利用顶空固相微萃取(HS-SPME)与气相色谱-质谱联用(GC-MS)技术,分析'波叶金桂'开花过程和一天不同时间点释放的花香变化。同时,利用扫描电镜和透射电镜对'波叶金桂'开花过程中花被片结构的变化进行研究。发现在'波叶金桂'开花过程中,花香的释放在香眼期较少,初花期与盛开期增加,初花期最高,末花期急剧下降。同时,桂花开花过程中花被片细胞内嗜锇酸基质颗粒的数量、形态及聚集部位存在明显变化。在一天当中花香释放从早到晚同样表现出先增加后减少的变化,9:00与15:00香气释放最多,而在7:00和19:00这两个时间段香气释放明显减少。得出'波叶金桂'花香释放量初花期最高的结论,并推测光强的变化是引起'波叶金桂'花香日变化的重要原因,嗜锇酸基质颗粒可能是桂花释放香气物质的基础。

关键词:'波叶金桂';花香成分;超微结构;顶空固相微萃取;气相色谱-质谱联用

  自然界中大部分植物都会释放香气,植物的天然香气成分可排斥杂草,防腐抗菌,抵御病虫害,还可传播信息,吸引授粉昆虫以繁衍后代。植物花香的释放伴随着花朵的开放而产生,且其释放量也会发生一定的变化。如黄兰[1]、紫丁香[2]、梅花[3]等植物的花香释放都随花期变化而表现出不同的变化规律,但总体来说都是半开期和盛开期的释放量较高。此外,植物花香成分的释放还受外界环境的影响,如光照、温度、水分,以及授粉昆虫等。如水仙[4]、珍珠梅[5]等植物的花香释放随着外界光照强度的变化而变化,而牡丹[6]、紫花醉鱼草和密蒙花[7]等各自的花香特征与其授粉者都有着密切的联系。

  目前,关于桂花(*Osmanthus fragrans*)花香的研究主要集中在不同品种、不同品种群间香气物质的种类和释放量的差异上,银桂中含有较高的芳樟醇以及芳樟醇氧化物,香气清幽淡雅;四季桂中芳樟醇、β-紫罗兰酮和芳樟醇化合物含量较高,香气清淡;丹桂甜香不够、清香不足,其香气中 α、β-紫罗兰酮的含量很低,且缺乏叶醇酯类、环己烯酯类、吡喃型芳樟醇氧化物和 4-酮基-β-紫罗兰酮;而金桂中含量较高的主要有 α-紫罗兰酮、β-紫罗兰酮、γ-癸内酯以及 4-酮基-β-紫罗兰酮,其香气甜润馥郁,最为浓郁[8-9]。故本文选择香气最为浓郁的金桂品种群中的'波叶金桂'进行其花香成分释放的研究,同时通过调查南京林业大学内绿化种植的金桂品种群桂花,发现校园内同时期种植的'波叶金桂'('Boye Jingui')树龄已达 25~30 年,正值壮年,适宜作本研究的实验材料。

  笔者利用顶空固相微萃取(HS-SPME)与气相色谱-质谱联用(GC-MS)技术分析'波叶金桂'不同开花期花香释放的变化情况,以及'波叶金桂'花香释放的日变化节律。同时。

---

* 原文发表于《南京林业大学学报(自然科学版)》,2018,42(2):97-104。
基金项目:"十二五"科技支撑"林木种质资源发掘与创新利用"(2013BA001B06)。

采用扫描电镜和投射电镜对'波叶金桂'释香过程(不同花期)中花被片细胞超微结构的变化进行了研究,旨在从细胞学角度了解桂花开花过程中香气的差异变化。

1 材料与方法

1.1 试验材料

新鲜的'波叶金桂'花朵采自南京林业大学校园,试验在南京林业大学现代分析测试中心完成。

仪器有固相微萃取仪(美国 SUPELCO 公司),包括手动固相微萃取进样器,65 μm PDMS/DVB 萃取头;气相色谱-质谱联用仪 Trace DSQ(美国 Thermo Electro-Finnigan);4 mL 顶空取样瓶(美国 SUPELCO 公司);QUANTA200 型环境扫描电镜;JGM-400 型透射电镜。

1.2 试验方法

1.2.1 样品采集

于 2014 年 10 月桂花开花期间(桂花花期一般可以持续 10 d 左右),晴朗无风、光照良好的上午 9:00—10:00 在校园内分别采取 4 个不同花期的花朵,即香眼期(开花前 1 d 左右,部分小花顶端出现孔隙)、初花期(约 10%花朵开放,大部分半闭合)、盛花期(70%以上的花朵开放,花被反卷)和末花期(盛花期后 1~2 d,且开始少量落花)[10]的桂花样品。

同时,采取初花期的'波叶金桂'花朵,从 7:00 开始到 20:00 结束,每隔 2 h 采样 1 次。选择生长位置基本一致的桂花花朵进行采样。采摘后立即带回实验室进行香气成分测定。

1.2.2 HS-SPME 分析

先将 65 μm PDMS/DVB 萃取头与 SPME 手动进样手柄进行组装,在气相色谱的进样口进行老化,老化温度为 250 ℃,老化时间 30 min。然后每次取长势、大小一致的花瓣 25 朵放置于 4 mL 萃取瓶中,并密封,于室温下(23 ℃左右)平衡 30 min。然后将经过老化的 SPME 纤维头通过聚四氟乙烯隔垫插入采样瓶中,萃取头置于花朵上方 1 cm 处,顶空萃取 40 min。最后将萃取头插入 GC 进样口,解吸 5 min,进行 GC-MS 分析。

1.2.3 GC-MS 分析

色谱条件:TTR-5MS 弹性石英纤维毛细管柱(30 m×0.25 mm×0.25 μm);载气为高纯度氦气(He),氦气流速为 1 mL/min,分流比为 10∶1,进样口温度 250 ℃;程序升温初始温度为 40 ℃,保持 2 min,以 2 ℃/min 升至 60 ℃,以 5 ℃/min 升至 100 ℃,以 10 ℃/min 升至 250 ℃,保持 5 min。

质谱条件:接口温度为 250 ℃,电子轰击源为 EI,离子电离能量为 70 eV,质量扫描范围 50~450 amu。

1.2.4 扫描电镜观察

将采集的桂花样品投入标准(FAA)固定液中,减压抽真空,使样品下沉,在 4 ℃冰箱中固定 48 h。然后梯度乙醇系列脱水,真空干燥,粘样,喷金镀膜[11],最后在扫描电镜下进行观察、照相。

1.2.5 透射电镜观察

将采集的桂花样品处理成长 5 mm,宽 1 mm 的长方形小块,固定于 0.2 mol/L 的磷酸缓冲液(pH=7.2)配制的 4%的戊二醛中,减压抽真空,直至样品下沉,置于 4 ℃冰箱中固定 48 h。然后用磷酸盐缓冲液冲洗 3 次,每次停留 10~20 min;加入 1%锇酸固定 3 h,磷酸缓冲液清洗 3 次,每次 10~20 min;系列乙醇梯度脱水;Epon812 渗透、包埋、聚合。将包埋块用超薄切片机切片,切片经醋酸双氧铀、柠檬酸铅双重染色[10]后,在透射电镜下进行观察、照相。

1.3 数据分析

对采集到的质谱图用计算机检索并与图谱库(NIST05)的标准质谱图进行对照分析,并查阅有关文献[9,12],确认香气物质的化学成分。根据离子流峰面积归一化法计算出样品中各香气物质的含量。

2 结果与分析

2.1 '波叶金桂'不同开花阶段花香成分分析

将'波叶金桂'不同开花阶段花香成分的总离子流色谱图信息经计算机检索和资料分析,扣除空气本底杂质,得到其香气的主要成分种类。

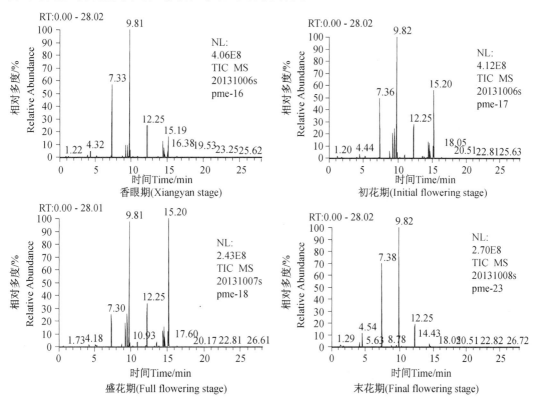

图 1 '波叶金桂'GC-MS 总离子流色谱图

Fig. 1 Total Ion chromatograms of *O. fragrans* 'Boye Jingui'

'波叶金桂'4个时期共检测出34种香气成分,主要包括萜烯类、醇类、醛类、酯类、酮类和烷烃类六大类,且各类香气成分在4个花期阶段表现出显著差异。香眼期共鉴定出20种化合物,主要是萜烯类和酯类化合物,未检测出醛类化合物。初花期花香成分略有增多,共检测出23种花香成分,主要成分与香眼期基本一致,包括顺式-芳樟醇氧化物、反式-芳樟醇氧化物、β-紫罗兰酮、α-紫罗兰酮、二氢-β-紫罗兰酮、γ-癸内酯等萜烯类、酯类化合物。盛花期的'波叶金桂'共鉴定出21种化合物,其主要成分也为β-紫罗兰酮、顺式-芳樟醇氧化物、反式-芳樟醇氧化物、α-紫罗兰酮、二氢-β-紫罗兰酮、γ-癸内酯等。到了末花期花香成分减少到17种,主要包括萜烯类、醛类以及醇类化合物。而从香眼期到末花期,花香成分的种类表现为先上升后下降,在初花期达到最多。但在各个开花阶段中顺式-芳樟醇氧化物、反式-芳樟醇氧化物、β-紫罗兰酮、二氢-β-紫罗兰酮以及γ-癸内酯等的含量均较高,是'波叶金桂'的主要成分,且大部分主成分到末花期会大幅度减少,甚至检测不到,其成分种类也与其他3个时期存在明显差异。

表1 '波叶金桂'不同开花阶段花瓣释放的香气成分及释放量
Tab. 1 The components and relative contents of aromatic compounds released from flowers of O. fragrans 'Boye Jingui' at different developmental stages

序号 No.	主要成分 Main component	质量分数/% Content			
		A	B	C	D
萜烯类	Terpenoids				
1	反式-β-罗勒烯 trans-β-Ocimene	—	0.07	0.15	
2	顺式-β-罗勒烯 cis-β-Ocimene	1.88	3.41	2	12.78
3	5-(1-羟基-1-甲基乙基)-2-甲基-2-环己烯 5-(1-Hydroxy-1-methylethyl)-2-methylcyclohex-2-en-1-one;	—	—	—	1.31
4	顺式-芳樟醇氧化物 cis-Linaloloxide	6.61	12.08	11.5	2.69
5	反式-芳樟醇氧化物 trans-Linaloloxide	7.94	14.19	15.21	11.41
6	芳樟醇 Linalool	0.71	0.22	0.21	1.31
7	(E,Z)-2,6-二甲基-2,4,6-辛三烯 2,4,6-Octatriene, 2,6-dimethyl-, (E,Z)-	—	—	0.38	
8	4,6(Z),8(E)大柱三烯 Megastigma-4,6(Z),8(E)-triene	0.94	1.12	2.29	
9	7(E),9,13 大柱三烯 Megastigma-7(E),9,13-triene		0.3	0.43	
10	环氧芳樟醇 Epoxy-linalooloxide				0.63
11	α-紫罗兰醇 α-Ionol	1.37	0.7	0.31	
12	β-紫罗兰醇 β-Ionol		0.23	0.58	
13	α-紫罗兰酮 α-Ionone	18.91	8	4.65	
14	二氢-β-紫罗兰酮 Dihydro-β-ionone	11.78	7.3	5.9	7.63
15	β-紫罗兰酮 β-Ionone	27.42	35.07	42.44	1.03
醛类	Aldehyde				
16	对异丙基苯甲醛 Benzaldehyde, 4-(1-methylethyl)-	—	—	—	47.54
酯类	Ester				
17	乙酸乙酯 Ethyl Acetate	1.9	0.72	0.35	—
18	2-丁烯酸乙酯 2-Butenoic acid, ethyl ester	2.62	2.1	2.27	—
19	异戊酸乙酯 Butanoic acid, 3-methyl-, ethyl ester	0.7	0.34		

续表

序号 No.	主要成分 Main component	质量分数/% Content			
		A	B	C	D
20	(E)-2-甲基-2-丁烯酸乙酯 Ethyl tiglate	0.3	—	—	—
21	3-羟基丁酸乙酯 Butanoic acid, 3-hydroxy-, ethyl ester	—	0.24	0.19	—
22	正己酸乙酯 Hexanoic acid, ethyl ester	0.2	0.11	—	—
23	乙酸叶醇酯 3-Hexen-1-ol, acetate, (Z)-	—	—	—	1.38
24	Z-3-甲基丁酸-3-己烯酯 cis-3-Hexenyl isovalerate	—	—	—	0.87
25	γ-癸内酯 γ-Decalactone	8.14	6.17	5.14	1.3
醇类 Alcohol					
26	2,2,6-三甲基-6-乙烯基四氢-2H-呋喃-3-醇 2H-Pyran-3-ol, 6-ethenyltetrahydro-2,2,6-trimethyl-	1.62	2.26	2.46	5.28
27	1-壬醇 1-Nonanol	—	0.24	0.18	—
28	顺-α,α-5-三甲基-5-乙烯基四氢化呋喃-2-甲醇 2-Furanmethanol, 5-ethenyltetrahydro-α,α-5-trimethyl-, cis-	—	—	—	0.44
29	对甲氧基苯乙醇 2-(4-Methoxyphenyl)ethanol	—	0.6	0.16	—
30	4-(2,6,6-Trimethyl-cyclohex-1-enyl)-butan-2-ol	4.08	3.49	2.91	1.73
烷烃类 Alkane					
31	辛烷 Octane	0.84	0.16	—	—
32	2,2,4,6,6-五甲基庚烷 Heptane, 2,2,4,6,6-pentamethyl-	—	—	—	0.51
33	正十三烷 Tridecane	—	—	—	0.49
酮类 Ketone					
34	4-(2,6,6-三甲基-2-环辛烯-1-基)-3-丁烯-2-酮 4-(2,6,6-Trimethyl-2(1)-cyclohexen-1-yl)-3-buten-2-one	0.18	0.25	0.32	—
	总计 Total	98.37	99.72	99.45	98.33

注：A—香眼期 Xiangyan stage；B—初花期 initial flowering stage；C—盛花期 full flowering stage；D—末花期 final flowering stage。

'波叶金桂'在不同开花阶段香气的释放量不同(表1)。香眼期检测出含量较高的香气物质有顺式-芳樟醇氧化物、反式-芳樟醇氧化物、α-紫罗兰酮、β-紫罗兰酮、二氢-β-紫罗兰酮以及γ-癸内酯等，其中β-紫罗兰酮含量最高(27.42%)，其次为α-紫罗兰酮、二氢-β-紫罗兰酮、γ-癸内酯。初花期检测出的香气物质以顺式-β-罗勒烯、顺式-芳樟醇氧化物、反式-芳樟醇氧化物、α-紫罗兰酮、β-紫罗兰酮、二氢-β-紫罗兰酮以及γ-癸内酯等为主，以β-紫罗兰酮含量最高(35.07%)，其次是反式-芳樟醇氧化物、顺式-芳樟醇氧化物、α-紫罗兰酮、二氢-β-紫罗兰酮、γ-癸内酯，分别为14.19%、12.08%、8.00%、7.30%、6.17%。盛花期所含的香气物质以顺式-芳樟醇氧化物、反式-芳樟醇氧化物、α-紫罗兰酮、β-紫罗兰酮、二氢-β-紫罗兰酮以及γ-癸内酯等为主，也是β-紫罗兰酮含量最高(42.44%)，其次是反式-芳樟醇氧化物、顺式-芳樟醇氧化物、二氢-β-紫罗兰酮、γ-癸内酯。末花期的香气成分主要为异丙基苯甲醛、顺式-β-罗勒烯、反式-芳樟醇氧化物、二氢-β-紫罗兰酮以及2,2,6-三甲基-6-乙烯基四氢-2H-呋喃-3-醇，分别为47.54%、12.78%、11.41%、7.63%、5.28%。

同时，'波叶金桂'的香气释放量在不同开花阶段的变化情况也有一定差别：顺式-芳樟醇氧化物、反式-芳樟醇氧化物、β-紫罗兰酮等芳香物质的总体变化趋势是先上升后下降；芳樟醇、二氢-β-紫罗兰酮物质的变化趋势是先下降后上升；而乙酸乙酯、α-紫罗兰醇、α-紫罗兰酮、γ-癸内酯等芳香物质的变化趋势是持续下降；2,2,6-三甲基-6-乙烯基四氢-2H-呋喃-3-醇等的变化趋势是持续上升；2-丁烯酸乙酯的总体变化趋势不大；其中，有些挥发性的芳香成分仅在某一个阶段出现，如(E)-2-甲基-2-丁烯酸乙酯仅在香眼期检测到，(E,Z)-2,6-二甲基-2,4,6-辛三烯仅出现在盛花期，对异丙基苯甲醛、乙酸叶醇酯、5-(1-羟基-1-甲基乙基)-2-甲基-2-环己烯、环氧芳樟醇等香气物质只在末花期检测到。

2.2 '波叶金桂'开花过程中花被片细胞结构的变化

在桂花释放香气的过程中，即不同花期桂花花被片的表皮细胞结构及细胞内含物有明显的规律性变化。香眼期花被片处于生长发育阶段，细胞排列紧密，细胞主要为大液泡所占据，少量圆形或椭圆形质体沿细胞壁紧密排列，质体中可见少量淀粉粒与大小、形状各异的嗜锇酸基质颗粒(图2A、B)；初花期花被片即将展开，细胞中质体数量较多，嗜锇酸基质颗粒变为形状大小一致的圆形，并开始溢出细胞壁外，于细胞壁外成串排列(图2C、D)；盛花期花被片完全展开，随着细胞增大，质体体积增大，但质体内嗜锇酸基质颗粒密度降低，细胞壁外嗜锇酸基质颗粒的数量也开始减少(图2E、F)，部分细胞器开始解体；末花期花被片用手触摸即脱落，质体数量减少，嗜锇酸基质颗粒甚少，细胞壁外几乎不见嗜锇酸基质颗粒(图2G、H)，此外，细胞间隙出现较大的缝隙，这与扫描电镜下看到的一致(图2I~L)，开花早期细胞间隙较小，到末花期细胞间隙增大。

由此可以发现，'波叶金桂'开花过程中，嗜锇酸基质颗粒的数量、形态，以及聚集部位均不相同。香眼期花被片细胞内嗜锇酸基质颗粒刚形成，初花期嗜锇酸基质颗粒形成大小一致的圆形颗粒，开始向细胞壁外分泌，在细胞壁外成串排列；盛花期时，嗜锇酸基质圆形颗粒的中间出现白色圆点，开始分解，细胞壁外聚集的数量也开始减少；末花期花被片细胞壁外几乎没有嗜锇酸基质颗粒，推测这些嗜锇酸基质颗粒与桂花香的释放存在一定关系。

2.3 '波叶金桂'香气成分的日变化规律

将'波叶金桂'一天内不同时间段花香成分的总离子流色谱图(图3)信息经计算机检索和资料分析，且扣除空气本底杂质，得到其香气的主要成分种类。

在'波叶金桂'桂花香气日变化中，共鉴定出29种香气成分，分别属于萜烯类、酯类、醇类、烷烃类、醛类、酮类、芳香烃类七大类化合物。其中萜烯类化合物种类最多，释放量最大，种类和释放量分别占到全部的40%~58%和49%~83%。其次为酯类、醇类化合物和醇类。而烷烃类、醛类、酮类、芳香烃类化合物的种类和释放量所占比例都较小。

'波叶金桂'香气成分的种类和释放量在一天中先增加后减少再增加再减少，早晨7:00共检查出19种香气成分，总释放量为85.51%。9:00和15:00香气成分种类达到最高(25种)，11:00、12:00、17:00和19:00香气成分种类分别为23种、24种、22种、20种。10:00香气成分的总释放量最大(99.70%)，与11:00、12:00以及15:00香气成分总释放量不存在明显差异(98.20%、98.93%、99.55%)，而17:00和19:00，香气成分总释放量开始大幅下降，仅为85.56%和75.81%。可以看出，'波叶金桂'花香的释放存在明显的日变化，早晨花香的释放逐渐增加，到中午略有下降，至下午又开始回升，而到傍晚释放出现大幅降低。

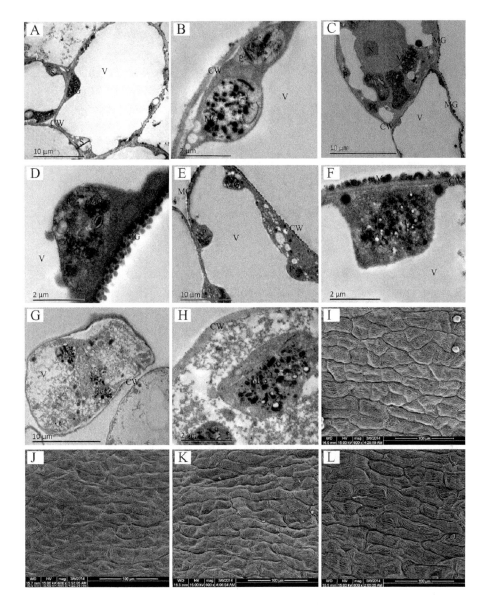

图2　不同花期桂花花被片超微结构的变化

Fig. 2　The variation in perianth microstructure during the blooming of *O. fragrans*

注:图 A~H 示透射电镜图片,图 A、B 示香眼期,图 C、D 示初花期,图 E、F 示盛花期,图 G、H 示末花期;图 I~L 示扫描电镜图片,图 I 示香眼期,图 G 示初花期,图 H 示盛花期,图 L 示末花期。细胞壁;细胞核;质体;线粒体;淀粉粒;嗜锇酸基质颗粒。Fig. A~H Transmission electron microscopy photograph:Fig. A、B Xiangyan stage, Fig. C、D initial flowering stage, Fig. E、F full flowering stage, Fig. G、H final flowering stage;Fig. I~L electronic scanning microscope photograph:Fig. I Xiangyan stage, Fig. G initial flowering stage, Fig. H full flowering stage, Fig. L final flowering stage. Cell wall(CW); Cell nucleus(N); Plastid(P); Mitochondria(M); Starch grain(S); Matrix granule(MG).

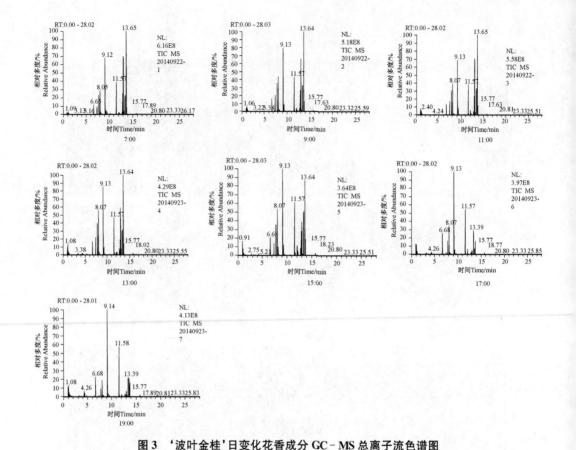

图 3 '波叶金桂'日变化花香成分 GC-MS 总离子流色谱图
Fig. 3 Diurnal variation of total Ion chromatograms of *O. fragrans* 'Boye Jingui'

在'波叶金桂'一天不同时间段释放的香气成分中,不同时段香气成分的种类差异比较大,如正己醛、2-己烯醛和正己醇仅存在于 17:00 和 19:00 这 2 个时段,其他时段没有检测到;而有些成分在其他时段中存在,在 17:00 和 19:00,特别是 19:00 却没检测到。同时,从表 2 中可以看出,顺式-芳樟醇氧化物、反式-芳樟醇氧化物、β-紫罗兰醇、α-紫罗兰酮、二氢-β-紫罗兰酮、β-紫罗兰酮、γ-癸内酯和 2,2,6-三甲基-6-乙烯基四氢-2H-呋喃-3-醇等化合物释放量较高。这些成分的释放量均表现出明显的日变化:从早晨开始释放量增加,到中午气温升高,光照变强,释放量略有下降,至下午 15:00 开始,释放量开始回升,到傍晚释放量大幅下降,但不同花香成分的变幅及达到最高值的时间存在差异(图 4)。其中二氢-β-紫罗兰酮和 β-紫罗兰酮释放量最高,分别从 7:00 的 16.39% 和 28.18%,增加到 9:00 的 17.60% 和 29.15%,然后到 11:00 释放量略有下降,随着光照强度的增加,13:00 释放量减小为 10.05% 和 22.88%,而 15:00 释放量又开始增加(11.2% 和 23.4%),至 17:00 和 19:00,释放量持续减少;反式-β-罗勒烯的释放量也表现出相似的变化规律;顺式-芳樟醇氧化物与反式-芳樟醇氧化物的释放量分别从早晨 7:00 的 6.33 和 7.92,至 9:00 增加到 7.31 和 10.17,而后 11:00 略有下降,13:00~17:00 变化不明显,变幅较小,19:00 急剧下降;β-紫罗兰醇和 2,2,6-三甲基-6-乙烯基四氢-2H-呋喃-3-醇的释放量从早晨逐渐增加(仅中午 11:00 略有减少),到 15:00 达最大值(2.24% 和 12.46%),17:00~19:00 开始下降;α-紫罗兰酮的释放量从 7:00~11:00 都较稳定,13:00 开始下降,到 17:00~19:00 急剧下降。

表 2 '波叶金桂'香气成分种类与释放量的日变化
Table 2 Diurnal variation of aroma compositions of *O. fragrans* 'Boye Jingui'

序号 No.	主要成分 Main component	质量分数/% Content						
		7:00	9:00	11:00	13:00	15:00	17:00	19:00
萜烯类 Terpenoids								
1	反式-β-罗勒烯 *trans*-β-Ocimene	2.49	3.57	3.29	2.9	4.05	1.56	1.58
2	顺式-β-罗勒烯 *cis*-β-Ocimene	—	0.11	0.14	0.24	0.33	—	—
3	顺式-芳樟醇氧化物 *cis*-Linalool oxide	6.33	7.31	7.03	8.27	8.8	6.03	3.75
4	反式-芳樟醇氧化物 *trans*-Linalool oxide	7.92	10.17	10.04	12.19	12.4	12.93	6.35
5	芳樟醇 Linalool	0.66	0.38	2.23	2.11	0.78	1.56	0.82
6	7(E),9,13 大柱三烯 Megastigma-7(E),9,13-triene	0.31	0.12	0.13	0.17	0.32	—	—
7	4,6(E),8(Z)大柱三烯 Megastigma-4,6(E),8(Z)-triene	0.59	0.08	0.10	0.25	0.22	—	—
8	α-紫罗兰醇 α-Ionol	0.51	0.43	0.47	1.29	0.77	0.55	—
9	β-紫罗兰醇 β-Ionol	0.77	1.51	1.02	1.86	2.24	0.70	1.32
10	α-紫罗兰酮 α-Ionone	6.49	6.07	6.70	5.96	4.59	2.05	2.37
11	二氢-β-紫罗兰酮 Dihydro-β-ionone	16.39	17.6	16.58	10.05	11.2	4.19	4.48
12	β-紫罗兰酮 β-Ionone	28.18	29.15	28.31	22.88	23.4	15.77	16.64
醛类 Aldehyde								
13	正己醛 Hexanal	—	—	—	—	—	6.96	8.23
14	2-己烯醛 2-Hexenal	—	—	—	—	—	3.04	4.62
酯类 Ester								
15	乙酸乙酯 Ethyl Acetate	0.91	1.73	1.59	1.56	1.66	—	—
16	2-丁烯酸乙酯 2-Butenoic acid, ethyl ester	—	0.63	—	1.89	2.04	2.87	—
17	3-羟基丁酸乙酯 Butanoic acid, 3-hydroxy-, ethyl ester	—	0.14	0.17	0.78	0.44	1.68	1.51

续表

序号 No.	主要成分 Main component	质量分数/% Content						
		7:00	9:00	11:00	13:00	15:00	17:00	19:00
18	乙-3-甲基丁酸-3-己烯酯 cis-3-Hexenyl isovalerate	0.49	—	0.23	—	—	—	—
19	γ-癸内酯 γ-Decalactone	7.24	9.52	9.13	10.19	12.46	11.76	9.03
醇类 Alcohol								
20	正己醇 1-Hexanol	—	—	—	—	—	2.94	4.61
21	1-壬醇 1-Nonanol	—	0.61	0.56	0.53	0.66	—	—
22	2,2,6-三甲基-6-乙烯基四氢-2H-吡喃-3-醇 2H-Pyran-3-ol, 6-ethenyltetrahydro-2,2,6-trimethyl-	1.88	3.15	3.13	4.49	3.74	4.18	3.08
23	对甲氧基苯乙醇 2-(4-Methoxyphenyl)ethanol	0.64	1.78	2.54	3.27	0.66	2.65	2.05
24	4-(2,6,6-Trimethyl-cyclohex-1-enyl)-butan-2-ol	2.95	4.03	3.38	6.48	6.74	1.43	1.13
烷烃类 Alkane								
25	二乙氧基甲烷 Methane, diethoxy-	0.27	0.68	0.67	0.87	0.92	0.74	0.43
酮类 Ketone								
26	2,2,6-三甲基-6-乙烯基四氢-2H-吡喃-3(4H)-酮 2H-Pyran-3(4H)-one, 6-ethenyldihydro-2,2,6-trimethyl-	0.49	0.44	0.58	0.42	0.32	0.75	0.62
27	4-(2,6,6-三甲基-2-环辛烯-1-基)-3-丁烯-2-酮 4-(2,6,6-Trimethyl-2(1)-cyclohexen-1-yl)-3-buten-2-one	—	0.24	0.21	0.13	0.38	—	—
芳香烃类 Arenes								
28	甲苯 Toluene	—	0.08	—	—	0.23	0.56	1.53
29	乙苯 Ethyl benzene	—	0.12	—	0.15	0.30	0.66	1.66
总计 Total		85.51	99.70	98.20	98.93	99.55	85.56	75.81

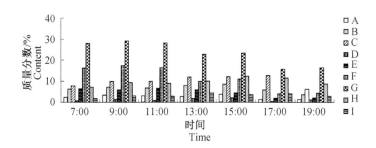

图4 '波叶金桂'主要花香成分释放的日变化

Fig. 4 Diurnal variation of major aroma compositions of *O. fragrans* 'Boye Jingui'

A. 反式-β-罗勒烯 *trans*-β-Ocimene；B. 顺式-芳樟醇氧化物 *cis*-Linaloloxide；C. 反式-芳樟醇氧化物 *trans*-Linaloloxide；D. β-紫罗兰醇 β-Ionol；E. α-紫罗兰酮 α-Ionone；F. 二氢-β-紫罗兰酮 Dihydro-β-ionone；G. β-紫罗兰酮 β-Ionone；H. γ-癸内酯 γ-Decalactone；I. 2,2,6-三甲基-6-乙烯基四氢-2H-呋喃-3-醇 2H-Pyran-3-ol, 6-ethenyltetrahydro-2,2,6-trimethyl-。

3 结论与讨论

多种芳香植物开花过程中香气成分的研究表明：随着植物花朵的开放和衰败，植物的花香释放会发生变化[13-14]，如玉簪花[15]的花香释放于盛花期达到顶峰，而后随着玉簪花的开放开始下降。本试验中，不同花期'波叶金桂'的花香成分的种类及其释放量随花期均表现出先增大后减小的规律，初花期达到最大。其中萜烯类化合物的种类和释放量最高，主要包括芳樟醇氧化物、α-紫罗兰酮、二氢-β-紫罗兰酮、β-紫罗兰酮等，是主要的花香成分。同时，试验发现，嗜锇酸基质颗粒在花被片开展的不同花期，数量、形态及聚集部位均不相同。香眼期花被片细胞内嗜锇酸基质颗粒刚形成，大小、形状各异，主要聚集在质体中；初花期嗜锇酸基质颗粒形成大小一致的圆形颗粒，并开始向细胞壁外分泌，在细胞壁外成串排列；盛花期时，嗜锇酸基质圆形颗粒的中间出现白色圆点，开始分解，细胞壁外聚集的数量也开始减少；末花期花被片细胞壁外几乎没有嗜锇酸基质颗粒，仅在细胞质体内少量存在。由此推测这些嗜锇酸基质颗粒与桂花香气的形成存在一定关系，是桂花释放香气的物质基础。这也与GC-MS分析的结果一致，初花期香气释放最大，盛花期开始下降，末花期急剧减少。有研究表明[16]，有些植物中的芳香气味来源于一种特殊腺体，称为芳香腺，这种芳香腺通常具有几层细胞厚的分泌组织，或在细胞内呈现为小滴状，但都会在较短时间内挥发掉。桂花细胞内嗜锇酸基质颗粒具备上述芳香腺的特征，认为桂花花被片分泌的嗜锇酸基质颗粒结构属于芳香腺，这与以往的研究结论相似[17]。但此次试验中没发现表皮细胞外壁外向外突起的表皮毛结构，推测是由于细胞壁外密集的嗜锇酸基质颗粒将毛粘合在一起，从而没有观察到。

此外，'波叶金桂'花香的释放表现出明显的日变化规律：清晨与傍晚花香成分少，释放量低，而在上午9:00与下午15:00香气成分的种类和释放量出现高峰，中午11:00和13:00香气成分的种类和释放量也略有减小。以往的研究表明许多植物的花香释放也都存在日变化，如'西伯利亚'百合[18]的花香释放呈现明显的日变化：清晨花香成分少，释放量低，随着光强的增加，百合花香的释放量表现出升高的趋势，而到了傍晚花香的释放明显减少。

综述上述,'波叶金桂'花香的释放随花朵的开放进程以及一天中不同的时间点,表现出规律性的变化,但引起变化的影响因素具有多样性,包括发育程度、花色、授粉者,以及地域分布和外界环境等。近年来相关的研究很多,比如孔滢等[19]研究发现8种常见的百合品种('Tresor''Ceb Dazzle''White Heaven''Marco Polo''Siberia''Sorbonne''Conca d'Or''Yelloween'),以及淡黄花百合、"Casa Blanca"百合和宜昌百合[20]的花香夜间释放比白天高,水仙[21]的花香也呈现明显的夜间挥发节律。而夹竹桃[22]的花香与其花色等因素有关,烟草[23]的花香与其授粉者等因素有关。此次试验中,'波叶金桂'花朵开放发育程度是影响花香释放的重要因素,外界环境也是引起'波叶金桂'花香释放日变化的主要因素。然而,此次试验是在室外采样(天气晴朗无风),光照、温度、水分等外界环境条件无法控制,且采样仅至20:00,'波叶金桂'花香释放是否存在夜间挥发节律,有待于进一步研究。

参考文献

[1] 蒋冬月,李永红,何昉,等.黄兰开花过程中挥发性有机成分及变化规律[J].中国农业科学,2012,45(6):1215-1225.
   JIANG D Y, LI Y H, HE F, et al. The components and changes of VOCs of *Michelia champaca* L. flower at different developmental stages[J]. Scientia Agricultura Sinica, 2012,45(6):1215-1225.

[2] LI Z G, LEE M R, SHEN D L. Analysis of volatile compounds emitted from fresh *Syringa oblate* flowers in different florescence by headspace solid-phase microextraction-gas chromatography-mass spectrometry[J]. Science Direct, 2006, 576: 43-49.

[3] 赵印泉,潘会堂,张启翔,等.梅花花朵香气成分时空动态变化的研究[J].北京林业大学学报,2010,32(4):201-206.
   ZHAO Y Q, PAN H T, ZHANG Q X, et al. Dynamics of fragrant compounds from *Prunus mume* flowers[J]. Journal of Beijing Forestry University, 2010,32(4):201-206.

[4] 窦雅君,翟娟,侯芳梅,等.不同光照强度对'金盏银台'水仙花香释放的影响[J].西北农业学报,2014,3(4):85-91.
   DOU Y J, ZHAI J, HOU F M, et al. Effect of different light intensities on the floral aroma emitted from Chinese daffodil (*Narcissus tazetta* L. var. *Chinensis Roem*)[J]. Acta Agriculturae Boreali-occidentalis Sinica, 2014, 3(4):85-91.

[5] 李海东,高岩,金幼菊.珍珠梅花挥发性物质日动态变化的研究[J].内蒙古农业大学学报(自然科学版),2004,25(2):54-59.
   LI H D, GAO Y, JIN Y J. The daily dymamic variances of the VOCs releasing from flower of *Siberia kirilowii* (Regel) Maxim[J]. Journal of Inner Mongolia Agricultural University (Natural Science Edition), 2004,25(2):54-59.

[6] LI S S, CHEN L G, XU Y J, et al. Identification of floral fragrances in tree peony cultivars by gas chromatography-mass spectrometry[J]. Scientia Horticulturae, 2012,142:158-265.

[7] GONG W C, CHEN G, LIU C Q, et al. Comparison of floral scent between and within *Buddleja fallowiana* and *Buddleja officinalis* (Scrophulariaceae)[J]. Biochemical Systematics and Ecology, 2014, 55: 322-328.

[8] 孙宝军,李黎,韩远记,等.上海桂林公园桂花芳香成分的HS-SPME-GC-MS分析[J].福建林学院学报,2012,32(1):39-42.
   SUN B J, LI L, HAN Y J, et al. HS-SPME-GC-MS analysis of different *Osmanthus fragrans*

cultivars from Guilin Garden in Shanghai[J]. Journal of Fujian College of Forestry,2012,32(1):39-42.

[9] 施婷婷,杨秀莲,王良桂. 桂花花朵香气成分的研究进展[J]. 化学与生物工程,2014,31(10):1-5.
SHI T T,YANG X L,WANG L G. Research progress of flavoring compositions in *Osmanthus fragrans* Lour. flower[J]. Chemistry and Bioengineering,2014,31(10):1-5.

[10] 杨康民,朱文江. 桂花[M]. 上海:上海科学技术出版社,2000.
YANG K M,ZHU W J. *Osmanthus fragrans*[M]. Shanghai:Shanghai Science and Technology Press,2000.

[11] 郭素枝. 电子显微镜技术与应用[M]. 厦门:厦门大学出版社,2008.
GUO S Z. Electron microscope technology and application[M]. Xiamen:Xiamen University Press,2008.

[12] 单晓丹. 气质联用技术在挥发性有机物分析中的初步应用[J]. 现代农业,2009,9:108-112.
SHANG X D. The primary application in analysis of volatile organic compounds by GC-MS[J]. Modern Agriculture,2009,9:108-112.

[13] 张莹,李辛雷,王雁,等. 文心兰不同花期及花朵不同部位香气成分的变化[J]. 中国农业科学,2011,44(1):110-117.
ZHANG Y,LI X L,WANG Y,et al. Changes of aroma components in *Oncidium sharry* baby in different florescence and flower parts[J]. Scientia Agricultura Sinica,2011,44(1):110-117.

[14] 周继荣,倪德江. 蜡梅不同品种和花期香气变化及其花茶适制性[J]. 园艺学报,2010,37(10):1621-1628.
ZHOU J R,NI D J. Changes in flower aroma compounds of cultivars of *Chimonanthus praecox*(L.)link and at different stages relative to *Chimonanthus* tea quality[J]. Acta Horticulturae Sinica,2010,37(10):1621-1628.

[15] LIU Q,SUN G F,WANG S,*et al*. Analysis of the variation in scent components of *Hosta* flowers by HS-SPME and GC-MS[J]. Scientia Horticulturae,2014,175:57-67.

[16] CAI X,ZHOU Y F,HU Z H. Ultrastructure and secretion of secretory canals in vegetative organs of bupleurum chinense DC[J]. Journal of Molecular Cell Biology,2008,41(2):96-106.

[17] 王丽梅. 桂花有效成分合成转化规律与药学研究[D]. 武汉:华中科技大学,2009.
WANG L M. The synthesis regularity of active components and pharmaceutical research of *Osmanthus fragrans* Lour[D]. Wuhan:Huazhong University of Science and Technology,2009.

[18] 张辉秀,冷平生,胡增辉,等. '西伯利亚'百合花香随开花进程变化及日变化规律[J]. 园艺学报,2013,40(4):693-702.
ZHANG H X,LENG P S,HU Z H,*et al*. The floral scent emitted from *Lilium*'Siberia' at different flowering Stages and diurnal variation[J]. Acta Horticulturae Sinica,2013,40(4):693-702.

[19] KONG Y,SUN M,PAN H T,*et al*. Composition and emission rhythm of floral scent volatiles from eight Lily cut flowers[J]. Journal of the American society for horticultural science,2012,137(6):376-382.

[20] KONG Y,SUN M,PAN H T,*et al*. Floral scent composition of *Lilium sulphureum*[J]. Chemistry of Natural Compounds,2013,49:362-364.

[21] RUíZ-RAMÓN F,ÁGUILA D J,EGEA-CORTINES M,*et al*. Optimization of fragrance extraction:Daytime and flower age affectscent emission in simple and double narcissi[J]. Industrial Crops and Products,2014,52:671-678.

[22] MAJETIC C J,LEVIN D A,Raguso R A. Divergence in floral scent profiles among and within

cultivatedspecies of Phlox[J]. Scientia Horticulturae,2014,172:285-291.

[23] RAGUSO R A, LEVIN R A, FOOSE S E, et al. Fragrance chemistry, nocturnal rhythms and pollination'Syndromes' in *Nicotiana*[J]. Phytochemistry,2003,63:265-284.

Study on the Aroma Component Emission Pattern of *Osmanthus fragrans* 'Boye Jingui'

SHI Tingting, Yang Xiulian, WANG Lianggui

(College of Landscape Architecture, Nanjing Forest University, Nanjing 210037, Jiangsu, China)

Abstract: The objective was to Clarify the aroma emission pattern of the *Osmanthus fragrans* 'Boye Jingui' in order to provide theoretical basis for the comprehensive development and utilization of *Osmanthus fragrans*. The aromatic components of 'Boye Jingui' at different flowering stages and different time points of one day was analyzed by using the method of headspace solid-phase micro-extraction (HS-SPME) and gas chromatography-mass spectrometry (GC-MS). And the variation in perianth microstructure during the blooming of *O. fragrans* was observed by electronic scanning microscope and transmission electron microscopy. In the 'Boye Jingui' flowering stages, the aroma releasing amount was lowest in the Xiangyan stage, but rose in the initial flowering stage and the full flowering stage, while arriving at the peak in the initial flowering stage, and then dropped rapidly in the final flowering stage. Meanwhile, the eosinophil matrix granule in the perianth cell of *O. fragrans* varied in morphology, number and location during the flowering period. The released amount of floral scent also showed the pattern of first increase and then decrease from dawn to dusk during one day, arriving at the peak at 9:00 and 15:00 and significantly decreasing at 7:00 and 19:00. At the 'Boye Jingui' flowering stages, for the released aroma amount was highest at initial flowering stage, it was speculated that the light intensity played a key role in the diurnal variation of floral scent amount, and that the eosinophil matrix granule could be an aroma releasing predecessor of *O. fragrans*.

Key words: *Osmanthus fragrans* 'Boye Jingui'; aroma component; ultrastructure; headspace solid-phase micro-extraction (HS-SPME); gas chromatography-mass spectrometry (GC-MS)

3个四季桂品种花瓣挥发性成分的 GC-MS 分析*

杨秀莲,施婷婷,文爱林,王良桂

(南京林业大学 风景园林学院,江苏 南京 210037)

摘 要：采用顶空固相微萃取结合 GC-MS 技术对 3 个四季桂品种盛花期花瓣的挥发性化学成分进行了分析,并比较了鲜花瓣与短期超低温冰箱保存后的花瓣挥发性成分的差异。结果表明：不同四季桂品种的香味主体成分比较接近,主要成分为 β-紫罗兰酮、二氢-β-紫罗兰酮、顺式-芳樟醇氧化物(呋喃型)、反式-芳樟醇氧化物(呋喃型)、γ-癸内酯、α-紫罗兰酮、正己醛、叶醛、顺式-芳樟醇氧化物(吡喃型)、芳樟醇、反式-芳樟醇氧化物(吡喃型)、顺-3-己烯醛。短时间的超低温保存对花瓣挥发性成分和含量影响不大,因此,在测定样品较多时可以在超低温冰箱中短期保存。

关键词：四季桂；挥发性成分；顶空固相微萃取；气相色谱质谱法

  花香是花卉的主要性状,是衡量花卉质量的重要内容,对提高花卉的观赏价值和经济价值有重要的意义。近年来,对花香的研究逐渐增多,很多学者对梅花[1]、蜡梅[2]、兰花[3-4]、百合[5-6]等植物的花香,以及花香成分的应用[7-8]都进行了研究。桂花(*Osmanthus fragrans* Lour.)为木犀科(Oleaceae)木犀属(*Osmanthus*)的常绿芳香花木,是一种享誉我国古今,集绿化、美化、香化于一体,观赏和实用兼备的优良园林景观树种[9-10]。栽培历史已达 2 500 多年,至今已有 160 多个栽培品种,分属于四季桂、银桂、金桂、丹桂 4 个品种群,且不同品种群的花香差异明显,四季桂和丹桂是淡香,银桂比四季桂更香,而金桂最为香浓[11-13]。桂花的香有别于兰花的幽香,梅花的淡香,水仙的清香,既浓郁,又有些香甜,被古人赞"清可绝尘,浓能溢远"的仙香,广泛应用于食品加工、医药保健、化妆品等行业。天然桂花香气异常,至今不能人工合成。近年来,对桂花花香成分的研究较多,从桂花挥发油成分的提取方法到花香成分和含量的分析等；同时,对花香成分的分析主要集中在桂花浸膏和精油的成分上,由于浸膏或精油的原料均是不同品种花瓣的混合物,且在制作过程中芳香成分的损失较多,不能真实反映某品种应有的香气和成分[14-17]。也有人采用顶空固相微萃取(HS-SPME)结合 GC-MS 技术对金桂、银桂和丹桂的活体植物头香成分进行了分析,但大部分均未涉及具体品种。由于不同地区不同品种的桂花花香成分差异很大,本研究比较 3 个四季桂品种鲜花和超低温冷冻花花瓣挥发性成分和释放量的差异,以确定桂花四季桂品种致香的关键成分和短时间低温保存对挥发性成分释放的影响,为进一步开展桂花花香研究提供指导。

---

* 原文载于《中南林业科技大学学报》,2015,35(10):127-133。
基金项目：国家林业局公益性行业专项(201204607)；"十二五"科技支撑"林木种质资源发掘与创新利用"(2013BA001B06)；江苏省高校自然科学基础研究项目(11KJB220002)；江苏高校优势学科建设工程资助项目。

1 材料与方法

1.1 试验材料及仪器

2009年9月底,在桂花开花期采集南京林业大学校园内的四季桂品种的花朵,品种包括'四季桂'('Sijigui')、'日香桂'('Rixiang Gui')、'大叶佛顶珠'('Daye Fodingzhu')。

试验仪器:手动固相微萃取进样器(美国SUPELCO公司),65 μm PDMS/DVB萃取头(美国SUPELCO公司),气相色谱质谱联用仪Trace DSQ(美国Thermo Electro-Finnigan公司),4 mL顶空取样瓶(美国SUPELCO公司)、水浴锅。

1.2 取样和测试方法

于桂花花朵开放的第一天,在上午露水褪尽后,每品种采集25朵花放入4 mL采样瓶内,其中'日香桂'取2份样品(一份用于测定鲜花花香成分,一份用于测定超低温-73 ℃保存一周后的冷冻花花香成分),密封瓶盖,在23 ℃左右室温下平衡10 min。将经老化(老化温度为250 ℃,老化时间为30 min)的SPME纤维头通过聚四氟乙烯隔垫插入放在50 ℃水浴的采样瓶中,萃取头置于花朵上方1~2 cm,吸附30 min,然后将萃取头插入GC进样口,解吸5 min,以后每次进样活化5 min。

色谱条件:TR-5MS毛细管色谱柱,柱长30 m,内径0.25 mm,膜厚0.25 μm,载气为高纯度氦气(He),氦气流速为1 mL/min,分流比为10∶1,进样口温度250 ℃。

升温程序:起始温度为40 ℃,保持2 min,以2 ℃/min升至60 ℃,以5 ℃/min升至100 ℃,以10 ℃/min升至250 ℃,保持5 min。

质谱条件:接口温度250 ℃,电离方式EI,离子电离能量70 eV,质量扫描范围(m/z)50~450 amu。

1.3 数据处理

所得谱图直接由该机数据处理系统进行检索,并查阅有关资料进行香气成分定性分析。桂花香味组分的相对含量根据总离子流色谱峰的峰面积归一化法计算。

2 结果与分析

2.1 不同保存方法对'日香桂'香气成分的影响

'日香桂'品种鲜花和超低温保存一周后的花朵香味成分的总离子色谱图见图1、图2。从图中可看出两者离子流色谱图的峰型契合度很高,个别物质保留时间略有不同,但重现性很好。使用计算机质谱数据库NBS进行检索,按峰面积归一化法计算得出各成分相对百分含量(表1)。从鲜花中分检出27种化合物,占总成分的97.92%,从超低温保存一周后的花朵中分检出25种化合物,占总成分的97.13%。在新鲜花瓣中检测出了0.05%的月桂烯和0.14%的柠檬醛,而在超低温保存的花瓣中未检测到。新鲜和冷冻花瓣的主要成分相同,均为β-紫罗兰酮、二氢-β-紫罗兰酮、顺式-芳樟醇氧化物(呋喃型)、反式-芳樟醇氧化物(呋喃型)、γ-癸内酯、α-紫罗兰酮、正己醛、叶醛、顺式-芳樟醇氧化物(吡喃型)、芳樟醇、反式-芳樟醇氧化物(吡喃型)等萜烯类和醛类化合物,共占到芳香物质总含量的90%以上。由此可见,短时间的超低温保存对花香成分的影响不大。

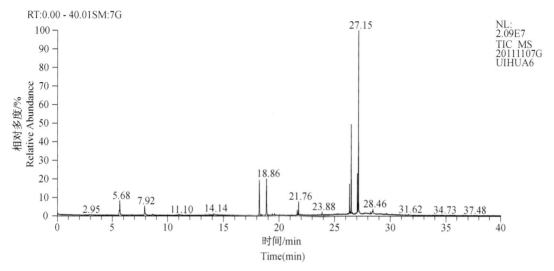

图 1　新鲜'日香桂'香味成分总离子图
Fig. 1　Total inonic current chromatogram of volantile components from the fresh flowers of *Osmanthus fragrans* 'Rixiang Gui'

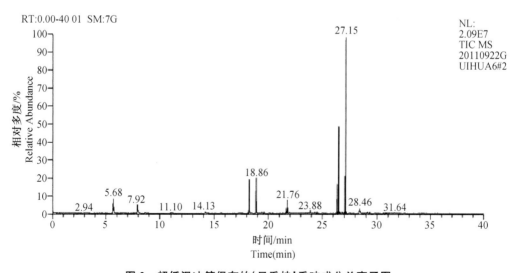

图 2　超低温冰箱保存的'日香桂'香味成分总离子图
Fig. 2　Total inonic current chromatogram of volantile components from the flowers of *Osmanthus fragrans* 'Rixiang Gui' preserved by ultra low temperature refrigerator

2.2　3个四季桂品种鲜花挥发性成分分析

3个四季桂花品种中共检测出37种挥发性化合物,可分为6大类,分别为萜烯类21种、醛类6种、酯类3种、醇类3种、烷烃类3种和酮类1种(表1)。品种间香气成分的种类及数量有明显差异,'四季桂'花香挥发成分中,萜烯类化合物种类最多(13种),醛类(5种)和烷烃类(3种)次之,醇类、酯类与酮类较少;'日香桂'花香挥发成分中萜烯类化合物种类最多(16种),醛类(5种)和酯类(3种)次之,醇类、酮类较少,没有检测到烷烃类化合物的释放;'大叶佛顶珠'花香挥发成分中也是萜烯类化合物种类最多(14种),醛类(6种)次之,醇类、酯类与烷烃类较少,且没有检测到酮类化合物的释放。对6大类花香挥发物进一步分析可以看出,3个四季桂品种香气成分中均以萜烯类化合物的相对含量最高,平均相对含量达76.56%,尤其是'日香桂'与'大叶佛顶珠'高达80%以上,'四季桂'相对较低,为68.16%。其次是醛类化合物,平均相对含量达15.22%,其中'日香桂'品种含量较低(9.65%)。而3个品种中酯类、醇类、烷烃类和酮类化合物的相对含量较低,均小于10%。

由表1可知,不同品种四季桂香气成分的种类及其相对含量也有一定差异,'四季桂'品种中共检测到花香挥发物种类26种,主体为顺-3-己烯醛、正己醛、叶醛、γ-2-己烯内酯、反式-芳樟醇氧化物(呋喃型)、顺式-芳樟醇氧化物(呋喃型)、芳樟醇、α-紫罗兰酮、二氢-β-紫罗兰酮、γ-癸内酯、β-紫罗兰酮,占香味成分总量的93.50%,其中紫罗兰酮类化合物占了将近60%的比例;而芳樟醇类化合物含量很低。从'日香桂'中共检测到27种花香挥发物,主体成分为正己醛、叶醛、顺式-芳樟醇氧化物(呋喃型)、反式-芳樟醇氧化物(呋喃型)、芳樟醇、反式-芳樟醇氧化物(吡喃型)、顺式-芳樟醇氧化物(吡喃型)、α-紫罗兰酮、二氢-β-紫罗兰酮、β-紫罗兰酮、γ-癸内酯,占香味成分总量的93.81%。其中紫罗兰酮类和芳樟醇类化合物含量都很高。在'日香桂'中检测到少量的罗勒烯、香叶醇、橙花醇、柠檬醛等物质,虽然含量少,但可能对'日香桂'的香味特性有重要的影响。'大叶佛顶珠'中仅检测到22种挥发性成分,主体成分顺-3-己烯醛、正己醛、叶醛、己醇、γ-2-己烯内酯、反式-芳樟醇氧化物(呋喃型)、顺式-芳樟醇氧化物(呋喃型)、芳樟醇、顺式-芳樟醇氧化物(吡喃型)、α-紫罗兰酮、二氢-β-紫罗兰酮、γ-癸内酯、β-紫罗兰酮,占到了总香味的93.80%。检测到其他2个品种中没有的α-松油醇、薄荷醇、4-酮基-β-紫罗兰酮等物质。

表1 3个四季桂品种花瓣香气成分比较
Tab. 1 Comparison of the main ingredients in O. fragrans

序号 No.	主要成分 Constituent	保留时间/min Retention time	质量分数/% Content			
			A	B	C	D
萜烯类 Terpene						
1	月桂烯 á-Myrcene	14.69	—	0.05	0.14	—
2	柠檬烯 Limonene	15.48	0.14	0.19	0.06	—
3	3-蒈烯 3-Carene	17.21	—	—	—	0.87
4	反式-罗勒烯 trans-Ocimene	17.23	0.37	0.12	0.04	0.07
5	反式-香叶醇 trans-Geraniol	17.83	0.07	0.15	—	0.09
6	反式-芳樟醇氧化物(呋喃型) trans-Linaloloxide(furan)	18.23	10.79	10.03	2.12	7.3
7	顺式-芳樟醇氧化物(呋喃型) cis-Linaloloxide(furan)	18.87	11.16	10.16	1.83	9.96
8	芳樟醇 Linalool	19.38	2.15	2.45	1.25	1
9	α-松油醇 α-Terpineol	19.43	—	—	—	0.06

续 表

序号 No.	主要成分 Constituent	保留时间/min Retention time	质量分数/% Content A	B	C	D
10	反式-芳樟醇氧化物(吡喃型) trans-Linaloloxide(pyran)	21.64	1.75	1.13	0.28	0.88
11	顺式-芳樟醇氧化物(吡喃型) cis-Linaloloxide(pyran)	21.77	3.16	2.86	0.58	1.55
12	橙花醇　Nerol	23.38	0.32	0.36	—	—
13	柠檬醛　Citral	23.79	—	0.14	—	—
14	薄荷醇　Mentho	24.22	—	—	—	0.23
15	α-紫罗兰醇　α-Ionol	25.59	—	—	0.15	—
16	β-紫罗兰醇　β-Ionol	26.19	0.08	0.16	—	—
17	α-紫罗兰酮　α-Ionone	26.33	5.17	5.41	5.92	6.17
18	二氢-β-紫罗兰酮　Dihydro-β-ionone	26.49	14.78	15.63	30.55	27.63
19	二氢-β-紫罗兰醇　Dihydr-β-ionol	26.62	0.44	0.44	0.56	—
20	β-紫罗兰酮　β-Ionone	27.15	28.34	30.56	22.87	21.84
21	4-酮基-β-紫罗兰酮　4-Oxo-β-ionone	27.28	—	—	—	0.14
	合计　Subtotal		78.72	79.84	66.35	77.79
醛类 Aldehyde						
22	顺-3-己烯醛　3-Hexenal,(Z)-	5.62	0.59	0.33	7.47	2.06
23	正己醛　Hexanal	5.69	4.67	5.22	5.22	4.2
24	叶醛　2-Hexenal,(E)-	7.94	3.21	3.6	6.82	7.75
25	苯甲醛　Benzaldehyde	13.75	0.16	0.09	0.21	0.17
26	辛醛　Octanal	17.35	—	—	—	0.23
27	壬醛　Nonanal	18.47	0.44	0.21	0.53	0.37
	合计　Subtotal		9.07	9.45	20.25	14.78
酯类 Ester						
28	γ-2-己烯内酯 2(5H)-Furanone,5-ethyl-	14.15	0.73	0.67	1.42	1.3
29	反式-丁酸-3-己烯酯 Butanoic acid,3-hexenylester,(E)-	24.95	0.28	0.16	—	—
30	γ-癸内酯　γ-Decalactone	27.05	7.18	6.78	8.03	1.77
	合计　Subtotal		8.19	7.61	9.45	3.07
醇类 Alcohol						
31	丁醇　1-Butanol	3.43	0.21	0.17	0.13	—
32	己醇　1-Hexanol	8.64	0.26	0.41	—	1.27
33	壬醇　1-Nonanol	18.56	—	—	0.08	—
	合计　Subtotal		0.47	0.58	0.21	1.27

续 表

序号 No.	主要成分 Constituent	保留时间/min Retention time	A	B	C	D
烷烃类 Alkane						
34	辛烷 1-octane	9.13	—	—	0.27	0.35
35	2,5-二甲基己烷 2,5-dimethylhexane	11.36	—	—	0.09	—
36	甲基环己烷 Cyclohexane, methyl-	22.83	—	—	0.04	—
	合计 Subtotal		0	0	0.4	0.35
酮类 Ketone						
37	2,2,6-三甲基-6 乙烯基-2H-四氢吡喃-3(4H)-酮 2H-Pyran-3(4H)-one, 6-ethenyldihydro-2,2,6-trimethyl-	19.59	0.68	0.44	0.69	—
	合计 Subtotal		0.68	0.44	0.69	0
	总计 Total		97.13	97.92	97.35	97.26

注释：A—超低温冰箱保存的'日香桂'花瓣；B—新鲜'日香桂'花瓣；C—新鲜'四季桂'花瓣；D—新鲜'大叶佛顶珠'花瓣。

Note: A—the flowers preserved by ultra low temperature refrigerator of 'Rixiang Gui'; B—the fresh flowers of 'Rixiang Gui'; C—the fresh flowers of 'Sijigui'; D—the fresh flowers of 'Daye Fodingzhu'.

比较 3 个四季桂品种的花香挥发性成分可以得知，正己醛、叶醛、芳樟醇、α-紫罗兰酮、二氢-β-紫罗兰酮、β-紫罗兰酮等为其共有成分，其中含量较高的为β-紫罗兰酮、二氢-β-紫罗兰酮、顺式-芳樟醇氧化物（呋喃型）、反式-芳樟醇氧化物（呋喃型）、γ-癸内酯、α-紫罗兰酮、正己醛、叶醛、顺式-芳樟醇氧化物（吡喃型）、芳樟醇、反式-芳樟醇氧化物（吡喃型）、顺-3-己烯醛、乙醇、γ-2-己烯内酯。图 3 对这 14 种主体成分的质量分数进行了比较。

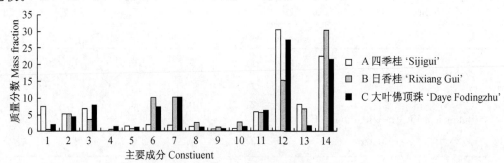

图 3 3 个四季桂品种主要成分含量对比

Fig. 3 Comparison figure of the main ingredients in three cultivars of *O. frgrans* Asiaticus Group

1：顺-3-己烯醛。2：正己醛。3：叶醛。4：己醇。5：γ-2-己烯内酯。6：反式-芳樟醇氧化物（呋喃型）。7：顺式-芳樟醇氧化物（呋喃型）。8：芳樟醇。9：反式-芳樟醇氧化物（吡喃型）。10：顺式-芳樟醇氧化物（吡喃型）。11：α-紫罗兰酮。12：二氢-β-紫罗兰酮。13：γ-癸内酯。14：β-紫罗兰酮。

1: 3-Hexenal, (Z)-; 2: Hexanal; 3: 2-Hexenal, (E)-; 4: 1-Hexanol; 5: 2(5H)-Furanone, 5-ethyl-; 6: *trans*-Linaloloxide(furan); 7: *cis*-Linaloloxide(furan); 8: Linalool; 9: *trans*-Linaloloxide(pyran); 10: *cis*-Linaloloxide(pyran); 11: α-Ionone; 12: Dihydro-β-ionone; 13: γ-Decalactone; 14: β-Ionone

从图3中可以看出，3个品种的β-紫罗兰酮和二氢-β-紫罗兰酮含量都很高，是香味中最主要的组成成分。日香桂中含量最高的是β-紫罗兰酮，达到了30.56%；而四季桂和大叶佛顶珠中含量最高的是二氢-β-紫罗兰酮，分别为30.55%和27.63%。3个品种中α-紫罗兰酮和正己醛的含量比较接近，其他主要成分的含量都有明显的差异。如四季桂中顺-3-己烯醛的含量比较高，达到了7.47%，而'日香桂'中的含量不足1%；'日香桂'中芳樟醇氧化物的含量要高于其他2个品种，顺式和反式呋喃型芳樟醇氧化物分别为10.16%和10.03%，而'四季桂'中的含量分别只有1.83%和2.12%，吡喃型芳樟醇氧化物含量也是'日香桂'较高，但'日香桂'的叶醛含量相对偏低。'四季桂'和'日香桂'中γ-癸内酯的含量相对较高，分别为8.03%和6.78%。

综上所述，'四季桂''日香桂'与'大叶佛顶珠'虽同属四季桂品种群，但其香气成分的类型、种类及其主要成分相对含量的差异还是很明显的。

3 结论与讨论

3.1 四季桂品种芳香成分的组成

综合前人的研究结果，植物花香主要由萜烯类、苯型烃类、脂肪酸及其衍生物以及一些含硫、含氮的化合物组成[18]。在本次检测的3个四季桂品种花瓣挥发性成分中，共检测出37种化合物，分属于萜烯类、醇类、醛类、酮类、酯类和烷烃类等6大类。主要为顺式-芳樟醇氧化物（呋喃型）、反式-芳樟醇氧化物（呋喃型）、正己醛、叶醛、顺式-芳樟醇氧化物（吡喃型）、芳樟醇、反式-芳樟醇氧化物（吡喃型）、顺-3-己烯醛、乙醇、γ-2-己烯内酯、α-紫罗兰酮、二氢-β-紫罗兰酮、γ-癸内酯、β-紫罗兰酮。曹慧等[12]研究表明金桂、银桂和丹桂的香气主要成分是芳樟醇及其氧化物、α-紫罗兰酮、β-紫罗兰酮和γ-癸酸内酯等，本试验3个四季桂品种主成分仅芳樟醇及其氧化物含量比较低，而γ-癸内酯、β-紫罗兰酮、二氢-β-紫罗兰酮的含量较高。这与孙宝军认为4个品种群的桂花芳香成分不存在根本性的差异的结论基本一致[13]。与四季桂其他品种相比，如'佛顶珠''九龙桂''天香台阁'[13]以及'圆叶四季桂''橙黄四季桂'[19]的香味主体成分基本一致。另外检测结果显示顺-3-己烯醛、正己醛、叶醛、γ-2-己烯内酯等的含量也很高，这与文献中的结果不同，猜测可能与试验材料、采集时间和提取方法有一定的关系。

3.2 四季桂品种香味特性

3个四季桂品种主要挥发性成分为萜烯类化合物，其中α-紫罗兰酮、二氢-β-紫罗兰酮和β-紫罗兰酮的含量高，香气甜润馥郁[20]；但芳樟醇含量低，甜香有余，清香不足，香味相对比较沉闷、不透发[21]。萜烯类化合物是其他许多香气浓郁植物（如百合[22]、姜花[23]、厚朴[24]）花香中普遍存在的重要组分。在同一桂花品种群内花瓣挥发性成分相差不大，因此花瓣挥发性成分组成可以作为桂花品种群划分的主要依据之一，而其对于桂花品种划分的作用还需要更多数据支持。

3.3 超低温保存方法对花香挥发性成分的影响

花香的组成和含量受到很多方面的影响，包括花发育、基因、内源生物钟和环境因素等，其中低温因素就是影响花香挥发物的环境因素之一。桂花花期短,一般为一个星期左右,花朵娇弱,香气极易损失,所以采摘后必须及时处理,但盛花期由于花量大而往往来不及处理,因此,探索桂花的适宜保藏方法有重要的现实意义。本试验对鲜花和冷冻桂花进行 GC-MS 分析,从总离子色谱图看出其峰型契合度很高,个别物质保留时间略有不同,但重现性很好。虽然个别物质没有在超低温保存的桂花中检测到,但主体成分没有受到太大影响,基本一致,所以认为短时间超低温保存桂花对桂花香味成分的影响在可控范围内。这一结果与其他芳香型观赏花卉的研究结果类似。白三叶的花朵在 10 ℃时的挥发物释放量明显低于 15 ℃和 20 ℃,即推测降低温度可以减缓挥发物的释放[25]。马希汉等研究了冷冻方法对玫瑰精油得率的影响,结果表明,能长时间地保藏玫瑰花,提取的精油亦不受影响[26]。Shang 等[27]报道白玉兰鲜花冷冻样品只含有几种主要挥发物,大多数含氧化合物及单萜烯类与倍半萜烯类还有烷烃类含量均为痕量或不存在,这可能是由于保存时间过长引起。本研究初步认为在分析桂花主体成分的实验中,超低温冰箱可以用于短时间保存桂花样品,为更好地研究利用桂花花香争取了时间。

参考文献

[1] 范正琪,李纪元,田敏,等. 三个山茶花种(品种)香气成分初探[J]. 园艺学报,2006,33(3):592-596.

[2] 熊敏,周明芹,向林,等. 蜡梅花挥发油成分的 GC-MS 分析[J]. 华中农业大学学报,2012,31(2):182-186.

[3] 张莹,王雁,田敏,等. 不同种兰花香气成分分析[J]. 分析科学学报,2012,28(4):502-506.

[4] 彭红明. 中国兰花挥发及特征花香成分研究[D]. 中国林业科学研究院,2009.

[5] Kong Ying, Sun Ming, Pan H T, et al. Comparison of different headspace gas sampling methods for the analysis of floral scent from *Lilium* 'Siberia' [J]. Asian Journal of Chemistry, 2012,24(12):1-4.

[6] Kong Ying, Sun Ming, Pan H T, et al. Composition and emission rhythm of floral scent volatiles from eight lily cut flowers[J]. Journal of the American Society for Horticultural Science, 2012,137(6):376-382.

[7] 李桂珍,梁忠云,陈海燕,等. 芳樟叶油中芳樟醇的单离工艺条件[J]. 经济林研究,2013,31(4):195-197.

[8] 刘万好,郑秋玲,张超杰,等. 无核化处理对'玫瑰香'葡萄果实香气的影响[J]. 经济林研究,2014,32(3):171-174.

[9] 廖建华,陈月华,覃事妮. 湖南烈士公园植物群落的观赏效果及生态效益比较分析[J]. 中南林业科技大学学报,2013,33(8):143-146.

[10] 彭其龙,曾思齐,肖柏松,等. 南方次生林珍贵乡土树种的价值评价体系研究[J]. 中南林业科技大学学报,2014,34(5):21-25.

[11] 向其柏,刘玉莲. 中国桂花品种图志(An Illustrated Monograph of the Sweet Osmanthus Cultivars in China)(双语本)[M]. 浙江:科学技术出版社,2008.

[12] 曹慧,李祖光,沈德隆. 桂花品种香气成分的 GC/MS 指纹图谱研究[J]. 园艺学报,2009(3):391-398.

[13] 孙宝军,李黎,韩远记. 上海桂林公园桂花芳香成分的 HS-SPME-GC-MS 分析[J]. 福建林学院学报,2012(1):39-42.

[14] 徐继明,吕金顺. 桂花精油化学成分研究[J]. 分析试验室,2007,26(1):37-41.

[15] Wang Chengzhong, Su Yue, Guo Yinlong. Analysis of the volatile components from flowers and leaves of *Osmanthus fragrans* Lour. by Headspace-GC-MS[J]. Chinese Journal of Organic

Chemistry, 2009, 29 (6): 948-955.
[16] Wang Limei, Li Maoteng. Variations in the components of *Osmanthus fraarans* Lour. Essential oil at different stages of flowering[J]. Food Chemistry, 2009, 114: 233-236.
[17] 杨宇婷,武晓红,田璞玉.桂花(晚银桂、贵妃红和窈窕淑女)挥发性成分分析[J].河南大学学报(医学版),2010,29(1):13-16.
[18] Pichersky E, Noel J P, Dudareva N. Biosynthesis of plant volatiles: Nature's diversity and ingenuity [J]. Science, 2006, 311: 808-811.
[19] 王晨.不同桂花品种的香气成分分析及红外光谱鉴定模型初步建立[D].南京林业大学,2012.
[20] 杨雪云,赵博光,刘秀华,等.金桂银桂鲜花挥发性成分的顶空固相微萃取GC/MS分析[J].南京林业大学学报(自然科学版),2008,32(4):86-90.
[21] 林翔云.天然芳樟醇与合成芳樟醇[J].化学工程与装备,2008(1):21-27.
[22] 张辉秀,胡增辉,冷平生,等.不同品种百合花挥发性成分定性与定量分析[J].中国农业科学 2013,46(4):790-799.
[23] 范燕萍,王旭日,余让才,等.不同种姜花香气成分分析[J].园艺学报,2007,34:231-234.
[24] 王洁,杨志玲,杨旭,等.不同花期厚朴雌雄蕊和花瓣香气组成成分的分析和比较[J].植物资源与环境学报,2011,20(4):42 48.
[25] Jakobsen H B, Olsen C E. Influence of climatic factors on emission of flower volatiles *in situ*[J]. Planta, 1994, 192(3): 365-371.
[26] 马希汉,王永红,尉芹,等.玫瑰花保藏方法与精油得率关系的研究[J].林产化学与工业,2005.25(1): 84-88.
[27] Shang Chunqing, Hu Yaoming, Deng Chunhui. *et al*. Rapid determination of volatile constituents of *Michelia alba* flowers by chromatography-mass spectrometry with solid-phase microextraction[J]. Journal of Chromatography A, 2002, 942: 283-288.

Analysis of Volatile Compounds from Petals of Three Species *Osmanthaus fragrans* Asiaticus Group Cultivars by Gas Chromatography-Mass Spectrometry

YANG Xiulian, SHI Tingting, WEN Ailin, WANG Lianggui

(College of Landscape Architecture, Nanjing Forestry University, Nanjing 210037, China)

Abstract: The volatile constituents from petals of the three species *Osmanthu fragrans* Asiaticus Group cultivars at stages of petal expansion were analyzed by GC-MS (gas chromatography-mass spectrometry) with headspace SPME (solid-phasemicro-extraction). Compared the differences between the volatile components of the fresh flower petals and short-term cryopreservative flower petals. The results showed that the main volatile components were closed, there were β-Ionone, Dihydro-β-ionone, *cis*-Linaloloxide (furan), *trans*-Linaloloxide (furan), γ-Decalactone, α-Ionone, Hexanal, 2-Hexenal, (E)-, Linalool, *cis*-Linaloloxide(pyran), *trans*-Linaloloxide(pyran), 3-Hexenal, (Z)-. And the results showed that little effect on volatile compounds of the short-term cryopreservation *Osmanthu fragrans*. The petal could be preserved in ultra-low temperature freezing for a short period When there are many samples.
Key words: *Osmanthaus fragrans* Asiaticus Group cultivars; volatile compounds; HS-SPME; GC-MS

11个桂花品种花瓣与叶片中矿质元素含量的比较[*]

林燕青,杨秀莲,凌 敏,王良桂

(南京林业大学 风景园林学院,江苏 南京 210037)

摘 要:矿质元素是动植物生长发育所必需的元素,为进一步开发富矿质元素、更具食品保健价值和园林应用价值的桂花新品种,用电感耦合等离子体发射光谱仪(ICP-MS)测定了11个桂花品种花瓣与叶片中的钙、镁、铝、锌、铜、铁、砷、铅元素含量,分析了花瓣与叶片中矿质元素的相关性。结果表明:11个桂花品种花叶中重金属元素砷和铅元素积累量未检出,有益矿质元素含量均较高;叶片中主要有益元素钙、锌、铁元素含量最高的分别为'速生金桂''苏州浅橙''红桂';花瓣中主要有益元素钙、锌元素含量最高的为'白洁',铁元素含量最高的是'苏州浅橙';叶片中的铁元素与花瓣中的铁元素呈极显著正相关,叶片中的锌、铜元素均与花瓣中的镁元素呈显著正相关,叶片中的镁元素与花瓣中的铜元素呈显著负相关。

关键词:桂花;叶片;花瓣;矿质元素;相关性

矿质元素是指除了碳、氢、氧外,由植物根系从土壤中吸收的元素,包括氮、磷、钾、钙、镁、铁、铜、锰、锌、铝、硼等。矿质元素是植物应用价值提高、产量形成和品质创新的物质基础,是植物生理代谢和生长发育所必需的元素[1]。重金属能抑制细胞分裂和致染色体畸变,导致出现染色体断裂、粘连、体细胞染色体不等交换、染色体环等畸变现象。此外,有毒的重金属元素会严重威胁人类的健康,是一类具有潜在危害的元素[2]。目前,对植物矿质元素的研究主要集中在果树及水稻(*Oryza sativa*)的叶片、果实和土壤矿质元素相关性方面,其中有关叶片和果实内矿质元素的变化在龙眼(*Dimocarpus longan*)[3]、枇杷(*Eriobotrya japonica*)[4]等中已有报道。有关花卉已有的营养物质研究发现,郁金香(*Tulipa gesneriana*)、杜鹃(*Rhododendron simisii*)[5]、百合花(*Lilium* spp.)、玫瑰花(*Rosa rugosa*)、木棉花(*Bombax malabaricum*)、牡丹(*Paeonia suffruticosa*)[6]、唐菖蒲(*Gladiolus hybridus*)[7]、山茶花(*Camellia japonica*)以及一些野生花卉[8]等花瓣含有较高的矿质元素。此外,还有部分桂花品种[9]的花瓣营养成分研究及梨花[10]花粉、花柱矿质元素研究,而对不同桂花品种花瓣与叶片中矿质元素含量的比较未见报道。桂花(*Osmanthus fragrans*)为木犀科(Oleaceae)木犀属(*Osmanthus* Lour.)植物,是著名的常绿观赏兼食用型植物,其花叶营养成分具有较大的研究价值,加强不同桂花品种矿质元素的研究,可为其植物营养诊断、合理施肥和培育植物新品种提供理论依据[11]。本实验分析了桂花11个品种叶片和花瓣中 Zn、Cu、Fe、Ca、Mg、Al、Pb、As 的含量,进行了品种间花瓣和叶片间的相关

---

[*] 原文载于《西部林业科学》,2015,44(4):79-83。
基金项目:国家林业局公益性行业科研专项(201204607);江苏高校优势学科建设工程资助项目(PAPD)。

性分析,旨在选出含矿质元素较高的品种,为桂花品质改善及筛选适合食用和园林环保不同用途的桂花品种提供科学依据。

1 材料和方法

1.1 实验材料

实验所用材料选自南京、杭州2地,共11个品种(表1),分属于4个品种群。

表1 实验用桂花品种
Tab. 1 Experimental *Osmanthus fragrans* cultivars

品种群 Cultivars group	品种名 Cultivars name	来源 Source
四季桂品种群	'四季桂''Sijigui'	南京林业大学校园
O. fragrans Asiaticus Group	'日香桂''Rixiang Gui'	南京林业大学校园
金桂品种群	'波叶金桂''Boye Jingui'	杭州园林绿化工程有限公司基地
O. fragrans Luteus Group	'速生金桂''Susheng Jingui'	南京林业大学校园
银桂品种群	'白洁''Baijie'	杭州园林绿化工程有限公司基地
O. fragrans Albus Group	'苏州浅橙''Suzhou Qiancheng'	杭州园林绿化工程有限公司基地
	'红桂''Honggui'	南京林业大学校园
丹桂品种群	'堰江桂''Yanjianggui'	杭州园林绿化工程有限公司基地
O. fragrans Aurantiacus Group	'橙江丹桂''Chengjiang Dangui'	杭州园林绿化工程有限公司基地
	'状元红''Zhuangyuan Hong'	南京林业大学校园
	'朱砂丹桂''Zhusha Dangui'	南京林业大学校园

1.2 测定方法

实验用桂花花瓣与叶片于2013年10月上旬取样,取样集中于盛花期(每花序有50%的盛花量),每株树随机采摘当年生叶片及新鲜花瓣60 g,每品种采样重量总约400 g,使用冰盒携带,之后烘干、粉碎后备用。

每个样品精密称取(0.20±0.02)g粉碎样于50 mL锥形瓶中,加浓硝酸与高氯酸(3:1)的混合酸10 mL,用保鲜膜进行封口,放置12 h。次日,在3 kW电热板上进行硝煮,如在蒸尽之前有机物仍未全部溶解,则补加浓硝酸,直至锥形瓶内溶液透明,冷却至常温。把硝煮后的溶液转移至25 mL容量瓶中,用超纯水稀释至刻度,摇匀之后上机测定,每处理重复3次。

使用PerkinElmer NexION300型电感耦合等离子体发射光谱仪(ICP-MS)(美国PE公司)进行测定,在仪器达到最佳工作状态时,按照选定的仪器工作条件,取样液直接测定元素含量,每测试重复3次。

1.3 统计分析方法

使用SPSS 19.0软件及Excel软件对实验数据进行相关分析。

2 结果与分析

2.1 不同桂花品种间叶片矿质元素含量比较

11个桂花品种叶片中矿质元素含量测定结果见表2。由表2可知,不同桂花品种叶片间各矿质元素含量存在差异。9号品种('朱砂丹桂')叶片中Mg元素含量最高,达2 709 mg/kg,11号品种('四季桂')叶片中Mg元素含量最低,为1 533 mg/kg;7号品种('速生金桂')叶片中Ca元素含量最高,为2 301 mg/kg,与其他品种叶片中Ca元素含量均呈极显著差异,4号品种('橙江丹桂')叶片中Ca元素含量最低,为1 081 mg/kg;10号品种('白洁')叶片中Al元素含量最高,达到528 mg/kg,与其他品种叶片中Al元素含量呈显著差异,11号品种('四季桂')叶片中Al元素含量最低,为87 mg/kg;2号品种('红桂')叶片中Fe元素含量最大,达到211 mg/kg,5号品种('波叶金桂')叶片中Fe元素含量最小,为51 mg/kg,各品种叶片中Fe元素含量存在差异;1号品种('苏州浅橙')叶片中Zn元素含量最高,为84 mg/kg,与其他品种叶片中Zn元素含量呈显著水平,其次为10号品种('白洁'),11号品种('四季桂')叶片中Zn元素含量最低,为16 mg/kg;Cu元素为叶片中含量最少的微量元素,1号品种('苏州浅橙')叶片中Cu元素含量最高,为27 mg/kg,与其他品种叶片中Cu元素含量呈显著水平,4号品种('橙江丹桂')与10号品种('白洁')叶片中的Cu元素含量最低,均为13 mg/kg;各品种叶片中有毒重金属元素As和Pb的测定值为0,很可能是其含量低于仪器检出限,仍有待于进一步研究确定。初步可知,11个桂花品种的叶片中所含有毒重金属成分较低。

表2 不同品种桂花叶片中矿质元素含量分析
Tab. 2 The analysis of mineral element content in leaves of different *Osmanthus fragrans* cultivars (mg/kg)

品种 Cultivars	Zn	Cu	Fe	Ca	Mg	Al
1	84±1.15 aA	27±0.49 aA	67±0.56 cC	1 495±11.88 bcB	2 616±9.40 aA	248±1.00 edC
2	43±1.70 bB	20±0.49 abAB	211±1.09 aA	1 267±17.56 cB	2 694±9.01 aA	200±0.35 dC
3	23±0.15 cB	21±0.90 abAB	180±0.47 abA	1 688±5.29 bB	1 705±5.00 bB	262±3.27 dC
4	31±0.35 bB	13±0.29 bAB	56±0.02 cC	1 081±27.00 cC	1 986±20.80 bB	246±1.25 dC
5	43±0.91 bB	17±0.33 abAB	51±1.40 cC	1 611±23.93 bB	1 956±3.92 bB	460±5.67 bA
6	26±0.25 bB	10±0.20 bB	124±3.44 bB	1 238±8.15 cC	1 861±18.18 bB	360±2.33 cB
7	25±0.55 bB	16±0.15 bAB	72±1.18 cC	2 301±24.19 aA	1 721±8.59 bB	115±1.12 eCD
8	47±4.99 bAB	22±0.77 abAB	58±0.85 cC	1 138±24.79 cC	1 842±3.41 bB	252±8.26 dC
9	28±0.38 bB	18±0.21 aAB	148±1.75 bB	1 290±14.13 cB	2 709±4.67 aA	178±1.62 edC
10	52±0.60 bAB	13±0.44 bAB	79±1.26 cC	1 587±14.77 bB	2 263±15.36 abAB	528±7.09 aA
11	16±0.62 cB	15±0.05 bAB	202±0.98 aA	1 445±16.80 bcB	1 533±15.52 bB	87±1.05 fD

注:1为'苏州浅橙';2为'红桂';3为'堰江桂';4为'橙江丹桂';5为'波叶金桂';6为'状元红';7为'速生金桂';8为'日香桂';9为'朱砂丹桂';10为'白洁';11为'四季桂';表中同列大小写字母分别表示1%和5%差异显著水平;有毒重金属元素As和Pb的测定值为0,其余表同。
Note: 1 is 'Suzhou Qiancheng'; 2 is 'Honggui'; 3 is 'Yanjianggui'; 4 is 'Chengjiang Dangui'; 5 is 'Boye Jingui'; 6 is 'Zhuangyuan Hong'; 7 is 'Susheng Jingui'; 8 is 'Rixiang Gui'; 9 is 'Zhusha Dangui'; 10 is 'Baijie'; 11 is 'Sijigui'; the same-case uppercase and lowercase letters in the table indicate significant differences between 1% and 5%, respectively; the toxic heavy metal elements As and Pb have a measured value of 0, and the rest are the same.

2.2 不同桂花品种间花瓣中矿质元素含量比较

11个桂花品种花瓣中矿质元素含量测定结果见表3。由表3可知,不同桂花品种花瓣间各矿质元素含量存在差异。

花瓣中Ca元素含量最高达2 658 mg/kg,含量最高的是10号品种('白洁'),其Ca元素含量与其他品种花瓣中Ca元素含量均呈极显著差异,4号品种('橙江丹桂')花瓣中Ca元素含量最低,为783 mg/kg。

11号品种('四季桂')花瓣中Mg元素含量最高,为1 488 mg/kg,其Mg元素含量与其他品种花瓣中Mg元素含量均呈显著水平,1号品种('苏州浅橙')花瓣中Mg元素含量最低,为989 mg/kg。

8号品种('日香桂')花瓣中Al元素含量最高为51 mg/kg,6号品种('状元红')与11号品种('朱砂丹桂')花瓣中Al元素含量最低,均为29 mg/kg。

花瓣的微量元素中,1号品种('苏州浅橙')花瓣中Fe元素含量最大,为168 mg/kg,6号品种('状元红')花瓣中Fe元素含量最小,为42 mg/kg。

10号品种('白洁')花瓣中Zn元素含量最高,为48 mg/kg,7号品种('速生金桂')花瓣中Zn元素含量最低,为21 mg/kg。

其中Cu元素为各品种花瓣中含量最少的微量元素,5号品种('波叶金桂')花瓣中Cu元素含量最高,为45 mg/kg。

11个桂花品种的花瓣中有毒重金属元素As和Pb的含量均为零。

表3 不同品种桂花花瓣中矿质元素含量分析
Tab. 3 The analysis of mineral element content in petals of different *Osmanthus fragrans* cultivars (mg/kg)

品种 Cultivars	Zn	Cu	Fe	Ca	Mg	Al
1	43±4.30 aAB	27±2.67 abA	168±3.77 abA	1 323±18.32 bcB	989±10.51 cB	50±1.75 aA
2	41±4.10 aAB	38±3.80 abA	54±2.76 dD	1 240±11.37 bcB	1 331±9.24 abdAB	30±0.55 bA
3	42±4.20 aAB	19±1.87 abA	51±4.16 dD	1 302±16.18 bcB	1 142±6.40 bcdB	42±0.31 abA
4	37±3.70 aAB	38±3.80 abA	45±0.05 dD	783±7.08 cC	1 096±1.80 bcdB	36±0.85 abA
5	27±2.70 bB	45±4.53 aA	44±1.29 dD	1 210±34.16 bB	1 085±2.41 bcdB	34±1.21 abA
6	27±2.73 bB	11±1.10 bA	42±1.69 dD	802±6.48 cC	1 082±10.98 bcdB	29±0.46 bA
7	21±2.07 bB	14±1.43 bA	73±3.97 cCD	1 132±10.86 cBC	1 207±19.87 ddB	36±1.15 abA
8	35±3.53 aAB	18±1.80 abA	52±5.89 dD	1 689±17.89 bB	1 234±6.19 bdB	51±1.93 aA
9	41±4.10 aAB	20±2.00 abA	116±9.69 bB	874±5.13 cC	1 132±11.58 bcdB	45±1.39 abA
10	48±4.80 aA	20±1.97 abA	87±3.65 cC	2 658±29.35 aA	1 137±16.14 bcdB	47±0.55 abA
11	23±2.27 bB	31±3.10 abA	130±2.97 bB	905±4.99 cC	1 488±8.28 aA	29±0.95 bA

2.3 叶片与花瓣矿质元素相关性分析

由表4可知,叶片与花瓣中6种元素间存在着复杂的相互关系。叶片中的Fe元素含量与花瓣中的Fe元素含量呈极显著正相关,相关系数为0.804;叶片中的Zn元素含量与花瓣中的Mg元素含量呈显著正相关,相关系数为0.710;叶片中的Cu元素含量与花瓣中的Mg

元素含量呈显著正相关性,相关系数为 0.675;叶片中 Mg 元素含量与花瓣中 Al 元素含量呈显著正相关,相关系数 0.633,证明这些存在正相关的元素间存在增效作用。叶片中的 Mg 元素含量与花瓣中 Cu 元素含量呈显著负相关,相关系数为 −0.652,即两元素间存在拮抗作用。其他元素间均呈不显著相关。

表 4　叶片与花瓣矿质元素相关性分析
Tab. 4　Correlation analysis of mineral elements in leaves and petals

叶片矿质元素 Leaf mineral element	花瓣矿质元素 Petal mineral element					
	Al	Zn	Cu	Fe	Ca	Mg
Zn	0.434	−0.011	−0.406	0.617	−0.458	0.710*
Cu	0.284	−0.068	−0.234	0.263	0.138	0.675*
Fe	0.321	0.162	−0.066	0.804**	−0.395	0.507
Ca	−0.335	−0.248	0.342	0.152	−0.033	−0.030
Mg	0.633*	0.092	−0.652*	−0.229	−0.195	0.343
Al	0.330	0.112	−0.349	0.555	−0.585	0.086

注:表中*,**分别表示 5% 和 1% 差异显著水平。
Note: * and ** in the table indicate significant differences between 5% and 1%, respectively.

3　结论与讨论

实验表明,4 个品种群,11 个不同品种桂花间叶片和花瓣中矿质元素含量存在差异。不同品种间矿质元素含量存在显著的差异,可能与品种生长势及生长环境有关[12]。品种对叶片中矿质元素含量变化的影响已有研究报道[13],结合之前杨秀莲[9,14]对 25 个不同桂花品种的研究,发现由于树体及土壤中营养成分积累不同,不同年份、不同地点,同一棵树上不同花瓣及叶片中的矿质元素含量差异也极大[9,15]。

本研究表明,'苏州浅橙'品种叶片中 Zn 元素含量最高,其次为'白洁'品种,'白洁'品种的花瓣中 Zn 元素含量最高,可为保健功效的桂花食品提供资源。'白洁'品种花瓣 Ca 元素含量极显著高于其他品种,可选'白洁'品种作为具有补钙功效的产品资源;由于 Ca 在植物体内移动性很小,Ca 缺乏时,植株早衰,不结实或少结实[16],可作为杂交育种的亲本,为选育抗早衰及提高结实率的品种提供资源。

不同桂花品种间叶片及花瓣矿质元素含量分析表明,同一元素在不同品种间含量差异很大,可见其蕴藏的遗传基础较广泛,利用和开发的潜力较大,可获得所需育种目标品种的概率也相应较大[17]。同理可知,这 11 个桂花品种对养分的吸收和利用存在一定的相关性和差异性。

本研究对 11 个桂花品种叶片和花瓣中 8 种矿质元素含量进行测定,结果表明,8 种矿质元素含量有不同程度差异性,不同桂花品种间叶片和花瓣的矿质元素含量均较高,具有很高的营养价值。选出了矿质元素含量高的品种如'白洁'花瓣中 Ca、Zn 元素含量较高,'速生金桂'叶片中 Ca 含量较高等。叶片与花瓣矿质元素相关性分析中探明了其间的关系,叶片中的 Fe 元素与花瓣中的 Fe 元素呈极显著正相关;Esmaeil Fallahi[11]等研究发现,叶片中的 Ca 元素与花瓣中的 Ca 元素呈显著负相关,与本研究结果有所不同,可能是与植物的种类相关,也可能是与植物种植环境相关,仍待进一步研究;叶片中的 Zn 元素含量与花瓣中的 Mg 元素含量呈显著正相关;叶片中的 Cu 元素含量与花瓣中的 Mg 元素含量呈显著正相关性;叶片中 Mg 元素含量与花瓣中 Al、Cu 元素含量分别呈显著正、负相关;这为杂交苗的早期筛选提供了依据。有关不同品种桂花花瓣与叶片中的矿质元素与栽培基质及其生长环境等的关系仍有待于进一步研究。

参考文献

[1] 卢桂宾,李春燕,郭晓东.外源钙肥对枣果实矿质营养元素含量的影响[J].经济林研究,2010,28(3):69-74.

[2] 陈久耿,晁代印.矿质元素互作及重金属污染的研究进展[J].植物生理学通讯,2014,50(5):585-590.

[3] 庄伊美,李来荣,江由,等.赤壳龙眼叶片与土壤常量元素含量年周期变化的研究[J].园艺学报,1984,11(3):79-87.

[4] 陆修闽,郑少泉,蒋际谋,等.'早钟6号'枇杷主要营养元素含量的年周期变化[J].园艺学报,2000,27(4):240-244.

[5] 幸宏伟.郁金香及杜鹃花花瓣营养成分分析[J].重庆工商大学学报(自然科学版),2005,22(1):53-55.

[6] 史国安,郭香凤,包满珠.不同类型牡丹花的营养成分及体外抗氧化活性分析[J].农业机械学报,2006,37(8):111-114.

[7] 刘晓辉,杨明,邓日烈,等.花瓣营养成分探析[J].北方园艺,2010(12):48-50.

[8] 许又凯,刘宏茂,刀祥生,等.五种野生花卉营养成分及其评价[J].营养学报,2005,27(5):437-438,440.

[9] 杨秀莲,赵飞,王良桂.25个桂花品种花瓣营养成分分析[J].福建林学院学报,2014,34(1):5-9.

[10] 贾兵,张绍铃.梨花粉、花柱与子房中激素和矿质元素含量的比较[J].园艺学报,2012,39(2):225-233.

[11] Esmaeil Fallahi, Fenton ELarsen. Root stock influence on leaf and fruit mineral status of 'Bartlett' and 'Anjou' pear[J]. Scientific Horticultural, 1984, 23(1):41-49.

[12] 安贵阳,范崇辉,杜志辉,等.苹果叶营养元素含量的影响因素分析[J].园艺学报,2006,33(1):12-16.

[13] 李港丽,苏润宇,沈隽.几种落叶果树叶内矿质元素含量标准值的研究[J].园艺学报,1987,14(2):81-89.

[14] 杨秀莲,王良桂,文爱林.桂花花瓣营养成分分析[J].江苏农业科学,2012,40(12):334-336.

[15] 徐文斌,王贤荣,石瑞,等.两个丹桂品种营养成分分析[J].林业科技开发,2010,24(6):117-119.

[16] 蒋淑丽.稻米矿质元素分析及其近红外测定技术的研究[D].杭州:浙江大学,2007.

[17] 汪禄祥,黎其万,王丽华,等.不同百合种球矿质元素含量的变异和相关性研究[J].北方园艺,2008(5):119-121.

Comparison of Mineral Element Content in Leaves And Petals of 11 *Osmanthus fragrans* Cultivars

LIN Yanqing, YANG Xiulian, LING Min, WANG Lianggui

(College of Landscape Architecture, Nanjing Forestry University, Nanjing 210037, China)

Abstract: Mineral elements are essential for plant growth. In order to develop the new *Osmanthus fragrans* cultivars with rich mineral elements, more health value and ornamental value, the mineral elements such as calcium, magnesium, aluminum, zinc, copper, iron, arsenic, lead in petals and leaves 11 *O. fragrans* cultivars were analyzed by using inductively coupled plasma optical emission spectrometry (ICP‑MS). The correlation of mineral elements in leaves and petals were also analyzed. The results showed that there were no heavy metals of arsenic and lead. The amounts of healthy elements were high, and the highest amounts of calcium, zinc, copper in leaves took place in 'Susheng Jingui', 'Suzhou Qiancheng', 'Honggui', that of calcium, zinc in petals was 'Baijie', and that of iron in petals was 'Suzhou Qiancheng'. Iron element in leaves and petals had a very significant positive correlation. Zinc, copper elements in leaves positively correlated with magnesium element in petals. Magnesium in leaves negatively correlated with copper element in petals.

Key words: *Osmanthus fragrans*; leaf; petal; mineral element; correlation

'紫梗籽银'桂种子内源抑制物质分析

施婷婷,杨秀莲,王良桂

(南京林业大学 风景园林学院,江苏 南京 210037)

摘 要：以'紫梗籽银'桂(*Osmanthus fragrans* 'Zigeng Zi Yin')种子为试材,对其内果皮、种实内源抑制物质进行提取、生物测定,并采用 GC-MS 进行鉴定。结果表明：①'紫梗籽银'桂种子内果皮、种实均含有抑制种子萌发和幼苗生长的物质,且内果皮、种实中各分离相的抑制效果均不相同。其中石油醚相抑制作用最弱,乙酸乙酯相和甲醇相抑制作用较强。②GC-MS 鉴定结果显示,'紫梗籽银'桂种子内果皮、种实中存在多种抑制物质,包括棕榈酸、油酸、9,12-十八碳二烯酸等发芽抑制物质,主要为脂肪酸类物质。

关键词：'紫梗籽银'桂;种子;内源抑制物质;发芽率;幼苗生长;气相色谱-质谱联用

许多植物的种子都存在休眠现象,而种子的休眠原因也是多种多样的,主要包括了种皮的吸水障碍、种胚的休眠以及种子中存在萌发抑制类物质等原因,其中由种子萌发抑制物引起的休眠是最常见的种子休眠类型之一。植物种子的内源抑制物质种类非常多,除脱落酸(ABA)外,还包括了有机酸、酚类、醛类、不饱和内酯类,以及一些气体类物质,如氢氰酸、乙烯、芥子油、氨等[1]。目前,关于青钱柳(*Cyclocarya paliurus*)[2]、四川牡丹(*Paeonia decomposita*)[3]、秦艽(*Gentiana macrophylla*)[4]等植物种子内源抑制物质的研究已取得了很大进展,而关于桂花种子内源抑制物质的研究少有研究报道。

'紫梗籽银'桂是桂花银桂品种群中的一个品种,其花香气清幽淡雅,且每年可开花 2~3 次,深受人们的喜爱。同时,'紫梗籽银'桂花后结实率高,果实丰满,性状稳定,是我国重要的园林绿化树种。然而其种子具有明显的休眠习性,这给'紫梗籽银'桂的苗木生产带来了很多不便。杨秀莲等[5]曾初步探讨了'紫梗籽银'桂种子的休眠原因,发现'紫梗籽银'桂种子坚硬的外种皮不是引起休眠的原因,推测种子的休眠是由种子中所含的抑制物质引起。故本研究以'紫梗籽银'桂为材料,从抑制种子萌发的内源抑制物质入手,分别比较不同有机溶剂相中抑制物质的作用,并通过气相色谱-质谱联用(GC-MS)技术对抑制物质成分进行初步分析。

1 材料与方法

1.1 试验材料

2007 年 5 月底,于南京林业大学校园内采集完全成熟的'紫梗籽银'桂种子;于南京市蔬菜种子公司购买'绿优 1 号'白菜种子(纯度 95%,净度 98%,发芽率 85%以上)。

---

* 原文发表于《福建林业科技》,2015,42(2):104-107。
基金项目：江苏省科技支撑项目(BE2011367);江苏省高校优势学科建设工程项目(PAPD)。

1.2 种子不同部位内源抑制物质的提取

取'紫梗籽银'桂种子内果皮和种实各 10 g 置于 250 mL 容量瓶中,用 80% 的甲醇溶液 200 mL,于 1~4 ℃的恒温条件下密闭浸提 48 h,然后将所得滤液在 35 ℃下减压浓缩后定容至 1 g·mL^{-1}(以每 mL 浓缩液中含有原材料的量计算),最后置于 1~4 ℃冰箱中保存备用。

1.3 种子不同部位内源抑制物质的初步分离

参照黄耀阁等[6]系统溶剂法,将'紫梗籽银'桂种子不同部位内源抑制物质的粗提液分为石油醚相、乙醚相、乙酸乙酯相、甲醇相和水相。即将'紫梗籽银'桂内果皮和种实 2 部分烘干(35 ℃),分别取磨碎后过 1 mm 筛的粉末各 15 g,各加入 5 倍体积的 80% 甲醇,在 1~5 ℃的恒温条件下密闭浸提,24 h 后过滤,重复浸提 3 次,混合 3 次所得的浸提液,再加入等量石油醚,相同条件下萃取 3 次,得到石油醚相和水相;然后在水相浸提液加入等量乙醚,相同条件下萃取 3 次,得到乙醚相和水相;再在水相浸提液中加入等量乙酸乙酯萃取,得到乙酸乙酯相与水相;再将水相浸提液在 65 ℃减压蒸干,除去水分得到干物质后,用甲醇溶解过滤,得到甲醇相和水相。最后将各有机相溶液分别置于旋转蒸发器 ZFQ85B 上进行浓缩蒸发,定容至 10 mL,置于 1~4 ℃冰箱中备用。

1.4 种子不同部位内源抑制物质各分离物活性的生物测定

将上述所得各分离相(0.2 g·mL^{-1})和稀释 1 倍后的各有机相溶液(0.1 g·mL^{-1})进行生物测定。在直径 9 cm 的培养皿内分别加入'紫梗籽银'桂种子内果皮和种实各分离相的提取液 3 mL,白菜籽 50 粒,待各分离相有机溶剂完全挥发(水相除外)后,每个培养皿内再加入蒸馏水 3 mL,并以同体积蒸馏水作为对照,每个处理 3 次重复。置于 25 ℃恒温光照培养箱内,进行白菜籽发芽试验,每天记录发芽数,48 h 后统计白菜籽的发芽率(以露出胚根为发芽的标准),72 h 后测其苗高和根长。

1.5 GC-MS 鉴定方法

对'紫梗籽银'桂种子内果皮和种实甲醇浸提液进行 GC-MS 鉴定,测定在中国林业科学院林产化学研究所分析测试中心质谱室进行鉴定。参照韩宝瑞等[7]鉴定方法:仪器型号为美国 Agelient 6890N/5973N GC-MS(气相色谱-质谱)联用仪;气相色谱条件为 HP-5MS 柱,30 m×0.32 mm×0.25 μm;升温程序 100~250 ℃(10 min);程序升温 10 ℃·min^{-1};进样 1 μL;分流比 100∶1;载气为氦气;汽化室温度 260 ℃。质谱条件:离子源 EI;源温 230 ℃;电离电压 70 eV。由计算机控制的库存信号检查各个成分的质谱图并与标准库谱图核对,以面积归一化法计算出各成分的相对百分含量。

2 结果与分析

2.1 内果皮和种实中各分离相对白菜种子发芽率和幼苗生长的影响

2.1.1 内果皮和种实中各分离相对白菜种子发芽率的影响

由图 1 可见,除石油醚相外,'紫梗籽银'桂内果皮和种实的乙醚相、乙酸乙酯相、甲醇相以及水相均对白菜籽的发芽有抑制作用,且方差分析结果表明,'紫梗籽银'桂内果皮和种实内源抑制物质浸提液中除石油醚相对白菜籽发芽率的影响不显著外,其他各相与对照相比均达到极显著差异。内果皮各分离相浸提液中乙酸乙酯相的抑制作用最强,浸提液 $0.1\ g\cdot mL^{-1}$、$0.2\ g\cdot mL^{-1}$抑制率分别达到 90.6%、94.8%;而以乙醚相浸提液的抑制活性最弱,仅 8.3%、21.67%;石油醚相对白菜籽发芽基本没有抑制作用,浸提液$0.1\ g\cdot mL^{-1}$处理的白菜籽发芽率比对照降低了 3.15%,浸提液 $0.2\ g\cdot mL^{-1}$处理的白菜籽发芽率比对照提高了 2.08%。相比内果皮,种实各分离相浸提液对白菜籽的发芽抑制作用更大,种实各分离相浸提液中也是乙酸乙酯相的抑制作用最强,浸提液$0.1\ g\cdot mL^{-1}$、$0.2\ g\cdot mL^{-1}$的抑制率分别达到 92.7%、95.8%;以水相浸提液的抑制活性最弱,仅为 20.8%、28.1%;石油醚相对白菜籽发芽基本没有抑制作用,其浸提液 $0.1\ g\cdot mL^{-1}$处理的白菜籽发芽率比对照降低了 6.25%,而浸提液 $0.2\ g\cdot mL^{-1}$处理的白菜籽发芽率比对照提高了 2.08%。据此初步证实'紫梗籽银'桂内果皮和种实均含有内源抑制物质,而且不同部位提取的不同有机相对白菜籽发芽率的抑制程度各有不同,种实各分离相对白菜籽发芽率的抑制作用的强弱顺序为:乙酸乙酯相>甲醇相>乙醚相>水相>石油醚相,而内果皮各分离相对白菜籽发芽率的抑制作用的强弱顺序则为:乙酸乙酯相>甲醇相>水相>乙醚相>石油醚相。

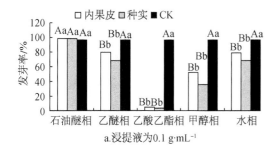 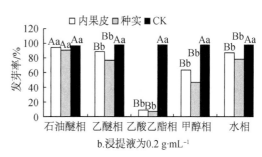

图1 '紫梗籽银'桂内果皮和种实中抑制物对白菜发芽率的影响

图1表明:'紫梗籽银'桂内果皮和种实浸提液的发芽测定结果大致相同,但各有机相对白菜籽发芽的抑制作用不同。'紫梗籽银'桂内果皮和种实中的抑制物质均主要集中在乙酸乙酯相和甲醇相中,对白菜种子发芽起到明显抑制作用;而其石油醚相均与对照无差异,说明石油醚相中也许不存在抑制物质。

2.1.2 内果皮和种实中各分离相对白菜籽幼苗生长的影响

从图2可以看出,'紫梗籽银'桂内果皮和种实各分离相对白菜籽幼苗生长的抑制作用差异达到极显著水平。'紫梗籽银'桂内果皮、种实各相对白菜籽苗高抑制作用的强弱顺序为甲醇相>乙酸乙酯相>水相>乙醚相>石油醚相,而对白菜根长的抑制作用的强弱顺序则为乙酸乙酯相>甲醇相>乙醚相>石油醚相>水相。特别是乙酸乙酯相和甲醇相处理的白菜籽苗高生长均为0,对根长的抑制率也都达到了 85%以上。同时,各相处理浓度越大,对白菜籽幼苗生长的抑制作用也越强。但通过方差分析和多重比较结果表明,除不同浓度间种实乙醚相对白菜苗高生长的抑制作用差异显著外,其他各相不同浓度间差异均不显著;除不同浓度间内果皮乙醚相、水相对白菜根系生长的抑制作用差异极显著外,其他各相不同浓度间差异也均不显著。而在相同浓度相同有机相的条件下,种实浸提物对白菜幼苗生长的影响均高于内果皮提取物。如内果皮水相$0.1\ g\cdot mL^{-1}$、

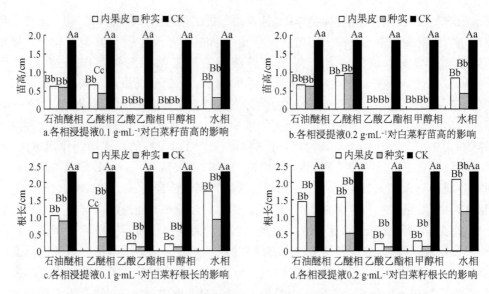

图 2　'紫梗籽银'桂内果皮和种实中抑制物对白菜籽幼苗生长的影响

0.2 g·mL^{-1} 处理的白菜苗高生长比对照分别减少了 54.3%、60%，白菜幼苗根长分别降低 9.7%、24.9%；而种实水相在相同浓度处理下白菜苗高生长量降低了 77.2%、81.8%，根长减少了 50.1%、59.7%。

由此可以说明，'紫梗籽银'桂内果皮和种实各分离相对白菜幼苗生长也均有抑制作用，且浓度越大，抑制作用越强。其中相同浓度条件下，种实浸提液对白菜幼苗生长的抑制作用比内果皮显著。但不管是种实还是内果皮，其甲醇相和乙酸乙酯相对白菜幼苗生长的抑制作用最强，表明内果皮和种实的甲醇相和乙酸乙酯相中存在着大量的抑制物质，且这些物质的抑制活性都较强。

2.2　种子不同部位提取的有机化合物种类和相对含量

'紫梗籽银'桂种子的内果皮、种实提取液经 GC-MS 分析得到的总离子色谱图见图 3、

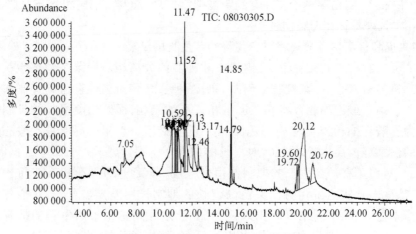

图 3　'紫梗籽银'桂种子种实中提取物的离子流程图

注：图上数字重叠处峰值从左到右分别是：10.50、10.59、10.86、10.60、10.70。

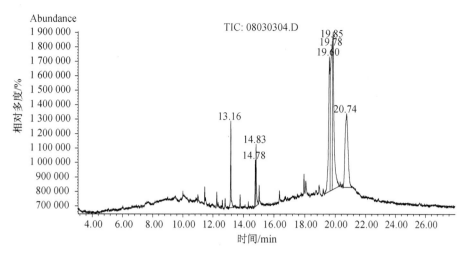

图 4 '紫梗籽银'桂种子内果皮中提取物的离子流程图

图4。从图中可看出两者离子流色谱图的峰型差异很大,个别物质保留时间也不尽相同,但重现性较好。通过质谱系统检索,按峰面积归一化法计算得出各成分相对百分含量,分析出离子流程图中峰面积和相似度较大、可能具有抑制作用的有机化合物共有9种(表1)。'紫梗籽银'桂种实中鉴定出的有机化合物主要为脂肪酸类和苯胺类物质,有油酸(4.208%)、棕榈酸(1.817%)、9,12-十八碳二烯酸(1.84%)和2,4-二甲苯胺(21.689%)、3,4-二甲苯胺(4.427%)等;内果皮浸提液中鉴定出的有机化合物也为脂肪酸类物质,可能有抑制作用的也是棕榈酸(4.691%)、油酸(4.022%)、9,12-十八碳二烯酸(2.264%)和2,4-二甲苯胺(24.248%)、3,4-二甲苯胺(18.506%)等。由此可知,'紫梗籽银'桂种子的内果皮和种实中提取的抑制物质种类比较少,主要是一些脂肪酸类和苯胺类物质,且种实中提取的有机物种类多于内果皮。

表 1 内果皮和种实甲醇浸提液中有机化合物种类及相对含量

序号	分子式	主要成分	质量分数/% 种实	质量分数/% 内果皮
1	—	Benzeoic caid,3-formyl-4,6-dihydroxy-2,5-dimethyl	6.967	—
2	—	Benzeoic caid,2-hydroxy-4-methoxy-6-methyl	9.213	—
3	—	1,1-Biphenyl,4-(hydroxyacetyl)-	4.609	—
4	$C_{16}H_{32}O_2$	n-Hexadecanoic acid 十六(烷)酸、棕榈酸	1.817	4.691
5	$C_{18}H_{32}O_2$	9,12-Octadecadienoic acid(Z,Z) -9,12-十八碳二烯酸	1.840	2.264
6	$C_{17}H_{33}COOH$	Oleic acid 油酸、顺式-9-十八(碳)烯酸	4.208	4.022
7	$C_8H_{11}N$	Benzenamine,2,4-dimethyl 2,4-二甲苯胺	21.689	24.248
8	$C_8H_{11}N$	Benzenamine,3,4-dimethyl 3,4-二甲苯胺	4.427	18.506
9	—	Pivalate,[1-nitro-3,4-dimethyl-3-cyclohexenyl]met	—	25.782

3 结论与讨论

'紫梗籽银'桂种子不同部位浸提液对白菜籽发芽率以及苗高、根长的生长均有显著的抑制作用。且内果皮和种实浸提液的白菜种子发芽以及幼苗生长测定结果变化趋势大致相同,但各有机相对白菜籽发芽及幼苗生长的抑制作用不同。在相同浓度处理下,'紫梗籽银'桂内果皮和种实各分离相对白菜籽发芽率以及苗高、根长生长的抑制作用的强度都存在一定差异。其中,无论是'紫梗籽银'桂的内果皮还是种实部分,其乙酸乙酯相和甲醇相对白菜籽发芽率以及苗高、根长生长的抑制作用均最强烈,尤其是对苗高生长的抑制率达到100%,故认为'紫梗籽银'桂内果皮和种实的乙酸乙酯相或甲醇相中存在着某些萌发抑制物质,这些物质对种子的萌发及其生长的抑制活性最强。同时,虽然不同浓度间抑制作用的差异不显著(可能是由于浓度梯度比较小),但还是可以看出抑制作用会随着各相处理浓度的增大而增强,这与其他植物[8-9]抑制物研究中的结论相符合。'紫梗籽银'桂内果皮和种实中各有机相的抑制作用或程度表现不同的原因可能由以下几方面引起:一是内果皮和种实的乙醚相、乙酸乙酯相、甲醇相和水相中萌发抑制物的种类及其含量存在差异;二是各抑制物的抑制能力不尽相同。许传莲等[10]也曾提出一种植物种子中可能存在多种抑制物质,但起关键作用的可能仅有一种或几种,对白菜种子起抑制作用的质量浓度对该植物种子可能无抑制作用,故'紫梗籽银'桂种子中所含抑制物对其自身萌发是否存在抑制作用,其自身所含抑制物的质量浓度效应等均有待进一步研究。

  植物种子中存在内源抑制物质是导致种子休眠的一个重要原因,相关研究也有许多。紫椴种子中存在壬酸、辛酸、棕榈酸、十五烷酸等发芽抑制物质[11]。青钱柳种子中鉴定到的抑制物中也包括了壬酸、辛酸、棕榈酸、邻苯二甲酸二乙酯和顺式-9,12-十八碳二烯酸等发芽抑制物质[2]。刘艳[12]指出梭梭种子中的萌发抑制物质主要是有机酸和酚类物质。于海莲等[13]从南方红豆杉种子甲醇浸提液中分离检测到的有机物质主要为有机酸类、酯类、胺类、醇类和酮类等。本试验中,经 GC-MS 分析鉴定结果显示,'紫梗籽银'桂内果皮和种实中存在多种有机化合物,这些化合物主要以脂肪酸类(油酸、棕榈酸和十八烯酸)、苯胺类(邻、间二甲苯胺)为主,它们对种子萌发都有抑制作用,但它们的作用机制和作用方式还不清楚。可见种子中的内源抑制物种类具有多样性,因此在考虑内源抑制物对种子休眠的影响时,不能简单地认为是某种抑制物质作用的结果,哪一种或几种真正起抑制作用还有待于深入研究。而且本次试验由于受气相色谱数据分析软件的限制,还有一部分含量较高但匹配度低于70%的化合物,其主要成分至今未能明确,其对于种子萌发是否具有抑制作用还有待进一步分析。

参考文献

[1] 孙佳,郭江帆,魏朔南.植物种子萌发抑制物研究概述[J].种子,2012,31(4):57-61.

[2] 尚旭岚,徐锡增,方升佐.青钱柳种子休眠机制[J].林业科学,2011,47(3):68-74.

[3] 宋会兴,刘光立,高素萍,等.四川牡丹种子浸提液内源抑制物活性初探[J].园艺学报,2012,39(2):370-374.

[4] 郭江帆,孙佳,魏朔南.秦艽种子内源萌发抑制物的初步探究[J].种子,2013,32(7):39-43,49.

[5] 杨秀莲,丁彦芬,甘习华.'紫梗籽银'桂种子休眠原因的初步探讨[J].江苏林业科技,2007,34(6):18-20,45.

[6] 黄耀阁,崔树玉,鲁歧,等.西洋参种子抑制物质的初步研究[J].吉林农业大学学报,1994,16(2):9-14.

[7] 韩宝瑞,黄耀阁,李向高.西洋参果肉中的六种发芽抑制物质[J].特产研究,2000(1):13-17.

[8] 史锋厚,沈永宝,施季森. 南京椴种子发芽抑制物研究[J]. 福建林学院学报,2007,27(3):222-225.
[9] 杨勇,刘光立,宋会兴,等. 四川牡丹胚乳浸提液对油菜种子萌发与幼苗生长的影响[J]. 西南农业学报,2013,26(1):89-92.
[10] 许传莲,崔树玉,张崇禧,等. 人参种子不同后熟期抑制物质的动态分析[J]. 人参研究,1996(3):26-28.
[11] 王海南. 紫椴种子休眠机理及休眠解除技术的研究[D]. 哈尔滨:东北林业大学,2012.
[12] 刘艳. 梭梭种子内源抑制物质及萌发生理生化变化研究[D]. 北京:北京林业大学,2007.
[13] 于海莲,李凤兰,赵翠格,等. 南方红豆杉种子发芽抑制物质的初步研究[J]. 北京林业大学学报,2009(5):78-83.

The Analysis of Inhibitor Substance for Seeds of *Osmanthus fragrans* 'Zigeng Zi Yin'

SHI Tingting, YANG Xiulian, WANG Lianggui

(College of Landscape Architecture, Nanjing Forestry University, Nanjing 210037, Jiangsu, China)

Abstract: With *Osmanthus fragrans* 'Zigeng Zi Yin' seeds as the test materials, its endocarp and endogenesis inhibitory substance was extracted, bioassayed and then identified by GC-MS. The result indicated that: ①There were some substance inhibiting seed germination and seedling growth in both endocarp and endogenesis of *O. fragrans* 'Zigeng Zi Yin' seeds, and the inhibitory effect of disperse phases in endocarp and endogenesis were different. Among these the inhibition of petroleum ether phase was weakest, while the ethyl acetate phase and methanol phase had higher inhibitory effects. ②The GC-MS results show that there were many inhibiting substances in both endocarp and endogenesis, including some germination inhibiting substances, such as Palmitic acid, oleic acid, 9,12-octadecaienoic acid, etc., fatty acid are the major ingredients.

Key words: *Osmanthus fragrans* 'Zigeng Zi Yin'; seeds; inhibitory substance; germination rate; seedling growth; GC-MS

3个桂花品种对NaCl胁迫的光合响应[*]

杨秀莲,母洪娜,郝丽媛,王良桂

(南京林业大学 风景园林学院,江苏 南京 210037)

摘 要:以3个1年生桂花(Osmanthus fragrans)品种幼苗为材料,设置低、中、高3个不同的NaCl浓度梯度,研究桂花盐害指数的变化、叶片光化学效率和叶绿素含量的影响。结果表明:盐害指数随着胁迫时间的延长、盐处理浓度的增大而增大;光化学效率和叶绿素含量随不同浓度处理时间的延长而下降,当盐胁迫浓度为40 mmol·L^{-1}时,3个桂花品种的光化学效率和叶绿素含量与对照相比差异不显著;当盐浓度分别为70、100 mmol·L^{-1}时,差异显著;但品种间差异不显著。

关键词:桂花;NaCl胁迫;盐害指数;实际光化学效率;叶绿素

土壤盐渍化不仅是沿海地区面临的主要问题,更是世界农林业面临的重要问题之一。随着城市化进程的推进,城市土壤的盐分、pH上升及其他污染物含量逐年增加,城市土壤结构和理化性质严重恶化,园林植物在城市的生存及其正常的生长发育面临着严峻的挑战。桂花(Osmanthus fragrans)作为中国十大传统名花之一,不仅具有很高的文化价值,而且集园林应用[1]、药用保健[2]等多种功能于一体,研究桂花在盐胁迫条件下的生理响应有助于筛选耐盐桂花品种在盐渍化地区的绿化美化,然而迄今为止,中国关于桂花耐盐胁迫的研究还未见报道。因此,本研究通过不同浓度梯度的NaCl胁迫处理后对桂花3个品种的盐害指数(SI)、实际光化学效率(ΦPSII)、最大光化学效率(F_v/F_m)、叶绿素(chl)含量进行测定,以期为进一步探讨盐胁迫对桂花的影响及其耐盐适应性研究,为城市园林中桂花养护管理提供理论依据。

1 材料与方法

1.1 材料

'潢川金桂'(O. fragrans 'Huangchuan Jingui')、'笑秋风'(O. fragrans 'Xiao Qiufeng')、'大叶银'桂(O. fragrans 'Daye Yin')盆栽1年生实生苗。在温室中将1年生实生苗(苗高10 cm)移栽于内径15 cm,高10 cm的盆内,基质为去离子水洗净的河沙,每品种240株,每盆种植3株。盆栽试验在南京林业大学温室进行,指标测定在南京林业大学风景园林学院实验室完成。

---

[*] 原文发表于《河南农业大学学报》,2015,49(2):105-108。
基金项目:江苏省科技支撑项目(BE2011367);国家林业局公益性行业专项(201204607);江苏高校优势学科建设工程项目(PAPD)。

试验采用 3 因素随机区组设计,模拟苏北土壤含盐量[3],采用含有 NaCl(40、70、100 mmol·L^{-1})的 Hoagland 溶液浇灌处理组,对照组浇灌不含 NaCl 的 Hoagland 溶液。为避免盐激效应,在正式处理前先进行预处理(40 mmol·L^{-1}),以后每 2 天浇灌预定浓度溶液 1 次,每次增加 30 mmol·L^{-1},每次浇 0.5 L·盆$^{-1}$(保证离子完全代换)。达到各处理浓度的当天为处理 0 d,处理 19 d 后选取植株测定荧光参数并进行破坏性取样,测定叶绿素含量,其间每天观察记录植株的受害症状。每处理 3 盆,3 次重复。

1.2 试验仪器

紫外可见分光光度计(岛津 UV-2450/2550)、电导仪(雷磁 DDS-308A)、精密天平、等离子体发射光谱仪(华科易通 HK-8100 型)、脉冲调制式荧光仪(英国 Hansatech FMS2 型)。

1.3 指标测定及数据处理

1.3.1 盐害指数的测定

对全部植株进行观察,根据叶片受害程度划分盐害等级并赋值,计算幼苗在不同 NaCl 胁迫条件下的盐害指数[4]。

$$SI = (1 \times n1 + 2 \times n2 + 3 \times n3 + 4 \times n4)/(4 \times m) \tag{1}$$

式中:$n1$、$n2$、$n3$、$n4$ 分别为不同盐害级别的受害株数;m 为总株数。0 级为全株无盐害症状;1 级为轻度盐害,少部分(小于 20%)叶片边缘有轻微发黄、枯焦、脱落症状;2 级为中度盐害,近 50% 叶片发黄、枯焦或脱落;3 级为重度盐害,50% 以上的叶片发黄、枯焦或脱落;4 级为极重度盐害,90% 以上叶片枯焦、脱落或植株死亡。

1.3.2 叶绿素含量的测定

桂花叶绿素的提取方法参考郝再彬[5]的试验方法。

以体积分数为 95% 的乙醇为空白对照,在波长 665、649、470 nm 下测定吸光度。

叶绿素浓度计算公式为:

$$C_a = 13.95A_{665} - 6.88A_{649} \quad C_b = 24.96A_{649} - 7.32A_{665} \tag{2}$$

叶绿素含量的计算公式为:

$$A = n \times C \times N \times W^{-1} \times 10^{-3} \tag{3}$$

式中:A 为叶绿体色素(mg·g^{-1});n 为提取液体积(mL);C 为色素的质量浓度(mg·L^{-1});N 为稀释倍数;W 为样品鲜重(g)。

1.3.3 荧光参数的测定

每个处理选择 5 株进行挂牌标记,每株选定中上部向光叶片;利用 FMS2 型脉冲调制式荧光仪,测定初始荧光 F_o、最大荧光 F_m、光下最大荧光 $F_m{'}$、稳态荧光 F_s 等荧光参数。测定暗适应下叶片的 F_o 和 F_m 时,需要把叶片夹入暗适应夹中暗适应 20 min;荧光参数的计算参照 Genty 等[6]及 FMS2 使用手册。

1.3.4 数据分析

用 Excel 2000 进行计算和作图,SPSS 19.0 分析软件分析数据。

2 结果与分析

2.1 盐胁迫下盐害指数的变化

3个桂花品种对盐胁迫的反应有差异,随着盐浓度的升高,桂花品种的盐害指数也呈现增加的趋势(表1)。观察发现,最早出现盐害症状的是'笑秋风'和'大叶银'桂,在100 mmol·L^{-1}的盐浓度下处理第5 d时出现下部叶片发黄、新叶边缘发焦的现象;70 mmol·L^{-1}处理的'潢川金桂'在第8 d时出现相同现象。70、100 mmol·L^{-1}处理的'笑秋风'在处理第10 d,出现下部叶片死亡枯萎现象,上部叶片长势良好;处理第19 d,100 mmol·L^{-1} NaCl浓度处理下,'笑秋风'保持相对较高的生长势,盐害系数较低。低浓度的盐分对'潢川金桂'的影响较小,胁迫第19 d,只表现为零星叶缘发焦,但高浓度(100 mmol·L^{-1})的盐胁迫处理下,'潢川金桂'植株全部受害。'大叶银'桂在处理第12 d时,出现个别植株死亡,处理中期,低浓度盐分对植株影响较小,只出现零星叶片掉落,生长势良好,中、高浓度处理与'笑秋风'的表现类似,处理后期,'大叶银'桂约2/3叶片发黄,但仍保持良好的生长状态。

表1 盐胁迫处理19 d后3个桂花品种的盐害指数
Tab. 1 Salt Injury Index of three cultivars of *Osmanthus fragrans* under salt stress after 19 days

NaCl 浓度/ (mmol·L^{-1}) NaCl Concentration	盐害指数/% Salt Injury Index		
	'笑秋风' 'Xiao Qiufeng'	'潢川金桂' 'Huangchuan Jingui'	'大叶银'桂 'Daye Yin'
0	0	0	0
40	21.59	25	33.33
70	45.97	27.5	52.08
100	51.14	70.83	61.81

2.2 盐胁迫对不同桂花品种光合特性的影响

随着盐胁迫的增强,3个桂花品种的实际光化学效率(ΦPSII)均呈现下降趋势,说明盐胁迫抑制了植物的光化学反应。在40、70、100 mmol·L^{-1}盐浓度胁迫下,'笑秋风'的ΦPSII比对照分别下降了14.08%、21.97%和49.69%,各浓度间差异显著;'潢川金桂'的ΦPSII比对照分别增加0.16%、下降8.13%和45.86%;'笑秋风'的ΦPSII分别下降8.27%、26.10%、39.24%。其中'潢川金桂'和'大叶银'桂的高浓度和中低浓度间差异显著,但不同浓度处理时品种之间的差异均不显著(图1)。

不同浓度盐胁迫下,3个桂花品种的最大光化学效率总体上随着处理浓度的升高呈下降趋势。在40、70、100 mmol·L^{-1}的盐浓度处理下,'笑秋风'的F_v/F_m值分别下降1.1%、0.92%、2.36%;'潢川金桂'的值分别下降了0.91%、7.95%、15.56%;'大叶银'桂的F_v/F_m值分别下降了0.38%、7.36%、1.89%。方差分析和多重比较显示,'笑秋风'和'大叶银'桂不同浓度间均无显著差异,仅'潢川金桂'在100 mmol·L^{-1}处理值与中低浓度处理值间有显著差异,并且与另2个桂花品种也有显著差异(图2)。试验表明在70 mmol·L^{-1}盐浓

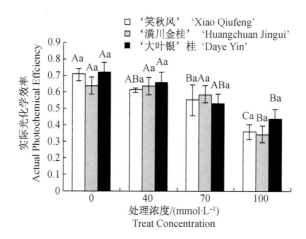

图 1　盐胁迫对 3 个桂花品种实际光化学效率的影响

Fig. 1　Effects of salt stress on the actual photochemical efficiency of three *O. fragrans* cultivars

注：小写字母表示相同浓度不同品种间差异显著（$P<0.05$）；大写字母表示同一品种不同处理浓度间差异显著（$P<0.05$）。下同。

Note：Lowercase letters indicate significant difference between different cultivars of the same concentration（$P<0.05$）；Uppercase letters mean significant difference between different concentrations to deal with the same cultivars（$P<0.05$）. The same as below.

度胁迫下，'笑秋风'的最大光化学效率受到的影响最小，与盐害指数值对应，因此认为'笑秋风'对中等浓度盐胁迫耐受力较强。

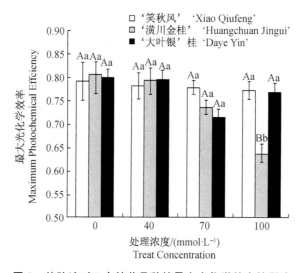

图 2　盐胁迫对 3 个桂花品种的最大光化学效率的影响

Fig. 2　Effects of salt stress on the maximal photochemical efficiency of three *O. fragrans* cultivars

2.3　盐胁迫对叶绿素总量和叶绿素 a/b 的影响

3 个桂花品种的叶绿素总量总体上随着盐胁迫的加强而呈下降趋势，在 40、70、100 mmol·L^{-1} 的盐浓度处理下，'笑秋风'的叶绿素总量比对照分别下降了 23.37%、32.48% 和 35.82%，处理和对照间差异显著；'潢川金桂'的叶绿素总量在 40 mmol·L^{-1} 处理时增加了1.03%（同PSII变化相似），在70、100 mmol·L^{-1}的盐浓度处理下比对照下降了4.65%和19.15%，其中100 mmol·L^{-1}处理值与中低浓度间差异显著；'大叶银'桂的叶绿素总量分别下降了7.48%、9.19%和8.03%，各浓度间无显著差异（图3）。

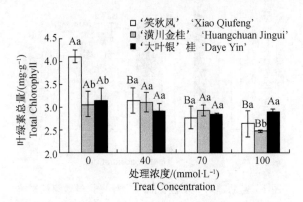

图3 盐胁迫19 d对3个桂花品种叶绿素总量的影响
Fig. 3　Effects of salt stress after 19 days on the chlorophyll content of three *O. fragrans* cultivars

3个桂花品种的叶绿素a和叶绿素b的比值（a/b）变化见图4。其中，'潢川金桂'的叶绿素a/b值在0、40 mmol·L^{-1}浓度处理时显著高于'笑秋风'和'大叶银'桂，且差异显著，不同盐浓度处理间也存在显著差异；而'笑秋风'和'大叶银'桂各浓度处理间差异不显著。100 mmol·L^{-1}的盐浓度处理时，3个品种间无显著差异。

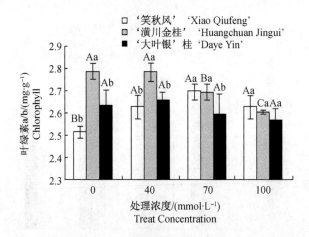

图4 盐胁迫处理19 d对3个桂花品种叶绿素a/b的影响
Fig. 4　Effects of salt stress after 19 days on the chlorophyll a/b of three *O. fragrans* cultivars

3　结论与讨论

在NaCl胁迫下，盐害指数可以反映NaCl胁迫对植物生长的综合伤害程度。桂花品种的盐害指数随着盐浓度的升高呈现增加的趋势，这与骆建霞等[7]人的盐害试验结果一致。综合考虑盐害症状出现的时间、盐害率、盐害指数等因素可以初步确定3个桂花品种的耐盐性强弱排序为：'笑秋风'＞'潢川金桂'＞'大叶银'桂。

盐胁迫植物降低光合作用的因素是多方面的,如盐胁迫影响 CO_2 扩散到结合部位,抑制同化产物转移,气孔因素、非气孔因素等。盐胁迫初期海滨锦葵光合降低的主要因素是气孔限制,在后期非气孔限制因素逐渐开始起作用,CO_2 补偿点随盐浓度增加的幅度越来越大[8]。因此,光合产物的损耗增加,实际光化学效率、最大光化学效率均降低;这与本研究得出的结论,即 NaCl 胁迫降低了试验中 3 个桂花品种的实际光化学效率和最大光化学效率相一致。

叶绿素含量的多少在一定程度上反映了植物光合作用强度的高低,从而影响植物的生长,因此,叶绿素含量不仅是直接反映植物光合能力的一个重要指标,而且是衡量植物耐盐性的重要生理指标之一。绿色植物叶绿素包括叶绿素 a、b 两种;盐胁迫条件下,叶绿素含量随盐浓度的升高而下降,并且高浓度处理的叶绿素总量与对照差异显著或极显著(图3);叶绿素 a 和叶绿素 b 的比值总体上也是随盐浓度升高而呈现先高后低的趋势。许祥明等[9]认为植物的叶绿素含量下降主要是由于叶绿素酶对叶绿素 b 的降解所致,对叶绿素 a 的影响较小,所以叶绿素 a/b 的比值上升。Santos[10] 的试验也表明随着 NaCl 浓度的增加,叶绿素含量和荧光性(F_m)下降,25 mmol·L^{-1} NaCl 浓度处理下叶绿素酶活性增强,高浓度(50、100 mmol·L^{-1})抑制叶绿素前体——氨基酮戊酸(5-aminolaevulinic acid)的合成,并且盐胁迫对叶绿素合成的影响远大于叶绿素酶降解。因此,叶绿素 a/b 值总体上也是随盐浓度升高而呈现先高后低的趋势。其原因可用许详明等[9]和 Santos[10] 的研究结果证实,即低浓度 NaCl 胁迫主要是叶绿素酶活性增强导致桂花叶绿素总量下降,并且叶绿 a/b 值较高;高浓度 NaCl 抑制叶绿素前体的合成,并且其效果远大于叶绿素酶。Mehta 等[11] 的研究发现,高盐胁迫后 PSII 受体侧完全可修复,但是供体侧只有不足 85% 可以修复。因此,高盐胁迫对 PSII 供体侧损害比受体侧更严重。虽然高盐胁迫对 PSII 供体的损害远远高于 PSII 的受体,但是这足以限制 PSII 的光合特性。

参考文献

[1] 汪小飞,史佑海,向其柏.中国古典园林与现代园林中桂花应用研究[J].江西农业大学学报 2006,5(2):85-87.
[2] 王丽梅.桂花有效成分合成转化规律与药学研究[D].武汉:华中科技大学,2009.
[3] 刘广明,杨劲松,姜艳.江苏典型滩涂区地下水及土壤的盐分特征研究[J].土壤,2005,37(02):163-168.
[4] 李倩中,苏家乐.4 种槭属植物耐盐性差异的研究[J].江苏农业科学,2009(06):227-228.
[5] 郝再彬.植物生理实验[M].哈尔滨:哈尔滨工业大学出版社,2004.
[6] GENTY B, BRIANTAIS J M, BAKER N R. The relationship between the quantum yield of photosynthetic electron transport and quenching of chlorophyll fluorescence [J]. Biochimicaet Biophysica Acta, 1989, 990(1):87-92.
[7] 骆建霞,史燕山,吕松.3 种木本地被植物耐盐性的研究[J].西北农林科技大学学报.2005,33(12):121-124,129.
[8] 林莺,李伟,范海,等.海滨锦葵光合作用对盐胁迫的响应[J].山东师范大学学报(自然科学版),2006(02):118-120.
[9] 许祥明,叶和春,李国凤.植物抗盐机理的研究进展[J].应用与环境生物学报,2000(4):379-387.
[10] SANTOS C V. Regulation of chlorophyll biosynthesis and degradation by salt stress in sunflower leaves[J]. Scientia Horticulturae,2004,103:93-99.

[11] MEHTA P, JAJOO A, MATHUR S, et. al. Chlorophyll a fluorescence study revealing effects of high salt stress on Photosystem II in wheat leaves [J]. Plant Physiology and Biochemistry, 2010, 48 (1): 16-20.

Photosynthetic Response of Three *Osmanthus fragrans* Cultivars to NaCl Stress

YANG Xiulian, MU Hongna, HAO Liyuan, WANG Lianggui

(College of Landscape Architecture, Nanjing Forestry University, Nanjing 210037, China)

Abstract: The effects of salt injury index, leaf $\Phi PSII$, F_v/F_m and Chlorophyll content of three one-year-old *Osmanthus fragrans* cultivars were investigated by using low, moderate and high NaCl concentrations. The results showed that the salt index increased with the increase of stress intensity and duration, while leaf $\Phi PSII$, F_v/F_m and Chlorophyll content were reduced. There was no significant difference when these seedlings were treated with 40 mmol · L^{-1} NaCl which was melted in Hoagland solution. But there was significant difference when treated by 70 mmol · L^{-1} and 100 mmol · L^{-1} NaCl respectively, whereas there was no significant difference among the three different cultivars.

Key words: *Osmanthus fragrans*; NaCl stress; salt injury index; $\Phi PSII$; chlorophyll

'朱砂丹桂'扦插技术及生根过程中生理生化分析*

杨秀莲,冯 洁,王良桂

(南京林业大学 风景园林学院,江苏 南京 210037)

摘 要:研究了激素种类、激素浓度、扦插基质对'朱砂丹桂'扦插生根的影响,找出了其生根最优组合,即 100 mg/L ABT 生根粉+珍珠岩+蛭石,并初步确定'朱砂丹桂'生根类型属皮部生根型。试验结果还表明,在生根过程中,可溶性糖含量与淀粉含量在不定根生出之前变化趋势呈负相关;可溶性蛋白在不定根形成之前持续下降,后上升;总氮值在不定根形成期下降,后上升;C/N 在不定根伸长期之前持续下降,后上升;过氧化物酶活性在根原基诱导期及不定根伸长期分别出现高峰。

关键词:'朱砂丹桂';扦插生根;营养物质;变化

桂花是我国十大传统名花之一,属于木犀科常绿小乔木,是一种绿化、美化、香化三者密切结合的优良园林树种[1]。'朱砂丹桂'(*Osmanthus fragrans* 'Zhusha Dangui')是丹桂的一种,其主干挺拔秀丽,枝叶婆娑多姿,清香四溢,花瓣橙红色,花量大,花色艳丽,观赏效果极佳,是优良的园林绿化和庭院观赏树种[2]。此外,'朱砂丹桂'食用、药用、经济价值也很高。近年来,市场需求不断提高,据调查,干径 8 cm 以上的丹桂稀缺,每株单价在 2 000 元以上,干径 3 cm 以上的植株每株也值数百至上千元。

桂花传统繁育多用嫁接、压条等方法,但是嫁接和压条会导致母树受损严重,成本高且繁殖量少,影响了桂花产业的发展。因而在生产上,以扦插繁殖最为普遍,但不同的品种有不同的扦插生根能力,有的品种扦插生根率高达 90% 以上,而有的却非常低。'朱砂丹桂'属扦插较难生根的树种,扦插成活率很低。因此,在本试验中研究了外源激素和扦插基质对插穗生根的影响,并试图了解其生根机理,旨在找出提高扦插生根率的最佳配方,为进一步加大'朱砂丹桂'的开发与推广奠定基础。

1 材料与方法

1.1 材料与场地

试验材料为'朱砂丹桂',树龄 30 年,生长在南京林业大学园林实训中心内,试验在该实训中心的温室内进行。

---

* 原文发表于《江苏农业科学》,2015,43(3):155-158。
基金项目:国家林业局公益性行业专项(编号:201204607)。

1.2 试验设计与方法

1.2.1 生根观察与最优化试验

6月初,在清晨温度较低时采集植物外围的1年生健康枝条,将采下的枝条放在清水中,插穗长度为8~10 cm,保留1~2片叶子。插穗上切口为平口,下端口斜切且尽量靠近叶节处。扦插前把插穗下端对齐捆好,插穗下端在提前配好的生长调节剂中速蘸30 s后扦插。插床用大号周转箱(58 cm×36 cm×18 cm),底部均匀打孔,以利于排水,扦插前2 d将基质装入容器内,用0.5%的高锰酸钾进行淋灌消毒。

扦插完毕后,周转箱上盖遮阴网,定期喷水,保持适宜扦插生根的温度与湿度。

(1)生根类型的观察以ABT生根粉(250 mg/L)处理插穗为对象,以清水作为对照处理。扦插后,每天观察插条愈伤组织形成和不定根出现的情况,至切口愈合,长出新根为止,记录每次观察的结果。每次随机抽取各处理5株,3次重复。

(2)生根率最优组合试验:采用$L_9(3^3)$正交设计,试验因素和水平见表1,共9个处理,每处理60株插穗,3次重复。生根结束后,对生根性状的5个指标:生根率、愈伤组织形成率、存活率、最长根长、根系效果指数进行测定,其中,根系效果指数=平均根长×根系数量/总插条数。

表1 丹桂嫩枝扦插正交试验因素和水平

水平	A:生长调节剂种类	B:生长调节剂浓度/(mg·L^{-1})	基质类型(C)
1	IAA	100	珍珠岩1:泥炭1
2	ABT生根粉	250	珍珠岩1:蛭石1
3	NAA	500	珍珠岩1:蛭石1:泥炭1

1.2.2 生根过程中韧皮部营养物质及过氧化物酶(POD)活性变化测定

选用珍珠岩+蛭石为基质,250 mg/L ABT生根粉处理插穗,进行扦插,每个处理50根,重复3次,以清水作对照。每隔一定的时间随机选取18~20根插穗,将所取材料用蒸馏水冲洗擦干后,置于冰盒内带回实验室,剥取插穗基部2 cm区域内的韧皮部,将其剪碎,置于超低温冰箱中保存备用。

采用考马斯亮蓝G-250染色法[3]测定可溶性蛋白含量;采用蒽酮比色法测定可溶性糖含量[4]和淀粉含量[5];用愈创木酚法[6]测定POD活性;采用H_2SO_4-H_2O_2消煮法[7]测定全氮含量。

试验数据处理:用Excel软件绘制图表,SPSS 17.0软件进行方差分析和多重对比。

2 结果与分析

2.1 生根类型观察结果

在5~26 d,处理的插穗切口开始有白色絮状愈伤组织产生,此期为愈伤组织诱导期。在27~42 d,处理的插穗切口开始出现白色突起,此期为根原基诱导期,而对照的插穗此时才出现愈伤组织、无白色突起产生。在42~57 d,大部分处理插穗已有不定根产生,此期为不定根形成期,而对照插穗中仅有小部分产生不定根。在57~64 d,有少数侧根形成,这个时期为不定根伸长期。64 d后,根成型,生根基本结束。对照组的生根时间比处理组平均晚12 d左右。此外,从插穗生根部位和生根时间看,其根原基发生在插穗的皮孔下部,与愈伤组织无关,因此朱砂丹桂是典型的皮部生根型植物。

2.2 丹桂扦插生根最优组合试验结果

从表2可以看出,各处理间扦插效果差异显著,其中,愈伤组织形成率为11.67%~71.67%,4号组合最高且与其他处理差异极显著,排序为4号>2号>3号>1号>6号>9号>(5号、8号)>7号。插穗存活率为6.67%~68.33%,4号最高,其次是3号,但二者之间差异不显著,排序为4号>3号>2号>1号>6号>9号>8号>(5号、7号)。生根率为10%~65%,4号生根率最高且与其他处理差异极显著,排序为4号>3号>1号>9号>6号>(2号、5号)>8号>7号。根系效果指数为0.31~1.44,4号最高,与1号差异不显著,排序为4号>1号>5号>3号>2号>6号>9号>8号>7号。最长根长2.43~9.03 cm,4号最高,与1号差异不显著,排序为4号>1号>3号>6号>5号>2号>9号>8号>7号。

表2 扦插生根性状试验结果

编号	组合	愈伤组织形成率/%	插条存活率/%	生根率/%	根系效果指数	最长根长/cm
1	$A_1B_1C_1$	40.00 Cd	40.00 C	38.33 C	1.36 Ab	8.72 A
2	$A_1B_2C_2$	56.67 Bb	55.00 B	13.33 E	0.93 Bcd	4.67 BCd
3	$A_1B_3C_3$	50.00 Bc	61.67 AB	48.33 B	1.09 ABc	5.73 B
4	$A_2B_1C_2$	171.67 Aa	68.33 A	65.00 A	1.44 A	9.03 A
5	$A_2B_2C_3$	13.34 Ef	6.67 E	13.33 E	1.13 BCd	4.80 BC
6	$A_2B_3C_1$	38.34 Cd	28.34 D	26.67 D	0.91 BCd	5.57 BC
7	$A_3B_1C_3$	11.67 Ef	6.67 E	10.00 E	0.31 Df	2.43 Cd
8	$A_3B_2C_1$	13.34 Ef	10.00 E	11.67 E	0.49 CDdf	3.20 BCd
9	$A_3B_3C_2$	28.34 De	25.00 D	28.33 D	0.67 BCD	3.60 BCd

注:各数值均指试验的3个重复的平均值。同列数据后不同大、小写字母分别表示在0.01、0.05水平上差异显著。

根据生根性状方差分析得知,正交试验的9个处理对生根性状的5个指标均有显著差异。由上述分析综合可知:在'朱砂丹桂'的夏季嫩枝扦插中,4号处理(100 mg/L ABT、珍珠岩+蛭石)的几个指标均比其他8个处理高,愈伤组织形成率达到71.67%,插穗存活率达到68.33%,生根率达到65.00%,根系效果指数为1.44,最长根长为9.03 cm。而7号处理(100 mg/L NAA、珍珠岩+蛭石+泥炭土)的生根指标相对较低,愈伤组织形成率仅为11.67%,插穗存活率6.67%,生根率10.00%,根系效果指数为0.31,最长根长为2.43 cm。

2.3 插穗皮部营养物质变化及POD活性变化

2.3.1 插穗皮部营养物质的变化

2.3.1.1 可溶性糖含量的变化 糖类物质是植物体内主要的营养储藏和运输形式,有研究表明,插穗生根与插穗内营养物质有关,可溶性糖是插穗生根和生存所必需的主要营养物质[8]。图1表明,'朱砂丹桂'插穗内可溶性糖含量的变化趋势是上升—下降—上升—下降,有较大波动。综合生根过程观察分析,250 mg/L ABT 生根粉处理过的插穗,初期由于淀粉类物质生物降解,含糖量有轻微上升,16~30 d 内,愈伤组织形成需要消耗营养,造成插穗内可溶性糖含量下降。扦插 30 d 后,可溶性糖含量达到最低,此后愈伤组织形成而使淀粉酶活性增强,促进淀粉水解,可溶性糖含量开始升高。57 d 后可溶性糖含量下降,此时是不定根伸长期,可溶性糖大量消耗。此外,处理与对照差别不大,可见 250 mg/L ABT 生根粉对插穗内部可溶性糖变化影响不大,这与前人研究结果不同,可能是由于试验材料的差异所致。

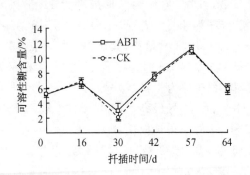

图 1　扦插过程中可溶性糖含量的变化　　　　图 2　扦插过程中淀粉含量的变化

2.3.1.2　淀粉含量的变化　淀粉是插穗中主要的储藏物质,淀粉通过水解转化为糖类从而供给生根所需的能量,插穗内淀粉含量越高,说明转化成可溶性糖的含量就越低[9]。由图 2 可见,扦插过程中插穗皮部淀粉含量的变化趋势基本是下降—上升。0~30 d,处理插穗皮部淀粉含量持续下降,对照则呈现先降后轻微上升的趋势。处理插穗皮部淀粉含量在30 d 时达到最低,可能是由于愈伤组织形成后,促使淀粉酶活力增强,部分淀粉被分解提供能量。之后含量上升,说明伴随着插穗不定根的形成,不定根开始吸收营养物质,淀粉得到积累。初步推断,250 mg/L ABT 生根粉能促进淀粉转化为可溶性糖的速度。

2.3.1.3　可溶性蛋白含量的变化　蛋白质在生物体内参与构成细胞、调节代谢等多种作用,因此测定可溶性蛋白含量变化也可反映出扦插生根过程中的复杂变化[10]。由图 3 知,可溶性蛋白是下降—上升—下降的趋势。ABT 生根粉处理的插穗,初始的 0~42 d 内,可溶性蛋白含量下降,蛋白质分解,为愈伤组织的形成提供营养物质。在 42~57 d 内,不定根诱导形成,储存了充足的营养物质,致使可溶性蛋白含量上升。虽然对照的变化趋势与处理的差不多,但是对照较处理更为和缓。这说明,250 mg/L ABT 生根粉刺激了插条,可能加快了可溶性蛋白运作的速度,从而促进生根。

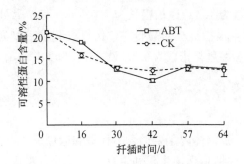
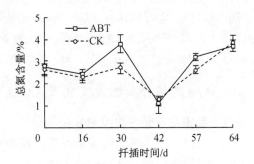

图 3　扦插过程中可溶性蛋白含量的变化　　　　图 4　扦插过程中总氮含量的变化

2.3.1.4 总氮含量的变化　氮元素也是植物生长的重要元素,与植物根部生长有着密切的关系[11]。由图4可知,总氮含量呈现下降—上升—下降—上升的趋势。在42 d不定根形成期,总氮含量达到最低值,可能是低氮的环境更有利于不定根的长出。在愈伤组织形成前期、不定根形成期,处理总氮的含量比对照高,初步推断250 mg/L ABT生根粉可能会加速总氮含量的生成。

2.3.1.5 碳氮比的变化　碳、氮营养是插穗生根前维持生命和生根所不可缺少的重要能源[12]。插穗生根与碳水化合物和含氮化合物的比率有关[13]。由图5可知:碳氮比值呈现下降—上升趋势,在愈伤组织和根原基的发育阶段,碳氮比值的下降,可能促进了根原基发端和发育,而在不定根伸长期碳氮比又开始上升。这说明'朱砂丹桂'不定根的形成产生需要较低的C/N值,直到不定根伸长期才需要较高的C/N值。

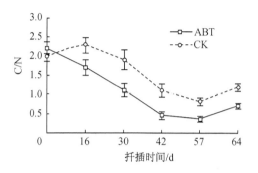

图5　扦插过程中碳氮比的变化

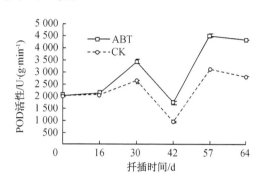

图6　扦插过程中POD的活性变化

2.3.2 插穗皮部内POD的活性变化

POD是植物体内酶促保护系统(即保护酶系统)的重要组成成分,能消除植物体内的内源IAA,促使诱导根原基[14]。POD活性与愈伤组织形成能力,以及与插穗不定根的诱导均有密切的关系[14]。图6显示了'朱砂丹桂'嫩枝扦插过程中插穗POD活性的变化,变化趋势为上升—下降—上升—下降。POD活性在愈伤组织的形成期和不定根伸长期均出现高峰,说明高活性的POD有助于消除体内的过氧化氢、酚类物质以及多余的IAA,从而有利于根原基的诱导。对照中POD活性与处理的变化趋势基本相同,但处理的插穗中POD活性比对照的高,且两者之间POD活性差异较大,相关分析表明,生根率与POD活性呈正相关,说明250 mg/L ABT生根粉能提高POD活性,从而提高生根率。

3 结论与讨论

3.1 '朱砂丹桂'生根类型与最优组合

本试验表明,'朱砂丹桂'扦插生根属于皮部生根型。激素的种类与浓度、基质种类对插穗的生根有显著的影响。100 mg/L ABT生根粉＋珍珠岩＋蛭石的组合处理效果最佳;而100 mg/L NAA＋珍珠岩＋蛭石＋泥炭土组合处理效果最差。

3.2 营养物质与生根的关系

插穗生根的过程是在很大程度上受碳水化合物充足供应所控制的过程。本研究中,

250 mg/L ABT 生根粉处理的插条中,其可溶性糖含量与淀粉含量在不定根生出之前变化趋势呈负相关,这证明其生根之前,淀粉水解转化为糖类,供给生根所需的能量,而生根后,根能吸收营养物质积累淀粉,淀粉含量呈上升趋势,可溶性糖也呈短暂上升。这与杜仲[8]、金露梅[16]的扦插生根过程中的变化基本一致。可溶性蛋白在不定根形成之前不断下降,说明蛋白质分解为愈伤组织的形成提供构成营养,与凹叶厚朴[17]扦插生根过程中蛋白质的变化曲线一致。总氮值在不定根形成期不断下降,说明低氮更利于生根。不定根的形成产生需要较低的 C/N 值,直到侧根生长期才需要较高的 C/N 值,这与落羽杉[18]扦插生根过程中 C/N 值的变化基本一致。但与前人对许多其他植物在这方面的研究结果不尽相同,因此笔者认为,高 C/N 值能促进生根这一理论不一定适合所有植物。POD 活性在根原基诱导期及侧根伸出期分别出现高峰,说明生根率与 POD 活性呈正相关,这与前人对欧榛[19]、马尾松[20]的扦插研究结果一致。

3.3 ABT 生根粉对营养物质变化的影响

虽然 ABT 生根粉对插穗内部可溶性糖的变化影响不大,但可溶性蛋白、淀粉、总氮的含量、C/N 值以及 POD 活性与对照组相比均发生了明显的变化:促进可溶性蛋白运作的速度,促进淀粉转化为可溶性糖的速度,加速总氮含量的生成,降低了 C/N 值。此外,还提高了其 POD 的活性,尤其是在愈伤组织形成期及不定根原基诱导的关键期。有研究证明,这些物质含量的动态变化与插穗生根相关,说明生长调节剂是通过调节插穗内代谢物质的含量来促进插穗生根[21]。本试验中'朱砂丹桂'嫩枝扦插的生根率还不是很高,本研究只是对影响扦插成活率的最佳组合以及相关因素进行了初步试验,要找到'朱砂丹桂'扦插的最佳时期和最佳生长调节剂配方、高效的育苗途径还有待深入研究。

参考文献

［1］向其柏,刘玉莲.桂花资源的开发与应用现状及发展趋势[J].南京林业大学学报(自然科学版),2004,28(增刊1):104-108.

［2］李士保,王长海,蔡凤林,等.丹桂苗木繁育技术研究[J].现代农业科技,2009(22):198,201.

［3］赵英永,戴云,崔秀明,等.考马斯亮蓝 G-250 染色法测定草乌中可溶性蛋白质含量[J].云南民族大学学报(自然科学版),2006,15(3):235-237.

［4］张志良.植物生理学实验指导[M].北京:高等教育出版社,1990:154-155.

［5］黄光文,沈玉平,李常健.甘薯淀粉含量测定的新方法[J].湖南农业科学,2010(17):109-111.

［6］李忠光,龚明.愈创木酚法测定植物过氧化物酶活性的改进[J].植物生理学通讯,2008,44(2):323-324.

［7］景丽洁,袁东海,王晓栋,等.水生植物总氮测定中两种消化方法的比较[J].环境污染与防治,2005,27(5):392-394.

［8］Rafiqul Hoque A T M. Rootability of *Dalbergia sissoo* Roxb. Cuttings from different clones at two different levels and their primary field growth performance[J]. Dendrobiology, 2008, 59(4): 9-12.

［9］吕明霞.梅花硬枝扦插繁殖与贮藏养分的关系[J].浙江农业科学,2000(4):48-50.

［10］徐丽萍,上官新晨,喻方圆.秤锤树嫩枝扦插过程中营养物质含量的变化[J].江西农业大学学报,2012,34(1):50-53.

［11］李振坚,陈俊愉,吕英民.木本观赏植物绿枝扦插生根的研究进展[J].北京林业大学学报,2001,23

(增刊2):83-85.
[12] 大山郎雄.植物扦插理论与技术[M].北京:中国林业出版社,1989:89.
[13] 郭素娟.林木扦插生根的解剖学及生理学研究进展[J].北京林业大学学报,1997,19(4):64-69.
[14] 原牡丹,侯智霞,翟明普,等. IAA分解代谢相关酶(IAAO、POD)的研究进展[J].中国农学通报,2008,24(8):88-92.
[15] Calderon Baltierra X V. Changes in peroxidase activity during root formation by *Eucalyptus globules* shoots raised in vitro[J]. Plant Perox Newlett,1994,46(4):27-29.
[16] 张忠微,石素霞,等.金露梅嫩枝插穗生根过程中营养物质含量变化研究[J].河北农业大学学报,2008,31(4):56-59.
[17] 马英姿,宋荣,宋庆安,等.凹叶厚朴扦插繁殖的生根机理研究[J].中国农学通报,2012,28(33):112-117.
[18] 吴落军.落羽杉的扦插繁殖技术与生根机理研究[D].南京:南京林业大学,2007.
[19] 扈红军,曹帮华,尹伟伦,等.不同处理对欧榛硬枝扦插生根的影响及生根过程中相关氧化酶活性的变化[J].林业科学,2007,43(12):70-75.
[20] 刘玉民,刘亚敏,马明,等.马尾松扦插生根过程相关生理生化分析[J].林业科学,2010,46(9):28-33.
[21] 王关林,吴海东,苏冬霞,等. NAA、IBA对樱桃砧木(*Prunus pseudocerasus* Colt)插条的生理、生化代谢和生根的影响[J].园艺学报,2005,32(4):691-694.

Physiological and Biochemical Analysis of Cutting Technique and Rooting Process of *Osmanthus fragrans* 'Zhusha Dangui'

YANG Xiulian, FENG Jie, WANG Lianggui

(College of Landscape Architecture, Nanjing Forestry University, Nanjing 210037, China)

Abstract: The effects of hormone species, hormone concentration and cutting matrix on the rooting of *O. fragrans* 'Zhusha Dangui' were studied. The optimal combination of rooting was found, namely 100 mg/LABT rooting powder + perlite + vermiculite, and the rooting type of *O. fragrans* 'Zhusha Dangui' was identified as rooting type of skin. The results also showed that during the rooting process, the soluble sugar content was negatively correlated with the starch content before the adventitious roots were born. The soluble protein continued to decrease before the adventitious root formation, and then increased; the total nitrogen value decreased during the adventitious root formation period and then increased; C/N continued to decrease before the adventitious root elongation period, and then increased; POD activity peaked in the root primordium induction period and the adventitious root elongation stage, respectively.

Key words: *O. fragrans* 'Zhusha Dangui'; cutting rooting; nutrient substance

不同桂花品种香气成分的差异分析*

杨秀莲,施婷婷,文爱林,王良桂

(南京林业大学,江苏 南京 210037)

摘 要:为研究不同桂花品种芳香物质的组分与质量分数,利用顶空固相微萃取(HS-SPME)与气相色谱-质谱联用(GC-MS)技术,对4个品种群8个桂花品种鲜花中的芳香物质进行分析。结果共检测出52种化合物,在四季桂品种群的2个品种和丹桂品种群的2个品种中分别都检测到32种化合物,在银桂品种群的2个品种和金桂品种群的2个品种中分别检测到34种和33种化合物。芳樟醇、芳樟醇氧化物、β-紫罗兰酮、二氢-β-紫罗兰酮、正己醛,以及叶醛等构成桂花香气的主要成分。总之,4个品种群的桂花香气成分及质量分数都存在差异,但同一桂花品种群内桂花的花香成分相差不大。

关键词:桂花;鲜花;香气成分;固相微萃取;气相色谱-质谱法

桂花是中国传统名花之一,花香清甜浓郁,被古人誉为"清可绝尘,浓能溢远"的仙香,有2 500多年的栽培历史[1]。同时,桂花还具有很高的食用、药用和观赏价值[2]。近些年,国内外关于桂花的研究主要集中在种质资源的调查收集与系统分类[3-5]、花瓣的营养成分[6]与药理作用[7-8]、繁殖与栽培[9-11]以及桂花花香[12-13]等方面,但桂花品种繁多,关于不同桂花品种香气成分的比较还较少。金荷仙等[14]分析了杭州满陇桂雨公园4个桂花品种的香气成分。曹慧等[15]根据不同桂花品种样本之间差异显著,同一桂花品种样本的相似度较高,提出了利用色谱指纹图谱相似度评价法和主成分投影分析法对不同桂花品种样本进行归类和鉴别。孙宝军等[16]对桂花4个品种群的8个品种盛花期的香气成分进行了鉴定。文中试图通过分析比较4个品种群8个桂花品种香气成分的差异,筛选出香气花香品质优良的品种,为桂花花香成分的进一步研究奠定基础。

1 材料与方法

2009年9月底,在南京林业大学校园内选择长势一致且健康的桂花采集花朵,四季桂品种群有'四季桂'(*Osmanthus fragrans* 'Sijigui')和'日香桂'(*O. fragrans* 'Rixiang Gui')2个品种;丹桂品种群有'雄黄'丹桂(*O. fragrans* 'Xionghuang')和'鄂橙'丹桂(*O. fragrans* 'E Cheng')2个品种;银桂品种群有'长梗白'(*O. fragrans* 'Changgeng Bai')和'晚银桂'(*O. fragrans* 'Wan Yingui')2个品种;金桂品种群有'波叶金桂'(*O. fragrans* 'Boye Jingui')和'速生金桂'(*O. fragrans* 'Susheng Jingui')2个品种。

---

* 原文发表于《东北林业大学学报》,2015,43(1):83-87。
基金项目:江苏省科技支撑项目(BE2011367);国家林业局公益性行业科研专项(201204607);江苏高校优势学科建设工程项目(PAPD)。

试验仪器：手动固相微萃取进样器（美国 SU-PELCO 公司），65 μm PDMS/DVB 萃取头（美国 SU-PELCO 公司），气相色谱-质谱联用仪 Trace DSQ（美国 Thermo Electro-Finnigan），4 mL 顶空取样瓶（美国 SUPELCO 公司）、水浴锅。

桂花芳香成分采集方法：于桂花花朵开放的第 1 天，在上午露水褪尽后，每品种采集 25 朵花放入 4 mL 采样瓶内，密封后插入 65 μmPDMS/ DVB 纤维头，于 50 ℃下顶空萃取 30 min。萃取完成后，取出纤维头，插入 GC-MS 进样口，解析 5 min 后，进样分析。

GC-MS 分析条件：气相色谱条件为 TR-5MS 毛细管色谱柱（30 m×0.25 mm×0.25 μm）；程序升温为起始温度 40 ℃保持 2 min，以 2 ℃·min^{-1}升至 60 ℃，再以 5 ℃·min^{-1}升至 100 ℃，再以 10 ℃·min^{-1}升至 250 ℃，保持 5 min；载气为高纯度氦气（He），氦气流速为 1 mL·min^{-1}，分流比为 10∶1；进样口温度 250 ℃。

质谱条件：质谱接口温度为 250 ℃；电离方式 EI，离子电离能量 70 eV；质量扫描范围（m/z）50～450 amu。

2 结果与分析

2.1 不同桂花品种芳香成分种类的比较

将不同品种群不同桂花品种芳香成分的总离子流色谱图（图 1）信息经计算机谱库检索及资料分析，且扣除空气本底杂质，得到其挥发物主要成分种类，并根据总离子流色谱峰的峰面积归一化法计算各香味组分的质量分数（表 1）。

由表 1 可见，4 个品种群的 8 个桂花品种共检测出 52 种花香化合物，可分为萜类、酯类、醛类、醇类、烷烃类、酮类和芳香烃类 7 类化合物。不同桂花品种群间花香成分差异明显，四季桂品种群香气成分中，萜类化合物种类最多（17 种），酯类（5 种）次之，醛、醇类、烷烃类、酮类较少，没有检测到芳香烃类化合物的释放；丹桂品种群花香成分中萜类化合物种类最多（15 种），酯类（7 种）和醛类（5 种）次之，醇类、酮类、芳香烃类较少，没有检测到烷烃类化合物的释放；银桂品种群花香成分中也是萜类种类最多（18 种），酯类（7 种）和醛类（6 种）次之，醇类、酮类较少，而烷烃类与芳香烃类化合物没有检测到；金桂品种群的花香成分中依旧是萜类种类最多（17 种），酯类（5 种）、醛类（5 种）和醇类（4 种）次之，芳香烃类与酮类较少，未检测出烷烃类化合物的释放。同时，这些成分的质量分数也存在较大差异，其中萜类化合物质量分数的差异不大，银桂品种群最高，丹桂品种群相对较低；醛类化合物以金桂品种群'波叶金桂'的质量分数最高，银桂品种群的'晚银桂'质量分数最低，前者是后者的 3.76 倍；酯类化合物质量分数以四季桂和丹桂品种群较金桂与银桂品种群高；醇类化合物以银桂品种群'长梗白'的质量分数最高，四季桂品种群中'四季桂'的质量分数最低，两者差异非常显著，前者是后者的 14.68 倍。而同一品种群内不同品种间香气成分的差异较小，如金桂品种群'波叶金桂'和'速生金桂'品种中检测到花香成分种类都为 28 种，主要香气成分的种类一致，仅这些成分的质量分数和微量成分的种类存在差异。

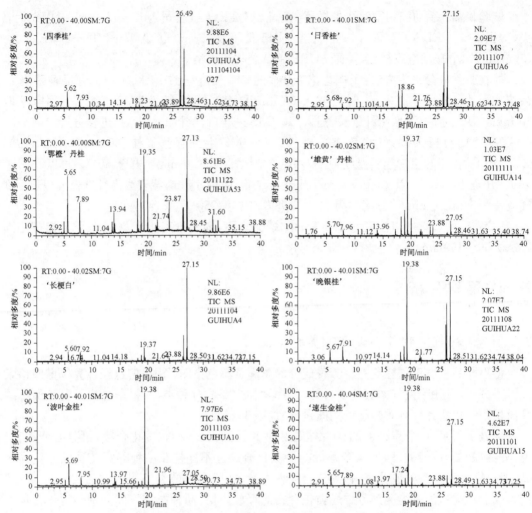

图1 不同桂花品种香气成分的GC-MS总离子流色谱图

表1 不同桂花品种芳香成分及质量分数(%)

桂花品种		萜类												
		1	2	3	4	5	6	7	8	9	10	11	12	13	14
四季桂	A	0.14	0.06	—	0.04	—	2.12	1.83	1.25	0.28	0.58	—	—	—	—
品种群	B	0.05	0.19	—	0.12	0.15	10.03	10.16	2.45	1.13	2.86	0.36	0.14	—	—
丹桂	C	—	—	—	2.5	—	9.08	10.53	36.89	1.81	2.43	3.4	—	0.28	—
品种群	D	0.42	—	—	0.86	—	8.03	9.86	15.25	0.82	1.97	—	—	0.38	—
银桂	E	0.26	—	0.06	0.33	—	1.8	3.57	9.51	0.8	0.91	0.12	—	0.07	—
品种群	F	0.68	—	—	0.29	—	3.35	5.3	29.5	0.94	1.54	0.33	—	—	—
金桂	G	0.23	—	—	9.22	0.25	3.51	4.64	27.58	0.25	0.55	0.25	—	—	0.1
品种群	H	0.26	—	0.18	5.63	0.26	2.52	3.35	41.2	0.26	0.5	0.37	0.07	—	—

续 表

桂花品种		萜 类									
		15	16	17	18	19	20	21	22	23	24
四季桂品种群	A	—	—	0.15	—	5.92	—	30.6	0.56	22.87	—
	B	—	—	—	0.16	5.41	—	15.6	0.44	30.56	—
丹桂品种群	C	—	—	—	—	0.98	—	1.25	—	6.15	0.42
	D	0.6	0.33	—	—	3.67	—	4.81	—	13.02	4.39
银桂品种群	E	—	0.23	—	—	1.4	—	9.56	0.38	35.01	0.62
	F	—	0.09	0.12	—	7.8	0.17	12.4	0.39	17.35	0.13
金桂品种群	G	—	—	—	—	1.85	—	0.72	—	14.4	0.28
	H	—	—	0.28	—	3.72	—	0.51	—	18.1	0.54

桂花品种		醛 类							
		25	26	27	28	29	30	31	32
四季桂品种群	A	7.47	5.22	6.82	0.2	—	0.53	—	—
	B	0.33	5.22	3.6	0.1	—	0.21	—	—
丹桂品种群	C	3.5	4.82	3.8	0.1	—	0.47	—	—
	D	11.4	8.13	8.13	0.3	—	0.95	1.25	0.25
银桂品种群	E	5.41	4.73	1.37	0.2	0.58	0.12	—	0.05
	F	—	5.63	0.77	0.3	—	0.71	—	—
金桂品种群	G	0.45	18.1	8.37	0.3	—	0.6	—	—
	H	3.88	5.22	3.68	0.2	—	0.48	—	—

桂花品种		酯类								醇类				
		33	34	35	36	37	38	39	40	41	42	43	44	45
四季桂品种群	A	1.42	—	—	8.03	—	—	—	—	0.13	—	0.08	—	—
	B	0.67	—	0.16	6.78	—	—	—	—	0.17	0.41	—	—	—
丹桂品种群	C	1.41	—	1.14	4.75	—	—	0.42	—	—	—	0.45	—	—
	D	1.99	—	—	6.49	—	0.53	—	—	—	0.84	—	0.37	—
银桂品种群	E	0.76	0.13	0.07	4.64	0.54	—	—	—	—	0.43	—	11.46	—
	F	0.95	—	0.81	0.54	0.24	—	0.18	—	—	1.32	—	6.21	—
金桂品种群	G	1.89	—	1.54	0.96	0.38	—	0.23	—	0.36	—	—	0.16	0.16
	H	1.08	—	0.85	2.63	—	—	0.13	—	—	—	0.08	0.82	—

续表

桂花品种		烷烃类			酮类		芳香烃	
		46	47	48	49	50	51	52
四季桂品种群	A	0.27	0.09	0.04	0.69	—	—	—
	B	—	—	—	0.44	—	—	—
丹桂品种群	C	—	—	—	0.56	—	0.19	—
	D	—	—	—	—	—	0.34	—
银桂品种群	E	—	—	—	2.26	—	—	—
	F	—	—	—	—	—	—	—
金桂品种群	G	—	—	—	—	—	—	0.08
	H	—	—	—	—	0.37	—	—

注:A '四季桂';B '日香桂';C '雄黄'丹桂;D '鄂橙'丹桂;E '长梗白';F '晚银桂';G '波叶金桂';H '速生金桂'。1.月桂烯;2.柠檬烯;3.3-蒈烯;4.反式-罗勒烯;5.反式-香叶醇;6.反式-芳樟醇氧化物(呋喃型);7.顺式-芳樟醇氧化物(呋喃型);8.芳樟醇;9.反式-芳樟醇氧化物(吡喃型);10.顺式-芳樟醇氧化物(吡喃型);11.橙花醇;12.柠檬醛;13.香叶醛;14.薄荷酮;15.5-三甲基-6-亚丁烯基-4-环己烯;16.顺-茉莉酮;17.α-紫罗兰酮;18.β-紫罗兰醇;19.α-紫罗兰醇;20.二氢-α-紫罗兰酮;21.二氢-β-紫罗兰酮;22.二氢-β-紫罗兰醇;23.β-紫罗兰醛;24.4-酮基-β-紫罗兰酮;25.顺-3-己烯醛;26.正己醛;27.叶醛;28.苯甲醛;29.辛醛;30.壬醛;31.癸醛;32.反,反-2,4-二己二烯醛;33.γ-2-己烯内酯;34.顺式-丁酸-3-己烯酯;35.反式-丁酸-3-己烯酯;36.γ-癸内酯;37.甲酸叶醇酯;38.戊酸叶醇酯;39.丁酸己酯;40.(E,Z)-2-丁烯酸-3-己烯酯;41.丁醇;42.正己醇;43.壬醇;44.顺式-叶醇;45.3,7-二甲基-1,5,7-辛三烯-三醇;46.辛烷;47.2,5-二甲基己烷;48.甲基环己烷;49.2,2,6-三甲基-6-乙烯基-2H-四氢吡喃-3(4H)-酮;50.3-羟基-2-丁酮;51.甲苯;52.丙苯。

2.2 桂花品种主要芳香成分分析

比较8个桂花品种的花香成分,从表1可以得知,芳樟醇、芳樟醇氧化物、β-紫罗兰酮、二氢-β-紫罗兰酮、正己醛,以及叶醛等为其共有的主要成分。进一步对其进行比较分析(图2)可知,芳樟醇、β-紫罗兰酮和二氢-β-紫罗兰酮的质量分数最高,是香味中最主要的组成成分。芳樟醇在金桂品种群'波叶金桂'和'速生金桂'以及'雄黄'丹桂、'晚银桂'中质量分数较高,分别达27.58%、41.20%、35.89%和29.50%;而四季桂品种群的两个品种中芳樟醇的质量分数仅为1.25%和2.45%,二氢-β-紫罗兰酮在四季桂品种中质量分数最高,为30.55%。其他主要成分的质量分数也都有明显的差异,如α-紫罗兰酮、叶醛、γ-癸内酯在4个品种群的8个桂花品种中的质量分数差异都比较大,其质量分数高的是质量分数低的10倍左右;金桂品种群'波叶金桂''速生金桂'和丹桂品种群'雄黄'丹桂的反式-罗勒烯质量分数比较高,而其他品种中的质量分数都不足1%;'鄂橙'丹桂和'波叶金桂'的正己醛的质量分数较高,分别为11.4%和18.08%,其他品种的质量分数差异不大,都在5%左右;γ-2-己烯内酯的质量分数差异不是很大,都在1%左右;而顺-3-己烯醛的质量分数却有很大差异,'四季桂'和'长梗白'的顺-3-己烯醛的质量分数都在5%以上,在'晚银桂'中却没有检测到顺-3-己烯醛的存在;银桂品种群2个品种中顺式-叶醇的质量分数较高,分别为11.46%和6.21%,而在丹桂品种群和金桂品种群中的质量分数很低,在四季桂品种群中都检测不到;芳樟醇氧化物普遍质量分数较高,品种间没有很大的差异。

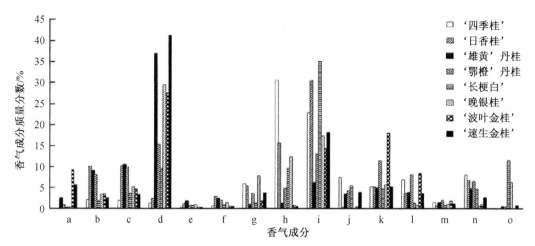

图 2 不同桂花品种主要香气成分质量分数对比图

a. 反式-罗勒烯;b. 反式-芳樟醇氧化物(呋喃型);c. 顺式-芳樟醇氧化物(呋喃型);d. 芳樟醇;e. 反式-芳樟醇氧化物(吡喃型);f. 顺式-芳樟醇氧化物(吡喃型);g. α-紫罗兰酮;h. 二氢-β-紫罗兰酮;i. β-紫罗兰酮;j. 顺-3-己烯醛;k. 正己醛;l. 叶醛;m. γ-2-己烯内酯;n. γ-癸内酯;o. 顺式-叶醇。

3 结束语

花香是观赏植物的重要品质指标之一,主要由烃类、烯类、醇类、酮类、醛类、醚类、酯类及芳香族化合物等组成[17]。在本次检测的 8 个桂花品种花香成分中,共检测到 52 种化合物,分为萜类、酯类、醛类、烷烃类、酮类和芳香烃类 7 大类。结果表明,在同一桂花品种群内花香成分种类相差不大,主体香味成分一致,但成分的质量分数或一些微量成分存在一定差异。然而,不同品种群桂花的香气成分差异明显,四季桂品种群中检测到了烷烃类物质的存在,而在其他 3 个品种群内均未检测到;在丹桂品种群和金桂品种群内检测到的芳香烃类化合物也未从四季桂品种群和银桂品种群中检测到。同时,这些成分的质量分数也存在较大差异,这些差异使得不同桂花品种群的香味各有不同。这与金荷仙[14]等认为不同桂花品种群的品种之间释放的挥发物在组分和质量分数上都存在差异的观点一致。孙宝军等[16]也发现 4 个品种群的桂花香气成分的质量分数明显不同,同时,认为 4 个品种群的桂花香气成分并不存在根本性的差异,与本试验认为 4 个品种群桂花的香味有一定的相似性、主体香味成分大部分一致的观点相似。顺-3-己烯醛、正己醛、叶醛、己醇、γ-2-己烯内酯、反式-芳樟醇氧化物(呋喃型)、顺式-芳樟醇氧化物(呋喃型)、芳樟醇、反式-芳樟醇氧化物(吡喃型)、顺式-芳樟醇氧化物(吡喃型)、α-紫罗兰酮、二氢-β-紫罗兰酮、γ-癸内酯和β-紫罗兰酮等物质在多数桂花中都具有较高的质量分数,因此,可确定为桂花香味的主要成分。其中,萜类化合物是最主要成分,在 8 个不同桂花品种香气成分中其质量分数都在 60% 以上,最高达到了 80.35%。这表明萜类化合物与桂花花香关系密切,相关香气浓郁的芳香植物研究[18-20]表明花香浓淡与挥发性的萜类化合物密切相关。

参考文献

[1] 陈洪国,汪华. 我国桂花种资资源的研究和利用及咸宁桂花发展现状[J]. 咸宁学院学报,2004,24(3):116-122.

[2] 高杰. 浅谈桂花在城市园林绿化中的应用[J]. 现代园林,2012(5):32-35.

[3] 向其柏. 中国桂花品种图志[M]. 杭州:浙江科学技术出版社,2008.

[4] 赵桂琴,王敬,冯忠武. 园林绿化树种桂花品种分类研究综述[J]. 四川林业科技,2013,34(1):105-106,40.

[5] 向民,段一凡,向其柏. 彩叶桂品种群的建立[J]. 南京林业大学学报(自然科学版),2014(1):封二-封三.

[6] 杨秀莲,赵飞,王良桂. 25个桂花品种花瓣营养成分分析[J]. 福建林学院学报,2014,34(1):5-10.

[7] Lee D G, Lee S M, Bang M H, et al. Lignans from the flowers of Osmanthus fragrans var. aurantiacus and their Inhibition effect on NO production[J]. Archives of Pharmacal Research, 2011, 34(12):2029-2035.

[8] 周秋霞,岳淑梅. 基于文献的桂花化学成分及药理作用研究现状分析[J]. 河南大学学报(医学版),2013,32(2):139-142.

[9] 时朝,王亚芝,刘国杰,等. 北方地区桂花嫁接繁殖技术研究[J]. 北方园艺,2013(3):70-72.

[10] 何礼军,孙连连,杨园园. 桂花种子发育与发芽研究综述[J]. 安徽农业科学,2013,41(11):4870,4948.

[11] 李林,韩远记,袁王俊,等. 桂花组织培养快繁体系的建立[J]. 河南大学学报(自然科学版),2013,43(6):667-671.

[12] 唐丽,唐芳,段经华,等. 金桂芳樟醇合成酶基因的克隆与序列分析[J]. 林业科学,2009,45(5):11-19.

[13] 杨宇婷,武晓红,田璞玉. 桂花(晚银桂、贵妃红和窈窕淑女)挥发性成分分析[J]. 河南大学学报(医学版),2010,29(1):13-16,20.

[14] 金荷仙,郑华,金幼菊,等. 杭州满陇桂雨公园4个桂花品种香气组分的研究[J]. 林业科学研究,2006,19(5):612-615.

[15] 曹慧,李祖光,沈德隆. 桂花品种香气成分的GC/MS指纹图谱研究[J]. 园艺学报,2009,36(3):391-398.

[16] 孙宝军,李黎,韩远记. 上海桂林公园桂花芳香成分的HS-SPME-GC-MS分析[J]. 福建林学院学报,2012,32(1):39-42.

[17] 李莹莹. 花香挥发物的主要成分及其影响因素[J]. 北方园艺,2012(6):184-187.

[18] 王洁,杨志玲,杨旭,等. 不同花期厚朴雌雄蕊和花瓣香气组成成分的分析和比较[J]. 植物资源与环境学报,2011,20(4):42-48.

[19] 张辉秀,胡增辉,冷平生,等. 不同品种百合花挥发性成分定性与定量分析[J]. 中国农业科学,2013,46(4):790-799.

[20] 魏丹,李祖光,徐心怡,等. HS-SPME-GC-MS联用分析3种兰花鲜花的香气成分[J]. 食品科学,2013,34(16):234-237.

Variance Analysis of Aromatic Components from Different Varieties of *Osmanthus fragrans*

YANG Xiulian, SHI Tingting, WEN Ailin, WANG Lianggui

(Nanjing Forestry University, Nanjing 210037, China)

Abstract: We studied the aromatic components and relative contents of *Osmanthus fragrans*, and analyzed the aromatic components of eight cultivars of *O. fragrans* from four *O. fragrans* groups by headspace solid-phase micro extraction (HS-SPME) and gas chromatography-mass spectrometry (GC-MS). A total of 52 aromatic compounds were identified from *O. fragrans*, 32 compounds were identified from two cultivars of *O. fragrans* Asiatic Group, 34 compounds were identified from two cultivars of *O. fragrans* Albus Group, 33 compounds were identified from two cultivars of *O. fragrans* Luteus Group, 32 compounds were identified from two cultivars of *O. fragrans* Aurantiacus Group, of which linalool, linalool oxide, β-io-none, dihydro-β-ionone, hexanal and 2-Hexenal (E) were the major aromatic components. Therefore, the aromatic com-pounds and their relative contents were different in four *O. fragrans* groups, but the differences of aromatic compounds among *O. fragrans* cultivars of same group were not obvious.

Key words: *Osmanthus fragrans*; flowers; aromatic component; HS-SPME; GC-MS

冷藏对5个桂花品种主要营养成分的影响*

杨秀莲,宋佳妮,赵 飞,王良桂

(南京林业大学 风景园林学院,江苏 南京 210037)

摘 要:以5个桂花品种为试验材料,研究冷藏时间对花瓣部分营养成分的影响。结果表明:(1)在2~5℃条件下冷藏5 d和10 d后,5个桂花品种花瓣的含水量、可溶性糖、可溶性蛋白质、维生素C、苹果酸和黄酮的含量都呈现下降的趋势,不同品种的下降幅度不同;(2)随着冷藏时间延长,营养物质损耗增加,冷藏5 d营养成分损耗比冷藏10 d的小。因此,短时间冷藏可提高桂花花瓣的利用率,提升桂花产业的经济效益。

关键词:桂花;冷藏;营养成分

植物的花是大自然的精华,许多植物的花中含有较高的矿质元素、蛋白质或氨基酸等,可食用、药用、保健用,是原生态的食用珍品[1-5]。桂花是我国传统十大名花之一,集药用、食用、观赏、绿化于一身。桂花花瓣中黄酮含量最高达28.9%,总游离氨基酸含量达15.11%,桂花中氨基酸配比合理,含有优质蛋白质[6]。桂花中还含有丰富的铁、锌、钙、镁等矿质元素,桂花香浓味甜,营养十分丰富,被称为"全营养食品"。

目前,食用桂花材料均是烘干或经盐、糖腌渍后使用,但由于桂花采花期比较集中,采下的桂花必须当天处理,否则容易褐化失去应有的食用价值。实际生产中常因为来不及处理而浪费了很多新鲜桂花,影响经济效益。本试验通过测定新鲜和冷藏一段时间后的桂花花瓣中部分营养物质的含量,了解冷藏对营养物质含量变化的影响,以期为延长桂花保存时间提供理论依据,提高桂花利用率,提升桂花的经济价值。

1 材料与方法

1.1 供试材料及处理

试验在南京林业大学风景园林试验示范中心实验室内进行。供试材料为南京林业大学校园内采摘的'四季桂'、'雨城丹'桂、'朱砂丹桂'、'紫梗籽银'桂品种的花瓣和在秦淮河河道管理处采摘的'小金铃'品种的花瓣。

将在盛花期采摘的新鲜桂花,处理干净后放入2~5℃冰箱中,分别于采摘当天和冷藏

---

* 原文发表于《西南农业学报》,2014,27(6):2720-2722。
基金项目:江苏省科技支撑项目(BE2011367);国家林业局公益性行业科研专项(201204607);江苏省高校自然科学基础研究项目(11KJB220002);江苏高校优势学科建设工程项目(PAPD)。

5 d、10 d 取样,测定花瓣含水量、可溶性糖、可溶性蛋白、Vc、有机酸、黄酮等营养物质的含量。

1.2 测定指标与方法

含水量的测定采用李合生的方法[7];可溶性糖的测定采用蒽酮比色法[7];可溶性蛋白含量的测定采用考马斯亮蓝染色法[7];Vc 含量的测定采用 1,10-菲咯啉法[8];苹果酸含量的测定采用酸碱中和滴定法[9];黄酮含量的测定采用 $Al(NO_3)_3$-$NaNO_2$ 方法[10]。

1.3 数据统计分析

用 Execl 软件作图,用 SPSS 19.0 统计软件进行单因素方差分析和 Duncan's 多重对比分析。

2 结果与分析

2.1 冷藏前后花瓣含水量的变化

采摘当天和冷藏 5、10 d 后花瓣中的含水量见图 1。不同品种花瓣含水量不同,'雨城丹'桂和'紫梗籽银'桂的含水量相对比其他 3 个品种低;随着冷藏时间延长,花瓣中含水量有一定程度减少,但差异不显著。

2.2 冷藏前后花瓣可溶性糖含量的变化

植物组织中可溶性糖含量的高低可以反映其在生长过程中光合作用的强弱,对营养有一定的指导意义。随着冷藏时间的延长,5 个桂花品种花瓣中可溶性糖含量呈下降趋势(图2),但不同品种表现出一定的差别,尤其是'雨城丹'桂,冷藏 5 d 时可溶性糖比新鲜时降低了 56.7%,而'小金铃'仅下降了 6.7%。方差分析和多重对比显示:'雨城丹'桂冷藏前后可溶性糖含量差异极显著,'四季桂'和'朱砂丹桂'新鲜和冷藏的花瓣中可溶性糖含量相比也有显著差异,但冷藏 5 和 10 d 花瓣中可溶性糖含量差异不显著;'小金铃'和'紫梗籽银'桂新鲜花瓣和冷藏 5 d 花瓣中可溶性糖含量无差异,与冷藏 10 d 花瓣中可溶性糖含量差异显著。

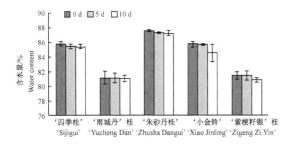

图 1 冷藏前后含水量的变化

Fig. 1 The change of the content of water before and after cold storage

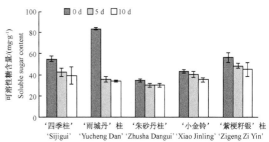

图 2 冷藏前后可溶性糖含量的变化

Fig. 2 The change of the content of soluble sugar before and after cold storage

2.3 冷藏前后花瓣可溶性蛋白质含量的变化

冷藏 0、5、10 d 时,5 个品种花瓣中可溶性蛋白质出现明显的下降趋势(图3)。方差分析和多重对比表明:'四季桂'、'雨城丹'桂和'朱砂丹桂'冷藏 0、5 d 时,花瓣可溶性蛋白质含量无显著性差异,但与冷藏 10 d 花瓣中可溶性蛋白质含量差异显著。'小金铃'在冷藏 0、5 和 10 d 之间花瓣中可溶性蛋白质含量均存在显著性差异。'紫梗籽银'桂冷藏 0 和 5、10 d 花瓣中可溶性蛋白质含量之间存在显著性差异,但冷藏 5 和 10 d 花瓣中可溶性蛋白质含量间无显著差异。

2.4 冷藏前后花瓣 Vc 含量的变化

5 个桂花品种新鲜花瓣中的 Vc 含量有一定的差异,其中 Vc 含量最高的'雨城丹'桂是'四季桂'的 2.55 倍。冷藏后各品种花瓣中的 Vc 含量均出现下降(图4)。其中'雨城丹'桂和'紫梗籽银'桂冷藏 5 d 时花瓣 Vc 含量下降最多,分别下降了 54.3% 和 65.8%,而'四季桂'和'朱砂丹桂'仅降低了 13.4% 和 3.7%。方差分析和多重对比显示:'朱砂丹桂'冷藏前后 Vc 含量无显著差异,'四季桂'和'雨城丹'桂冷藏 0 d 与冷藏 5、10 d 花瓣 Vc 含量间差异显著,但冷藏 5 和 10 d 花瓣 Vc 含量间差异不显著。'小金铃'和'紫梗籽银'桂 0、5 和 10 d 花瓣 Vc 含量间均存在显著差异。研究表明:冷藏期间花瓣 Vc 含量的降低主要是由于花瓣呼吸作用的消耗使得有机物逐渐减少,合成 Vc 的底物不足;同时 Vc 化学性质不稳定,易被分解和氧化[11-12]。

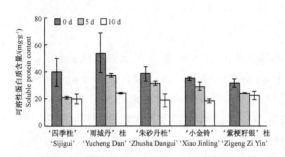

图3 冷藏前后可溶性蛋白质含量的变化
Fig. 3 The change of the content of soluble protein before and after cold storage

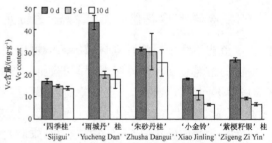

图4 冷藏前后维生素C含量的变化
Fig. 4 The change of the content of Vc before and after cold storage

2.5 冷藏前后花瓣内苹果酸含量的影响

桂花花瓣经冷藏 5、10 d 后,花瓣中的苹果酸含量逐渐下降(图5)。其中降幅较大的是'小金铃',花瓣冷藏 10 d 后,比新鲜时苹果酸含量下降了 62.6%,降幅最小的是'雨城丹'桂,仅下降 7.6%。方差分析和多重对比表明:'四季桂'、'雨城丹'桂和'朱砂丹桂'冷藏前后苹果酸的含量未发生显著变化;'小金铃'和'紫梗籽银'桂冷藏 0 和 5、10 d 花瓣中苹果酸含量间有显著差异,冷藏 5 和 10 d 花瓣中苹果酸含量间无差异。

2.6 冷藏前后花瓣内黄酮含量的变化

5 个桂花品种冷藏前后花瓣中的黄酮含量呈下降趋势(图6)。其中降幅最大的是'紫梗籽银'桂,冷藏 10 d 后黄酮含量下降了 54.7%;而'四季桂'和'小金铃'的黄酮含量降幅较小,分别为 26.2%和 25.8%。方差分析和多重对比显示:'四季桂'和'小金铃'3 个时间段黄酮含量变化无显著差异;'雨城丹'桂冷藏 0 和 5 d 间黄酮含量无显著差异,但 10 d 后黄酮含量出现显著差异;'朱砂丹桂'和'紫梗籽银'桂的黄酮含量在冷藏 0 和 5、10 d 间存在显著差异,但冷藏 5 和 10 d 间无显著差异。

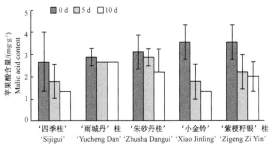

图 5　冷藏前后苹果酸含量的变化

Fig. 5　The change of the content of malic acid before and after cold storage

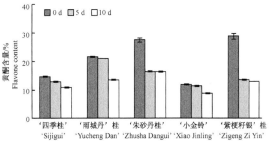

图 6　冷藏前后黄酮含量的变化

Fig. 6　The change of the content of flavone before and after cold storage

3　讨论与结论

新鲜的植物组织在采摘后许多生理变化仍在继续进行,如呼吸加强,贮藏物质消耗加快等,从而导致衰老进程加快,品质劣变等。经研究,短时间的低温处理可降低贮藏过程中的生理代谢速度,延缓贮藏物质的消耗,使植物组织原有的营养成分得到较好的保持[13-15]。本试验中随着冷藏时间的延长,不同品种花瓣中营养物质含量有所下降,但品种间差别比较大,其中冷藏 5 d 的花瓣和新鲜花瓣相比,'四季桂'花瓣中可溶性蛋白质和苹果酸含量,'雨城丹'桂花瓣中可溶性糖和 Vc 含量,'朱砂丹桂'花瓣中黄酮含量,'紫梗籽银'桂花瓣中 Vc、苹果酸、黄酮含量以及'小金铃'花瓣中 Vc 和苹果酸含量降幅较大,其他物质降幅相对较小。冷藏 10 d 和冷藏 5 d 的花瓣相比,各物质含量降幅相对较小。

由此可知,不同品种冷藏后营养物质含量变化有差异,冷藏虽降低了部分营养物质的含量,但相比来不及处理而使桂花褐化造成浪费,短时间的冷藏仍不失为一种有效的贮藏方法。

参考文献

[1] 向昌国,李文芳,陈阳波,等.南瓜茎、叶、花的营养成分分析[J].天然产物研究与开发,2010,22(01):68-71.

[2] 李秀芬,张德顺,王少鸥,等.木槿两变型花蕾的营养成分分析[J].植物资源与环境学报,2009,18(4):85-87.

[3] 魏学智,刘少华,师学琴,等.红柄白鹃梅营养成分的分析[J].西北植物学报,2005,25(8):1657-1660.

[4] 金丽丽,杨武,孙向丽,等.百合花中微量金属元素及其营养成分分析[J].西北师范大学学报(自然科学版),2008,44(6):58-61.

[5] 金潇潇,陈发棣,陈素梅,等.20 个菊花品种花瓣的营养品质分析[J].浙江林学院学报,2010,27(1):22-29.

[6] 杨秀莲,赵飞,王良桂.25个桂花品种花瓣营养成分分析[J].福建林学院学报,2014,34(1):1-5.

[7] 李合生.植物生理生化实验原理和技术[M].北京:高等教育出版社,2000:167-169.

[8] Nobuhiko A N, Tsutsumi K, Sanceda, N G, et al. A rapid and sensitive method for the determination of ascorbic acid using 4,7-diphenyl-1,10-phenanthroline[J], Agricultural Biology and Chemistry, 1981,45(5):1289-1290.

[9] 黄松,吴月娜,刘梅,等.茚三酮比色法测定青天葵中总游离氨基酸的含量[J].中国中医药信息杂志,2010,12(17):50-52.

[10] 元晓梅,蒋明蔚,胡正芝,等.聚酰胺吸附-硝酸铝显色法测定山楂及山楂制品中的总黄酮含量[J].食品与发酵工业,1996(04):27-31.

[11] 郭香凤,向进乐,李秀珍.贮藏温度对西兰花净菜品质的影响[J].农业机械学报,2008,39(2):201-203.

[12] 邱贺媛,曾宪锋.漂烫处理对两种蔬菜中硝酸盐亚硝酸盐及 V_C 含量的影响[J].农产品加工,2005(2):65-66.

[13] 林本芳,鲁晓翔,李江阔,等.冰温贮藏对西兰花保鲜的影响[J].食品工业科技,2012,19:312-316.

[14] 张胜珍.不同贮藏方式对长蕊石头花主要营养成分含量的影响[J].北方园艺,2010(21):61-63.

[15] 焦岩,李保国,张岩,等.冰箱冷藏条件对果蔬贮藏质量的影响[J].食品科学,2002,25(3):150-153.

Effects of Cold Storage on Major Nutrients of Five *Osmanthus fragrans* Cultivars

YANG Xiulian, SONG Jiani, ZHAO Fei, WANG Lianggui

(College of Landscape Architecture, Nanjing Forestry University, Nanjing 210037, China)

Abstract: Taking five cultivars of *Osmanthus fragrans* as experimental materials, the impacts of cold storage time on some nutrients of petals were studied. The results show as follows: (1) the content of water, soluble sugar, soluble protein, vitamin C, malic acid and flavone show downward trends under storage condition of 2~-5 ℃ after 5 and 10 days, which falling range varied as different cultivars. (2) As prolonging the cold storage time, the losing nutritions of 5 days were less than that of 10 days. In consequence, cold storage of short time can improve the utilization of the petals and economic benefit of Osmanthus industry.

Key words: *Osmanthus fragrans*; cold storage; nutrient

桂花花朵香气成分的研究进展

施婷婷,杨秀莲,王良桂

(南京林业大学 风景园林学院,江苏 南京 210037)

摘 要:桂花花朵的香味由一系列低分子量的挥发性化合物组成,并形成其独有的特征。对桂花花朵香气成分的物质种类、影响因素进行了简单介绍,并对其提取与分析方法的研究与应用进展进行了综述。

关键词:桂花花朵;香气成分;影响因素;分析方法

桂花(Osmanthus fragrans Lour.)是我国传统的十大名花之一,集绿化、美化、香化于一体,属观赏和实用兼具的优良园林树种,深受人们喜爱。同时,桂花花朵内含物丰富,含有人体所需的蛋白质、碳水化合物、矿物质、维生素等成分,营养价值很高,被称为"全营养食品"[1]。早在两千多年前,中国人就以桂花为配料制作各种美食,如桂花糕、桂花麻糖、桂花月饼、桂花鸭等,深受人们喜爱。目前,桂花的应用十分广泛,除了在食品中添加外,在酿酒、制茶、饮料等工业生产中,桂花也是一种常用的重要调香剂,桂花酒、桂花茶在市场上广受赞誉;此外,桂花花朵的香气清新高雅、幽雅留长,常被作为高级名贵的天然香料。因此,花用桂花市场前景广阔,关于桂花花朵香气成分开发利用方面的研究成为热点。作者在此对桂花花朵香气成分的物质种类、影响因素、提取与分析方法等进行综述,拟为桂花花朵香气成分的开发利用提供参考。

1 桂花花朵香气成分的物质种类

综合前人的研究结果,植物花香主要由萜烯类、苯型烃类、脂肪酸及其衍生物以及一些含硫、含氮的化合物组成[2]。桂花芳香成分的研究始于20世纪60年代[3-4],其香气成分包括几百种挥发性化合物,目前已经鉴定出的主要香气成分大约有60多种,按照桂花中香气成分的化学结构,可将其分为萜烯类、醇类、醛类、酮类、酯类和烷烃类等化合物[5]。而且,每一类、每一种香味物质对桂花花朵香味的贡献都不一样。其中,一些重要的香气物质及其感官特性如表1所示。

---

* 原文发表于《化学与生物工程》,2014,31(10):1-5。

基金项目:"十二五"国家科技支撑计划项目(2013BA001B06),江苏省高校自然科学基础研究项目(11KJB220002),国家林业局公益性行业专项(201204607);江苏高校优势学科建设工程项目(PAPD)。

表 1 桂花花朵中重要的香气物质及其感官特性
Tab. 1 Odour and content of some important flavoring compositions in *Osmanthus fragrans* flowers

种类 Type	名称 Name	气味特征 Odour characteristics	参考文献 References
萜烯类 Terpenes	二氢-β-紫罗兰酮 Dihydro-β-ionone	木香、紫罗兰花香 Woody, violet orchid	[6]
	α、β-紫罗兰酮 α,β-ionone	木香、紫罗兰花香、果香 Woody, violet, and fruity	[7-8]
	芳樟醇 Linalool	木香,玫瑰花香,果香 Woody, rose, fruity	[6,9-10]
	芳樟醇氧化物 Linalool Oxide	樟脑木香 Camphor wood incense	[6,9-10]
	松油醇 Terpineol	海桐花香、铃兰花香 Sea paulownia, lily of the vauey	[11]
	香叶醇 Geraniol	玫瑰花香气、天竺葵花香 Rose, geranium, floral	[9]
	橙花醇 Neroli	玫瑰花香、柠檬味 Rose and lemon	[12]
	柠檬醛 Citral	柠檬味,薄荷味 Lemon, mint	[6]
	3-蒈烯 3-decene	木香、松木香 Woody, pinewood	[6]
醛类 Aldehyde	叶醛 Leafaldehyde	杏草、果香 Vaniwa and fruity	[9,13]
	壬醛 Furfural	玫瑰花香、甜橙味 Rose, sweet, orange	[12]
	正己醛 N-hexanal	青草味、苹果味 Grass, apple	[9,14]
酯类 Ester	γ-癸内酯 γ-decalactone	椰子、桃子香 Coconut, peach	[9,15]
	甲酸叶醇酯 Formic acid ester	青草味 Grass	[16]
酮类 Ketones	茉莉酮 Jasminone	茉莉花香 Jasmine	[6]
	6-戊基-2-吡喃酮 6-pentyl-2-pyrone	乳酪味 Cheese	[6]
醇类 Alcohol	顺-3-己烯醇 (Z)-3-hexenol	青草味、苹果味 Grass, apple	[17]

2 桂花花朵香气成分的影响因素

桂花花朵的香气首先取决于桂花的品种,同时,也与花期、研究对象、采集方法等有很大的关联。通常,在鲜花作原材料的前提下,桂花品种对香气成分起决定作用,但只有在适宜的花期条件下,才能使桂花花朵发挥其馥郁芳香。

2.1 品种

早在20世纪80年代我国学者就对桂花香气的成分进行了研究,结果表明,丰富的桂花品种间香气物质的种类和含量有很大的差异,但大多数桂花品种的主要香气成分基本相同。金桂中β-紫罗兰酮、α-紫罗兰酮及γ-癸内酯含量较高并含有4-酮基-β-紫罗兰酮,香气甜润馥郁;银桂中含有较高的芳樟醇及芳樟醇氧化物,香气清幽淡雅;丹桂中α、β-紫罗兰酮的含量很低,故甜香不够,且缺乏叶醇酯类、环己烯酯类、吡喃型芳樟醇氧化物和4-酮基-β-紫罗兰酮,故清香不足[18-20]。

巫华美等[21]研究发现同一产地的银桂净油和金桂净油的化学成分及其含量也存在明显差异。Deng等[22]也利用HS/SPME分析比较了金桂和银桂的挥发性化合物,发现这2

种桂花的化合物总量十分接近,但每种化合物的含量还是有很大不同。金荷仙等[23]分析了杭州满陇桂雨公园4个桂花品种的香气成分,确定了芳樟醇氧化物、芳樟醇、β-紫罗兰酮、2H-β-紫罗兰酮、α-紫罗兰酮、香叶醇、罗勒烯等为桂花香味的主体成分,认为不同桂花品种群的品种之间释放的挥发物在组分和相对含量上都存在差异。曹慧等[17]根据不同品种桂花样本之间差异显著,同一品种桂花样本的相似度较高,提出了利用色谱指纹图谱相似度评价法和主成分投影分析法对不同品种的桂花样本进行归类和鉴别。孙宝军等[24]采集桂花4个品种群的8个品种盛花期的香气成分,认为4个品种群的桂花香气成分并不存在根本性的差异,但这些成分的含量明显不同。

2.2 花期

花香的释放伴随着花朵的开放而产生,且发生一定的变化,文心兰[25]、菊花[26]、栀子[27]、百合[28]等观赏植物在花发育过程中香气成分的种类和含量均发生显著的变化。许多研究表明,桂花香气成分的差异,可能与桂花采集花期也有密切关系。李祖光等[29]曾分析比较3种桂花(金桂、银桂和丹桂)5个不同开花阶段的头香成分,发现不同花期桂花芳香成分差异明显,半开期及盛开期阶段桂花的鲜花花色鲜艳,头香成分含量较多。Wang等[15]将桂花花期分为4个阶段,分析其芳香成分,认为桂花不同花期的香气成分差异显著,初花期的精油产量最高,各主体香气成分的含量也最高,这不同于李祖光等人的研究结果,可能是由于采集方法及外界条件等不同造成的。丁成斌等[20]研究发现栽培50年内的桂花花朵香气馥郁,100年以上香气成分则下降。

2.3 研究对象

目前,对桂花花朵香气成分的研究大多以桂花浸膏或净油为研究对象[8,21,30],但提取的桂花浸膏或净油会受到高温或萃取溶剂等的干扰,从而影响桂花的真实香气。张晓林等[31-32]比较了桂花头香和净油的香气成分,发现头香与净油的成分有较大的差别:头香的主要成分是单萜化合物,低沸点含氧化合物相当多,而净油中这些低沸点化合物大部分损失掉了,因此,以桂花浸膏或净油为研究对象会影响桂花香气的真实性。杨志萍等[33]比较了β-D-葡萄糖苷酶酶解前后桂花鲜花的香气成分,发现酶解后的主要香气成分质量分数均明显升高,认为β-D-葡萄糖苷酶可水解桂花鲜花香气的前体物质,释放出桂花鲜花中潜在的香气成分。

综上所述,桂花花朵香气的主要成分为紫罗兰酮、γ-癸内酯、芳樟醇及其氧化物等,但不同品种、不同花期以及不同研究对象对桂花花朵的香气成分都有一定的影响,造成成分的不同或各种成分含量的差异。

3 桂花花朵香气成分的提取与分析方法

传统的植物芳香成分的提取方法主要有:水蒸气蒸馏法、榨磨法、挥发性溶剂浸提法和吸附法[34],这些方法会破坏天然香味中某些热敏性或不稳定性的成分,使香味失去真实性。随着科技的发展,许多提取方法得到改进,新技术也广泛用于植物芳香成分的提取中,使提取得到的植物芳香成分更接近天然。近年来桂花花朵香气成分的提取方法主要有:水蒸气蒸馏法[30]、挥发性有机溶剂萃取法[20]、动态顶空法[23,32]、超临界萃取法[21,35-36]、固相微萃取法(SPME)[22-33]以及活性炭吸附丝吸附法[37]等,这些方法各有利弊,不同的研究方法研究相同样品的结果差别也会很大,故应根据研究香气的目的和分析物质的特性选择不同的方法。

3.1 水蒸气蒸馏法

早在16世纪的欧洲,水蒸气蒸馏技术就已普遍应用于植物香精油的提取。水蒸气蒸馏法操作简单、成本低、提取量大,至今仍被人们所沿用,是批量提取精油的主要方法。徐继明等[30]利用水蒸气蒸馏法提取桂花精油,发现桂花精油中的主要香气成分含量与先前的报道相比都偏低,而邻苯二甲酸酯及其衍生物等对身体有害的物质含量较高,推测可能是由于桂花鲜花内的一些热敏性或不稳定性的芳香物质被水蒸气蒸馏的高温破坏所致。微波-同时蒸馏萃取法是利用基体物质的某些区域或萃取体系中的某些组分因吸收微波能力的差异而被选择性加热,从而进行物质分离的方法。具有高效率、高选择性、不会破坏天然热敏性物质等优点。与常规蒸馏法和萃取法比较,微波-同时蒸馏萃取法得到的精油提取率高、色泽浅、质地纯,但存在香味失真、非香气成分含量较高[38]等问题。

挥发性有机溶剂萃取法

挥发性有机溶剂萃取法[39]是利用挥发性有机溶剂将原料中某些成分萃取出来,过去的几十年在天然香料加工中得到了迅速的应用和推广。目前,桂花就常采用有机溶剂浸提的方法获得浸膏,再用乙醇溶解浸膏,低温脱蜡得到净油[40]。麦秋君[8]曾用乙醇萃取了桂花浸膏,并采用气相色谱-质谱联用(GC-MS)技术分析了桂花净油的香气成分,共鉴定出了35种化合物,主要为酮类、醇类、高级脂肪酸及其酯类等,造成桂花净油幽清香韵的甜清花香。挥发性有机溶剂萃取法由于加工受热温度低,能保持原有植物原料的香气,化学稳定性强,可以提高精油质量。但挥发性有机溶剂萃取法也会造成浸提溶剂的残留,从而影响浸膏或者精油的真实香气。

动态顶空法

动态顶空又称吹扫捕集技术,即高效富集自然状态下释放的挥发性物质。张晓林等[32]曾用动态顶空法研究新鲜桂花的香气成分,但桂花的用量很大(10 kg),且洗脱时需用大量溶剂也有可能会带入或损失某些组分。金荷仙等[23]用活体植株动态顶空套袋捕集法,即在循环密闭的条件下高效富集自然状态下的桂花挥发性成分,发现各挥发性成分与以往研究结果存在一定差异,但一些主要的化合物,如氧化芳樟醇、芳樟醇、β-紫罗兰酮、2H-β-紫罗兰酮等物质,在色谱保留时间的先后顺序上,与以往研究具有一致性。较其他方法,如蒸馏、压榨或溶剂萃取等方法得到的精油或浸膏更能逼真地反映桂花自然释放的香气物质。

静态顶空法

静态顶空法是指在已达平衡的密闭容器中直接抽取样品顶空气体,并与GC结合对样品进行分析的一种技术,主要应用于复杂样品挥发性成分的定性分析检测。王呈仲等[41]采用静态顶空-气相色谱-质谱联用方法分析检测了新鲜桂花的挥发性成分,包括顶空-气相色谱-四极质谱、顶空-气相色谱-飞行时间色谱以及顶空-气相色谱-串联质谱联用技术对桂花芳香成分进行检测,并采用多维定性分析对检出的挥发性成分进行了鉴定。从桂花香气中检测出了41种挥发性成分,以单萜类和倍半萜类物质为主,含量高的组分有β-芳樟醇、β-

紫罗兰酮、反式-香叶醇、呋喃型芳樟醇氧化物等,与相关研究结果基本一致。

超临界 CO_2 萃取法

超临界 CO_2 萃取技术是以超临界 CO_2 做萃取剂,从液体或固体物料中萃取、分离有效成分。随着科技的发展,超临界 CO_2 萃取技术已成为一种新型萃取分离技术,被广泛应用于香精香料[42]、色素[43]、医药[44]等生产部门。利用超临界 CO_2 萃取技术提取天然香精香料,如从桂花、菊花[45]、含笑[46]中提取花香精,不仅可以有效地提取芳香组分,还可以提高纯度,保持其天然香味。施云海等[35]使用超临界 CO_2 萃取技术提取桂花精油,鉴定其主要成分相同与挥发性溶剂浸提得到的精油相同,但两种精油中的主要挥发性成分在含量上的差异明显。同时,以超临界 CO_2 进行萃取制得的桂花浸膏比用有机溶剂石油醚浸取制得的桂花浸膏特征香气完整,天然感好,品质高。刘虹等[36]利用 GC-MS 对超临界 CO_2 萃取的桂花浸膏的净油进行分析,发现超临界 CO_2 萃取中溶剂残留量少且可以富集具有烯丙基苯结构和紫罗兰酮结构的化合物,使净油更接近天然桂花香气。巫华美等[47]研究发现超临界 CO_2 萃取精油含较多酮类、醇类、酯类和醛类化合物,香气完全,新鲜感好,品质较高。与水蒸气蒸馏法相比,超临界 CO_2 萃取技术产物提取及分离效率高、提取时间短、蒸发潜热低[48],避免了高温对萃取物的破坏;与溶剂法相比,超临界 CO_2 萃取技术无化学溶剂残留、无污染,且其萃取成分更完全,香气、色泽更自然。但超临界 CO_2 萃取技术装置费时、操作复杂,不适于工业化生产。

活性炭吸附丝吸附法

活性炭吸附丝吸附法具有操作方便、吸附效率高等优点,广泛应用于医学研究[49]、香气成分分析[50]等领域。桂花香气是由挥发性有机物组成,活性炭吸附丝对其具有很强的吸附能力。冯建跃等[37]采用活性炭吸附丝累积采集桂花鲜花的香气成分,通过热解脱附结合 GC-MS 技术进行定性定量分析,鉴定结果与动态顶空法研究鲜桂花的香气成分基本相同。用活性炭吸附丝色谱法分析鲜花的香气成分可直接进行活体分析,采样比动态顶空法、超临界 CO_2 萃取法方便。而且香气成分被吸附在活性炭上,能够方便地被热解吸出来进行 GC-MS 分析,避免了溶剂萃取等带来的成分损失或溶剂的二次污染。

固相微萃取法

SPME 属于非溶剂型萃取法,集样品的萃取、解析和进样于一体,几乎不产生二次污染,但不同萃取条件会影响测定结果。常用的萃取头有 50/30 μm DVB/CAR/PDMS、50 μm PDMS/DVB、65 μm PDMS/DVB 以及 100 μm PDMS/DVB,不同萃取头性质不同。曹慧等[17]用 50/30 μm DVB/CAR/PDMS 的萃取头,从金桂中鉴定出 20 中香气成分,从银桂中鉴定出 13 种香气成分,从丹桂中鉴定出 9 种香气成分;李光祖等[29]也用 50/30 μm DVB/CAR/PDMS 的萃取头,从金桂的头香成分中分析出了 22 种化合物。杨雪云等[6]对 50 μm PDMS/DVB、65 μm PDMS/DVB 和 100 μm PDMS/DVB 三种萃取头进行了对比研究,发现不同的萃取头检测到的化合物数目和含量不同,65 μm PDMS/DVB 萃取头吸附的香味成分最多;Wang 等[15]、孙宝军等[24]、杨宇婷等[51]用的均是 65 μm PDMS/DVB 萃取头。萃取的温度也影响着各香味成分的挥发量,进而影响吸附在萃取头上的香味成分的含量和比例。温度太低,不利于桂花香气成分的挥发;温度太高则使得某些分子量较大或半挥发性成分比例提高,从而不能真实地反映自然条件下桂花的香味。Wang 等[15]在 50 ℃下萃取 40 min,从桂花花朵香气中分析出了 11 种分子量在 250 以上的化合物。李祖光等[29]在(25±5) ℃

室温条件下测定了桂花花朵的香气成分,其中分子量在250以上的化合物很少。

综上所述,不同的萃取方法对桂花浸膏或精油的萃取率及花香成分的鉴定影响较大,浸膏或精油的部分香气成分会被高温或溶剂萃取破坏,如水蒸气蒸馏法因温度过高,会对鲜花内很多不稳定的芳香物质造成破坏,但因其操作简单、成本低、产量大,适合大规模工业化生产;利用挥发性有机溶剂(如石油醚、乙醇)虽能保留桂花花朵的主要香气成分,但残留的有机溶剂不易去除,影响花香成分的真实性;微波-同时蒸馏萃取法提取率较高,但有效成分比例较低,大大影响了桂花花朵的香气成分;超临界CO_2萃取的精油中芳香成分的相对含量高于其他萃取方法,且香气成分更浓郁逼真,但对设备要求过高,难以产业化。

4 展 望

香气作为花朵重要品质指标之一越来越受到人们的关注,桂花作为著名的香花植物也成了近年来研究的热点。目前,关于桂花花朵香气的研究在不同种类和品种间香气成分的组成和含量差异、特征性香气成分的鉴定、花朵成熟过程中香气物质的变化及花香的释放规律等方面已取得一定的研究成果。但桂花花朵香气物质的合成、释放是一个复杂多样的问题。桂花的香气物质不但种类居多,而且同一种类包括的香气物质更多,如桂花香气中常见的萜烯就有20多种,因此还有待深入研究。

近年来,利用基因工程的方法改良植物花香已有报道,如蜡梅的法呢基焦磷酸合成酶基因 *CpF-PPS*[52]、白姜花的倍半萜合成酶基因 *Hc-Sesqui*[53] 等花香基因相继被克隆。然而,桂花花朵香气成分相关的酶或基因等分子生物学方面的研究较少。同时,桂花精油的提取率非常低,提取工艺也还不完善[30,38]。因此,今后应加强桂花花朵香气合成关键酶基因的克隆研究,在分子水平上对桂花花朵香气成分的合成进行调控,改善桂花花朵的香气成分;完善桂花精油的提取工艺,提高桂花精油的得率及品质,更好地指导相关生产实践。

参考文献

[1] 杨康民,朱文江. 桂花[M]. 上海:上海科学技术出版社,2000.

[2] PICHERSKY E, NOEL J P, DUDAREVA N. Biosynthesis of plant volatiles:Nature's diversity and ingenuity[J]. Science,2006,311:808-811.

[3] ISHIGURO T, KOGA N, NARA K. Components of the flowers of *Osmanthus fragrans* II. odorous component osmane and acids[J]. Yakugaku Zasshi,1957,77:566-567.

[4] SISIDO K S, KUROZUMI K, UTINIOT O K, et al. Fragrant flower constituents of *Osmanthus fragrans* Peif. Ensent[J]. Oil Rec,1966,5:557-560.

[5] LI Yingying. The main composition and affecting factors of aroma volatiles in flowers[J]. Northern Horticulture,2012(06):184-187.
李莹莹. 花香挥发物的主要成分及其影响因素[J]. 北方园艺,2012(06):184-187.

[6] YANG Xueyun, ZHAO Boguang, LIU Xiuhua, et al. Analysis of volatile compounds from the fresh flowers of *Osmanthus fragrans* var. *thunbergii* and *O. fragrans* var. *latifolius* by SPME and GC-MS[J]. Journal of Nanjing Forestry University(Natural Sciences Edition),2008,32(4):86-90.
杨雪云,赵博光,刘秀华,等. 金桂银桂鲜花挥发性成分的顶空固相微萃取GC-MS分析[J]. 南京林业大学学报(自然科学版),2008,32(4):86-90.

[7] GAO Qian, XIANG Nengjun, WU Yi-qin. Infuence of α-Ionone on the volatile compounds of cigarette mainstream smoke[J]. Flavour Fragrance Cosmetics, 2010, 2(4):21-25.
高茜,向能军,吴亿勤.卷烟中添加 α-紫罗兰酮对主流烟气挥发性成分影响的研究[J].香料香精化妆品,2010,2(4):21-25.

[8] MAI Qiujun. Analysis of chemical constituents of osmanthus absolute oil[J]. Journal of Guangdong University of Technology,2000,17(1):73-75.
麦秋君.桂花净油化学成分分析[J].广东工业大学学报,2000,17(1):73-75.

[9] MU Jinmei. The compounding process of osmanthus fragrance[J]. Flavour Fragrance Cosmetics, 2008(2):40-41.
牟进美.浅谈桂花香精的调配[J].香料香精化妆品,2008(2):40-41.

[10] LIN Xiangyun. Natural linalool and synthesis of linalool[J]. Chemical Engineering & Equipment, 2008(1):21-27.
林翔云.天然芳樟醇与合成芳樟醇[J].化学工程与装备,2008(1):21-27.

[11] CHEN Lijun, GAO Jianhong, WANG Shen, et al. Identification of the aroma compounds in orange oil by gas chromatography-mass spectrometry/olfactometry[J]. Fine Chemicals, 2012, 29(2): 143-146.
陈丽君,高建宏,王申,等.气相色谱-质谱与嗅觉测量法联用分析橙油中致香物质[J].精细化工,2012,29(2):143-146.

[12] WANG Qiushuang, CHEN Dong, XU Yongquan, et al. Investigation and comparison of the aroma components in guangdong black tea[J]. Journal of Tea Science,2012,32(1):9-16.
王秋霜,陈栋,许勇泉,等.广东红茶香气成分的比较研究[J].茶叶科学,2012,32(1):9-16.

[13] LIU Jinyun, WEI Jie, HUANG Jing, et al. GC-TOFMS analysis of the volatile components from leaves of *Melissa officinalis* L. by two extraction methods[J]. Journal of Anhui Agricultural Sciences,2012, 40(5):2621-2623.
刘劲芸,魏杰,黄静,等.2种方法提取香蜂花叶挥发性成分的 GC-TOFMS 分析[J].安徽农业科学,2012,40(5):2621-2623.

[14] LI Guopeng, JIA Huijuan, WANG Qiang, et al. Changes of aromatic composition in Youhongli (*Pyrus ussuriensis*) fruit during fruit ripening[J]. Journal of Fruit Science, 2012, 29(1):11-16.
李国鹏,贾惠娟,王强,等.油红梨(*Pyrus ussuriensis*)果实后熟过程中香气成分的变化[J].果树学报,2012,29(1):11-16.

[15] WANG Limei, LI Maoteng. Variations in the components of *Osmanthus fragrans* Lour. Essential oil at different stages of flowering[J]. Food Chemistry, 2009,114:233-236.

[16] XUE Chaoqun, WANG Jianwei, XI Jiaqin, et al. Relationship between physical-chemical indexes and full flavor style degree of flue-cured tobacco leaves[J]. Tobacco Science & Technology,2012(1):56-56.
薛超群,王建伟,奚家勤,等.烤烟烟叶理化指标与浓香型风格程度的关系[J].烟草科技,2012(1):56-56.

[17] CAO Hu, LI Zuguang, SHEN Delong. GC/MS Finger print analysis of *Osmanthus fragrans* Lour. in different varieties[J]. Acta Horticulturae Sinica,2009,36(3):391-398.
曹慧,李祖光,沈德隆.桂花品种香气成分的 GC/MS 指纹图谱研究[J].园艺学报,2009,36(3):391-398.

[18] WEN Guangyu, YU Fenglan, WANG Huating, et al. Studies on chemical constituents of the absolute from Flower of *Osmanthus fragrans*(Guihua)[J]. Journal of Integrative Plant Biology,

1983,25(5):467-471.

文光裕,丁风兰,王华亭,等.桂花净油的成分研究[J].植物学报,1983, 25(5):467-471.

[19] ZHU Meili, DING Desheng, HUANG Zuxuan, et al. Head space constituents of different varieties of *Osmanthus fragrans*[J]. Journal of Integrative Plant Biology,1985,27(4):412-418.

祝美莉,丁德生,黄祖萱,等.桂花不同变种的头香成分研究[J].植物学报,1985,27(4):412-418.

[20] DING Chengbin, XIONG Guangtong, WANG Qiang. Studies on chemical constituents of the absolute oils from *Osmanthus fragrans* flowers[J]. Guizhou Science,1993,11(3):41-45.

丁成斌,熊光同,王强.贵州桂花净油的成分研究[J].贵州科学,1993,11(3):41-45.

[21] WU Huamei, CHEN Xun, HE Xiangyin, et al. The chemical constituents of the absolute oils from *Osmanthus fragrans* flowers[J]. Acta Botanica Yunnanica,1997,19(2):213-216.

巫华美,陈训,何香银,等.贵州桂花净油的化学分析[J].云南植物研究,1997,19(2):213-216.

[22] DENG C H, SONG G X. Application of HS-SPME and GC-MS to characterization of volatile compounds emitted from osmanthus flowers[J]. Annalidi Chimic,2004,94(12):921-927.

[23] JIN Hexian, ZHENG Hua, JIN Youju, et al. Research on major volatile components of *Osmanthus fragrance* cultivars in Hangzhou Manlong Guiyu Park[J]. Forest Research,2006,19(5):612-615.

金荷仙,郑华,金幼菊,等.杭州满陇桂雨公园4个桂花品种香气组分的研究[J].林业科学研究,2006, 19(5):612-615.

[24] SUN Baojun, LI li, HAN Yuanji, et al. HS-SPME-GC-MS analysis of different *Osmanthus fragrans* cultivars from Guilin Garden in Shanghai[J]. Journal of Fujian College of Forestry,2012, (1):39-42.

孙宝军,李黎,韩远记.上海桂林公园桂花芳香成分的HS-SPME-GC-MS分析[J].福建林学院学报,2012,(1):39-42.

[25] ZHANG Ying, LI Xinlei, WANG Yan, et al. Changes of aroma components in *Oncidium* Sharry Baby in different florescence and flower parts[J]. Scientia Agricultura Sinica,2011,44(1):110-117.

张莹,李辛雷,王雁,等.文心兰不同花期及花朵不同部位香气成分的变化[J].中国农业科学,2011, 44(1):110-117.

[26] XU Jin, LI Yingying, ZHENG Chengshu, et al. Studies of aroma compounds in Chrysanthemum in different florescence and inflorescence parts and aroma releasing[J]. Acta Botanica Boreali-Occidentalia Sinica,2012,32(4):722-730.

徐瑾,李莹莹,郑成淑,等.菊花不同花期及花序不同部位香气成分和挥发研究[J].西北植物学报, 2012,32(4):722-730.

[27] TAN Yitan, XUE Shan, TANG Huizhou. Analysis of aroma constituents in *Gardenia jasminoides* at different flowering stages[J]. Food Science,2012,33(12):223-227.

谭谊谈,薛山,唐会周.不同花期栀子花的香气成分分析[J].食品科学,2012,33(12):223-227.

[28] ZHANG Huixiu, LENG Pingsheng, HU Zenghui, et al. The floral scent emitted from *Lilium* 'Siberia'at different flowering stages and diurnal variation[J]. Acta Horticulturae Sinica, 2013,40 (4):693-702.

张辉秀,冷平生,胡增辉,等.'西伯利亚'百合花香随开花进程变化及日变化规律[J].园艺学报, 2013,40(4):693-702.

[29] LI Zuguang, CAO Hu, ZHU Guohua, et al. Study on chemical constituents of fragrance released from fresh flowers of three different *Osmanthus fragrans* Lour. during different florescences[J]. Chemistry and Industry of Forest Products,2008,28(3):75-80.

李祖光,曹慧,朱国华,等.三种桂花在不同开花期头香成分的研究[J].林产化学与工业,2008,28(3):

75-80.
- [30] XU Jiming, LU Jinshun. A study on chemical composition of the essential oils from *Osmanthus fragran* flowers[J]. Chinese Journal of Analysis Laboratory,2007,26(1):37-41.
徐继明,吕金顺.桂花精油化学成分研究[J].分析试验室,2007,26(1):37-41.
- [31] ZHANG Xiaolin, LIN Zuming, JIN Sheng, et al. A Study on the composition of top note of *Osmanthus* in Hangzhou[J]. Chemistry,1984(9):20-21.
张晓林,林祖铭,金声,等.杭州栽培桂花头香化学成分的研究[J].化学通报,1984(9):20-21.
- [32] ZHANG Xiaolin, LIN Zuming, JIN Sheng, et al. A Study on the composition of top note of *Osmanthus* in Hangzhou[J]. Chemical Journal of Chinese Universities,1986,7(8):695-700.
张晓林,林祖铭,金声,等.杭州桂花头香成分的研究[J].高等学校化学学报,1986,7(8):695-700.
- [33] YANG Zhiping, YAO Weirong, QIAN He. Effect of β-D-Glucosidase on aromatic compounds of sweet *Osmanthus*[J]. Fine Chemicals,2005,22(12):924-926.
杨志萍,姚卫蓉,钱和.β-D-葡萄糖苷酶对桂花香气成分的影响[J].精细化工,2005,22(12):924-926.
- [34] YANG Kangming, ZHANG Jing, PENG Tingting. The Chinese renowned ancient sweet osmanthus [J]. Garden,2002(10):54-55.
杨康民,张静,彭婷婷.享誉今古的中国桂花[J].园林,2002(10):54-55.
- [35] SHI Yunhai, YANG Zhongwen, ZHOU Zhanyun. Study on the extraction of *Osmanthus fragrans* extract by supercritical CO_2[J]. Journal of Guangdong University of Technology,1993(4):5-8.
施云海,杨中文,周展云.超临界二氧化碳提取桂花浸膏的研究[J].香精香料化妆品,1993(4):5-8.
- [36] LIU Hong, HE Zhenghong, SHEN Meiying. Study on the oils of *Osmanthus fragrans* extract by supercritical CO_2[J]. Guangxi Forestry Science,1996,25(3):127-131.
刘虹,何正洪,沈美英.超临界二氧化碳萃取桂花净油化学成分的研究[J].广西林业科学,1996,25(3):127-131.
- [37] FENG Janyue, ZHAO Jing, HUANG Qiaoqiao, et al. Study on aroma components of osmanthus by absorption wire gas chromatography/mass spectrometry[J]. Journal of Zhejiang University(Science Edition),2001,28(6):672-675.
冯建跃,赵菁,黄巧巧,等.吸附丝色谱-质谱法用于桂花香气研究[J].浙江大学学报(理学版),2001,28(6):672-675.
- [38] ZHANG Jian. Study on extraction and analysis of *Osmanthus fragrans* Lour. essential oil[D]. Hangzhou Zhejiang University of Technology,2006.
张坚.桂花精油的提取与成分分析的研究[D].杭州:浙江工业大学,2006.
- [39] JIA Guiru, YANG Haiyan. Mechanism of special material extraction from cells[J]. Transactions of the Chinese Society of Agricultural Engineering,1998,14(2):68-71.
贾贵儒,杨海燕.从细胞中萃取特定物质机理的研究[J].农业工程学报,1998,14(2):68-71.
- [40] MA Xihan, WANG Yonghong, HU Yayun, et al. Advances in the research of oil-bearing roses[J]. Journal of Northwest Forestry University,2004,19(4):138-141.
马希汉,王永红,胡亚云,等.精油玫瑰研究[J].西北林学院学报,2004,19(4):138-141.
- [41] WANG Chengzhong, SU Yue, GUO Yinlong. Analysis of the volatile components from flowers and leaves of *Osmanthus fragrans* Lour. by Headspace-GC-MS[J]. Chinese Journal of Organic Chemistry,2009,29(6):948-955.
王呈仲,苏越,郭寅龙.顶空-气相色谱-质谱联用分析桂花和叶中挥发性成分[J].有机化学,2009,29(6):948-955.

[42] CHEN Jianhua, WEI Maoshan, LI Zhong, et al. Study on cumin essential oil and roasted cumin essential oil[J]. Science and Technology of Food Industry, 2012, 33(7):322-326.
陈建华,韦茂山,李忠,等. 孜然精油及其熟制精油的研究[J]. 食品工业科技,2012,33(7):322-326.

[43] LIANG Yexing, XIONG Jiayan. Research progress in supercritical CO_2 extraction of natural pigments [J]. The Beverage Industry, 2013, 16(7):1-7.
梁叶星,熊家艳. 超临界 CO_2 技术应用于天然色素萃取的研究进展[J]. 饮料工业,2013,16(7):1-7.

[44] LUO Shunian, WANG Jin, LI Moxin, et al. Study on extraction of blackcurrant seed oil by supercritical carbon dioxide[J]. Cereals & Oils, 2010(2):22-24.
罗淑年,王瑾,李默馨,等. 超临界 CO_2 萃取黑加仑籽油研究[J]. 粮食与油脂,2010(2):22-24.

[45] JIN Jianzhong, TONG Jianying. Supercritical carbon dioxide extraction of volatile oil from *Chrysanthemum morifolium*[J]. Food Science, 2010, 31(14):125-127.
金建忠,童建颖. 超临界 CO_2 萃取杭白菊挥发油的工艺研究[J]. 食品科学,2010,31(14):125-127.

[46] WANG Hongwu, LIU Yanqing, LU Xiange, et al. Supercritical CO_2 extraction GC-MS analysis of *Michelia Multifloliate* for its volatile constituent[J]. Physical Testing and Chemical Analysis Part B (Chemical Analysis), 2007, 43(7):537-539.
汪洪武,刘艳清,鲁湘鄂,等. 含笑花挥发油的超临界二氧化碳萃取及气相色谱-质谱分析[J]. 理化检验(化学分册),2007,43(7):537-539.

[47] WU Huamei, HE Xiangyin. *Osmanthus fragrans* concrete extraction by supercritical CO_2 fluid[J]. Guizhou Sciencce, 1997, 15(1):32-35.
巫华美,何香银. 超临界 CO_2 提取桂花浸膏的工艺研究[J]. 贵州科学,1997,15(1):32-35.

[48] CAO Mingxia, XU Yi, ZHAO Tianming, et al. Application progress of supercritical fluid extraction in the natural plant extracts active ingredients[J]. Guangzhou Chemical Industry, 2010, 38(8):23-25, 37.
曹明霞,徐溢,赵天明,等. 超临界萃取在天然植物成分提取中的应用进展[J]. 广州化工,2010,38(8):23-25,37.

[49] CAI Yaoxin, ZHANG Renzheng, SUN Lan, et al. Nanometer activated carbon adsorption docetaxel effect on proliferation and apoptosis of A549 cells[J]. Journal of Zhengzhou University (Medical Sciences), 2012(6):759-761.
蔡要欣,张韧铮,孙岚,等. 纳米活性炭吸附多烯紫杉醇对 A549 细胞增殖及凋亡的影响[J]. 郑州大学学报(医学版),2012(6):759-761.

[50] HUANG Qiaoqiao, FENG Jianyue. Study on the variation in Narcissus aroma composition during blossoming[J]. Journal of Instrumental Analysis, 2004, 23(5):110-113.
黄巧巧,冯建跃. 水仙花开放期间香气组分变化的研究[J]. 分析测试学报,2004,23(5):110-113.

[51] YANG Yuting, WU Xiaohong, TIAN Puyu, et al. On the volatile components in *Osmanthus fragrans* (Wanyinggui, Guifeihong and Yaotiaoshunv)[J]. Journal of Henan University (Medical Science), 2010, 29(1):13-16.
杨宇婷,武晓红,田璞玉. 桂花(晚银桂、贵妃红和窈窕淑女)挥发性成分分析[J]. 河南大学学报(医学版),2010,29(1):13-16.

[52] LI Ruihong, FAN Yanping, YU Rangcai, et al. Molecular cloning and expression of sesquiterpenoid synthase gene in *Hedychium coronarium* Koenig[J]. Acta Horticulturae Sinica, 2008, 35(10):1527-1532.
李瑞红,范燕萍,余让才,等. 白姜花倍半萜合成酶基因的克隆及表达[J]. 园艺学报,2008,35(10):1527-1532.

[53] LIN X,ZHAO K G,CHEN L Q. Molecular cloning and expression of *Chimonanthus praecox* farnesyl pyrophosphate synthase gene and its possible involvement in the biosynthesis of floral volatile sesquiterpenoids[J]. Plant Physiology Biochnology ,2010,48 (10/11): 845-850.

Research Progress in Flavoring Compositions of *Osmanthus fragrans* Lour.

SHI Tingting, YANG Xiulian, WANG Lianggui

(College of Landscape Architecture, Nanjing Forestry University, Nanjing 210037, China)

Abstract: Floral scent of *Osmanthus fragrans* is a complex mixture of low molecular weight volatile compounds, which gives *O. fragrans* flower its unique fragrance. Flavoring compositions for *O. fragrans* flower have been comprehensively studied, which is focused on the changes caused by different varieties、times、regions and analytical methods of *O. fragrans*. In this article, the types of flavoring compositions in *O. fragrans* flower, influence factors, analytical methods and so on are summarized.

Key words: *Osmanthus fragrans* flower; flavoring compositions; affecting factors; analytical methods

保存方法对桂花精油提取及香气成分的影响[*]

施婷婷,杨秀莲,赵林果,王良桂

(南京林业大学 风景园林学院,江苏 南京 210037)

摘 要:通过水蒸气蒸馏法分别对冷冻、氯化钠、柠檬酸及自然干燥保存的桂花进行精油提取,并采用 GC-MS 对所得精油的有效成分进行分析,比较不同保存方法处理及保存 30、60 和 90 d 后桂花的精油得率及香气成分的差别。结果表明:冷冻保存、氯化钠保存及柠檬酸保存是保存桂花的较好方法,自然干燥保存法不可取;氯化钠、柠檬酸保存桂花较合适的时间为 60 d。

关键词:桂花;精油;保存方法;香气成分

精油,商业上称"芳香油",医药上称"挥发油",是一种具有一定挥发性的油状液态物质[1-2],具有经济、卫生、使用快速、香味均匀等优点,广泛应用在食品、烟酒工业和化妆品等领域[3-5]。近年来的一系列科学研究发现,植物精油还具有十分明显的保健功效,在抗菌和预防、辅助治疗慢性疾病等方面尤为引人注目[6-7]。桂花精油香气清新高雅、幽雅留长,深受人们的喜爱,是一种高级天然香料。国内外关于桂花精油提取以及化学成分分析已有报道[8-12]。

要实现桂花精油的工业化生产,就需要大量的桂花原材料,但桂花花期极短,仅有 1 个星期左右,花朵娇弱,香气极易损失,采摘后需及时处理。然而大量的桂花往往来不及处理,所以必须对桂花进行预处理,将其保存起来,然后再加工生产桂花精油。因此,如何使用合适的方法保存桂花鲜花,尽可能地减少精油的损失有着重要的现实意义。桂花保存可采用糖渍、柠檬酸保存、食盐保存等方法[13-14]。笔者通过水蒸气蒸馏法分别对冷冻、氯化钠、柠檬酸及自然干燥保存的桂花进行精油提取,并采用 GC-MS 对所得精油进行成分鉴定,比较不同保存方法处理的桂花精油产率及香气成分的差别,以确定合适的保存方法,为桂花精油的工业化生产提供理论基础。

1 材料与方法

1.1 试验材料及仪器

试材:桂花鲜花,采摘于湖北省咸宁市咸安区桂花基地,品种为'小金球'。将新鲜桂花中花梗及杂物去除,待用。乙醚、无水硫酸钠试剂均为分析纯。

---

[*] 原文发表于《南京林业大学学报(自然科学版)》,2014,38(增刊):105-110。
基金项目:国家林业公益性行业科研专项项目(201204607)。

试验仪器:气相色谱质谱联用仪 Trace DSQ(美国 Thermo Electro-Finnigan),水蒸气蒸馏装置。

色谱条件:TR-5MS(30 m × 0.25 mm × 0.25 μm)毛细管色谱柱。

升温程序:起始温度为 50 ℃,保持 2 min,以 10 ℃/min 升至 100 ℃,以 5 ℃/min 升至 250 ℃,保持 10 min;载气为高纯度氦气(He),氦气流速为 1 mL/min;分流比为 10∶1;进样量 1 μL。

质谱条件:接口温度 250 ℃,电离方式 EI,离子电离能量 70 eV,质量扫描范围 40～700 amu。

1.2 桂花保存方法

冷冻保存:将去除花梗及杂物的新鲜桂花置于超低温冰箱(−20 ℃)中冷冻保存。

氯化钠储藏法:称取桂花 1 000 g、氯化钠 400 g,在 10 L 广口瓶内撒一层氯化钠铺一层桂花,分层压实,加盖密封保存。

柠檬酸储藏法:称取桂花 1 000 g、柠檬酸 20 g,混合均匀,12 h 后滤去卤水,再加上 180 g 氯化钠和 5 g 白矾封盖表面,加盖密封保存。

自然干燥:将去除花梗及杂物的新鲜桂花置于阴凉处自然阴干,密封常温保存。

1.3 桂花精油的提取

称取一定量的采用不同方式保存的桂花,用蒸馏水洗涤后进行常规水蒸气蒸馏 3 h,蒸馏液用 1 000 mL 乙醚萃取 3 次,萃取液用无水 Na_2SO_4 干燥,过滤,水浴蒸去乙醚,得到淡黄色油状液体,称质量,计算桂花精油的得率。同时,采用 GC-MS 对所得精油进行成分分析。

2 结果与分析

2.1 不同保存方法对桂花精油得率和香气成分的影响

试验结果表明,冷冻干燥的桂花精油得率为 0.13%,利用柠檬酸和氯化钠保存的桂花精油得率分别为 0.16%、0.15%,与新鲜桂花精油得率(0.15%)相近,而自然干燥处理后的桂花精油产率仅为 0.07%,相比鲜花有很大程度的下降,这可能是由于各香气成分在自然阴干过程中散失造成。

采用 GC-MS,对新鲜桂花及冷冻、氯化钠、柠檬酸及自然干燥保存处理的桂花精油香气成分进行分析,分析结果如表 1 所示。

表 1 不同保存方法桂花精油香气成分分析

Tab. 1 The fragrant components of essential oils in *O. fragrans* from different preservation methods

序号 No.	主要成分 Main component	质量分数/%				Content
		A	B	C	D	E
1	乙酸乙酯 ethyl acetate	0.71	1.51	2.59	2.75	1.45
2	顺式-氧化芳樟醇 *cis*-Linalool oxide	0.79	2.48	—	—	0.14
3	反式-氧化芳樟醇 *trans*-linalool oxide	2.50	8.71	4.40	1.78	0.34

续表

序号 No.	主要成分 Main component	质量分数/% Content				
		A	B	C	D	E
4	芳樟醇 Linalool	0.17	0.23	0.53	0.89	—
5	壬醛 Nonanal	0.29	—	0.35	0.24	
6	2,2,6-三甲基-6-乙烯基四氢-2H-吡喃-3-醇 2H-Pyran-3-ol, 6-ethenyltetrahydro-2,2,6-trimethyl-	22.98	30.51	2.90	0.71	
7	香叶醇 trans-Geraniol	0.62	0.54	0.62	1.70	0.73
8	α-松油醇 α-Terpineol	—	—	2.13	1.69	
9	胡薄荷酮 Pulegone	0.45	0.18			
10	甲基丁香酚 Benzene,1,2-dimethoxy-4-(2-propenyl)-	4.11	0.72	—	—	
11	对甲氧基苯乙醇 2-(4-Methoxyphenyl)ethano	3.13	1.31	1.98		
12	4-(2,6,6-三甲基-2-环己烯-1-炔基)-3-丁烯-2-酮 3-Buten-2-one,4-(2,6,6-trimethyl-2-cyclohexen-1-yl)-	—			0.27	
13	壬酸 Nonanoic acid	7.65	—	4.21	2.33	
14	十三烷 Tridecane	—	—	3.05	1.64	5.72
15	4-叔丁基苯酚 Phenol, p-tert-butyl-	—	—	0.87	—	0.28
16	4-(1-甲基乙烯基)-1-环己烯-1-甲醇乙酸酯 p-Mentha-1(7),8(10)-dien-9-ol	—	—	—	0.27	0.13
17	薄荷醇 Menthol	—	—	—	0.22	
18	α,α-4-三甲基环己基甲醇 Cyclohexanol,1-methyl-4-(1-methylethyl)-	—	—	0.95	—	1.47
19	2,6,6-三甲基-1-环己烯-1-乙醇 1-Cyclohexene-1-ethanol, 2,6,6-trimethyl-	—	—	—	0.49	2.45
20	巨豆-4,6(Z),8(E)-三烯 Megastigma-4,6(Z),8(E)-triene	—	—	—	0.29	
21	脱氢乙酸 Dehydroacetic Acid	—	—	1.26	3.77	3.45
22	4-(2,6,6-三甲基-2-环己烯-1-基)-2-丁烯酮 2-Butanone,4-(2,6,6-trimethyl-2-cyclohexen-1-ylidene)-	—	—	—	0.76	
23	β-紫罗兰醇 β-Ionol	—	—	—	0.38	—
24	6-甲基-5-(1-甲基亚乙基)-6,8-壬二烯-2-酮 6,8-Nonadien-2-one,6-methyl-5-(1-methylethylidene)-	—	—	—	0.75	
25	4-(2,6,6-三甲基-1,3-环己二烯-1-基)-2-丁酮 2-Butanone,4-(2,6,6-trimethyl-1,3-cyclohexadien-1-yl)-	0.17	0.11	0.75	—	
26	α-紫罗兰酮 α-Ionone	3.61	2.49	2.15	3.21	1.09
27	二氢-β-紫罗兰酮 Dihydro-β-ionone	11.82	16.28	13.71	14.52	5.42
28	4-(2,6,6-三甲基环乙烯基)-2-丁醇 4-(2,6,6-Trimethyl-cyclohex-1-enyl)-butan-2-ol	0.14	0.27	4.18	3.46	
29	6-戊基-2H-吡喃-2-酮 2H-Pyran-2-one,6-pentyl-	0.83	0.53	0.38	0.51	—
30	γ-癸内酯 γ-Decalactone	5.91	3.64	5.79	6.29	1.41
31	β-紫罗兰酮 β-Ionone	15.01	16.44	20.84	38.97	5.35
32	N-乙酰基酪胺 N-Acetyltyramine	—	1.57	1.45	0.97	3.91

续 表

序号 No.	主要成分 Main component	质量分数/%				Content
		A	B	C	D	E
33	十七烷 Heneicosane	1.34	1.89	2.86	—	7.20
34	十九烷 Nonadecane	0.15	0.35	0.48	—	10.37
35	二氢猕猴桃内酯 2(4H)-Benzofuranone,5,6,7,7a-tetrahydro-4,4,7a-trimethyl-,(R)-	0.32	0.44	0.64	0.19	—
36	十二烯基丁二酸酐 2-Dodecen-1-yl(-) succinic anhydride	—	—	0.77	0.75	—
37	黑蚁素 Furan,3-(4,8-dimethyl-3,7-nonadienyl)-,(E)-	—	—	2.82	1.37	1.49
38	十五烷 Pentadecane	—	—	4.38	1.38	10.56
39	己基癸醇 1-Decanol,2-hexyl-	—	—	0.51	—	1.54
40	愈创蓝油烃 Guaiazulene	1.50	—	0.90	—	0.91
41	棕榈酸 Hexadecanoic acid	6.21	5.88	6.47	3.25	3.33
42	亚麻酸 Linolenic acid	1.34	0.58	0.98	0.61	1.69
43	3-酮-β-紫罗兰酮 3-Keto-β-ionone	0.59	—	1.51	1.35	0.14
44	3-氧-β-紫罗兰酮 3-Oxo-β-ionone	3.86	0.54	—	—	—
45	邻苯二甲酸二丁酯 Dibutyl phthalate	0.30	0.25	—	—	13.11
46	邻苯二甲酸二异丁酯 Phthalic acid diisobutyl ester	1.69	0.78	0.57	—	10.93
	总计 Total	98.19	98.23	97.98	97.76	94.61

注:A. 新鲜桂花;B. 冷冻保存;C. 柠檬酸保存;D. 氯化钠保存;E. 自然干燥保存。

采用水蒸气蒸馏法提取的新鲜桂花精油中主要含有28种香气成分,其中含量较高的有芳樟醇氧化物、2,2,6-三甲基-6-乙烯基四氢-2H-呋喃-3-醇、壬酸、α-紫罗兰酮、二氢-β-紫罗兰酮、γ-癸内酯、β-紫罗兰酮、棕榈酸等。同时,经冷冻干燥的桂花精油与新鲜桂花精油的成分无本质差异,仅缺少壬醛、壬酸、愈创蓝油烃、3-酮-β-紫罗兰酮4种成分,而且相对于鲜花,冷冻干燥处理后的桂花精油中各主要香气组分的相对质量分数变化并不太大,说明各香气组分的挥发损失程度相差不大。经柠檬酸、氯化钠保存的桂花精油分别含有34、32种香气成分,但其主要香气成分为芳樟醇氧化物、壬酸、α-紫罗兰酮、二氢-β-紫罗兰酮、γ-癸内酯、β-紫罗兰酮、棕榈酸等,与新鲜桂花精油的主要香气成分相似,仅存在一些微量成分的差异。同时还发现,经氯化钠保存处理的桂花精油中没有检测到邻苯二甲酸二丁酯与邻苯二甲酸二异丁酯。而经自然干燥处理的桂花精油香气成分发生了较大的变化,其中芳樟醇氧化物、α-紫罗兰酮、二氢-β-紫罗兰酮、γ-癸内酯、β-紫罗兰酮等主要香气成分的含量都明显减少(图1),而且在新鲜桂花精油中含量较高的2,2,6-三甲基-6-乙烯基四氢-2H-呋喃-3-醇、甲基丁香酚、对甲氧基苯乙醇等物质在自然干燥处理后都没有检测出来,同时邻苯二甲酸二丁酯与邻苯二甲酸二异丁酯含量也明显增大。

通过对冷冻、氯化钠、柠檬酸、自然干燥保存,以及新鲜桂花进行精油提取,比较4种不同方法处理后桂花精油的产率和香气成分,结果表明:冷冻干燥的桂花精油总体能保留鲜花的原有香气成分,精油得率基本不变。经柠檬酸、氯化钠保存的桂花精油得率及其主要香气成分的种类和相对含量与新鲜桂花相似,但一些微量的香气成分与新鲜桂花存在一定差异。而自然干燥提取的桂花精油的主要香气成分已基本散失,邻苯二甲酸二丁酯与邻苯二甲酸二异丁酯的含量也明显升高,且精油得率也大幅下降。因此,冷冻干燥保存、柠檬酸、氯化钠保存方法为较合适的保存桂花方法。

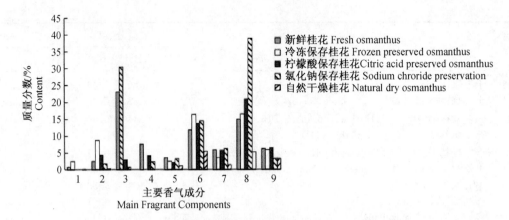

图1 不同保存方法提取的桂花精油主要成分含量对比
Fig. 1 Comparison of the main ingredients in different storage methods of essential oils of *O. fragrans*

1.顺式-氧化芳樟醇;2.反式-氧化芳樟醇;3.2,2,6-三甲基-6-乙烯基四氢-2H-呋喃-3-醇;4.壬酸;5.α-紫罗兰酮; 6.二氢-β-紫罗兰酮;7.γ-癸内酯;8.β-紫罗兰酮;9.棕榈酸。

2.2 不同保存时间对桂花精油主要香气成分的影响

由于冷冻干燥保存存在成本较高、保存时间短等问题,在此主要研究氯化钠、柠檬酸两种方法分别保存桂花30、60和90 d,分析不同保存时间的桂花精油香气成分变化(表2)。

从表2可以看出,虽然保存方法及保存时间不同,但6个样品中含量最高的香气物质基本一致,大部分都是萜烯类化合物,主要有反式-氧化芳樟醇、芳樟醇、壬酸、α-紫罗兰酮、二氢-β-紫罗兰酮、4-(2,6,6-三甲基环乙烯基)-2-丁醇、γ-癸内酯、β-紫罗兰酮、棕榈酸等。虽然两种不同保存方法的桂花精油香气成分有一定差异,但主要成分基本不变,而且在保存的过程中,主成分含量的变化也比较一致,呈现规律性变化。比如,α-紫罗兰酮的含量在保存过程中逐渐减少,α-紫罗兰酮在鲜花中的质量分数为3.61%,氯化钠保存30 d后质量分数为3.21%,保存60 d后含量减少至2.77%,可以看出变化范围都在0.4%左右。而当保存至90 d时,α-紫罗兰酮的含量开始骤减,仅有0.33%;柠檬酸保存时α-紫罗兰酮的含量变化趋势基本一致,保存30 d时减少至2.15%,保存60 d后含量为2.06%,变化相对较小,保存至90 d时,α-紫罗兰酮的含量也开始骤减(0.58%),相比鲜花减少了6倍以上。桂花精油中主要成分大部分都呈现此类变化,也有少部分的主要成分(如反式-氧化芳樟醇、芳樟醇、壬酸等)在保存过程中的含量基本没有变化,或变化不明显。有些成分在保存过程中含量略有增加,如二氢猕猴桃内酯的含量由氯化钠保存30 d的0.19%到保存60 d时的0.28%,再到保存90 d时,质量分数高达2.26%;柠檬酸保存时二氢猕猴桃内酯的含量变化与氯化钠一致,保存30 d时含量为0.64%,保存60 d后为0.55%,略有减少,当保存90 d时,质量分数又大幅提高至1.94%。此类物质还有N-乙酰基酪胺、亚麻酸、邻苯二甲酸二异丁酯等。还有些成分在鲜花及保存初期未被检测出,而在保存后期中存在,如棕榈醛、肉豆蔻醛、硬脂醛等。

不管是氯化钠保存,还是柠檬酸保存,在保存60 d前,桂花精油的香气成分变化不大,但到第3个月(60~90 d)其精油成分的含量及种类都发生了较大变化。因此,认为桂花用于精油提取可在氯化钠或柠檬酸条件下保存60 d。

表 2 不同保存时间对桂花精油主要香气成分的影响
Tab. 2 Effect of preservation time on the main fragrant components of essential oil

序号 No.	主要成分 Main component	质量分数/% Content					
		柠檬酸保存 Citric acid preservation			氯化钠保存 NaCl preservation		
		30 d	60 d	90 d	30 d	60 d	90 d
1	顺式-氧化芳樟醇 cis-Linalool oxide	—	—	1.61	—	—	—
2	反式-氧化芳樟醇 trans-Linalool oxide	4.40	4.97	4.82	1.78	3.69	3.21
3	芳樟醇 Linalool	0.53	1.80	0.27	0.89	1.51	0.76
4	壬醛 Nonanal	0.35	0.51	—	0.24	0.45	—
5	2,2,6-三甲基-6-乙烯基四氢-2H-吡喃-3-醇 2H-Pyran-3-ol,6ethenyltetrahydro-2,2,6-trimethyl-	2.90	2.09	—	0.71	2.70	0.18
6	香叶醇 trans-Geraniol	0.62	0.51	—	1.70	1.45	0.48
7	α-松油醇 α-Terpineol	2.13	2.77	0.93	1.69	1.46	0.99
8	壬酸 Nonanoic acid	4.21	3.82	4.33	2.33	2.06	2.79
9	脱氢乙酸 Dehydroacetic acid	1.26	2.02	1.98	3.77	3.36	4.67
10	α-紫罗兰酮 α-Ionone	2.15	2.06	0.58	3.21	2.77	0.33
11	二氢-β-紫罗兰酮 Dihydro-β-ionone	13.71	10.93	9.23	14.52	11.02	5.34
12	4-(2,6,6-三甲基环乙烯基)-2-丁醇 4-(2,6,6-Trimethyl-cyclohex-1-enyl)-butan-2-ol	4.18	3.07	3.32	3.46	4.16	2.64
13	6-戊基-2H-吡喃-2-酮 2H-pyran-2-one,6-pentyl-	0.38	0.30	—	0.51	0.22	0.29
14	γ-癸内酯 γ-Decalactone	5.79	4.43	3.73	6.29	9.37	5.38
15	β-紫罗兰酮 β-Ionone	20.84	21.61	11.36	38.97	29.78	11.23
16	N-乙酰基酪胺 N-Acetyltyramine	1.45	2.67	3.16	0.97	0.72	11.6
17	二氢猕猴桃内酯 2(4H)-Benzofuranone,5,6,7,7a-tetrahydro-4,4,7a-trimethyl-,(R)-	0.64	0.55	1.94	0.19	0.28	2.26
18	黑蚁素 Furan,3-(4,8-dimethyl-3,7-nonadienyl)-(E)-	2.82	0.94	1.00	1.37	1.31	—
19	2-甲基-4-(2,6,6-三甲基-1-环己烯-1-基)-2-丁烯醛 2-Butenal,2-methyl-4-(2,6,6-trimethyl-1-cyclohexen-1-	—	1.24	2.85	—	0.88	1.91
20	肉豆蔻醛 Tetradecanal	—	0.61	0.85	—	0.19	0.74
21	棕榈酸 Hexadecanoic acid	6.47	4.71	0.37	3.25	1.94	1.71
22	亚麻酸 Linolenic acid	0.98	1.74	9.27	0.61	0.41	1.63
23	硬酯醛 Octadecanal	—	2.89	5.96	—	5.57	7.38
24	油酸酰胺 9-Octadecenamide,(Z)-	—	0.87	1.33	—	0.18	1.53
25	3-酮-β-紫罗兰酮 3-Keto-β-ionone	1.51	1.69	1.62	1.35	0.76	—
26	邻苯二甲酸二异丁酯 Phthalic acid diisobutyl ester	0.57	1.63	5.47	—	—	—

3 结论

(1) 通过对冷冻保存、氯化钠保存、柠檬酸保存及自然干燥保存的桂花进行精油提取，比较4种不同方法处理后的桂花精油的得率和香气成分。结果表明，冷冻保存的桂花总体能保留鲜花的原有香气成分，精油得率基本不变；氯化钠、柠檬酸保存的桂花精油得率和主要香气成分也基本一致，仅一些微量香气成分的种类和含量存在差异；自然干燥处理的桂花精油香气成分及得率都有很大程度下降。因此，冷冻保存、氯化钠保存、柠檬酸保存方法为较合适的保存桂花方法。同时，氯化钠、柠檬酸保存的桂花精油虽然其香气成分存在差异，但桂花精油的主要香气成分都存在，还检测到了一些其他芳香类物质，而且在氯化钠保存处理的桂花精油中没有检测到邻苯二甲酸二丁酯与邻苯二甲酸二异丁酯，其被视为生殖分泌干扰物质在桂花精油的研究中早有报道[15]，这些有待进一步的验证和利用。

(2) 试验利用氯化钠、柠檬酸保存法对不同保存时间桂花精油香气成分的变化进行研究。研究发现保存的桂花在前60 d其主要香气成分变化不大，60 d后的主要香气成分的含量大幅减少，香气成分的种类也有所减少。此次试验中冷冻保存也是较为合适的方法，但−20 ℃的冷冻处理保存时间不长，保存1个月后桂花花瓣即开始褐化；同时，此次试验也没有比较不同保存时间桂花精油提取率的变化，故尚需对冷冻处理（超低温）保存时间对桂花精油的得率和花香成分的影响进行研究，了解冷冻处理是否可以更长久地保存桂花。

参考文献

［1］毛海舫,李琼.天然香料加工工艺学[M].北京:中国轻工业出版社,2006.

［2］江志利,张兴,冯俊涛.植物精油研究及其在植物保护中的利用[J].陕西农业科学,2002(1):32-36.
Jiang Z L, Zhang X, Feng J T. Study of plant essential oils and its application in plant protection[J]. Shanxi Journal of Agricultural Sciences, 2002(1): 32-36.

［3］杨秀莲,赵飞,王良桂.25个桂花品种花瓣营养成分分析[J].福建林学院学报,2014,34(1):5-10.
Yang X L, Zhao F, Wang L G. Analysis of nutrients in the petals of 25 cultivars of *Osmanthus fragrans*[J]. Journal of Fujian College of Forestry, 2014, 34(1): 5-10.

［4］边名鸿,叶光斌,陈欲云,等.新型桂花米酒的酿造工艺研究[J].酿酒科技,2013,8:19-22.
Bing M H, Ye G B, Chen Y Y, et al. Study on brewing technology of new *Osmanthus fragrans* wine [J]. Liquor-Making Science & Technology, 2013, 8:19-22.

［5］王丹,谢小丽,胡璇,等.天然香料在化妆品中的应用现状[J].现代生物医学进展,2013,31:6189-6193.
Wang D, Xie X L, Hu X, et al. Application of natural fragrance in cosmetics[J]. Progress in Modern Biomedicine,2013,31:6189-6193.

［6］Lee Do-Gyeong, Lee Sang-Min, Bang Myun-Ho, et al. Lignans from the flowers of *Osmanthus fragrans* var. *aurantiacus* and their inhibition effect on NO production[J]. Archives of Pharmacal Research, 2011, 34(12):2029-2035.

［7］周秋霞,岳淑梅.基于文献的桂花化学成分及药理作用研究现状分析[J].河南大学学报(医学版),2013,32(2):139-142.
Zhou Q X, Yue S M. Analysis of the research status of *Osmanthus fragrans* Lour. in chemical constituents and pharmacological actions based on literature[J]. Journal of Henan University(Medical

Science),2013,32(2):139-142.

[8] 徐继明,吕金顺.桂花精油化学成分研究[J].分析试验室,2007,26(1):37-41.
Xu J M, Lu J S. A study on chemical composition of the essential oils from *Osmanthus fragrans* flowers[J]. Chinese Journal of Analysis Laboratory, 2007, 26(1):37-41.

[9] 李发芳,胡西亮.不同提取方法对桂花精油品质的影响[J].氨基酸和生物资源,2012,34(2):59-62.
Li F F, Hu X L. Effect of different extraction methods on Osmanthus oil quality[J]. Amino Acids & Biotic Resources, 2012,34(2):59-62.

[10] 巫华美,陈训,何香银,等.贵州桂花净油的化学分析[J].云南植物研究,1997,19(2):213-216.
Wu H M, Chen X, He X Y, et al. The chemical constituents of the absolute oils from *Osmanthus fragrans* flowers[J]. Acta Botanica Yunnanica, 1997,19(2):213-216.

[11] 赵俊宏,薛宝玉,李文锋,等.桂花精油β-环糊精包合物的制备工艺研究[J].河南科学,2013,31(2):158-160.
Zhao J H, Xue B Y, Li W F, et al. The inclusion compound preparation of essential oil of *Osmanthus fragrans* with β- cyclodextrin [J]. Henan Science,2013,31(2):158-160.

[12] 原玲芳,敖明章,金文闻,等.香兰素延缓桂花精油香气释放的作用[J].食品科学,2013,34(19):73-75.
Yuan L F, Ao M Z, Jin W W, et al. Application of vanillin in sustained aroma release from essential oil from *Osmanthus fragrans*[J]. Food Science,2013,34(19):73-75.

[13] 邱丽颖.桂花保鲜贮藏技巧[J].农村新技术,2009,15:37.
Qiu L Y. *Osmanthus fragrans* preservation techniques [J]. New Rural Technology,2009,15:37.

[14] 何礼军,王珊崇,杨园园.桂花的育花与贮藏及加工技术[J].现代农业科技,2013,17:189,191.
He L J, Wang S C, Yang Y Y. Technology of breeding and processing flowers of *Osmanthus fragrans* (Thurb.) Lour. [J]. Modern Agricultural Science and Technology,2013,17:189,191.

[15] 麦秋君.桂花净油化学成分分析[J].广东工业大学学报,2000,17(1):73-75.
Mai Q J. Analysis of chemical constituents of *Osmanthus* absolute oil[J]. Journal of Guangdong University of Technology,2000,17(1):73-75.

Influence of Storage Method on the Fragrant Components of Essential Oils of *Osmanthus fragrans*

SHI Tingting, YANG Xiulian, ZHAO Linguo, WANG Lianggui

(College of Landscape Architecture, Nanjing Forestry University, Nanjing 210037, China)

Abstract: The influence of the storage method of *Osmanthus fragrans*, which were kept under frozen, citric acid preservative, NaCl preservative and dryness conditions, respectively, on the quality of essential oils were analyzed by GC-MS. The results show that the yield and the fragrant components of essential oils from the frozen, citric acid preservative and NaCl preservative flower were very similar to those from the fresh flower. As the citric acid preservative, NaCl preservative time increased, their main fragrant components of O. fragrans kept unchanged, but their contents showed a nonlinear decreasing trend which accelerated significantly after two months of cryopreservation, suggesting that O. fragrans should not be stored for longer than 60 d under citric acid or NaCl preservative conditions.

Key words:: *Osmanthus fragrans*; essential oils; storage method; fragrant components

两种植物生长延缓剂对盆栽'日香桂'的矮化效应

王 萍,杨秀莲,王春君,王良桂

(南京林业大学 风景园林学院,江苏 南京 210037)

摘 要:采用不同浓度的植物生长延缓剂多效唑与比久对盆栽2年生'日香桂'进行叶面喷施处理,研究不同处理对'日香桂'矮化的影响。结果表明:在适宜浓度梯度范围内,随着植物生长延缓剂施用量的加大,矮化效果增强,新枝高生长量明显受到抑制,节间长度缩减,地径增粗,根冠比增加,根系活力提高,叶色变浓绿,叶绿素含量提高。但过高浓度矮化剂处理会产生药害现象。经药效综合评分法确定,500 mg/L 多效唑和3 000 mg/L 比久对'日香桂'的矮化效果最好。

关键词:'日香桂';矮化效应;植物生长延缓剂;多效唑;比久

'日香桂'(*Osmanthus fragrans* 'Rixiang Gui')是经人工选育出的桂花新品种,因不断开花且香气浓郁而得名[1]。属常绿灌木,适应性强,适生区广,矮化后的桂花可做盆栽欣赏。目前生产上常用修剪技术对其进行矮化,但单纯使用修剪技术,在生长旺盛季节'日香桂'还是会出现徒长现象,且规模化生产时,需要耗费大量人力物力,频繁修剪还会造成大量养分流失。为探讨'日香桂'盆栽矮化栽培技术,采用修剪结合施用生长延缓剂的方法,观测多效唑(PP_{333})和比久(B_9)对'日香桂'的矮化效应。

1 材料与方法

1.1 试验材料及处理

试验材料为2年生'日香桂',选择田间生长健壮、长势一致的植株移栽到塑料盆中,选用1:1的园土和泥炭土混合物作为栽培基质。于2012年5月对'日香桂'进行重短截,去除侧枝,留25 cm左右主干,为生长延缓剂处理做准备。

植物生长延缓剂为多效唑(PP_{333})50%可湿性粉剂和比久(B_9)50%可湿性粉剂。将两种植物生长延缓剂分别设置4个质量浓度梯度:PP_{333}为200、500、700和1 500 mg/L;B_9为1 000、2 000、3 000、6 000 mg/L;以清水喷施处理作为对照(CK)。

采用叶面喷施的方法,喷药量以全部叶片表面湿润为原则。在新枝长出2片新叶后进行第1次药剂处理,之后每15 d用药剂处理1次,连续处理3次。药剂处理于9月下旬加施2次。每个处理5盆,重复3次。停止施药1个月后,测定各形态指标和生理生化指标。

---

* 原文发表于《南京林业大学学报(自然科学版)》,2014,38(增刊):30-34。
基金项目:江苏省林业"三新"工程项目(LYSX[2012]23);江苏省"六大人才高峰"项目。

1.2 形态指标和生理生化指标的测定

新枝高生长量:用卷尺测定新抽生枝条的高度。

地径增长量:施药前用游标卡尺测定距地面 2 cm 处的粗度,记为 M;停止施药 1 个月后,再次测定距地面 2 cm 处的粗度,记为 N,则地径增长量= $N-M$。

节间长度:测量枝条长度,记录叶片对数,即节数,节间长度=枝条长度/节数。

根冠比:挖取整株植株,冲洗干净,在根部剪断,将植株分为根、茎两部分,再将植株放烘箱内 105 ℃ 杀青 15 min,在 90 ℃ 恒温烘干至恒量,最终的质量即为干质量。地下部分与地上部分干质量比即为根冠比。

畸形叶数:记录每个新生枝条上表现为扭曲、发黄、变干的畸形叶片数量。

药害程度:延缓剂药害反映'日香桂'叶的异常变化(以 CK 为参考标准)。将药害程度分为 5 级:1 级,畸形叶片数在 0.5 片及以内;2 级,畸形叶片数 0.5~1.0 片;3 级,畸形叶片数 1.0~2.0 片;4 级,畸形叶片数 2.0~3.0 片;5 级,畸形叶片数 3.0 片以上。分别给予权重分 0.1、0.2、0.3、0.4、0.5。

药效综合评分(Y)方法:参照姜英等[2]的方法适当调整,将各处理的新枝高生长量、地径增长量、节间长度、根冠比分别与 CK 相比,换得两者的相对减少或增长量。分别赋予新枝高相对减少量(X_1)、地径相对增加量(X_2)、节间长度相对减少量(X_3)、根冠比相对增长量(X_4)、药害等级(X_5)适合的权重(依次为 0.3、0.2、0.3、0.2、0.3),计算抑制效果分 $Q = (0.3X_1 + 0.3X_2 + 0.3X_3 + 0.1X_4)$;计算药效综合评分 $Y = Q - 0.3X_5$。

根系活力:采用氯化三甲苯基四氮唑(TTC)法[3]测定。

叶绿素含量:采用李合生(丙酮乙醇混合液)的方法[4]测定。

1.3 数据处理

采用 Excel 2003 软件处理数据,采用 SPSS 19.0 软件进行方差分析及 Duncan 新复极差法进行显著性差异检验。

2 结果与分析

2.1 植物生长延缓剂对'日香桂'形态指标的影响

2.1.1 新枝高生长量

随着多效唑和比久浓度的增加,'日香桂'新枝高生长量均呈现递减趋势(表1)。由表1可知,用这两种植物生长延缓剂处理盆栽'日香桂',在减少新枝高生长量方面,均与 CK 达到差异显著水平($P<0.05$)。用 700 mg/L 多效唑和 3 000 mg/L 比久进行叶喷处理,新枝高生长量分别比 CK 缩短 47.05% 和 43.72%,与处理 1、2 之间均达到差异显著水平,但与处理 4 之间差异不显著,说明用 700 mg/L 多效唑和 3 000 mg/L 比久叶喷处理,矮化效果最好。但是浓度过高会发生药害现象,表现为叶片先端变黄、变干等症状。

两种生长延缓剂均能抑制'日香桂'枝条的生长,但两者达到最佳效果的浓度不同,'日香桂'对多效唑较比久敏感。

表1 植物生长延缓剂处理对盆栽'日香桂'植株形态变化的影响
Tab. 1 Effects of plant growth retardants treatments on morphologic changes of Osmanthus fragrans 'Rixiang Gui'

药剂 Agentia	质量浓度/ (mg·L^{-1}) Mass concentration	新枝高生长量/cm Increment of new branches height	地径增长量/cm Ground diameter increment	节间长度/cm Internodes length	根冠比 Root shoot ratio	畸形叶片数/片 Deformed leaf number	综合评分 Synthetical mark
多效唑 PP$_{333}$	0	28.46 a	0.341 bc	3.71 a	1.012 c	0 a	0
	200	23.53 b	0.337 c	3.53 a	1.080 b	0 a	0.069 7 C
	500	18.87 c	0.362 ab	3.33 bc	1.247 a	0.5 b	0.143 5 A
	700	15.07 d	0.373 a	3.35 bc	1.151 b	1.3 c	0.122 1 B
	1 500	14.26 d	0.323 c	3.29 c	1.078 b	4.2 d	0.024 3 D
比久 B$_9$	0	28.57 a	0.340 ab	3.72 a	1.012 e	0 a	0
	1 000	24.17 b	0.321 b	3.62 a	1.124 d	0 a	0.048 6 D
	2 000	19.57 c	0.342 ab	3.45 b	1.217 b	0.2 b	0.108 3 B
	3 000	16.08 d	0.358 a	3.44 b	1.263 a	0.5 b	0.164 4 A
	6 000	14.75 d	0.338 ab	3.40 b	1.206 c	1.2 c	0.098 3 C

注：同列不同小写字母代表用 Duncan 新复极差法测验的处理间在 0.05 水平上存在显著性差异。下同。综合评分列数字后 A、B、C、D 分别表示获得等级。

2.1.2 地径增长量

'日香桂'地径随生长延缓剂浓度的增加处理间呈现先增加后减少的趋势(表1)。由表1看出，用 700 mg/L 多效唑处理的地径增长量和 CK 差异显著，各比久处理结果与 CK 差异均不显著，说明适宜浓度的多效唑和比久促进了'日香桂'地径的增长，但幅度不大。就促进地径增长来说，500~700 mg/L 多效唑和 3 000 mg/L 比久效果最好，过量浓度则会抑制地径增粗。

2.1.3 节间长度

施用生长延缓剂后，'日香桂'的节间长度与 C 相比均不同程度地减小(表1)。由表1可以看出，在多效唑和比久不同浓度处理下，除处理1以外，'日香桂'节间长度均显著低于CK，处理1与处理2、3、4差异显著，而处理2、3、4之间差异不显著。说明处理2(500 mg/L 的多效唑和 2 000 mg/L 比久)对抑制'日香桂'节间生长的效果最好。

2.1.4 根冠比

随着生长延缓剂浓度的增大，'日香桂'根冠比呈现先增大后减小的趋势(表1)。多效唑在 500 mg/L 时根冠比达到最大值，且均与其他处理差异显著，比 CK 增加 23.22%；比久在 3 000 mg/L 时根冠比达到最大值，且均与其他处理差异显著，比 CK 增加 24.80%。

2.1.5 畸形叶数

低浓度生长延缓剂处理(200 mg/L 多效唑和 1 000 mg/L 比久)没有发现畸形叶片，但随着植物生长延缓剂浓度的增加，畸形叶数显著增加，处理2、3、4均与 CK 和处理1达到显著差异(表1)。

2.1.6 药效综合评分

在'日香桂'的矮化过程中，评价生长延缓剂的矮化效果不能单纯以对'日香桂'的抑制效果为标准,抑制过轻达不到植株矮化、株型紧凑的目的,抑制过重则叶片会产生异常变化,达不到美观的要求,另外还要考虑药害问题,因此,需要多指标综合评价。该试验选择新枝高相对减少量、地径相对增加量、节间长度相对减少量、根冠比相对增加量和药害等级5个指标对生长延缓剂的效果做评价,结果如表1所示。评价结果表明:获得A级的有500 mg/L多效唑、3 000 mg/L比久,这两组处理抑制效果明显,且无明显药害;700 mg/L多效唑的抑制作用稍显过度,2 000 mg/L比久抑制作用稍弱,均获得B级;200 mg/L多效唑处理抑制作用过轻获得C级,6 000 mg/L比久对'日香桂'生长抑制过度,产生比较多的畸形叶,影响美观,也获得C级;1 500 mg/L多效唑对'日香桂'生长的抑制作用最强烈,产生最多的畸形叶片,影响盆栽观赏性,获得D级,1 000 mg/L比久处理,对'日香桂'生长的抑制作用最弱,获得D级。

2.2 植物生长延缓剂对'日香桂'生理生化指标的影响

2.2.1 根系活力

不同浓度多效唑和比久处理下,'日香桂'根系活力均呈先增后减的趋势,多效唑在700 mg/L处理下达到最大值,比久在2 000 mg/L处理下达到最大值(表2)。多效唑处理浓度由低到高,'日香桂'根系活力分别比CK增加了11.25%、13.44%、47.19%和-8.75%;比久处理浓度由低到高时,'日香桂'根系活力分别比CK增加了22.88%、59.38%、55.63%和0.31%。多重比较可知:多效唑在500~700 mg/L质量浓度处理下,'日香桂'根系活力与CK呈显著差异,700 mg/L处理时,其效果最好;比久在2 000~3 000 mg/L处理下,'日香桂'根系活力与其他处理呈显著差异,说明2 000~3 000 mg/L比久能显著提高'日香桂'的根系活力。

表2 植物生长延缓剂处理对盆栽'日香桂'植株生理生化变化的影响
Tab. 2 Effects of plant growth retardants treatments on physiology and biochemistry changes of *O. fragrans* 'Rixiang Gui'

药剂 Agentia	质量浓度/(mg·L^{-1}) Mass concentration	根系活力/ (μg·g^{-1}·h^{-1}) Root vigor	叶绿素含量(FW)/ (mg·g^{-1}) Chlorophyll content
多效唑 PP$_{333}$	0	3.20 cd	1.211 d
	200	3.56 bc	1.330 c
	500	3.63 b	1.413 b
	700	4.71 a	1.460 a
	1 500	2.92 d	1.079 e
比久 B$_9$	0	3.20 c	1.208 c
	1 000	3.90 b	1.291 b
	2 000	5.10 a	1.484 a
	3 000	4.98 a	1.201 c
	6 000	3.21 c	0.940 d

2.2.2 叶绿素含量

在多效唑处理下,'日香桂'叶片叶绿素含量随着多效唑和比久浓度的增加,均呈现先增大后减小的变化趋势,多效唑在700 mg/L处理时达到最大值,比久在2 000 mg/L处理时达到最大值。多重比较分析发现,多效唑各处理均与 CK 差异显著,处理 1、2、3 光合色素含量均高于 CK,处理 3 均高于其他处理,比 CK 增加了 20.56%,且与其他处理达到显著差异,而处理 4 光合色素含量低于 CK,比 CK 减少了 10.90%。比久除处理 3 外,其他处理均与 CK 差异显著,处理 2 光合色素含量均高于其他处理,比 CK 增加 22.85%,且与其他处理达到显著差异,而处理 4 光合色素含量低于 CK,比 CK 减少了 22.19%(表 2)。说明 700 mg/L 多效唑和 2 000 mg/L 比久处理能显著提高'日香桂'光合色素的含量,有助于提高'日香桂'的光合作用,促进有机物的生产,进而促进'日香桂'的生长;而过高浓度延缓剂处理会抑制'日香桂'光合色素的产生,不利于其光合作用,进而抑制'日香桂'的生长。

3. 讨论

多效唑和比久是两种常用的生长延缓剂,研究表明,使用多效唑和比久能抑制植物茎的伸长,缩短节间长度,促进地径增粗、根冠比增大,这与前人的研究[5-11]一致。对'日香桂'施用多效唑和比久后,其植株新枝高度随处理浓度的增大而减小,达到了矮化的目的,但施用过高浓度的多效唑会造成'日香桂'叶片畸形,施用过高浓度的比久会造成'日香桂'叶片边缘枯黄变干。叶片是植物重要的组织器官,直接影响植物的光合作用,进而影响植株正常的生长发育。因此,笔者建议施用生长延缓剂处理'日香桂'时,多效唑浓度要低于 1 500 mg/L,比久低于 6 000 mg/L。在本试验中,700 mg/L 多效唑及 3 000 mg/L 比久对'日香桂'的矮化效果与 CK、处理 1、2 呈显著差异($P<0.05$),矮化效果最好,且能使植株紧凑美观。通过对'日香桂'节间长度的变化分析可以发现低浓度的多效唑(200~500 mg/L)和比久(1 000~2 000 mg/L)能显著控制'日香桂'节间长度的增长,而随着浓度的提高,多效唑和比久对'日香桂'节间长度的抑制作用并不显著。有研究认为多效唑与比久可降低植物内源吲哚乙酸的含量,抑制顶端生长,促使侧芽生长,使株型矮壮[12]。该研究认为低浓度的多效唑和比久对'日香桂'生长的抑制作用主要体现在抑制节间长度的生长,而随着浓度的升高,枝高还在减少,节间长度变化不大,说明节间减少了,叶片也相应减少,所以高浓度矮化剂是通过抑制新叶的萌发抑制了其生长。

地径随两种植物生长延缓剂处理浓度的增大呈先增大后减小的趋势,地径在 700 mg/L 多效唑处理与 3 000 mg/L 比久处理时效果较好,过高或过低浓度处理差异不显著,1 500 mg/L 多效唑处理'日香桂'地径比 CK 降低,这与前人关于生长延缓剂能使一串红[13]、菊花[14]等地径显著增加以及使大白杜鹃[15]的地径极显著减少的效果不完全一致。虽然使用多效唑和比久后,能提高吲哚乙酸氧化酶的活性,从而降低植物内源吲哚乙酸的含量,延缓作物的纵向伸长,增强横向生长[16],但地径的增长受多方面因素的影响,随着生长延缓剂浓度的增大,也会给植物带来其他方面的影响,如叶片变异、生理指标的变化。

施用生长延缓剂,对'日香桂'生长的影响是多方面的。评价生长延缓剂的效果也应从多个方面综合考虑,而不能单纯以对'日香桂'高度的抑制效果为标准,抑制过轻达不到矮化植株的目的,抑制过重则叶片会产生药害,影响美观,反而降低其观赏价值。所以还要考虑药害问题,因此需要对多效唑和比久所引起的外部形态的变化进行综合评估,从植株的新枝高生长量、地径增长量、节间长度、根冠比及药害程度 5 个指标的影响综合考虑,结果显示 500 mg/L 多效唑和 3 000 mg/L 比久处理的综合评分最高,说明其对'日香桂'的综合效果最好;浓度过低,效果不明显,浓度过高抑制过度,出现药害现象。因此,从多效唑和比久对'日香桂'外部形态指标的影响上来看,500 mg/L 多效唑和 3 000 mg/L 比久是合适的处理浓度。

生长延缓剂对'日香桂'根部产生了影响,有研究表明,多效唑和比久能促进植物根系活力[17-18],促进地上部分同化物向地下部分运输。该试验结果显示,700 mg/L 多效唑和 2 000~3 000 mg/L 比久能显著提高'日香桂'根系活力,根系活力提高,根系发达,则能促进地上部分的生长。根冠比随多效唑和比久浓度的增加呈先增大后减小(仍高于对照)的趋势,说明施用生长延缓剂能调整'日香桂'有机物的分配,促进有机物向根部的转移。'日香桂'根系活力提高,根系发达,有利于其抗旱、抗倒伏能力的提高。

叶片光合色素含量是反映植物光合能力的一个重要指标。色素含量的增加不仅有利于叶片净光合速率的提高,而且同时还能有效防止膜脂过氧化,保障光合作用的正常进行。大量研究表明,施用生长延缓剂能够促进植物光合作用的变化[19-20],叶片光合色素的含量与植物光合作用密切相关,叶绿素是植物进行光合作用的主要色素,该研究中,700 mg/L 多效唑和 2 000 mg/L 比久处理能显著提高'日香桂'光合色素的含量,有助于提高'日香桂'的光合作用,促进有机物的生产,进而促进'日香桂'的生长。

参考文献

[1] Xiao L. Rare flowers and trees—*Osmanthus fragrans* 'Rixiang Gui'[J]. Chinese Native Produce, 1995(6):35.

[2] 姜英,彭彦,李志辉,等.多效唑、烯效唑和矮壮素对金钱树的矮化效应[J].园艺学报,2010,37(5):823-828.
Jiang Y, Peng Y, Li Z H, *et al*. Effects of paclobutrazol, uniconazole and chlorcholinchlorid on dwarfing of *Zamioculcasz amiifolia*[J]. Acta Horticulturae Sinica, 2010,37(5):823-828.

[3] 张志良,瞿伟菁,李小芳.植物生理学实验指导[M].4 版.北京:高等教育出版社,2009.

[4] 李合生.植物生理生化实验原理和技术[M].北京:高等教育出版社,2000.

[5] 汪良驹,孙文全,李友生.PP_{333}对水仙花的矮化效应及其生理机制初探[J].园艺学报,1990,17(4):313-319.
Wang L J, Sun W Q, Li Y S. The research of dwarfing effect and physiological mechanism of PP_{333} on *Nacissus tazetta* var. *chinensis*[J]. Acta Horticulturae Sinica, 1990, 17(4):313-319.

[6] 裘文达,刘克斌.PP_{333}和 B_9 对菊花茎伸长和开花期的影响(简报)[J].植物生理学通讯,1989(6):31-33.
Qiu W D, Liu K W. Effect of PP_{333} and B_9 on stem elongation and florescence of *Dendranthema grandiflora*[J]. Plant Physiology Communications, 1989(6):31-33.

[7] 刘碧桃,官凤英,范少辉,等.两种生长延缓剂对绿竹矮化作用的研究[J].西北林学院学报,2011,26(4):142-147.
Liu B T, Guan F Y, Fan S H, *et al*. Dwarfing effect of two rdtardants on *Dendrocalamopsis oldhami*[J]. Journal of Northwest Forestry University, 2011, 26(4):142-147.

[8] 梁根桃,沈锡康,方星.多效唑对菊花株型和开花的影响[J].浙江林学院学报,1993,10(1):97-108.
Liang G T, Shen L K, Fang X. Effect of paclobutrazol on morphology and florescence of *Dendranthema grandiflora*[J]. Journal of Zhejiang Forestry College,1993,10(1):97-108.

[9] 武荣花,李东东,张晶,等.植物生长延缓剂对盆栽月季矮化效果的研究[J].河南农业科学,2013,42(5):141-145.
Wu R H, Li D D, Zhang J, *et al*. The dwarfing effect of plant growth retardants on potted Chinese rose[J]. Journal of Henan Agricultural Sciences,2013,42(5):141-145.

[10] 汤勇华,张栋梁,顾俊杰.不同生长延缓剂对盆栽玫瑰的矮化效果[J].江苏农业科学,2013,41(3):138-140.
Tang Y H, Zhang D L, Gu J J. The dwarfing effect of different plant growth retardants on potted rose[J]. Jiangsu Agricultural Sciences, 2013,41(3):138-140.

[11] 黄艳花,梁萍,石爱莲.植物生长延缓剂对两种绿篱植物矮化效应研究[J].南方农业学报,2011,42(3):284-287.
Huang Y H, Liang P, Shi A L. Effects of plant growth retardants on *Hibiscus rosa-sinensis* L. and *Carmona microphylla* (Lam.) G. Don[J]. Journal of Southern Agriculture,2011,42(3): 284-287.

[12] Lv C P, Chen H X. Effect of paclobutrazol on basin-cultured Canna generalis' reduction[J]. Journal of Hunan Agricultural University(Natural Science Edition),2003,29(2): 129-130.
吕长平,陈海霞.多效唑对盆栽花叶美人蕉的矮化效果[J].湖南农业大学学报(自然科学版).2003, 29(2):129-130.

[13] 毛龙生,高勇.PP_{333},B_9,CCC对盆栽一串红矮化效应研究[J].园艺学报,1991,18(2):177-179.
Mao L S, Gao Y. The dwarfing effect of PP_{333}, B_9 and CCC on potted *Salvia splendens* [J]. Acta Horticulturae Sinica,1991,18(2): 177-179.

[14] 史益谦.氯丁唑不同施用方法对菊花生长发育的影响[J].湖南农学院学报,1991,2:151-157.
Shi Y Q. Effect of different fertilization of paclobutrazol on the growth and development of chrysanthemum [J]. Journal of Hunan Agricultural, 1991, 2:151-157.

[15] 赵云龙,李朝婵,陈训.植物生长延缓剂对大白杜鹃的矮化效果[J].贵州农业科学,2013,41(4):143-146.
Zhao Y L, Li Z C, Chen X. Effect of plant growth retardant on dwarfing of *Rhododendron decorum* [J]. Guizhou Agricultural Sciences,2013,41(4): 143-146.

[16] 朱佳文,洪亚辉,萧浪涛,等.PP_{333}和B_9对白兰花生长的影响[J].湖南农业大学学报(自然科学版),2003,29(6):482-484.
Zhu J W, Hong Y H, Xiao L T, *et al*. Effects of plant growth retardants on the growth of *Michelia alba*[J]. Journal of Hunan Agricultural University(Natural Science Edition), 2003, 29(6): 482-484.

[17] 赵秀芬,房增国,高祖明.多效唑对稻麦苗期根系活力和叶片IAA氧化酶、过氧化物酶活性的影响[J].广西农业科学,2006,37(4):379-381.
Zhao X F, Fang Z G, Gao Z M. Effects of paclobutrazol (PP333) on root vigor and IAA oxidase and peroxidase activities in leaf of rice and wheat seedlings [J]. Guangxi Agricultural Sciences, 2006, 37(4): 379-381.

[18] 张剑,张志国,隋艳晖.植物生长延缓剂对万寿菊穴盘苗生长的控制作用研究[J].中国生态农业学报,2007,15(6):101-103.
Zhang J, Zhang Z G, Sui Y H. Effects of plant growth regulators on *Tagetes erecta* L. seedling growth[J]. Chinese Journal of Eco-Agriculture, 2007,15(6): 101-103.

[19] 罗栋,王雁,刘秀贤,等.植物生长延缓剂浸种对地涌金莲的矮化效应[J].林业科学研究,2009,21(6):847-851.
Luo D, Wang Y, Liu X X, *et al*. Study on dwarfing effects of seed soaking with plant growth retardantson *Musella lasiocarpa*[J]. Forest Research, 2009, 21(6):847-851.

[20] 董运斋,王四清.生长延缓剂在观赏植物中应用的研究进展[J].北方园艺,2004(6):14-16.
Dong Y Z, Wang S Q. Advance in growth retardant applications of ornamental plants research [J]. Northern Horticulture, 2004(6): 14-16.

The Dwarfing Effects of Paclobutrazol and Daminozide on Potted *Osmanthus fragrans* 'Rixiang Gui'

WANG Ping, YANG Xiulian, WANG Chunjun, WANG Lianggui

(College of Landscape Architecture, Nanjing Forestry University, Nanjing 210037, China)

Abstract: This experiment studied the dwarfing effect of different concentration of paclobutrazol (PP_{333}) and daminozide on two-year-old potted *Osmanthus fragrans* 'Rixiang Gui'. The result showed that in a proper concentration range, the dwarfing result of these two plant growth retardants at a high concentration was better than at a low concentration. In a proper concentration range, increment of new branches height was inhibited, the internodes length was shortened, at the same time, the ground diameter, root shoot ratio and root vigor were increased, also, the leaves turn dark green, and the chlorophyll contents was added. But too high concentration leads to phytotoxicity phenomena, bad to ornamental value of 'Rixiang Gui'. By using comprehensive evaluation method, a conclusion was drawn that 500 mg/L paclobutrazol and 3 000 mg/L daminozide treatments could effectively dwarf the growth of 'Rixiang Gui'.

Key words: *Osmanthus fragrans* 'Rixiang Gui'; dwarfing effect; plant growth retardant; paclobutrazol; daminozide

'波叶金桂'扦插生根过程中营养物质和激素含量变化

杨秀莲,李春意,王良桂

(南京林业大学,江苏 南京 210037)

摘　要:研究不同浓度的生长调节剂处理下'波叶金桂'插穗的生根性状及营养物质、内源激素含量变化。结果表明:生长调节剂处理可显著提高'波叶金桂'插穗的生根率及不定根数量,以500 mg/L 脱落酸(NAA)处理最有效。在不定根发生过程中,'波叶金桂'插穗皮部可溶性糖含量呈先降后升趋势;前期淀粉含量下降较快,后期趋缓;整个生根过程中全氮含量总体呈下降趋势;500 mg/L NAA 处理下'波叶金桂'插穗插后 15 d 生长素(IAA)含量最高,随后大幅下降;ABA 的变化趋势与 IAA 正相反,500 mg/L NAA 处理下'波叶金桂'插穗插后 15 d ABA 含量最低,随后逐渐上升。

关键词:'波叶金桂';扦插;生根;营养物质;内源激素

桂花(*Osmanthus fragrans* Lour.)是我国传统的十大名花之一,具有很高的食用价值及药用价值。桂花品种繁多,根据花色可分为四季桂品种群、银桂品种群、金桂品种群、丹桂品种群,其中以金桂品种群的观赏价值最高。'波叶金桂'是金桂品种群中的优异品种。扦插是桂花的主要繁殖方式,但不同品种桂花的扦插生根能力有很大差异。自然条件下,'波叶金桂'的插穗不易生根,这也是目前该品种在园林中应用较少的主要原因。本试验采用不同浓度的生长调节剂处理'波叶金桂'的插穗,并研究插穗内部营养物质及内源激素含量变化,旨在为繁殖'波叶金桂'提供理论依据。

1 材料与方法

1.1 材料及处理

试验所用插条于 2009 年 5 月采自南京林业大学校园内 30 年生'波叶金桂'。扦插试验在南京林业大学园林示范中心进行。用塑料周转箱(58 cm×36 cm×18 cm)做插床,底部均匀打孔方便排水,采用蛭石、珍珠岩(体积比 1∶1)混合基质。扦插前 3 d 用 0.5% 高锰酸钾溶液浸泡插条 24 h,随后用清水冲洗干净,待用。选取母树中部外围生长健壮、无病虫害的当年生半木质化枝条,剪成长约 10 cm 的小段。插前用不同浓度的脱落酸(NAA)、生长素(IAA)、ABT 生根粉等 3 种生长调节剂处理插条 30 s,以清水作对照。采取随机区组设计,

---

* 原文发表于《江苏农业科学》,2014,42(6):202-205。
基金项目:国家林业公益性行业科研专项(201204607);江苏省普通高校自然科学研究项目(11KJB220002);江苏省高校优势学科建设工程项目(PAPD)。

每处理 90 根插条,每小区 30 根插条,每处理重复 3 次,随机排列,插条间距 10 cm。扦插后保持适宜湿度。每隔 7 d 用 0.5% 多菌灵消毒 1 次。扦插后 71 d 统计生根率、根数、根长等指标。选择 500 mg/L NAA 溶液处理的插穗按相同方法进行扦插,清水处理作对照。分别在插后 15、24、44、57、71 d 取样。每处理随机抽取 4 根插穗,每处理 3 次重复。将插穗立即放入冰盒中带回实验室,用清水冲洗干净并擦干,剥取基部 2 cm 处的皮部测定内含物、激素含量。

1.2 测定方法

取置于超低温冰箱(−70 ℃)中保存备用的皮部,加 2 mL 内含 1 mmol/L 2,6-二叔丁基对甲基苯酚的 80% 甲醇溶液,冰浴下研磨成匀浆,转入 10 mL 试管中,再用 2 mL 提取液分次将研钵冲洗干净,一并转入试管中,摇匀后置于 4 ℃ 冰箱中提取 4 h,1 000 r/min 离心 15 min,取上清液。用酶联免疫吸附分析法测定 IAA、ABA、赤霉素(GA)、细胞分裂素(ZR) 4 种内源激素的含量,重复 3 次[1]。采用蒽酮比色法[2]测定可溶性糖及淀粉含量,采用考马斯亮蓝 G-250 法[2]测定可溶性蛋白质含量,参照高俊凤的方法[3]测定全氮含量,皮部 C/N 值计算公式如下:

$$C/N = 可溶性糖质量分数/植物全氮质量分数$$

1.3 数据处理

用 SPSS 13.10 软件进行方差分析,用 Duncan's 新复极差法检验差异显著性。

2 结果与分析

2.1 不同生长调节剂对'波叶金桂'扦插生根的影响

由表 1 可知,3 种生长调节剂处理下'波叶金桂'插穗生根率都在 60% 以上,500 mg/L NAA 处理下插穗生根率最高,是对照的 3.6 倍。500 mg/L NAA 处理下插穗不定根数量

表 1 不同生长调节剂处理下'波叶金桂'插穗的生根率、根长、根数

生根调节剂	浓度(mg·L^{-1})	生根率/%	总根数/条	总根长/cm	平均根长/cm
NAA	100	60 Bb	40 Cc	108.40 Cc	2.71 BCc
	300	70 BCbc	66 Ff	114.84 Dd	1.74 Aa
	500	90 Dd	84 Gg	136.08 Ff	1.62 Aa
IAA	100	60 Bb	23 Bb	75.00 Bb	3.26 Ee
	300	70 BCbc	42 Cc	121.38 Ee	2.89 CDd
	500	85 Dd	41 Cc	121.00 Ee	2.95 Dd
ABT	100	85 Dd	57 Ee	121.35 Ee	2.13 Bb
	300	80 CDcd	54 Ee	146.88 Gg	2.72 BCc
	500	85 Dd	49 Dd	136.71 Ff	2.79 BCc
清水	CK	25 Aa	3 Aa	6.63 Aa	2.21 Bb

注:同列数据后不同大写字母表示差异极显著,不同小写字母表示差异显著。

也最多,达84条,对照仅有3条。由此可知,3种生长调节剂均可有效提高'波叶金桂'插穗的生根率,500 mg/L NAA可显著提高'波叶金桂'插穗生根率及不定根数量,IAA对根长的促进作用较好。

2.2 扦插生根过程中皮部营养物质含量

2.2.1 可溶性糖、可溶性淀粉含量 由图1可知,'波叶金桂'插穗皮部可溶性糖含量呈先降后升趋势。NAA处理下插穗皮部可溶性糖含量在扦插后第24 d最低。扦插后第15 d至第24 d,NAA处理下插穗已经形成较多的愈伤组织,对照此时变化不明显。扦插后期,随着不定根的出现及生长,光合产物积累量增加,'波叶金桂'插穗根部可溶性糖含量不断上升。淀粉水解转化为糖类,供给插穗生根所需的能量。插穗内淀粉含量越高,说明可溶性糖含量越低[3]。由图2可知,NAA处理及对照处理的淀粉含量变化趋势均是前期下降较快,后期趋缓。但NAA处理下'波叶金桂'插穗皮部淀粉含量比对照低很多,表明用NAA处理'波叶金桂'插穗能显著提高淀粉的降解速度,前期'波叶金桂'插穗愈伤组织及不定根生长需要消耗大量可溶性糖,淀粉含量下降表明有更多的淀粉转化为糖类。

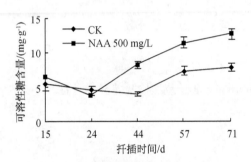 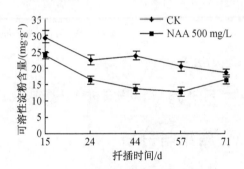

图1 不同扦插时间'波叶金桂'插穗　　图2 不同扦插时间'波叶金桂'插穗
　　　皮部可溶性糖含量　　　　　　　　　　皮部可溶性淀粉含量

2.2.2 全氮含量、C/N值 从图3可知,NAA处理、对照处理下'波叶金桂'插穗在整个生根过程中全氮含量总体呈下降趋势,且NAA处理下'波叶金桂'插穗皮部总氮含量低于对照。这也证实了高氮条件不利于插穗生根的结论[4]。从图4可知,不同扦插时间'波叶金桂'插穗皮部C/N值变化趋势与可溶性糖含量变化趋势相似,C/N值的变化主要是由可溶性糖含量变化造成的。NAA处理下'波叶金桂'插穗皮部C/N值升高,从而提高了生根率,证明了C/N值高有利于生根的结论[4]。

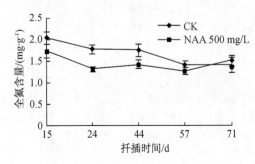 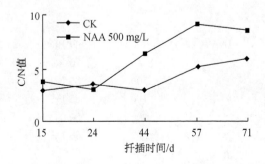

图3 不同扦插时间'波叶金桂'插穗皮部全氮含量　　图4 不同扦插时间'波叶金桂'插穗皮部C/N值

2.2.3 可溶性蛋白质含量 可溶性蛋白质与植物的形态发生有很大关系,为细胞的生长提供物质基础。从图 5 可知,前 24 d,NAA 处理下'波叶金桂'插穗皮部可溶性蛋白质含量比对照低,44 d 后高于对照。结合外部形态观察可知,前 24 d 是'波叶金桂'插穗皮部愈伤组织及根原基形成期,要消耗较多的蛋白质,不定根形成后,可溶性蛋白含量又逐渐升高。对照处理下'波叶金桂'插穗皮部可溶性蛋白质含量前 24 d 也呈下降趋势,44 d 到达低谷。

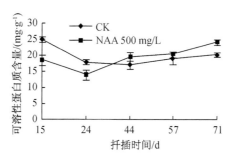

图 5 不同扦插时间'波叶金桂'插穗皮部可溶性蛋白质含量

2.3 扦插生根过程中插穗皮部内源激素含量

激素在不定根形成过程中起着核心作用,它可以诱导根原基产生,影响新根形成。由图 6 可知,NAA 处理下'波叶金桂'插穗皮部 IAA 含量呈现先降低后升高再降低的趋势。扦插后 15 d IAA 含量最高,此时正是愈伤组织及根原基大量形成期,较高的 IAA 含量有利于根原基形成,44 d 达到低谷,随后缓慢上升。对照 IAA 含量变化比较平缓。由此可知,NAA 处理显著提高了插穗皮部的 IAA 含量。愈伤组织生长要消耗部分内源 IAA,新根生出后'波叶金桂'体内合成内源 IAA,从而使内源 IAA 含量上升。本试验结果与前人研究结论一致[4-6]。ABA 是插穗生根的抑制剂。由图 7 可知,NAA 处理下'波叶金桂'插穗皮部 ABA 含量在前 44 d 升高,后逐渐降低,整个生根过程中 NAA 处理下 ABA 含量始终低于对照。低浓度 ABA 有利于 IAA 及其他有利生根的物质向插穗基部运输[6]。ZR 是细胞分裂素的一种,由图 8 可知,NAA 处理下'波叶金桂'插穗皮部 ZR 含量呈先升后降再升的趋势,对照处理下'波叶金桂'插穗皮部 ZR 含量总体呈缓慢下降趋势,前期下降较快,后期比较缓慢。不定根长出后,ZR 含量逐渐上升,这可能是新生根重新合成 ZR 所致[6-9]。由图 9 可知,NAA 处理下'波叶金桂'插穗皮部 GA 含量呈现先上升后下降趋势,44 d 达到峰值,此时正是不定根大量形成时期。NAA 处理下 GA 值始终低于对照,说明较低水平的 GA 对生根有促进作用,这与长蕊杜鹃的变化相反,与珍珠黄杨变化相似[7-9]。

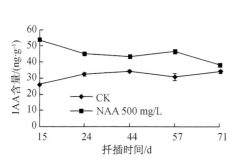 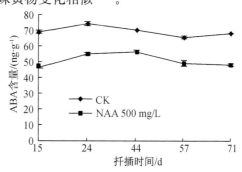

图 6 不同扦插时间'波叶金桂'插穗皮部 IAA 含量 **图 7 不同扦插时间'波叶金桂'插穗皮部 ABA 含量**

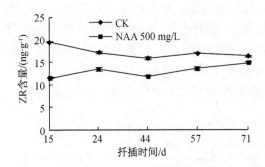
图8 不同扦插时间'波叶金桂'插穗皮部 ZR 含量

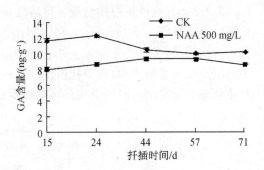
图9 不同扦插时间'波叶金桂'插穗皮部 GA 含量

3 结论与讨论

3.1 不同生长调节剂与生根的关系

应用植物生长调节剂可促进难生根树种生根[9-11]。本试验结果表明,500 mg/L NAA 处理下'波叶金桂'插穗生根率可达 90%,且不定根数量较多。IAA、ABT 处理下'波叶金桂'插穗生根率也极显著高于对照。由此可知,外源生长调节剂可以有效促进'波叶金桂'的插穗生根,这与前人的研究结果相同[12]。

3.2 营养物质与生根的关系

营养物质是插穗生根的基本条件之一,碳水化合物与氮素化合物不仅是插穗生根生长不可或缺的物质,也是插穗在生根前维持生命的重要能源。可溶性糖类是生根必需的营养物质,为大量不定根的形成提供养分。插穗内的可溶性糖积累于基部,用于诱导根原基及不定根的形成。本研究结果表明,'波叶金桂'插穗可溶性糖及蛋白质含量均呈先降后升的趋势,并且都高于对照,而淀粉含量则持续降低,说明根原基的形成及不定根的出现需要消耗大量碳水化合物及蛋白质。500 mg/L NAA 处理下'波叶金桂'插穗总氮含量呈现出先降低后升高的趋势,证实了高氮条件不利于插穗生根的结论[4]。500 mg/L NAA 处理下'波叶金桂'插穗 C/N 值除 24 d 时略低于对照外,其余时间均高于对照且持续上升,这与张玉臣等的研究结果一致[13]。一定的碳水化合物及氮素化合物是'波叶金桂'不定根形成的基础,插穗能否大量生根与插穗体内的营养物质含量高低存在密切关系[14]。

3.3 内源激素与生根的关系

植物不定根的发生虽然受多种因素的影响,但内源激素及外源激素在根原基的形成中起关键作用。高浓度的内源 GA 会抑制不定根的形成[15-16],本研究表明,500 mg/L NAA 处理下'波叶金桂'插穗插后 15 d IAA 含量最高,随后大幅下降,结合形态观察,此期为根原基发育高峰期,细胞在生长素的作用下大量分裂并分化不定根原基,44 d 降到谷值,此时大量不定根长出。ABA 的变化趋势与 IAA 相反,500 mg/L NAA 处理下'波叶金桂'插穗插后 15 d ABA 最低,随后逐渐上升。IAA 是促进插穗生根的主要内源激素,ABA 是抑制生根的主要内源激素。GA含量变化趋势与ABA相似,ZR含量则呈先上升后下降再上升的趋势,它们对生根影响小。ABA、GA、ZR含量与生根率呈负相关,IAA含量与生根率呈正相关,说明IAA与生根率关系最密切。外源NAA对内源IAA的合成有促进作用,高浓度的IAA有利于根原基的分化及形成。因此,外源NAA对'波叶金桂'不定根的形成有促进作用[9]。外源激素对插穗生根效果的影响可能不是通过直接改变某种内源激素含量而影响生根,而是使内源激素之间达到了一定的平衡,从而促进生根。另外,施用外源激素在愈伤形成期降低了生根部位的GA含量,低浓度的GA很可能有利于愈伤组织形成。本研究中由于第一次取样时间是在插后15 d,结合形态结构观察,此期已是愈伤组织根原基大量形成期,因此前期的很多变化均未能体现出来。

参考文献

[1] 俞良亮,乔瑞芳,季孔庶.不同外源激素对杂交鹅掌楸扦插生根过程中内源激素变化的影响[J].东北林业大学学报,2007,35(9):24-26.

[2] 李合生.植物生理生化实验原理与技术[M].北京:高等教育出版社,2006.

[3] 高俊风.植物生理学实验指导[M].2版.北京:高等教育出版社,2006.

[4] 吕明霞.梅花硬枝扦插繁殖与贮藏养分的关系[J].浙江农业科学,2000,1(4):200-202.

[5] 张忠微,石素霞,张彦广,等.金露梅嫩枝插穗生根过程中营养物质含量变化研究[J].河北农业大学学报,2008,31(4):56-59.

[6] 齐永顺,张志华,王同坤,等.同源四倍体玫瑰香葡萄嫩枝扦插不定根发生过程中内源激素的变化[J].园艺学报,2009,36(4):565-570.

[7] 李朝婵,赵云龙,张冬林,等.长蕊杜鹃扦插内源激素变化及解剖结构观察[J].林业科学研究,2012,25(3):360-365.

[8] 高洁,曾笑菲,刘晓辉,等.滇杠柳扦插繁殖及生根相关理化特性的动态分析[J].中药材,2011,34(6):841-845.

[9] 黄焱,季孔庶,方彦,等.珍珠黄杨春季扦插生根性状差异及内源激素变化[J].浙江林学院学报,2007,24(3):284-289.

[10] 袁利利,张林,王厚新,等.华北五角枫'京2'插穗生根过程中内源激素变化[J].中国农学通报,2012,28(13):61-64.

[11] 安三平,王丽芳,石红,等.蓝云杉不同品种扦插生根能力和生根特性研究[J].林业科学研究,2011,24(4):512-516.

[12] 朴楚炳,张有富,苗锡臣,等.促进红松插穗生根能力的研究[J].林业科技,1996,21(6):5-8.

[13] 张玉臣,周再知,梁坤南,等.不同植物生长调节剂对白木香扦插生根的影响[J].林业科学研究,2010,23(2):278-282.

[14] 詹亚光,杨传平,金贞福,等.白桦插穗生根的内源激素和营养物质[J].东北林业大学学报,2001,29(4):1-4.

[15] 王新建,何威,张秋娟,等.豫楸1号扦插生根过程中营养物质含量及氧化酶类活性的变化[J].林业科学,2009,45(4):156-161.

[16] 郭英超,杜克久,贾哲.兴安圆柏扦插生根过程中相关内源激素特征分析[J].中国农学通报,2012,28(1):44-48.

Changes of nutrient and hormone contents during Cutting Rooting of *Osmanthus fragrans* 'Boye Jingui'

YANG Xiulian, LI Chunyi, WANG Lianggui

(Nanjing Forestry University, Nanjing 210037, China)

Abstract: The effects of different concentrations of growth regulators on the rooting traits, nutrients and endogenous hormone contents of *O. fragrans* 'Boye Jingui' cuttings were studied. The results showed that the rooting rate and adventitious root number of cuttings were significantly increased by growth regulator treatment, and 500 mg/L NAA treatment was the most effective. During adventitious rooting, the content of soluble sugar in the cuttings' bark decreased first and then increased; the content of starch decreased rapidly in the early stage and slowed down in the late stage; the total nitrogen content decreased in the whole rooting process; the content of IAA was the highest on the 15th day after the cuttings were treated with 500 mg/L NAA, and then decreased sharply; On the contrary to IAA, the content of ABA was the lowest on the 15th day and then gradually increased under 500 mg/L NAA treatment.

Key words: *O. fragrans* 'Boye Jingui'; cuttage; rooting; nutrient; endogenous hormones

25个桂花品种花瓣营养成分分析*

杨秀莲,赵 飞,王良桂

(南京林业大学 风景园林学院,江苏 南京 210037)

摘 要:为选择营养品质优良的食用桂花品种,对25个桂花品种盛花期花瓣中的可溶性糖、可溶性蛋白质、水分、维生素C、有机酸、游离氨基酸、黄酮以及6种矿质元素含量进行了测定,并用主成分分析法,对其营养成分进行比较,筛选出营养品质优良的品种。结果表明:不同桂花品种可溶性糖含量为10.85~91.28 mg·g^{-1},可溶性蛋白质含量为14.70~59.28 mg·g^{-1},水分含量为79.09%~87.45%,维生素C含量为3.99~46.30 mg·g^{-1},有机酸含量为1.56~3.78 mg·kg^{-1},黄酮含量为61.94~288.98 mg·g^{-1},总游离氨基酸含量为53.73~151.05 mg·g^{-1};各矿质元素含量分别为锌19.78~72.67 mg·kg^{-1},铁52.47~207.98 mg·kg^{-1},镁705.19~1 541.31 mg·kg^{-1},钙631.71~3 679.83 mg·kg^{-1},钠54.85~220.70 mg·kg^{-1},钾9 002.17~16 979.13 mg·kg^{-1}。桂花品种间营养成分存在很大差别,'堰虹桂'、'雨城丹'桂、'青山银桂'、'醉肌红'4个品种的综合营养价值较高,'波叶金桂''镰叶金桂''天女散花'3个品种的综合营养价值较低。

关键词:桂花;花瓣;营养品质;主成分分析

近年来将花卉作为一种新型的美食原料逐渐成为时尚,对花卉营养品质的研究也在逐渐开展,已有的研究发现,郁金香(*Tulipa gesneriana*)、杜鹃(*Rhododendron simii*)[1]、牡丹(*Paeonia suffruticosa*)[2]、玫瑰花(*Rosa rugosa*)、百合花(*Lilium* spp.)、木棉花(*Bombax malabaricum*)、山茶花(*Camellia japonica*)、唐菖蒲(*Gladiolus hybridus*)[3]以及一些野生花卉[4]等的花瓣均含有较高矿质元素、蛋白质和氨基酸等,可食用、药用和保健用等,是原生态的食用珍品。桂花(*Osmanthus fragrans*)隶属木犀科(Oleaceae)木犀属(*Osmanthus*),是著名的观赏植物,在中国长江中下游普遍栽培,目前已确定的有160多个品种[5]。桂花是传统的香花植物,花虽小,但香味持久稳定。中国拥有悠久的食用桂花的传统,早在2000年前,就有桂花制酒、桂花制茶的记载,至今桂花酒、桂花茶仍负盛名。此外,现在在市场上还陆续出现了桂花糕、桂花糖、桂花月饼、桂花露饮料等食品50多种。目前,以桂花为原料制作的食品层出不穷,但对其营养成分的研究并不多,仅徐文斌等[6]对福建省'浦城丹桂'和'小叶丹桂'的营养成分进行了初步研究。本文试图通过测定不同品种桂花花瓣的可溶性糖、可溶性蛋白质、水分、氨基酸、有机酸、黄酮以及6种矿质元素的含量,用主成分分析法,比较各品种间营养品质差异,从观赏桂花中筛选出营养品质优良的品种,为桂花花瓣的合理利用提供理论基础。

---

* 原文发表于《福建林学院学报》,2014,34(1):5-10。
基金项目:江苏省高校自然科学基础研究基金资助项目(11KJB220002);国家林业局公益性行业专项基金资助项目(201204607);江苏省高校优势学科建设工程项目(11KJB220002)。

1 材料与方法

1.1 试验材料

试验中所用桂花品种采自南京、杭州两地,共 25 个品种,分属于 4 个品种群(表 1)。

表 1 用于测试的桂花品种*

Tab. 1 Clutivars of *O. fragrans* used for analysis

品种群 Cultivars group	品种名 Cultivars name	树龄/a Tree age	来源 Source
四季桂品种群 *O. fragrans* Asiaticus Group	'四季桂'(*O. fragrans* 'Sijigui')	15	南京林业大学校园
	'天女散花'(*O. fragrans* 'Tiannü Sanhua')	15	杭州园林绿化工程有限公司基地
银桂品种群 *O. fragrans* Albus Group	'波叶银桂'(*O. fragrans* 'Boye Yin')	25	南京林业大学校园
	'速生银桂'(*O. fragrans* 'Susheng Yingui')	15	杭州园林绿化工程有限公司基地
	'紫梗籽银'桂(*O. fragrans* 'Zigeng Zi Yin')	20	南京林业大学校园
	'青山银桂'(*O. fragrans* 'Qingshan Yingui')	15	江苏淳盛农业科技有限公司基地
	'晚银桂'(*O. fragrans* 'Wan Yingui')	25	南京林业大学校园
	'卷叶黄'(*O. fragrans* 'Juanye Huang')	25	南京林业大学校园
金桂品种群 *O. fragrans* Luteus Group	'圆瓣金桂'(*O. fragrans* 'Yuanban Jingui')	20	南京情侣园
	'墨叶金'桂(*O. fragrans* 'Moye Jin')	20	南京情侣园
	'桃叶黄'(*O. fragrans* 'Taoye Huang')	20	南京林业大学校园
	'波叶金桂'(*O. fragrans* 'Boye Jingui')	15	杭州园林绿化工程有限公司基地
	'镰叶金桂'(*O. fragrans* 'Lianye Jingui')	15	杭州园林绿化工程有限公司基地
	'小金铃'(*O. fragrans* 'Xiao Jinling')	20	南京秦淮河河道管理处办公区
	'速生金桂'(*O. fragrans* 'Susheng Jingui')	15	南京林业大学校园
	'多芽金'桂(*O. fragrans* 'Duoya Jin')	30	南京林业大学校园
	'金球'桂(*O. fragrans* 'Jinqiu')	15	杭州园林绿化工程有限公司基地
丹桂品种群 *O. fragrans* Aurantiacus Group	'雨城丹'桂(*O. fragrans* 'Yucheng Dan')	15	南京林业大学校园
	'平脉红'(*O. fragrans* 'Pingmai Hong')	15	杭州园林绿化工程有限公司基地
	'橙红丹桂'(*O. fragrans* 'Chenghong Dangui')	15	杭州园林绿化工程有限公司基地
	'状元红'(*O. fragrans* 'Zhuangyuan Hong')	15	杭州园林绿化工程有限公司基地
	'堰虹桂'(*O. fragrans* 'Yanhong Gui')	15	杭州园林绿化工程有限公司基地
	'朱砂丹桂'(*O. fragrans* 'Zhusha Dangui')	30	南京林业大学校园
	'满条红'(*O. fragrans* 'Mantiao Hong')	15	杭州园林绿化工程有限公司基地
	'醉肌红'(*O. fragrans* 'Zuiji Hong')	15	杭州园林绿化工程有限公司基地

* 所测试品种均为经木犀属国际登录中心专家鉴定的品种,树龄为 15~30a,正值桂花的成年期,长势旺盛,开花量大而稳定。

1.2 试验方法

因桂花花期较短,前后仅 1 周时间,故取样集中在花后的第 2、3 天(植株中 50%以上的花序上花朵开放的时期),采集各品种花瓣,冰盒保存带回实验室,进行营养物质含量的测定。矿质元素含量只测定1次,其余各个指标含量重复3次测定,取有效结果的平均值。

1.2.1 测定方法 含水量的测定采用烘干法[7];可溶性糖含量测定采用蒽酮法[7];可溶性蛋白质含量用考马斯亮蓝G-250染色法测定[7];矿质元素含量测定参照裴仁济等[8]方法;维生素C含量采用1,10-菲咯啉法测定[9];有机酸含量测定采用酸碱中和滴定法[10];黄酮含量测定采用$Al(NO_3)_3$-$NaNO_2$方法[11];采用茚三酮法测定桂花中游离氨基酸的含量[12]。

1.2.2 统计分析方法 用SPSS 13.0软件进行方差分析和主成分分析。

2 结果与分析

2.1 25个桂花品种的营养成分

25个桂花品种的主要营养成分含量见表2。其中可溶性糖含量最高的是'桃叶黄',为912.8 mg·g^{-1},'状元红'最低,为10.85 mg·g^{-1};'雨城丹'桂的可溶性蛋白含量最高,达到了59.28 mg·g^{-1},而最低的'镰叶金桂'仅为14.70 mg·g^{-1};'波叶金桂'的维生素C含量最高,达到46.30 mg·g^{-1},'堰虹桂'只有3.99 mg·g^{-1},最高和最低品种相差达10倍之多;有机酸(以苹果酸计)含量最高的是'平脉红',为3.78 mg·kg^{-1},含量相对较低的是'满条红'和'醉肌红',为1.56 mg·kg^{-1},品种间差异不是很大;'紫梗籽银'桂的黄酮含量为288.98 mg·g^{-1},是所有品种中最高的,'晚银桂'的黄酮含量仅为61.94 mg·g^{-1};总游离氨基酸含量最高的是'紫梗籽银'桂,为151.05 mg·g^{-1},含量最低的是'晚银桂',为53.73 mg·g^{-1}。

表2 25个桂花品种部分营养成分含量
Tab. 2 Contents of some nutrient compositions of 25 *O. fragrans* cultivars

品种 Cultivars	水分含量/% Moisture content	可溶性糖含量/ (mg·g^{-1}) Soluble sugar content	可溶性蛋白质含量/ (mg·g^{-1}) Soluble protein content	维生素C含量/ (mg·g^{-1}) Vitamin C content	有机酸含量/ (mg·g^{-1}) Organic acid content	黄酮含量/ (mg·g^{-1}) Flavonoid content	总游离氨基酸含量/ (mg·g^{-1}) Total free amino acid content
'圆瓣金桂'	82.41 lh	37.02 fg	20.31 hijk	9.30 gh	2.66 ab	165.39 f	73.66 i
'墨叶金'桂	79.90 j	58.39 d	15.65 ijk	10.97 g	2.22 ab	111.57 k	103.39 fg
'雨城丹'桂	81.11 i	83.12 b	59.28 a	43.09 b	2.89 ab	216.93 d	68.11 ij
'波叶银'桂	79.57 j	65.00 c	24.79 fjhi	5.02 i	2.44 ab	182.77 e	108.02 fg
'桃叶黄'	79.30 j	91.28 a	21.70 hijk	31.41 c	3.56 a	125.38 hi	85.68 h
'平脉红'	85.22 bcde	43.99 e	15.40 jk	11.37 g	3.78 a	160.41 f	128.42 d
'橙红丹桂'	87.44 a	35.10 gh	15.10 ijk	26.28 d	2.44 ab	146.65 g	115.63 e
'状元红'	86.97 a	10.85 l	22.46 hijk	18.61 f	2.44 ab	114.05 jk	64.08 jk
'波叶金桂'	85.07 bcde	38.39 fg	23.29 ghi	46.30 a	1.78 b	127.21 h	57.64 l
'堰虹桂'	83.41 g	34.58 gh	21.80 hijk	3.99 i	3.11 ab	122.74 jk	101.11 g
'朱砂丹桂'	87.45 a	34.99 gh	37.71 bcd	31.19 c	3.11 ab	275.35 b	134.62 cd
'满条红'	83.96 fg	35.08 gh	33.96 cdef	40.87 b	1.56 b	254.79 c	148.53 a
'速生银桂'	84.78 cdef	44.30 e	30.83 defg	16.54 f	2.44 ab	120.03 ij	81.92 h
'金球'桂	84.71 def	21.81 k	28.95 efgh	22.42 e	2.45 ab	144.07 g	146.51 ab
'紫梗籽银'桂	81.42 i	56.4 d	31.30 defg	26.40 d	3.56 a	288.98 a	151.05 a

品种 Cultivars	水分含量/% Moisture content	可溶性糖含量/(mg·g⁻¹) Soluble sugar content	可溶性蛋白质含量/(mg·g⁻¹) Soluble protein content	维生素C含量/(mg·g⁻¹) Vitamin C content	有机酸含量/(mg·g⁻¹) Organicacid content	黄酮含量/(mg·g⁻¹) Flavonoid content	总游离氨基酸含量/(mg·g⁻¹) Total free amino acid content
'青山银桂'	85.45 bcd	13.84 l	40.49 bc	29.40 cd	2.22 ab	146.82 g	117.41 e
'四季桂'	85.75 b	54.87 d	38.92 bcd	16.90 f	2.67 ab	146.37 g	108.74 f
'醉肌红'	85.69 bc	26.00 ijk	45.86 ab	11.74 g	1.56 b	113.81 ij	150.63 a
'天女散花'	84.44 ef	28.19 ij	23.36 ghij	27.40 d	2.44 ab	129.38 h	58.24 kl
'小金铃'	85.66 bc	43.28 e	34.96 cde	17.94 f	3.56 a	119.33 ij	57.92 kl
'镰叶金桂'	84.50 ef	28.09 ij	14.70 k	18.52 f	2.45 ab	145.94 g	60.58 kl
'速生金桂'	80.74 i	44.30 e	25.16 fghi	6.12 hi	3.11 ab	87.69 l	58.78 kl
'卷叶黄'	81.19 i	40.25 ef	22.00 hijk	11.32 g	2.22 ab	129.76 h	138.46 bc
'晚银桂'	79.09 j	24.45 jk	48.10 ab	7.37 hi	3.11 ab	61.94 m	53.73 l
'多芽金'桂	81.62 hi	30.71 hi	40.43 bcd	4.48 i	2.76 ab	94.03 l	56.11 l

注:同列数据后附不同字母表示差异达0.05显著水平。

2.2 25个桂花品种的矿质元素

25个桂花品种的矿质元素含量见表3。由表3可知,钙元素含量为631.71~3 679.83 mg·kg⁻¹,含量较高的品种有'堰虹桂''桃叶黄''状元红''醉肌红';铁元素含量为52.47~207.98 mg·kg⁻¹,含量较高的品种有'堰虹桂'、'青山银桂'、'状元红'、'紫梗籽银'桂;镁元素含量为705.19~1 541.31 mg·kg⁻¹,含量较高的品种有'堰虹桂'、'晚银桂'、'金球'桂、'波叶金桂';钠元素含量为54.85~220.70 mg·kg⁻¹,含量较高的品种有'青山银桂''金球'桂、'满条红'、'波叶银'桂;锌元素含量为19.78~72.67 mg·kg⁻¹,含量较高的品种有'堰虹桂''圆瓣金桂''状元红''橙红丹桂';钾元素含量为9 002.17~16 979.13 mg·kg⁻¹,含量较高的品种有'状元红''晚银桂''速生银桂''堰虹桂'。

表3 25个桂花品种部分矿质元素含量/(mg·kg⁻¹)
Tab. 3 Contents of mineral elements of 25 *O. fragrans* cultivars/(mg·kg⁻¹)

品种 Cultivars	钙元素含量 Calcium content	铁元素含量 Iron content	镁元素含量 Magnesium content	钠元素含量 Sodium content	锌元素含量 Zinc content	钾元素含量 Potassium content
'圆瓣金桂'	1 050.53	92.09	1 150.56	77.92	65.64	14 838.11
'墨叶金'桂	1 116.10	67.14	1 047.40	104.05	33.91	11 723.20
'雨城丹'桂	1 412.30	81.48	1 166.12	98.04	30.98	13 603.07
'波叶银'桂	1 551.43	140.98	911.70	181.52	49.91	11 762.65
'桃叶黄'	2 679.32	96.62	705.19	159.98	45.06	13 311.83
'平脉红'	1 165.29	80.51	1 018.36	154.25	55.25	9 465.64
'橙红丹桂'	1 590.20	103.44	1 136.22	148.68	60.62	15 670.43
'状元红'	2 302.14	146.38	1 342.66	124.79	65.05	16 979.13
'波叶金桂'	905.06	65.43	1 363.01	114.42	34.55	15 720.25
'堰虹桂'	3 679.83	207.98	1 541.31	124.07	72.67	16 295.90

续 表

品种 Cultivars	钙元素含量 Calcium content	铁元素含量 Iron content	镁元素含量 Magnesium content	钠元素含量 Sodium content	锌元素含量 Zinc content	钾元素含量 Potassium content
'朱砂丹桂'	1 897.71	115.11	1 206.01	68.96	41.54	14 719.81
'满条红'	1 219.40	55.80	1 048.72	193.03	47.76	9 002.17
'速生银桂'	828.11	74.98	1 220.32	127.04	47.91	16 449.06
'金球'桂	1 563.69	84.02	1 411.00	198.99	53.45	12 831.88
'紫梗籽银'桂	1 309.71	142.43	967.36	100.47	41.21	13 078.65
'青山银桂'	1 114.02	191.61	1 115.72	220.70	51.10	13 465.32
'四季桂'	853.47	118.33	1 223.84	70.30	19.78	13 973.80
'醉肌红'	1 993.99	91.81	1 295.76	174.22	60.42	15 766.90
'天女散花'	631.71	59.45	1 047.80	136.05	35.52	13 621.02
'小金铃'	1 337.97	52.47	1 294.80	54.85	39.38	13 661.29
'镰叶金桂'	1 104.76	81.91	1 005.47	67.19	42.09	12 590.81
'速生金桂'	1 574.92	75.63	1 183.16	76.10	41.10	12 328.93
'卷叶黄'	1 585.90	73.59	996.14	95.79	54.54	11 733.27
'晚银桂'	1 517.50	117.42	1 470.01	91.96	48.40	16 659.40
'多芽金'桂	1 482.14	75.52	1 197.78	128.45	31.13	16 026.70

2.3 25个桂花品种的营养品质主成分分析

2.3.1 主成分特征值、贡献率及累计贡献率分析 采用主成分分析法,对25个桂花品种所测的营养成分进行综合评价。由表4可知,前5个主成分贡献率分别为21.93%、17.48%、14.20%、12.07%、7.84%,5个主成分的累计贡献率达到73.52%,表明前5个主成分能够代表全部性状的73.52%的综合信息。因此,选取前5个主成分作为桂花营养品质的重要主成分。

表 4 主成分特征值、贡献率及累计贡献率
Tab. 4 Eigenvalue, contribution ratio and cumulative contribution ratio of principal components

主成分	特征值	贡献率/%	累计贡献率/%
第1主成分	2.851	21.93	21.93
第2主成分	2.272	17.48	39.41
第3主成分	1.846	14.20	53.61
第4主成分	1.569	12.07	65.68
第5主成分	1.019	7.84	73.52

2.3.2 主成分分析因子载荷阵分析 由表5可以看出,前5个主成分反映的各指标在桂花营养品质测定中所起到的作用存在差异。第1主成分中,钾、钙、镁、铁、锌的负载权数高,表明第1主成分主要反映的是桂花花瓣中矿质元素指标;第2主成分主要反映游离氨基酸、黄酮和钠等3个指标;第3主成分主要反映可溶性糖含量;第4主成分主要反映可溶性蛋白和有机酸含量;而第5主成分主要反映维生素C含量。

表5 主成分分析因子载荷阵
Tab. 5 Load factors in principal component analysis

因素 Factor	第1主成分 First principal component	第2主成分 Second principal component	第3主成分 3rd principal component	第4主成分 4th principal component	第5主成分 5th principal component
含水量	0.273	0.314	−0.704	0.039	−0.429
可溶性糖	−0.598	−0.115	0.568	0.313	0.071
可溶性蛋白质	−0.006	−0.073	−0.287	0.648	0.180
维生素C	−0.438	0.330	−0.442	0.380	0.610
游离氨基酸	−0.075	0.833	0.082	0.049	−0.030
黄酮	−0.440	0.616	−0.035	0.447	−0.267
有机酸	0.063	−0.237	0.393	0.521	−0.418
钾元素	0.595	−0.383	−0.217	0.379	0.018
钙元素	0.574	0.191	0.229	0.215	0.017
钠元素	0.131	0.658	0.032	−0.212	0.480
镁元素	0.691	−0.162	−0.330	0.199	0.061
铁元素	0.601	0.336	0.304	0.337	0.032
锌元素	0.697	0.402	0.241	−0.210	−0.101

2.3.3 各主成分值与营养品质得分值 表6列出25个品种桂花的5个主成分值,根据公式 $\sum \lambda_i u_i$(λ 为主成分贡献率;u 为主成分值;$i=1,2,\cdots,5$)求出桂花各品种营养品质得分以及排名。从表6中可以看出,'堰虹桂'、'雨城丹桂'、'青山银桂'、'醉肌红'4个品种的花瓣营养品质较高,可作为优良的食用品种;'天女散花''镰叶金桂'等品种的营养价值较低。

表6 各主成分值与营养品质得分值
Tab. 6 Scores of principal components and comprehensive scores of nutritional quality evaluation in different *O. fragrans* cultivars

品种 Cultivars	第1主成分 First principal component	第2主成分 Second principal component	第3主成分 3rd principal component	第4主成分 4th principal component	第5主成分 5th principal component	得分 Score	排名 Ranking
'堰虹桂'	2.675 7	0.427 2	1.657 5	0.616 3	−0.399 3	0.939 9	1
'雨城丹'桂	0.112 6	0.288 9	1.372 9	0.725 2	1.399 7	0.467 5	2
'青山银桂'	0.673 7	1.459 7	−0.809 4	0.214 1	1.280 9	0.414 1	3
'醉肌红'	1.091 6	0.853 6	−0.675 1	−0.148 8	1.475 5	0.390 4	4
'波叶银'桂	−0.438 0	0.630 2	1.798 7	−0.436 5	0.897 0	0.287 2	5
'状元红'	1.879 8	0.013 1	−0.534 6	−0.199 8	−0.734 3	0.256 8	6
'桃叶黄'	−0.943 6	−0.018 8	2.384 3	0.491 8	0.504 9	0.227 4	7
'紫梗籽银'桂	−0.972 7	1.105 5	0.837 5	1.551 9	−0.821 8	0.221 8	8
'金球'桂	0.509 3	1.131 5	−0.621 5	−0.499 5	0.528 0	0.202 3	9
'晚银桂'	1.173 2	−1.686 0	0.170 1	0.452 8	1.546 1	0.162 7	10
'朱砂丹桂'	−0.119 0	0.933 7	−0.824 6	1.850 7	−1.476 5	0.127 6	11
'橙红丹桂'	0.566 7	0.773 7	−0.516 1	−0.447 6	−1.075 0	0.047 8	12

续 表

品种 Cultivars	第1主成分 First principal component	第2主成分 Second principal component	第3主成分 3rd principal component	第4主成分 4th principal component	第5主成分 5th principal component	得分 Score	排名 Ranking
'满条红'	−1.303 6	2.046 7	−0.784 7	−0.353 9	0.600 8	−0.035 3	13
'卷叶黄'	−0.313 4	0.255 1	0.781 6	−1.113 0	−0.177 4	−0.061 4	14
'平脉红'	−0.444 8	0.803 7	0.637 3	−1.008 9	−1.420 0	−0.099 7	15
'多芽金'桂	0.245 7	−1.423 4	−0.145 9	−0.074 3	1.209 0	−0.129 8	16
'圆瓣金桂'	0.299 2	−0.492 4	0.278 6	−0.623 2	−0.952 8	−0.130 8	17
'速生银桂'	0.114 2	−0.609 4	−0.722 2	−0.172 6	0.185 6	−0.190 3	18
'四季桂'	−0.592 2	−0.584 3	−0.710 8	0.891 9	−0.122 5	−0.234 8	19
'速生金桂'	−0.106 6	−1.392 3	0.772 0	−0.668 9	−0.086 7	−0.244 6	20
'墨叶金'桂	−0.961 5	−0.640 8	0.840 2	−1.201 3	0.151 9	−0.336 6	21
'小金铃'	−0.151 4	−1.313 5	−0.611 6	0.374 6	−0.901 6	−0.375 1	22
'波叶金桂'	−0.382 6	−0.615 4	−1.742 9	0.008 7	0.051 8	−0.434 6	23
'镰叶金桂'	−0.399 0	−0.636 6	−0.355 6	−0.989 3	−1.379 9	−0.476 0	24
'天女散花'	−0.648 8	−0.572 6	−1.064 9	−0.785 1	−0.122 7	−0.498 0	25

注：综合得分在0.300 0以上的为营养价值较高品种；得分在−0.300 0以下的为营养价值较低品种；得分在−0.300 0～0.300 0之间的为营养价值中等品种。

3 讨论

文中测定了4个桂花品种群25个品种的营养成分，结合曾经对19个品种的分析[13]，发现同一品种在不同年份采集的花瓣其营养成分含量差异较大，如'紫梗籽银'桂、'晚银桂'和'朱砂丹桂'都是采自同一棵树，但其含量变化很大，这可能和树体当年的营养成分积累有关。此外，同一品种在不同地点采集的花瓣其营养成分含量也有很大差异，如'桃叶黄'金桂分别采自无锡梅园和临安青山苗圃，其矿质元素的含量显著不同，这与生长地土壤中的矿质元素含量有很大的关系。

两次研究均选择了4个品种群的不同品种，虽然不同品种群的花瓣颜色有差异，香味也不同，'四季桂'品种群色淡味浅，银桂品种群色渐深味较浓，金桂品种群色深味浓，而丹桂品种群色深味浅。但从测定结果看颜色香味与营养物质含量并非成正比，很多金桂品种如'速生金桂''墨叶金'桂、'小金玲'、'波叶金桂'、'镰叶金桂'等总得分和排名均在部分四季桂品种、银桂品种和丹桂品种之后，由此可见，花瓣营养物质含量与花瓣的颜色和香味相关性较小。

总体来看，桂花品种中的矿质元素以及可溶性糖、可溶性蛋白和黄酮等的含量均较高，具有很好的营养价值[6]。通过对不同品种的营养成分含量进行分析，可加以合理利用，如'雨城丹'桂、'晚银桂'和'醉肌红'等3个品种的可溶性蛋白质含量较高，作为食材可以满足人体对植物蛋白的需求；'朱砂丹桂'、'紫梗籽银'桂等的黄酮含量较高，而黄酮类化合物对心血管系统、消化系统等有显著药效活性，且具有抗炎、免疫调节、抗肿瘤、镇痛及保护肝脏等药理作用[14],这些桂花品种的花瓣可以用于一些保健品中;另外桂花中还含有丰富的矿质元素:钾能调节细胞内液的渗透压,调节 pH 值,维持神经、肌肉的兴奋性;钙是骨骼的主要组成成分,直接影响人体的生长发育;镁具有维持核酸结构的稳定性,激活人体各种酶的作用。因此,将桂花作为食材、药材等加以利用,具有较高营养价值和良好的保健作用。

主成分分析法常被用于分析复杂综合性状和优良品质选择,如肖泽鑫等[15]通过主成分分析进行台湾相思(Acacia confusa)优树选择,金潇潇等[16]通过主成分分析确定了 20 种菊花(Dendranthema × grandiflorum)的营养品质,最终筛选出 3 种营养品质较高可食用的菊花种。本文对 25 个桂花品种花瓣中的营养成分进行主成分分析,其中'堰虹桂'、'雨城丹'桂、'青山银桂'、'醉肌红'4 个品种的综合得分较高,比较适合作为食用桂花品种。

参考文献

[1] 幸宏伟.郁金香及杜鹃花花瓣营养成分分析[J].重庆工商大学学报(自然科学版),2005,22(1):53-55.

[2] 史国安,郭香凤,包满珠.不同类型牡丹花的营养成分及体外抗氧化活性分析[J].农业机械学报,2006,37(8):111-114.

[3] 刘晓辉,杨明,邓日烈,等.花瓣营养成分探析[J].北方园艺,2010(12):48-49,50.

[4] 许又凯,刘宏茂,刀祥生,等.五种野生花卉营养成分及其评价[J].营养学报,2005,27(5):437-440.

[5] 向其柏,刘玉莲.中国桂花品种图志[M].杭州:浙江科学技术出版社,2008:6.

[6] 徐文斌,王贤荣,石瑞,等.两个丹桂品种营养成分分析[J].林业科技开发,2010,24(6):117-119.

[7] 邹琦.植物生理生化实验指导[M].北京:中国农业出版社,1997.

[8] 裴仁济,陈小强,孙宁,等.火焰原子吸收光谱法测定不同花色非洲紫罗兰金属元素[J].江苏农业科学,2010(6):444-445.

[9] Nobuhiko A N, Tsutsumi K, Sanceda N G, et al. A rapid and sensitive method for the determination of ascorbic acid using 4,7-diphenyl-1,10-phenanthroline[J]. Agricultural Biology and Chemistry, 1981,45(5):1289-1290.

[10] 李合生.植物生理生化实验原理和技术[M].北京:高等教育出版社,2000:167-169.

[11] 元晓梅,蒋明蔚,胡正芝,等.聚酰胺吸附-硝酸铝显色法测定山楂及山楂制品中的总黄酮含量[J].食品与发酵工业,1996(4):27-31.

[12] 黄松,吴月娜,刘梅,等.茚三酮比色法测定青天葵中总游离氨基酸的含量[J].中国中医药信息杂志,2010,17(12):50-51,52.

[13] 杨秀莲,王良桂,文爱林.桂花花瓣营养成分分析[J].江苏农业科学,2012,40(12):334-336.

[14] 程秋月,郭菁,张成义.黄酮类化合物药理作用的研究[J].北华大学学报(自然科学版),2011,12(2):180-183.

[15] 肖泽鑫,彭剑华,詹潮安,等.主成分分析在台湾相思优树选择标准和方法中的应用研究[J].中南林业科技大学学报,2012,32(5):54-58.

[16] 金潇潇,陈发棣,陈素梅,等.20 个菊花品种花瓣的营养品质分析[J].浙江林学院学报,2010,27(1):22-29.

Analysis of Nutrients in the Petals Of 25 Cultivars of *Osmanthus fragrans*

YANG Xiulian, ZHAO Fei, WANG Lianggui

(College of Landscape Architecture, Nanjing Forestry University, Nanjing 210037, China)

Abstract: To select edible cultivars of *Osmanthus fragrans* with excellent nutritional quality from 25 cultivars, nutritional components in flowers were studied at full-bloom stage, including content of soluble sugar, soluble protein, water, vitamin C, organic acid, free amino acid, flavonoids and 6 mineral elements. The principal component analysis was used to screen cultivars with excellent quality by comparing nutritional components in the flowers. The results showed that soluble sugar content was $10.85 \sim 91.28$ mg $\cdot$ g^{-1}, soluble protein $14.70 \sim 59.28$ mg $\cdot$ g^{-1}, water $79.09\% \sim 87.45\%$, vitamin C $3.99 \sim 46.30$ mg $\cdot$ g^{-1}, organic acid $1.56 \sim 3.78$ mg $\cdot$ g^{-1}, flavonoids $61.94 \sim 288.98$ mg $\cdot$ g^{-1} and free amino acid $53.73 \sim 151.05$ mg $\cdot$ g^{-1}. Also contents of 6 mineral elements were as follows: zinc at $19.78 \sim 72.67$ mg $\cdot$ kg^{-1}, iron at $52.47 \sim 207.98$ mg $\cdot$ g^{-1}, magnesium at $705.19 \sim 1\,541.31$ mg $\cdot$ kg^{-1}, calcium at $631.71 \sim 3\,679.83$ mg $\cdot$ kg^{-1}, sodium at $54.85 \sim 220.70$ mg $\cdot$ kg^{-1}, and potassium at $9\,002.17 \sim 16\,979.13$ mg $\cdot$ kg^{-1}. There was large difference between cultivars, cultivars like *O. fragrans* 'Yanhong Gui', *O. fragrans* 'Yucheng Dan', *O. fragrans* 'Qingshan Yingui', *O. fragrans* 'Zuiji Hong' exhibited higher nutritional quality, however, *O. fragrans* 'Boye Jingui', *O. fragrans* 'Lianye Jingui', *O. fragrans* 'Tiannü Sanhua' showed lower nutritional quality.

Key words: *Osmanthus fragrans*; petal; nutritional quality; the principal component analysis

桂花花瓣营养成分分析*

杨秀莲,王良桂,文爱林

(南京林业大学,江苏 南京 210037)

摘要：为选择营养品质优良的食用桂花品种,对19个桂花品种盛花期花瓣中的可溶性糖、可溶性蛋白质、水分、维生素C、花青素以及6种矿质元素含量进行了测定,并用主成分分析法对桂花品种的营养品质进行比较,筛选出营养品质优良的品种。结果表明:不同品种桂花花瓣的可溶性糖含量在30.4～114.7 g/kg之间,可溶性蛋白质含量在6.6～31.5 g/kg之间,维生素C含量为0.271～0.473 g/kg,花青素相对含量为0.98～5.50,含水量为811.8～884.8 g/kg,矿质元素含量分别为锌57.34～89.73 mg/kg、铁132.9～173.7 mg/kg、镁1.85～2.47 g/kg、钙1.94～2.63 g/kg、钠226.5～324.5 mg/kg、钾28.34～42.78 g/kg,桂花品种间营养成分存在很大差别,'桃叶黄'金桂、'红十字'丹桂、'竹叶银'桂、'波叶金桂'4个品种的营养价值较高,'日香桂''长梗白'和'四季桂'等品种的营养价值较低。

关键词：桂花;品种;花瓣;营养品质;主成分分析

近年来将花卉作为一种新型的美食原料逐渐成为时尚,对花卉营养品质的研究也在逐渐开展,经测定:郁金香、杜鹃[1]、牡丹[2]、玫瑰花、百合花、木棉花、山茶花、唐菖蒲[3]以及一些野生花卉[4]等的花瓣均含有较多的矿质元素和蛋白质或氨基酸等,可食用、药用、保健用,是原生态的食用珍品。桂花隶属木犀科木犀属,是著名的观赏植物,在我国长江中下游普遍栽培,目前已确定就有160多个品种[5]。桂花是传统的香花植物,花虽小,但香味持久稳定。我国拥有悠久的食用桂花传统,早在2 000年前,就用桂花制酒、制花茶。现在,桂花酒、桂花茶仍负盛名,同时,还开发了桂花糕、桂花糖、桂花月饼、桂花露饮料等食品50多种。但对桂花营养成分的研究不多,仅徐文斌等对'浦城丹桂'和'小叶丹桂'的营养成分进行了初步研究[6]。本研究对不同品种桂花花瓣中的可溶性糖、可溶性蛋白质、水分、维生素、花青素以及6种矿质元素含量进行了测定,并用主成分分析法,比较品种间营养品质差异,从观赏桂花中筛选出营养品质优良的品种,为桂花花瓣的合理利用提供试验数据。

1 材料与方法

1.1 供试材料

本试验所用的桂花品种共19个,分别采自南京林业大学(12个品种)和无锡梅园(7个

---

* 原文发表于《江苏农业科学》,2012,40(12):334-336。
基金项目:国家林业局公益性行业科研专项(201204607);江苏省高校自然科学基础研究项目(11KJB220002);江苏省高校优势学科建设工程项目(PAPD)。

品种),具体品种如下:'四季桂'、'日香桂'、'大叶佛顶珠'、'旱银桂'、'长梗白'、'金粟'、'紫梗籽银'桂、'晚银桂'、'串银球'、'竹叶银桂'、'金满楼'、'金狮桂'、'波叶金桂'、'桃叶黄'金桂、'速生金桂'、'早红'丹桂、'朱砂丹桂'、'红十字'丹桂、'墨叶丹'桂。

1.2 试验方法

因桂花花期较短,前后仅1周时间,故取样集中在花后的第2、3天(植株中50%以上的花序上花朵开放的时期),采集各品种花瓣进行营养物质含量的测定。矿质元素只测定1次,其余各个指标重复3次测定,取有效结果的平均值。

1.2.1 测定方法

含水量的测定采用烘干法[7];可溶性糖含量采用蒽酮法[7];可溶性蛋白质采用考马斯亮蓝G-250染色法[7];矿质元素测定参照裴仁济等的方法[8];花青素含量的测定采用胡亚东等的方法[9];维生素C含量测定采用1,10-菲咯啉法[10]。

1.2.2 统计分析方法

用Excel软件绘制柱状图,用SPSS 13.0软件进行方差分析和主成分分析。

2 结果与分析

2.1 桂花部分营养成分含量

19个桂花品种可溶性糖含量以'红十字'丹桂最高,为114.7 g/kg,'四季桂'最低,只有30.4 g/kg。可溶性蛋白质含量差异极大,其中以'桃叶黄'金桂最高,达到了31.5 g/kg;'紫梗籽银'桂的含量较低,为6.6 g/kg,仅为前者的20.95%。'波叶金桂'的维生素C含量最高,达到了0.47 g/kg,'金粟'银桂含量较低,只有0.27 g/kg。丹桂品种群4个品种的花青素相对含量比较高,'墨叶丹'桂、'红十字'丹桂、'朱砂丹桂'和'早红'丹桂的花青素相对含量分别为5.05、5.50、4.20和2.37,其他品种群桂花的花青素相对含量则比较低,'金满楼'的含量最低,只有0.98(表1)。

表1 19个桂花品种的部分营养成分和矿质元素含量

品种	部分营养成分含量/(g·kg⁻¹)				花青素相对含量	矿质元素含量/(mg·kg⁻¹)					
	可溶性糖	含水量	可溶性蛋白质	维生素C		铁	钠	钾	钙	镁	锌
'日香桂'	84.4 g	811.8 a	7.4 a	0.357 e	1.10 b	157.3	274.3	34 672	2 336	2 028	63.38
'长梗白'	64.2 de	827.8 d	6.6 a	0.384 g	1.22 c	132.9	294.7	31 372	1 937	2 436	61.47
'金满楼'	63.4 cde	828.1 e	12.8 be	0.417 h	0.98 a	164.2	248.3	37 836	2 436	2 387	70.83
'墨叶'丹桂	59.6 be	830.8 f	12.9 be	0.433 j	5.05 i	146.3	276.4	29 453	2 135	1 974	59.49
'串银球'	80.8 fg	843.4 i	14.9 ed	0.468 l	1.10 b	168.2	268.3	32 822	2 274	2 346	65.92
'金粟'银桂	84.9 gh	843.1 i	15.9 ede	0.271 a	1.48 e	146.7	226.5	33 557	2 036	2 473	62.34
'金狮桂'	62.9 cde	817.8 e	9.3 ab	0.285 b	1.30 eb	172.3	293.5	38 229	2 462	2 312	73.51
'紫梗籽银'桂	70.9 de	812.8 ab	6.6 a	0.335 e	1.36 d	165.9	236.5	35 728	2 190	2 453	68.53
'大叶佛顶珠'	52.9 b	837.8 g	15.8 ede	0.375 f	1.04 ab	171.8	302.7	42 777	2 442	2 119	77.26

续表

品种	部分营养成分含量/(g·kg⁻¹)				花青素相对含量	矿质元素含量/(mg·kg⁻¹)					
	可溶性糖	含水量	可溶性蛋白质	维生素C		铁	钠	钾	钙	镁	锌
'晚银桂'	67.5 de	824.5 d	18.3 de	0.423 i	1.26 cd	159	285.3	38 252	2 632	2 142	72.45
'早银桂'	70.7 ef	814.8 b	26.9 f	0.354 e	1.03 ab	154	253.7	4 633	2 436	2 357	89.73
'波叶金桂'	78.9 fg	831.0 f	18.4 de	0.473 l	1.73 f	142.8	324.5	37 416	2 573	2014	78.36
'红十字'丹桂	114.7 j	856.6 j	15.3 ed	0.445 k	5.50 i	156.3	274.5	35 274	2 468	2 421	80.92
'速生金桂'	54.5 be	838.0 g	15.9 ede	0.386 g	1.07 ab	173.7	266.5	36 592	2 536	2 396	75.43
'早红'丹桂	67.1 de	837.6 g	15.3 de	0.354 e	2.37 g	142.4	215.2	31 511	2 035	2 167	63.82
'竹叶银桂'	93.6 hi	860.9 k	20.6 e	0.413 b	1.48 e	164.4	274.3	37 837	2 144	2018	74.37
'桃叶黄'金桂	107.9 i	884.8 l	31.5 g	0.445 k	1.56 e	153.2	257.8	38 625	2 363	1 937	77.59
'四季桂'	30.4 a	835.6 g	12.3 be	0.348 d	1.26 ed	164.3	247.7	28 336	2 073	1 846	57.34
'朱砂丹桂'	69.8 ef	839.7 b	12.4 be	0.375 f	4.20 h	157.4	248.3	35 821	2 149	2 173	67.54

注：同一列内不同字母表示在 0.05 水平上有显著差异；可溶性糖、可溶性蛋白和矿质元素为干质量比，其余为鲜质量比。

2.2 部分矿质元素含量

19 个桂花品种部分矿质元素的含量见表 1。各矿质元素含量分别为：铁 132.9～173.7 mg/kg，其中'紫梗籽银'桂、'速生金桂'、'大叶佛顶珠'、'金狮桂'、'串银球'较高；钠 226.5～324.5 mg/kg，含量最高的是'波叶金桂'；钾 28.34～42.78 g/kg，含量最高的是'大叶佛顶珠'；钙 1.94～2.63 g/kg，含量最高的是'晚银桂'；镁 1.85～2.47 g/kg，含量最高的是'金粟'银桂；锌 57.34～89.73 mg/kg，含量最高的是'早银桂'。

2.3 桂花营养品质主成分分析

2.3.1 不同桂花品种主成分特征值、贡献率及累计贡献率分析

采用主成分分析法，对 19 个桂花品种所测的营养成分进行综合评价。由表 2 可知，前 5 个主成分贡献率分别为 32.59%、21.33%、12.79%、11.02%、7.17%，5 个主成分的累计贡献率达到 84.90%，表明前 5 个主成分已经代表了全部性状的 84.90% 的综合信息，因此选取前 5 个主成分作为桂花营养品质的重要主成分。

表 2 桂花营养品质主成分特征值、贡献率及累计贡献率

主成分	特征值	贡献率/%	累计贡献率/%
第 1 主成分	3.585	32.59	32.59
第 2 主成分	2.346	21.33	53.92
第 3 主成分	1.407	12.79	66.71
第 4 主成分	1.212	11.01	77.73
第 5 主成分	0.789	7.17	84.90

2.3.2 不同桂花品种主成分分析因子载荷阵分析

由表3可以看出,前5个主成分反映的各指标在桂花营养品质测定中所起到的作用存在差异。第1主成分中,可溶性糖、可溶性蛋白质、花青素、锌、镁、钙的负载权数高,表明第1主成分主要反映的是桂花花瓣中这几个指标;第2主成分主要反映的是含水量和维生素C两个指标;第3主成分主要反映的是铁的含量;第4主成分主要反映的是钠的含量;而第5主成分主要反映的是钾的含量。

表3 桂花品质主成分分析因子载荷阵

因素	主成分				
	第1主成分	第2主成分	第3主成分	第4主成分	第5主成分
可溶性糖	0.526	0.516	0.351	0.364	−0.125
可溶性蛋白质	0.571	0.418	0.274	0.456	−0.230
水分含量	0.473	0.678	0.255	−0.234	0.189
维生素C	0.019	0.564	−0.210	0.511	0.501
花青素	0.580	0.413	−0.457	0.066	0.031
锌	0.934	−0.146	0.134	0.161	0.027
铁	0.496	−0.246	−0.695	0.187 5	−0.211
镁	0.782	−0.433	0.185	−0.016	−0.043
钙	0.768	−0.444	−0.136	0.071	−0.053
钠	−0.129	−0.336	0.548	0.656	−0.147
钾	0.295	−0.626	0.192	−0.235	0.604

2.3.3 不同桂花品种各主成分值及品质得分比较

表4列出的19个品种桂花的5个主成分值,根据公式$\sum \lambda_i u_i$(λ:主成分贡献率,u:桂花各品种主成分值,$i=1,2,\cdots,5$)求出各桂花品种营养品质得分。从表4中可以看出,'桃叶黄'金桂、'红十字'丹桂、'竹叶银桂''波叶金桂'4个品种的花瓣营养品质较高,可作为优良的食用品种;'四季桂''长梗白'和'日香桂'等品种的营养价值较低。

表4 各桂花品种的主成分值与营养品质得分

品种	第1主成分	第2主成分	第3主成分	第4主成分	第5主成分	得分	排名
'桃叶黄'金桂	3.137	2.906	1.400	1.170	0.592	2.031	1
'红十字'丹桂	2.680	1.642	0.072	1.791	0.694	1.949	2
'竹叶银桂'	1.289	1.388	0.913	0.931	0.293	0.909	3
'波叶金桂'	2.645	0.177	−2.328	0.438	0.091	0.530	4
'串银球'	0.191	0.322	0.104	0.199	−0.789	0.245	5
'速生金桂'	0.882	−1.620	0.711	0.270	−1.100	0.181	6
'晚银桂'	1.466	−0.955	−0.682	0.493	−0.497	0.155	7
'大叶佛顶珠'	1.562	−1.762	−0.052	0.672	−0.511	0.106	8
'早银桂'	0.214	−0.762	0.602	0.288	−1.432	0.068	9
'朱砂丹桂'	−0.762	0.850	0.124	0.302	1.423	0.063	10
'金满楼'	0.414	−1.340	0.457	−0.421	−0.882	−0.059	11
'金粟'银桂	−1.772	0.277	1.739	−1.445	−0.568	−0.261	12

续表

品种	第1主成分	第2主成分	第3主成分	第4主成分	第5主成分	得分	排名
'早红'丹桂	−1.890	1.459	0.588	−1.049	0.967	−0.360	13
'墨叶丹'桂	−1.793	2.046	−2.249	0.252	2.393	−0.550	14
'金狮桂'	−0.148	−2.691	0.377	0.322	−1.070	−0.578	15
'紫梗籽银'桂	−1.465	−1.697	1.075	−0.659	−0.836	−0.702	16
'日香桂'	−1.214	−0.625	−0.545	−0.227	0.166	−0.906	17
'长梗白'	−1.816	−0.160	−1.968	−2.457	−0.474	−1.148	18
'四季桂'	−3.620	0.543	−0.384	−0.870	1.541	−1.673	19

3 讨论

本试验仅测定了4个桂花品种群19个品种的营养成分,其含量差异较大。桂花品种众多,根据不同品种的营养成分加以合理利用,进行食用桂花品种的选择很有必要。如'桃叶黄'金桂、'早银桂'和'竹叶银桂'3个品种的可溶性蛋白质含量较高,可以满足人体对植物蛋白的需求;'波叶金桂'、'串银球'、'红十字'丹桂和'桃叶黄'金桂的维生素C含量较高,维生素C具有清除体内自由基、预防癌症的作用,这些品种的桂花可以用于一些保健品中;'墨叶丹'桂、'红十字'丹桂、'朱砂丹桂'和'早红'丹桂的花青素含量较高,多食用这些品种的桂花能发挥花青素抗氧化、预防心脑血管疾病、抗突变、保护肝脏等多种生理功能。另外桂花中还含有丰富的矿质元素,钾能调节细胞内液的渗透压,调节pH值,维持神经、肌肉的兴奋性;钙是骨骼的主要组成成分,直接影响人体的生长发育;镁具有维持核酸结构的稳定性,激活人体各种酶的作用。

主成分分析法常被用于分析复杂综合性状和优良品质选择,如肖泽鑫等通过主成分分析进行台湾相思优树选择[11];金潇潇等通过主成分分析确定了20种菊花的营养品质,最终筛选出3种营养品质较高的可食用的菊花品种[12]。本试验对19个桂花品种花瓣中的营养成分进行主成分分析,其中'桃叶黄'金桂、'红十字'丹桂、'竹叶银桂'、'波叶金桂'等4个品种的综合得分较高,适合作为食用桂花品种。

桂花不同花期花瓣内营养物质会发生变化,朱诚等研究认为可溶性蛋白质含量盛花期达到最大[13],而陈洪国则认为桂花的可溶性蛋白质和可溶性糖含量在初花期最高[14]。由于桂花的采收一般靠人工在谢花期摇落,因此对于谢花期花瓣的营养成分需进一步研究,从而确定桂花的最佳采收时间和采收方式。另外研究中选取的指标不够全面,能更好体现花瓣营养品质的指标如氨基酸、黄酮等应在今后的研究中加以应用。

参考文献

[1] 幸宏伟.郁金香及杜鹃花花瓣营养成分分析[J].重庆工商大学学报(自然科学版),2005,22(1):53-55.

[2] 史国安,郭香凤,包满珠.不同类型牡丹花的营养成分及体外抗氧化活性分析[J].农业机械学报,2006,37(8):111-114.

[3] 刘晓辉,杨明,邓日烈,等.花瓣营养成分探析[J].北方园艺,2010(12):48-50.

[4] 许又凯,刘宏茂,刀祥生,等.五种野生花卉营养成分及其评价[J].营养学报,2005,27(5):437-440.
[5] 向其柏,刘玉莲.中国桂花品种图志[M].杭州:浙江科学技术出版社,2008.
[6] 徐文斌,王贤荣,石瑞,等.两个丹桂品种营养成分分析[J].林业科技开发,2010,24(6):117-119.
[7] 邹琦.植物生理生化实验指导[M].北京:中国农业出版社,1997.
[8] 裴仁济,陈小强,孙宁,等.火焰原子吸收光谱法测定不同花色非洲紫罗兰金属元素[J].江苏农业科学,2010(6):444-445.
[9] 胡亚东,贾惠娟,孙崇德,等.桃果实中花青苷的提取、检测及应用[J].果树学报,2004,21(2):167-169.
[10] Nobuhiko A N, Tsutsumi K, Sanceda N G, et al. A rapid and sensitive method for the determination of ascorbic acid using 4,7-diphenyl-1,10-phenanthroline[J]. Agricultural Biology and Chemistry, 1981, 45(5):1289-1290.
[11] 肖泽鑫,彭剑华,詹潮安,等.主成分分析在台湾相思优树选择标准和方法中的应用研究[J].中南林业科技大学学报,2012,32(5):54-58.
[12] 金潇潇,陈发棣,陈素梅,等.20个菊花品种花瓣的营养品质分析[J].浙江林学院学报,2010,27(1):22-29.
[13] 朱诚,曾广文.桂花花衰老过程中的某些生理生化变化[J].园艺学报,2000,27(5):356-360.
[14] 陈洪国.桂花开花进程中花瓣色素、可溶性糖和蛋白质含量的变化[J].武汉植物学研究,2006,24(3):231-234.

Analysis of Nutritional Components of *Osmanthus fragrans* Petals

YANG Xiulian, WANG Lianggui, Wen Ailin

(College of Landscape Architecture, Nanjing Forestry University, Nanjing 210037, China)

Abstract: In order to select varieties of *Osmanthus fragrans* with excellent nutritional quality, the contents of soluble sugar, soluble protein, water, vitamin C, anthocyanin and six mineral elements in the petals of 19 *Osmanthus fragrans* cultivars at full bloom stage were determined. And the nutritional quality of *Osmanthus fragrans* was compared by principal component analysis, and the varieties with good nutritional quality were selected. The results showed that the soluble sugar content of different varieties of *Osmanthus fragrans* petals was between 30.4 and 114.7 g/kg, the soluble protein content was between 6.6 and 31.5 g/kg, the vitamin C content was 0.271 to 0.473 g/kg, the relative content of anthocyanins is 0.98—5.50, the water content is 811.8～884.8 g/kg, the contents of mineral elements are zinc 57.34～89.73 mg/kg, iron 132.9～173.7 mg/kg, magnesium 1.85～2.47 g/kg, calcium 1.94～2.63 g/kg, sodium 226.5～324.5 mg/kg, potassium 28.34～42.78 g/kg. There are great differences in nutrient composition between *Osmanthus fragrans* varietis, and 4 varieties('Taoye Huang'、'Hong Shizi'、'Zhuye Yin'、'Boye Jingui') have higher nutritional value, the nutritional value of 'Rixiang Gui'、'Changgeng Bai'、'Sijigui'is low.

Key words: *Osmanthus fragrans*; cultivars; petals; nutritional quality; principal component analysis

桂花种子对赤霉素处理的生理生化响应

杨秀莲,王良桂

(南京林业大学 风景园林学院,江苏 南京 210037)

摘　要:赤霉素处理并经低温层积后测定'紫梗籽银'桂种子萌发过程中胚和胚乳内各种激素和营养物质含量的变化,结果表明:随着层积时间的延长,1 000 mg/L 赤霉素处理下种子其胚和胚乳中的脱落酸(ABA)含量逐渐减少,至 45 d 时比对照减少 51.6%,赤霉素(GA)、生长素(IAA)、细胞分裂素(ZR)的含量逐渐增加,至 45 d 时分别比对照增加 76.66%、88.34%、80.9%;GA/ABA、IAA/ABA、ZR/ABA,以及(GA+IAA)/ABA 比值也呈先增后减的趋势。进一步证实了 ABA 抑制种子的萌发,而 GA 促进萌发;同时,IAA、ZR 在解除种子休眠过程中也起到一定的促进作用。层积期间,种子可溶性糖和淀粉含量逐渐降低,可溶性蛋白质则呈现先升后降再升又降的趋势。

关键词:'紫梗籽银'桂;种子;胚和胚乳;内源激素;营养物质

桂花(*Osmanthus fragrans*)是一种享誉我国古今,集绿化、美化、香化于一体,观赏和实用兼备的优良园林树种,至今已有 160 多个品种。由于大多数桂花的栽培品种因子房退化而不能结实,致使桂花的种子不易获得;且桂花种子具有一定的脱水敏感性和休眠特性,当年 4—5 月份成熟的种子需层积至第 2 年 2—3 月才能萌发,所以目前桂花繁殖主要采用扦插法、嫁接法、高压法,很少用播种繁殖,这影响了桂花杂交育种的开展。张义等[1]利用赤霉素(GA)浸种和低温层积相结合的方法进行桂花种子的催芽试验,证明低温层积和赤霉素浸种可有效促进种子的萌发。笔者以'紫梗籽银'桂种子为例,测定了赤霉素处理结合低温层积过程中种子内源激素和营养物质含量的动态变化,为揭示外源激素和层积处理对桂花种子休眠的解除和萌发机制提供科学依据。

1　材料与方法

1.1　供试材料及处理

试验所用的桂花种子 2007 年 5 月初采于南京林业大学的'紫梗籽银'桂。'紫梗籽银'桂属银桂品种群(*O. fragrans* Albus Group),结实量大,种子采收容易。将种子去除外、中果皮,仅留内果皮(文中均称种皮),种粒饱满,质量较好。采用蒸馏水(记为 G0)和质量浓度为 1 000 mg·L^{-1} 的 GA 溶液(记为 G10)浸种 48 h,然后分别与湿沙按 1∶3 混合放入塑料

---

* 原文发表于《南京林业大学学报(自然科学版)》,2012,36(1):63-67。
基金项目:江苏省科技支撑计划(BE2011367);江苏省林业三项工程项目(LYSX [2007]01);南京林业大学科技创新基金项目。

盆中,上面覆盖保鲜膜并用剪刀剪几个小洞,利于种子呼吸。将塑料盆置于 2～4 ℃ 冰箱中进行低温层积,间隔一定时间翻动并适时补充水分,层积 93 d 结束试验。层积期间每半个月随机取 G0、G10 处理的种子各 20 粒,剥取胚乳和胚,用刀片切成厚约 1 mm 的薄片,再纵横各切一刀成 4 小片,混合后取一定量进行种子营养物质和内源激素含量的测定,重复 3 次。同时各处理取 4×30 粒种子,在 25 ℃ 恒温、24 h 光照条件下做种子发芽试验,统计种子的发芽率。

1.2 生理生化指标的测定

测定内源激素含量时,取 0.5 g 胚和胚乳鲜样,加 2 mL 内含 1 mmol·L^{-1} 2,6-二叔丁基-4-甲基苯酚(BHT)的 80% 甲醇溶液,在冰浴下研磨成匀浆,转入 10 mL 试管,再用 2 mL 提取液分次将研钵冲洗干净,一并转入试管中,摇匀后放置在 4 ℃ 冰箱中提取 4 h,1 000 r/min 离心 15 min,取上清液。沉淀中加 1 mL 提取液,搅匀,置 4 ℃ 下再提取 1 h,离心,合并上清液并记录体积,残渣弃去。上清液过 C-18 固相萃取柱。具体步骤是:80% 甲醇(1 mL)平衡柱→上样→收集样品→移开样品后用 100% 甲醇(5 mL)洗柱→100% 乙醚(5 mL)洗柱→100% 甲醇(5 mL)洗柱→循环。将过柱后的样品转入 5 mL 塑料离心管中,真空浓缩干燥,除去提取液中的甲醇,用样品稀释液定容。采用中国农业大学作物化学控制实验室建立的间接酶联免疫吸附法[2]进行内源激素测定。可溶性糖及淀粉含量的测定采用邹琦的蒽酮比色法[2],可溶性蛋白质含量的测定采用邹琦的考马斯亮蓝 G-250 法[2],稍做修改。数据统计分析与处理采用 Excel 和 SPSS 13.0 软件完成。

2 结果与分析

2.1 桂花种子内源激素含量的变化

2.1.1 细胞分裂素(ZR)

不同处理'紫梗籽银'桂种子胚和胚乳中的 ZR 含量变化如图 1A 所示。由图 1 可知,随着层积时间的延长,G0 处理中 ZR 含量有升降但变化不显著,至层积结束 ZR 含量比层积前仅下降 1.48%;G10 处理的 ZR 含量在层积 45 d 时出现一较大峰值,含量比对照增加 80.9%,至 60 d 又比对照降低 59.6%,此后又有所回升。

方差分析结果表明,两处理不同层积时间的 ZR 含量均达差异极显著水平。多重对比显示,除层积 30～93 d 之间 ZR 含量没有差异外,其余各时间段 ZR 含量差异均达极显著水平。说明在层积期内,GA 处理和低温层积对'紫梗籽银'桂胚和胚乳中 ZR 含量变化的作用较明显。

2.1.2 生长素(IAA)

两种不同处理 IAA 含量的变化如图 1B 所示,G0 处理下在 15 d 即达到峰值,比层积前增加了 135.63%;在以后的 1 个月时间内逐渐下降,至层积 45 d 后基本保持不变。层积结束时比对照下降了 59.83%。G10 处理下则先降后升再降,在层积 45 d 时达到峰值,比对照增加 88.34%,后急速下降,60 d 后基本保持不变,层积结束时 IAA 含量比对照下降了 41.76%。

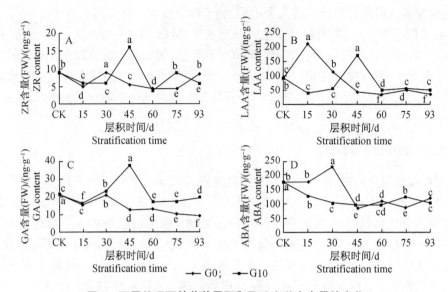

图1 不同处理下桂花种子胚和胚乳中激素含量的变化
Fig. 1 Changes of hormone contents of different treatments in embryo and endosperm

方差分析结果表明:两处理层积时间的IAA含量差异均达极显著水平。多重对比结果表明,G0处理除60.93 d,G10处理除75.93 d外,其余时间段的IAA含量差异均达极显著水平。

2.1.3 赤霉素(GA)

随着层积时间的延长,G0处理中GA含量在层积15 d时略有下降,后又缓慢回升,至30 d时达到峰值,随后一直处于缓慢下降的过程中,层积结束时共下降了56.15%。G10处理下则于15 d后快速上升,至45 d时达到最大值,比对照增加了76.66%,也是G0处理的2.99倍,至60 d急剧下降,后又缓慢上升(图1C)。比较两种处理可以看出,经外源GA处理的种子层积过程中内源GA含量始终比未处理的高,说明外源GA的处理可显著增加层积期间种子内源GA的含量,从而有利于休眠的解除。

方差分析和多重对比结果表明:从两处理层积时间的GA含量看,G0处理除45.60 d,G10处理除60.75 d外,其余均达到极显著水平。

2.1.4 脱落酸(ABA)

随着层积时间的延长,G0处理下桂花种子ABA的含量先是逐渐上升,至30 d时,达到峰值,以后迅速下降,45 d后保持基本稳定,此时的ABA含量比层积前下降了51.6%,至层积结束,ABA含量比对照下降了32.8%。而G10处理下种子ABA含量在层积前期快速降低,30 d后基本保持稳定,至层积结束时ABA含量比层积前降低了41%(图1D)。

方差分析结果表明:两处理下不同层积时间的ABA含量差异均达到极显著水平。多重对比表明,除75 d和93 d时ABA含量变化差异不显著外,其余时间段ABA含量差异均达极显著水平。

2.1.5 各激素含量之间比值的变化

有研究认为种子休眠和萌发不仅与植物内源激素的绝对含量有关,还与各类激素之间的平衡、特别是促进生长的激素与抑制生长的激素之间的比例及平衡有关[3-6]。以G10处理的种子中 ABA、GA、IAA、ZR 含量为例,各激素之间的比值结果如图 2 所示,可以看出 GA、IAA、ZR 与 ABA 含量的比值在层积 45 d 达到峰值,此后有所下降,但基本维持稳定。这一结果表明种子的休眠受控于 ABA 类抑制物质和 GA、IAA 类促进物质的相互作用,根据发芽试验观察结果,种子层积 45 d 以后开始萌发。这说明外源激素对桂花种子的萌发具有一定的调控作用。

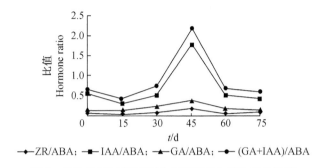

图 2　G10 处理下桂花种子内源激素含量比值的变化

Fig. 2　The changes of hormone ratio in embryo and endosperm of G10

2.2　桂花种子营养物质含量的变化

2.2.1　可溶性糖

可溶性糖含量呈先增后降的趋势(图 3A)。方差分析结果表明:G0 处理在 0 d 与其他时间段的差异均显著,而其他时间段之间差异不显著;G10 处理下种子可溶性糖含量在 30、45 和 75 d 差异显著,其他时间段差异不显著。

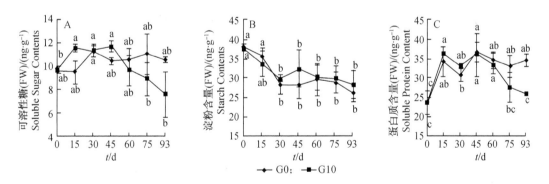

图 3　不同处理种子中的营养物质含量变化

Fig. 3　The changes of nutrient substance contents of different treatments

注:小写字母表示在 0.05 水平下差异显著,相同字母之间表示差异不显著。

2.2.2　淀粉

随着层积时间的延长,两处理下桂花种子淀粉含量在层积期间均逐渐减少(图 3B)。这说明随着层积时间的延长,种子中的代谢活动逐渐加强,生理生化反应加剧,种子对碳水化合物的需求也不断增加,与此相对应,淀粉含量逐渐下降。

2.2.3 蛋白质

不同处理下桂花种子的可溶性蛋白质含量变化结果如图 3C 所示。随着层积时间的延长,两处理的可溶性蛋白质呈现先升后降再升又降的趋势,至层积结束(93 d),G10 处理比 G0 处理低 24.8%。

方差分析表明,不同层积时间差异达显著水平。经多重对比可知,G0 处理中,0、30、45 d 间可溶性蛋白质含量差异显著,而其他各时间段差异不显著;G10 处理中,0、15、75 d 间差异显著,其他各时间段差异不显著。

2.3 不同层积时间下桂花种子发芽率变化

对不同层积时期的'紫梗籽银'桂种子进行发芽试验,结果见图 4。由图 4 可知,G10 处理的种子在 45 d 后大部分种子出现裂口,层积 60 d 后开始发芽,发芽率为 14.45%,而层积 75 d 时显著提高至 71.11%,至 105 d 时为 72.37%,即 75 d 后发芽率基本保持稳定。而 G0 处理的种子在层积 75 d 才开始发芽,发芽率仅为 4.44%,层积 95 d 和 105 d 发芽率分别为 8.89%和 26.67%。

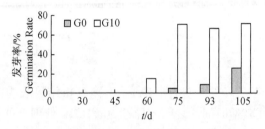

图 4 不同层积时间下桂花种子的发芽率
Fig. 4 Germination rate of sweet *Osmanthus* seeds during different stratification time

3 讨论

3.1 内源激素与种子休眠

由于 ABA 广泛存在于休眠种子中,因此对引起种子休眠的内源激素的研究也主要集中在 ABA 上,一般认为高浓度的 ABA 是造成植物种子生理休眠的重要原因[7],其主要作用是抑制核酸的正常代谢,干扰核糖核酸 RNA 的合成,从而阻抑了蛋白质和有关酶的合成,使种子不能进行正常的代谢活动。在苹果[8]、桃[9]、水曲柳[10]等种子中,内源 ABA 含量与其休眠深度之间呈明显的正相关。GA 能有效地打破种子休眠,加速不休眠种子的萌发[11]。关于 IAA 在解除种子休眠方面的报道不多,而且结论也各不相同。Kochanko 等[12]认为 IAA 参与了休眠调节,杨万霞等[13]认为在层积处理中,休眠种子中 IAA 含量总体呈上升趋势,对解除休眠起到一定的作用;另有一些研究则认为种子休眠解除过程与 IAA 含量关系不大,过高的 IAA 含量甚至起到负面作用[14-15]。细胞分裂素是促进细胞分裂和扩大的激素,具有抵消发芽抑制物质的作用,调控种子发育中的物质和能量代谢[16]。本试验中,层积前种子内 ABA 含量显著高于其他激素的含量,层积期间 ABA 含量不断降低,说明桂花种子的休眠主要是由体内过多的ABA造成的。试验中,用GA浸种,人为地增加GA的含量,打破促进生长的激素与抑制生长的激素之间的平衡,与未经GA处理的G0相比,随着层积时间的增长,种子ABA含量下降显著加快,GA、IAA、ZR的含量则明显增加。GA/ABA、IAA/ABA、ZR/ABA,以及GA+IAA/ABA的比值也呈现出前期明显增加的趋势。试验结果进一步证实了ABA对种子萌发有抑制作用,而GA能促进种子的萌发。同时,IAA、ZR在种子解除休眠、萌发的过程中也起到一定的促进作用。因此,种子休眠和萌发不仅与植物内源激素的绝对含量有关,还与各类激素之间的平衡,特别是促进生长的激素与抑制生长的激素之间的比例及平衡有关。这与山楂种子中ABA含量降低,GA、IAA、ZR含量增加,以及GA、IAA、IR与ABA的比值增大、休眠解除的结论一致[17]。

3.2 低温层积与种子营养物质含量的变化

种子萌发时,种子中贮藏的淀粉等大分子物质会在酶的作用下转化为小分子可溶性物质,以便为胚的生长所利用,促进幼苗的建成和生长。可溶性蛋白质是植物体建构物质,种子可溶性蛋白质为种子萌发提供氮素营养,种胚分化生长与蛋白质代谢密不可分。'紫梗籽银'桂种子在层积过程中,内部的营养物质也发生了相应的变化,其中可溶性糖含量总体表现为先升后降,随后趋于稳定,而淀粉的含量总体上均呈现逐渐降低的趋势。糖与淀粉之间的消长变化与种子生长发育密切相关,表明种子中的代谢活动逐渐加强,种子对碳水化合物的需求也不断增加,淀粉发生水解导致含量逐渐下降;可溶性糖含量的变化比较复杂,因为种子中的淀粉可转化为可溶性糖,而可溶性糖是呼吸作用的直接底物,部分可溶性糖转化为ATP等直接供种子发芽使用。可溶性蛋白质含量呈先上升后下降,然后趋于较缓慢下降的趋势。这可能是由于新采种子在层积初期体内可溶性淀粉和蛋白质仍处于合成阶段。15 d后,随着层积的继续,蛋白质含量逐渐降低。这说明蛋白质在层积过程中不断地进行分解代谢和转化,为种子萌发提供能量来源。本试验中,两处理在层积前期3种营养物质含量的变化趋势比较相似,但60 d后G10处理下种子的可溶性糖和蛋白质的下降速度要比G0处理下降快,结合发芽试验结果,说明此时种子开始萌发,需要消耗较多的营养物质。

参考文献

[1] 张义,宋春燕.赤霉素浸种与低温层积对桂花种子发芽的影响[J].中国林副特产杂志,2005(6):9-10.

[2] 邹琦.植物生理生化实验指导[M].北京:中国农业出版社,1997.

[3] Khan A A. 种子休眠和萌发的生理生化[M].王沙生,洪铁宝,高荣孚,等,译.北京:农业出版社,1989.

[4] Femell A. Systems and approaches to studying dormancy: Introduction to the workshop[J]. Hortiscience, 1999, 34(7):1172-1173.

[5] Foley M E. Genetic basis for dormancy in wild oat[J]. Hortiscience, 1999, 34(7):1174-1176.

[6] 曹帮华,蔡春菊.银杏种子后熟生理与内源激素变化的研究[J].林业科学,2006,42(2):32-37.

[7] 郑光华.种子生理研究[M].北京:科学出版社,2004.

[8] Rudnicki R. Studies on abscisic acid in apple seeds[J]. Planta, 1969, 86(1):63-68.

[9] Bonamy P A, Donnis F G. ABA levels in seeds of peach II. Effects of stratification temperature[J]. J Amer Soc Hort Sci, 1988, 102(1):26-28.

[10] 郭维明,李日韦,郭廷翘,等.水曲柳种子后熟期间内源抑制物的特点及其与更新的关系[J].东北林业大学学报,1991,19(6):44-53.

[11] 党海山,张燕君,江明喜,等.濒危植物毛柄小勾儿茶种子休眠与萌发生理的初步研究[J].武汉植物学研究,2005,23(4):327-331.

[12] Kochanko V G, Gitzesik M, Chojnowsk M, et al. Effect of temperature, growth regulators and other chemicals on *Echinacea purpurea* (L.) Moench seed germination and seedling survival[J]. Seed Sci & Technol, 1998, 26:547-554.

[13] 杨万霞,方升佐.青钱柳种子综合处理过程中内源激素的动态变化[J].南京林业大学学报(自然科学版),2008,32(5):85-88.

[14] 李金克,郑然,王沙生,等.脱落酸和酚酸在红松种子休眠中的作用[J].河北林学院学报,1996,11(3/4):189-193.

[15] 高红兵,吴榜华,孙振良.东北红豆杉种子层积过程中内源生长素和脱落酸含量的变化[J].吉林林学院学报,1998,14(4):187-189

[16] 王三根.细胞分裂素与植物种子发育和萌发[J].种子,1999,105(4):35-37,42.

[17] 李秉真,乌云,田瑞华,等.山楂种子休眠和后熟期间内源激素的变化[J].植物生理学通讯,1998,34(4):254-256.

The Physiological and Biochemical Responses of sweet Osmanthus Seeds to Gibberellic Acid Treatment

YANG Xiulian, WANG Lianggui

(College of Landscape Architecture, Nanjing Forestry University, Nanjing 210037, China)

Abstract: In this paper, the changes of endogenous hormones and nutrient substances of 'Zigeng Zi Yin' seeds were studied by low temperature stratification after gibberellic acid treatment. The results showed that the abscisic acid (ABA) reduced gradually as the stratification time increased, and it decreased 51.6% compared to CK 45 days after treatment, but the gibberellic acid (GA), indole-3-acetic acid (IAA) and cytokinin (ZR) increased to 76.66%, 88.34%, 80.9% respectively 45 days later. The ratios of GA/ABA, IAA/ABA, ZR/ABA and (GA+IAA)/ABA first increased and then decreased. The experimental results showed that ABA restrained from germination. By contrast, GA promotes seeds germination, IAA and ZR were also found positive in breaking the seed dormancy. During cold stratification treatment, contents of soluble sugar and starch decreased gradually; contents of soluble protein displayed a "up-down-up-down" change.

Key words: 'Zigeng Zi Yin'; seeds; embryo and endosperm; endogenous hormones; nutrient substances

淹水对2个桂花品种生理特性的影响

王良桂,杨秀莲

(南京林业大学 风景园林学院,江苏 南京 210037)

摘 要:以'天香台阁''朱砂丹桂'的2年生扦插幼苗为试材,通过盆栽试验,研究其在不同程度淹水条件下的生理响应。结果表明,不同淹水胁迫对2个品种生理特性的影响趋势相同,均表现为叶绿素含量持续缓慢下降;电导率呈现先降后升趋势;游离脯氨酸、丙二醛、过氧化物酶、超氧化物歧化酶总体上均呈现先升后降趋势,但对2个品种生理特性产生的影响大小各异。通过分析,认为'朱砂丹桂'的耐淹力比'天香台阁'强。

关键词:桂花;淹水胁迫;生理特性

涝害在长江中下游及沿海地区是一种常见的自然灾害,每年都对农作物造成巨大的经济损失,过去20多年来,关于植物对涝渍的适应性的研究逐渐成为植物逆境研究的热点之一。对于农作物如玉米[1]、花生[2]等以及绿化树种如枫杨[3]、鹅掌楸[4-5]、银杏[6]等都做了很多研究,取得了较大的进展。桂花($Osmanthus\ fragrans$ Lour.)作为重要的亚热带园林树种,集绿化、美化、香化、观赏、实用于一身,却有着不耐涝渍,尤忌积水的怕水特性,遭受涝渍危害后根系发黑腐烂,叶尖先是焦枯,其后全部脱落,导致全株死亡。目前我国共有桂花品种162个,这些品种观赏特性、生态习性各异。其中四季桂品种群中的'天香台阁'以花朵大(直径1~2.5 cm)、花期长(秋季9—11月和春季3—5月为盛花期,冬季在室内可开花,夏季有少量花开)著称;丹桂品种群中的'朱砂丹桂'以花色艳(橙红色,极为鲜艳)而极具观赏价值,此两品种在园林中应用极广。作者试图以'天香台阁'和'朱砂丹桂'为材料,开展在不同涝渍胁迫下桂花的生理反应研究。这对进一步了解不同桂花品种生态习性以及为沿江沿海园林绿地选择耐湿桂花品种,扩大桂花应用等做一些有益的尝试,为今后进一步的研究提供重要的借鉴意义。

1 材料与方法

1.1 试验材料与处理

供试材料为2年生'天香台阁'和'朱砂丹桂'的扦插幼苗,苗高约40~50 cm,由浙江金华复兴苗圃提供。幼苗于春季上盆栽培,盆径15 cm,园土作基质,于2006年7月8日开始淹水处理。试验设对照(正常管理)、浅淹(水与盆土面齐平)和深淹(水与盆边齐平)3个处理,每盆3株,每处理3个重复,随机区组排列。自开始处理第4 d取第1次样,后每隔6 d

---

* 原文发表于《安徽农业大学学报》,2009,36(3):382-386。
基金项目:江苏省林业三项工程项目(LYSX[2007]01)。

定期采自顶端以下第3至5片叶。测定叶片内主要生化指标为:叶绿素含量、超氧化物歧化酶(SOD)、丙二醛(MDA)、过氧化物酶(POD)、游离脯氨酸、相对电导率。

1.2 测定方法

叶绿素含量测定采用丙酮乙醇混合液法[7];游离脯氨酸含量采用茚三酮比色法测定;SOD活力测定根据邹琦方法[8],稍做改动;POD活性、MDA含量、细胞膜透性均根据邹琦方法[8]。

2 结果与分析

2.1 淹水胁迫对不同品种桂花幼苗叶绿素含量的影响

2个桂花品种在淹水胁迫过程中的叶绿素含量变化如图1所示。叶绿素a、叶绿素b和总叶绿素的含量均呈逐渐下降趋势,但下降的程度不同。至淹水处理结束时,'天香台阁'总叶绿素含量比对照下降33%,叶绿素a含量下降32%,叶绿素b含量下降35.1%;'朱砂丹桂'总叶绿素比对照下降24.7%,叶绿素a含量下降22.5%,叶绿素b含量下降28.6%。由上可知叶绿素b含量的下降程度要比叶绿素a大,'天香台阁'叶绿素含量的变化比'朱砂丹桂'大。

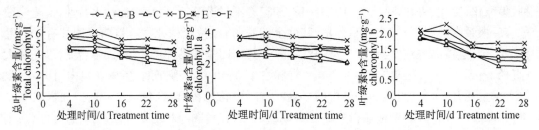

A.'天香台阁'对照;B.'天香台阁'浅淹;C.'天香台阁'深淹;D.'朱砂丹桂'对照;E.'朱砂丹桂'浅淹;F.'朱砂丹桂'深淹。下同。
A. Control of 'Tianxiang Taige'; B. 'Tianxiang Taige' light water logging; C. 'Tianxiang Taige' severe water logging; D. Control of 'Zhusha Dangui'; E. 'Zhusha Dangui' light water logging; F. 'Zhusha Dangui' severe water logging. The same below.

图1 淹水胁迫下桂花叶片总叶绿素、叶绿素a和叶绿素b含量的变化
Fig.1 Changes of total chlorophyll, chlorophyll a and b in seedling leaves under different water logging stress

2.2 淹水胁迫对不同桂花品种各生理指标的影响

2.2.1 对电导率的影响

随着淹水胁迫时间的延长,2个品种处理的相对电导率值均比对照要高,呈现先降后升的趋势(图2),其中'天香台阁'2种处理相对电导率高于'朱砂丹桂',且深淹处理高于浅淹处理,说明2个品种膜系统都受到了不同程度的伤害。

2.2.2 对游离脯氨酸含量的影响

桂花在淹水胁迫下,处理的叶片游离脯氨酸含量均呈现先升后降趋势,前期增幅较小,

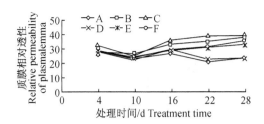

图 2 淹水胁迫下桂花叶片电导率的变化

Fig. 2 Changes on relative electric conductivity of seedling leaves under the different waterlogging stress

16 d 后增幅较大,至 22 d 时达到最大,其中'天香台阁'深淹处理比对照增加了 5 倍,'朱砂丹桂'深淹处理比对照增加近 2 倍,22 d 后又有所下降(图 3)。

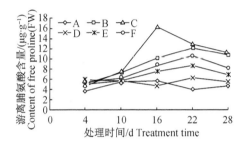

图 3 淹水胁迫下桂花叶片游离脯氨酸含量的变化

Fig. 3 Changes of proline content in seedling leaves under different waterlogging stress

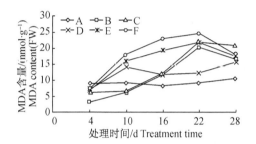

图 4 淹水胁迫下桂花叶片 MDA 含量的变化

Fig. 4 Changes of MDA content in seedling leaves under different waterlogging stress

2.2.3 对 MDA 含量、POD 和 SOD 活性的影响

MDA 是植物细胞膜脂过氧化作用的产物之一,其含量的高低可以反映逆境胁迫下植物受伤害的程度。从图 4 可知,淹水胁迫使桂花品种叶片中 MDA 含量随着处理时间的延长而呈现先增后降的趋势,淹水第 22 d 处理的 MDA 含量达最大值,后又逐渐下降。深淹处理 22 d 时'天香台阁'MDA 含量比对照增加了 136.9%,'朱砂丹桂'增加了 100.6%。

由图 5 可知,2 个品种不同处理的 POD 活性变化有所差异:'天香台阁'浅淹时呈现先降后升的变化趋势,16 d 时降到最低,而后逐渐上升,22 d 后才开始有较快的上升,至 28 d 时比对照高出 26.2%;深淹时则前期逐渐上升至 16 d 时达最大值,比对照高出 38.5%,后快速下降;'朱砂丹桂'浅淹时 10 d 前和 22 d 后上升较快,10～22 d 之间变化比较平缓,而深淹时则呈现先升后降再升的趋势,在 22 d 时达到最大值,比对照高出 47.8%。

淹水胁迫下,桂花 2 个品种 SOD 活性的变化表现为先升后降趋势,前期的上升可能是植物对水分胁迫的一种适应能力,以维持体内各种代谢之间的平衡,但在后期这种平衡遭到破坏,SOD 活性逐渐降低,造成细胞内活性氧积累而产生毒害,使植株遭受伤害。由图 6 可知,22 d 是淹水胁迫的一个重要时间,此时间以后,SOD 活性开始下降,尤其是'天香台阁'深淹处理的 SOD 活性下降最快。

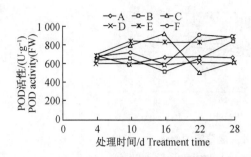

图 5 淹水胁迫下桂花叶片 POD 活性的变化
Fig. 5 Changes of POD activity in seedling leaves under different waterlogging stress

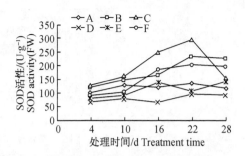

图 6 淹水胁迫下桂花叶片 SOD 活性的变化
Fig. 6 Changes of SOD activity in seedling leaves under different waterlogging stress

2.3 各指标之间的相关性分析

对桂花淹水处理的各指标进行相关性分析(见表1)表明,除了POD活性与游离脯氨酸含量、POD活性、SOD活性、MDA含量、电导率、游离脯氨酸含量与总叶绿素含量、叶绿素a含量、叶绿素b含量呈负相关外,其余均呈正相关。其中SOD活性与MDA含量、电导率、游离脯氨酸含量,MDA含量与电导率、电导率与游离脯氨酸含量、总叶绿素含量与叶绿素a含量和叶绿素b含量、叶绿素a含量与叶绿素b含量呈极显著正相关;而SOD活性、MDA含量、电导率、游离脯氨酸含量与总叶绿素含量、叶绿素a含量、叶绿素b含量呈极显著负相关。

表 1 各指标间的相关性
Tab. 1 The correlation coefficients of comparison among physiology indicates

项目 Item		POD 活性	SOD 活性	MDA 含量	电导率 Electric Conductivity	游离脯氨酸含量 Proline	总叶绿素含量 Total Chlorophyll	叶绿素a含量 Chlorophyll a
SOD 活性	I	0.200						
	II	0.059						
MDA 含量	I	0.222*	0.297**					
	II	0.035	0.004					
电导率 Electric Conductivity	I	0.162	0.536**	0.391**				
	II	0.127	0.000	0.000				
游离脯氨酸 Proline	I	−0.241**	0.349**	0.127	0.298**			
	II	0.022	0.001	0.233	0.004			
总叶绿素含量 Total Chlorophyll	I	−0.111	−0.636**	−0.333**	−0.588**	−0.336**		
	II	0.296	0.000	0.001	0.000	0.001		
叶绿素a含量 Chlorophyll a	I	−0.090	−0.557**	−0.136	−0.526**	−0.307**	0.929**	
	II	0.401	0.000	0.202	0.000	0.003	0.000	
叶绿素b含量 Chlorophyll b	I	−0.115	−0.596**	−0.520**	−0.538**	−0.299**	0.870**	0.625**
	II	0.280	0.000	0.000	0.000	0.004	0.000	0.000

* 表示5%差异显著水平。** 表示1%差异显著水平。
* Correlation insignificant at 0.05 level(2-tailed). ** Correlation insignificant at 0.01 level(2-tailed). Pears on correlation. Sig. (2-tailed).

3 讨论

本试验结果表明,在淹水处理过程中,保护性酶系都表现出积极的响应,在处理前期活性都表现出上升趋势,在处理 22 d 达到最大值,随着处理时间延长,酶活性逐渐降低。植物在逆境条件下,植物体内活性氧的产生能力大于清除能力,相对过量的活性氧影响生物膜和其他生物功能分子的结构或功能。SOD、POD 等活性氧清除剂具有维持活性氧代谢平衡、保护膜结构的功能,因此胁迫初期两者活性均有一定的增加。随着胁迫时间的延续,涝渍胁迫较重的处理其 POD 活性先于 SOD 活性下降,其间出现一定的反复,但 SOD 活性在淹水 22 d 降低。

细胞膜的稳定性是植物细胞执行正常生理功能的基础。植物在逆境条件下体内通常会产生活性氧,从而启动膜脂过氧化或膜脂脱脂作用,或改变细胞质膜透性,进而影响植物代谢,表现为 MDA 的积累与相对电导率的增加。本试验中随着胁迫时间的延长,桂花叶片的 MDA 含量和相对电导率都增加了,这与陈洪国[9]的研究结果一致,其他植物如中华蚊母[10]、银杏[6]、苹果[11]以及烟草[12]等水涝研究中均出现相同情况。

逆境条件下植物体内游离脯氨酸含量会增加,游离脯氨酸含量的提高是在逆境条件下植物的自卫反应之一。桂花品种在淹水胁迫时,体内游离脯氨酸含量随淹水时间的延长不断积累,处理 22 d 后,游离脯氨酸含量达到峰值,随后有所下降。这与任雪萍[13]等研究小麦叶片脯氨酸含量在湿涝害时的变化规律相似。脯氨酸含量不断增加,维持了细胞的膨胀,同时保护酶和膜系统免受伤害。但是由于植物体能够承受的胁迫有限,所以在后期表现出不同程度的下降。

叶绿素含量是植物叶片光合性能、营养状况和衰老程度的直观表现。大量研究结果表明,在各种逆境胁迫下,植物体内叶绿素均会受到不同程度破坏,使其含量降低。在水涝胁迫后,2 个桂花品种的叶绿素含量均表现为持续下降,且与其他生理指标呈极显著的负相关,这与张晓平等对鹅掌楸的研究结果一致[4]。李林通过研究花生品种间耐湿涝的差异,表明叶绿素含量及其变化状况是衡量花生品种耐湿涝能力高低的重要指标之一,叶绿素含量高低与叶色深浅密切相关。因此凭借在湿涝前后叶色的变化就能初步判断植物品种的耐湿涝性差异[2]。

综上所述,淹水处理的第 22 d 是一个重要的转折点,此时植物体内一些生理指标出现较明显的变化,表明此时植物对胁迫的防御系统遭到破坏。

参考文献

[1] 刘晓忠,李建坤,王志霞,等.涝渍逆境下玉米叶片超氧化物歧化酶和过氧化氢酶活性与抗涝性的关系[J].华北农学报,1995,10(3):29-32.
[2] 李林.花生品种间耐湿涝性差异及其机理研究[D].长沙:湖南农业大学,2004.
[3] 李纪元,饶垅兵,潘德寿,等.淹水胁迫下枫杨种源 MDA 含量的地理变异[J].浙江林业科技,1999,19(4):22-27.
[4] 张晓平,方炎明,陈永江.淹涝胁迫对鹅掌楸属植物叶片部分生理指标的影响[J].植物资源与环境学报,2006,15(1):41-44.
[5] 潘向艳,季孔庶,方彦.淹水胁迫下杂交鹅掌楸无性系几种酶活性的变化[J].西北林学院学报,2007,

22(3):43-46.
[6] 陈颖,谢寅峰,沈惠娟. 银杏幼苗对水分胁迫的生理响应[J]. 南京林业大学学报(自然科学版),2002,26(2):55-58.
[7] 张宪政. 植物叶绿素含量测定——丙酮乙醇混合液法[J]. 辽宁农业科学,1986(3):26-28.
[8] 邹琦. 植物生理生化实验指导[M]. 北京:中国农业出版社,1997.
[9] 陈洪国. 桂花幼苗对不同程度水分胁迫的生理响应[J]. 华中农业大学学报,2006,25(2):190-193.
[10] 彭秀,肖千文,罗韧,等. 淹水胁迫对中华蚊母生理生化特性的影响[J]. 四川林业科技,2006,27(2):17-20.
[11] 杨宝铭,吕德国,秦嗣军,等. 淹水对'寒富'苹果保护酶系和根系活力的影响[J]. 沈阳农业大学学报,2007,38(3):291-294.
[12] 曾淑华,刘飞虎,覃鹏,等. 淹水对烟草生理指标的影响[J]. 烟草科技,2004(1):36-38.
[13] 任雪萍,吴诗光,牛明功,等. 湿涝害对灌浆期冬小麦叶片脯氨酸含量的影响——脯氨酸含量的变化与小麦的抗性关系[J]. 周口师专学报,1998,15(5):72-74.

Effects of Water Logging Stress on the Physiological Characteristics of Two Cultivars of Sweet Osmanthus Flower

Wang Lianggui, Yang Xiulian

(College of Landscape Architecture, Nanjing Forestry University, Nanjing 210037, China)

Abstract: In this paper, two-year-old cuttings of 'Tianxiang Taige' and 'Zhusha Dangui' of sweet osmanthus flower cultivars were selected to study the physiological responses to different water logging stress. The results showed that the changing trend of physiological characteristics of two cultivars was consistent, but the degree of effect was dissimilar under the different water logging stress. With the extension of water logging time, the chlorophyll content decreased gradually, the relative electric conductivity decreased at first and then increased subsequently, and MAD content, the proline content, the activities of POD and SOD increased firstly, then decreased 22 days later. The results indicated that water logging resistance of 'Zhusha Dangui' was higher than that of 'Tianxiang Taige'.

Key words: sweet osmanthus flower; water logging stress; physiological responses

'紫梗籽银'桂种子休眠原因的初步探讨*

杨秀莲，丁彦芬，甘习华

(南京林业大学　风景园林学院,江苏　南京　210037)

摘　要：通过对'紫梗籽银'桂种子种皮结构的电镜扫描、种子发芽率、生活力、种皮吸水量测定以及种子不同部位浸提液对白菜种子萌发影响的试验,初步探讨了'紫梗籽银'桂种子的休眠原因。结果表明,'紫梗籽银'桂种子坚硬的外种皮不是引起休眠的原因,种子的休眠可能是由于种皮和种仁内含有发芽抑制物质引起的。

关键词：'紫梗籽银'桂；种子休眠；抑制物质

'紫梗籽银'桂是银桂品种群中的一个品种,小乔木,高达4～5 m,树冠广卵形,紧密,长势旺盛,分枝粗壮,枝条斜展。1年生花枝长8～10 cm,每节具花芽2～4对,多3对,花梗随果实生长,逐渐由绿变为紫色、深紫色,故名。花黄白色,香味浓郁,每年开花2～3次。花后结实丰满。品种性状稳定,园林应用较广。但由于其种子具有一定的休眠特性,当年4—5月成熟的种子需沙藏到第二年2—3月种子裂口露白时才能播种,如能在当年使其萌发,进一步缩短育苗时间,将有助于桂花品种的苗木生产。张义等[1]利用赤霉素浸种和低温层积相结合的方法进行桂花种子的催芽试验,证明桂花种子具有休眠特性,低温层积和赤霉素浸种可有效促进种子的萌发；袁王俊等[2]利用离体培养技术对层积3个月的种子进行胚培养,打破了种子的休眠,促进了萌发。作者以'紫梗籽银'桂为例,通过对桂花种子的形态结构、种子发芽率、休眠抑制物质等的研究,旨在探讨桂花种子的休眠原因,以期寻找解除休眠的最佳方法,有效地缩短休眠时间,提高播种繁殖率。

1　材料与方法

1.1　材料

试验所用'紫梗籽银'桂种子,于2007年5月采自南京林业大学校园内。

白菜种子:南京市蔬菜种子公司的'绿优1号'白菜种子,纯度95％,净度98％,发芽率85％以上。

1.2　方法

1.2.1　种子发芽率、生活力的测定

随机选取颗粒饱满、大小均一的种子30粒,重复4次,在25 ℃的光照培养箱中做萌发

---

* 原文发表于《江苏林业科技》,2007,34(6):18-20,45。
基金项目：江苏省农业三项工程项目(BK2005133)。

试验,30 d后统计其发芽率。生活力测定采用四唑法(TTC法)鉴定:取新鲜种子30粒,去除种子的硬质种皮,小心拨开胚乳,取出胚,放入0.5%的四唑溶液中,在30 ℃的恒温箱中进行染色,每小时观察记录1次,染色时间以着色稳定不再染色为止,重复3次。染色结束后,沥去溶液,用清水冲洗,根据胚染色的部位,染色面积的大小和染色程度,逐个判断种子的生活力。其中胚根、子叶全部染色,胚根染色,子叶部分染色的种子记为有生活力的种子;胚根染色、子叶未染色,胚根未染色、子叶染色和胚根、子叶均未染色的种子记为无生命力种子。将经判断有生活力的种子数除以供试种子总数即为种子的生活力。

1.2.2　种子的外表皮结构和吸水量测定

随机选取10粒正常大小的种子,用钳子撕裂先端置于2.5%戊二醛中固定,经系列丙酮脱水,转入乙酸异戊酯后,进行临界点干燥、喷金,扫描电镜下观察种子表面和横断面结构,并进行记录和拍照。种子的吸水量测定参照沈永宝等[3]的方法。

1.2.3　种子内含发芽抑制物质的生物检测

各取种子的中种皮(种壳)和种仁(包括胚、胚乳和内种皮)25.0 g,用小型粉碎机粉碎后放入盛有250 mL蒸馏水的容量瓶中,在40 ℃恒温水浴中浸提24 h,过滤定容后得到水提液。在培养皿中放置滤纸,分别加入一定量的浸提原液、60%原液、30%原液,每种处理放置100粒白菜种子,于25 ℃恒温全光照条件下培养,48 h测定发芽率,72 h测量苗高和根长,重复3次,用蒸馏水处理的白菜种子作对照。

2　结果与分析

2.1　种子的生活力和发芽率测定结果

经测定,新鲜种子的生活力为79.5%,置床1个月后观察种子的发芽率为0,因此种子的发芽率低不是因为种子的生活力低所导致,而是种子本身具有休眠的特性。

2.2　种子的形态特征、解剖结构和吸水量结果

'紫梗籽银'桂种子为核果,长椭圆形,成熟时肉质果皮(外种皮)为紫黑色,贮藏前已将其去除。中种皮土黄色,木质化,稍硬,有多条突起的纵条纹,每一纵条纹又向两边伸出很多细的条纹,形如枝杈状(见图1:中部表面),内种皮膜质。种子有胚乳,胚完整,位于胚乳中间(见图1:内部胚和胚乳)。种子中种皮纵剖面扫描电镜图显示,中种皮由3～4层横向排列的细胞和4～5层纵向排列的长管状细胞组成(见图1:头部断面),细胞排列较疏松,可见明显的细胞腔,细胞壁上有多数小孔,可能与透水有关。通过对种子头部纵切观察,可见明显的发芽孔(见图1:头部发芽孔),这一结构的存在,既可以减小胚根突破中种皮时的机械阻力,又可改善种子的透气透水性。通过定时测定完整种子和破壳种子的水分变化情况,绘出种子吸水进程的变化曲线(见图2),从图2中可见完整种子和破壳种子的吸水速率没有明显差异,随着时间的延长,两种处理的种子吸水量呈逐步上升趋势,种子吸胀72 h后,吸水量达88%,趋于饱和状态。反映了种子有良好的透水性。

由此可见,'紫梗籽银'桂种子较硬的木质化中种皮不存在透水障碍,不是造成种子休眠的原因。

中部表面

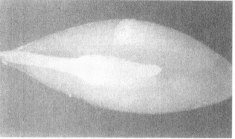
内部胚和胚乳

头部断面

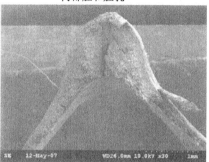
头部发芽孔

图1 '紫梗籽银'桂种子的解剖结构

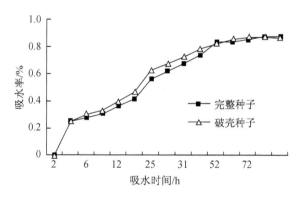

图2 '紫梗籽银'桂种子吸水曲线

2.3 种子不同部位浸提液对白菜种子发芽率和幼苗生长的影响

种子中种皮和种仁浸提液对白菜种子萌发和生长有一定的抑制作用(见表1)。方差分析结果表明,不同部位浸提液的不同配比度对白菜种子的发芽率和对白菜幼苗苗高与根长的影响均达到了显著水平,并且随着浸提液配比度的增大,这种抑制作用也逐渐增强。如经种皮浸提液不同配比度处理后苗高由 1.87 cm 下降到 0.97 cm,根长由 1.99 cm 下降到 1.06 cm,而种仁浸提液不同配比度处理后苗高下降到了 0.54 cm,根长降到 0.23 cm,且幼苗在萌发后大多表现出弯曲、黄化、瘦弱现象。其中种仁浸提液的抑制作用比种皮浸提液要强。而未经处理的对照幼苗生长迅速,根系粗壮。

上述结果表明'紫梗籽银'桂种子的种皮和种仁内均含有一定活性的抑制物质,且种仁内的抑制物质含量要比种皮内的高。

表1 不同配比度'紫梗籽银'桂种子浸提液对白菜种子萌发和幼苗生长的影响

浸提液配比度/%		发芽率/%	苗高/cm	根长/cm
外种皮	30% 原液	75.7	1.04	1.48
	60% 原液	64.5	1.01	1.23
	100% 原液	54.7	0.97	1.06
种仁	30% 原液	72.3	0.80	0.41
	60% 原液	53.7	0.56	0.25
	100% 原液	31.0	0.54	0.23
	CK	86.0	1.87	1.99

3 结论与讨论

3.1 种子基本形态、解剖结构、透水性和休眠的关系

很多研究表明,种皮的构造与休眠有直接的关系,有些植物种子就是由于种皮坚硬、透水性差而导致休眠。如珍珠花种子的外种皮由多层石细胞组成,不仅起着机械限制作用,而且妨碍了种子胚乳和胚对水分的吸收,从而导致种子不能萌发[4]。而核桃种子[5]和南方红豆杉[6]种子皆因其坚硬种皮所导致的机械障碍和透性障碍而休眠。通过对桂花种子解剖结构和透水性实验,表明桂花种子中种皮虽然木质化且较硬,但其头部具有一萌发孔,极大地提高了种皮的透水性,吸水72 h即可达到饱和状态,故种皮坚硬不是造成种子休眠的原因。

3.2 抑制物质与种子休眠的关系

Albertus Magus(1206—1280)最早提出种子休眠与抑制物质有关。研究发现许多种子中存在着抑制发芽的物质,主要分布于果皮(如油菜)、果肉(如西洋参)[7]、外种皮(如青钱柳)[8]、胚乳(如日本女贞)以及胚(如向日葵)等部位[9]。本文通过对'紫梗籽银'桂种皮和种仁浸提液的白菜籽萌发试验结果推断,桂花种子的不同部位,尤其是种仁浸提液中含有发芽抑制物质,随着浸提液配比度由低到高,对白菜种子的发芽率、苗高和根长的影响越来越明显,这说明种子中所含的这些物质能抑制种子的萌发,但这些内源抑制物质是否是导致桂花种子休眠的原因,仍有待进一步研究。

参考文献

[1] 张义,宋春燕.赤霉素浸种与低温层积对桂花种子发芽的影响[J].中国林副特产,2005(6):9-10.
[2] 袁王俊,董美芳,尚富德.利用离体培养技术打破桂花种子休眠的试验[J].南京林业大学学报(自然科学版),2004,28(增刊):91-93.
[3] 沈永宝,郭永清,喻方圆.北美鹅掌楸种子外种皮发芽抑制物研究[J].江苏林业科技,2004,31(6):10-11.
[4] 刘幼琪,洪艳艳,罗颖.珍珠花种子发芽条件的研究[J].湖北大学学报(自然科学版),1999,21(1):

81-83.
[5] 李近雨. 核桃种子烂种原因及发芽条件的研究[J]. 林业科学, 1994, 20(1): 18-24.
[6] 史忠礼, 周菊华. 南方红豆杉种子休眠与萌发的研究[J]. 浙江林业科技, 1991, 11(5): 164.
[7] 韩东, 黄耀阁, 李向高, 等. 西洋参果实中发芽抑制物质——二苯胺的分离鉴定[J]. 吉林农业大学学报, 2001, 23(4): 60-63, 68.
[8] 杨万霞, 方升佐. 青钱柳种皮甲醇浸提液的生物测定[J]. 植物资源与环境学报, 2005, 14(4): 11-14.
[9] 孔祥海. 抑制物质和种子休眠[J]. 龙岩师专学报, 2002, 20(6): 50-52.

Primary Research on Dormancy Mechanism of Seeds of 'Zigeng Zi Yin' Osmanthus

YANG Xiulian, DING Yanfen, GAN Xihua

(College of Landscape Architecture, Nanjing Forestry University, Nanjing 210037, China)

Abstract: In order to discuss the dormancy cause of 'Zigeng Zi Yin' osmanthus seed, the seed coat structure, seed germination ratio, life energy, water permeability, and the germination of *Brassica pekinensis* seed incubated in the soaking fluid of different parts of osmanthus seed were probed into. The results showed that the hardy testa had no effect on the seed dormancy, and there were some contents in the testa and kernel, which maybe restrain the seed's germination.

Key words: 'Zigeng Zi Yin' osmanthus; seed dormancy; germination inhibitor

桂花品种切花瓶插衰老生理研究

杨秀莲,姜晓装,丁彦芬,向其柏

(南京林业大学,江苏 南京 210037)

摘 要:通过试验研究了不同桂花品种切花在开花和衰老过程中花朵的保护酶活性、丙二醛、细胞质膜透性及可溶性蛋白质等的变化规律。结果表明:3个桂花品种花朵的可溶性蛋白质的含量均呈下降趋势,细胞质膜透性和丙二醛含量随花朵的衰老而增大;超氧化物歧化酶活性前期上升后期下降,呈现出明显的峰值;过氧化物酶表现为上升趋势。综合比较各项指标测定结果,3个品种的瓶插寿命长短为:'晚籽银桂'>'多芽金桂'>'四季桂'。保鲜液对延长桂花花期有一定的效果。

关键词:桂花;切花;衰老;生理变化

桂花作为我国传统的名花名木之一,其适应性广,是常绿阔叶树种中较耐寒的一种,而且其开花正值传统佳节——中秋节期间,万点金黄,芬芳飘逸,沁人肺腑。但由于桂花花期较短,一般为5~7 d,最佳观赏期和采花期只有2~4 d,极大地影响了桂花的观赏和采收。

目前,有关桂花开花和衰老机理的研究较多,朱诚等对桂花的乙烯及膜脂过氧化水平和生长调节物质等的变化进行了探讨[1-2],周媛等研究了乙烯与$CaCl_2$对桂花切花衰老的影响[3],陈洪国等研究了不同品种桂花开花进程中花瓣色素、可溶性糖和蛋白质含量的变化[4]。本研究以'多芽金桂''晚籽银桂''四季桂'等品种为材料,从采后瓶插期间花朵内在生理变化入手,分析测定了桂花花朵在衰老过程中的过氧化物酶(POD)及超氧化物歧化酶(SOD)的活性变化、丙二醛(MDA)含量和蛋白质的变化规律,以了解不同品种的瓶插寿命和保鲜液对延长桂花花期的作用,以求为延缓桂花衰老、增加其观赏价值和开发切花资源提供理论依据。

1 材料与方法

1.1 材料

'多芽金桂''晚籽银桂''四季桂'均为南京林业大学校园内栽植,采样树高3~4 m,树龄20~30a,生长健壮。于2006年9月19日清晨剪取位于树体中上部向阳面含苞欲放的花枝,带回实验室。

---

* 原文发表于《林业科技开发》,2007,21(6):29-32。
基金项目:江苏省农业三项工程项目(BK2005133)。

1.2 方法

1.2.1 瓶插方法

剪取长约 45 cm 的花枝,为便于采集花朵将叶片剪除后分别插入装有蒸馏水(文、图中称对照)和保鲜液(30 g·L^{-1}蔗糖＋200 mg·L^{-1} 8-羟基喹啉,保鲜液配方经预试验遴选后确定,文、图中称处理)的花瓶,每品种两种处理各 3 瓶。在室内自然条件下(室温为 25 ℃左右,相对湿度 50%～60%)进行瓶插水养,瓶插液隔天更换。每天中午选取代表当天开花特征的花朵作为测定材料。

1.2.2 生理生化测定方法

可溶性蛋白质含量测定采用考马斯亮蓝法[5]。SOD 活力测定根据邹琦方法[5],稍做改动。POD 活性测定采用愈创木酚法[5]。MDA 的含量根据硫代巴比妥酸的三氯乙酸溶液与 MDA 的显色反应测定[5]。细胞膜透性根据相对电导率测定法[5](雷磁-DSS120 型电导仪):取花朵 0.2 g,适当剪碎,置于 20 mL 双蒸水中,静置 2 h,测其电导值 C_1,然后将其煮沸 15 min,自然冷却到室温后测电导值 C_2,每处理重复 3 次。计算相对电导率:相对电导率＝$(C_1-C)/(C_2-C)\times 100\%$($C$ 为双蒸水的电导率值)。

2 结果与分析

2.1 可溶性蛋白质含量的变化

图 1 显示,3 个品种的蛋白质含量在瓶插期间总体上呈稳定下降趋势,其中'多芽金桂'和'晚籽银桂'在 2 d 内下降较快,而'四季桂'在瓶插第 1 天就出现急剧下降,以后则呈缓慢下降趋势。蛋白质含量的下降表明花朵在瓶插后蛋白质水解作用明显加强。蛋白质的变化趋势与水养寿命有一定的关系,3 个品种中,'晚籽银桂'的蛋白质含量的变化较'多芽金桂'和'四季桂'缓慢,表明'晚籽银桂'的瓶插寿命相对较长。而 3 个品种的对照值均比处理下降快,表明保鲜液可适当改善花朵营养状况,在一定程度上延长花朵的瓶插时间。

2.2 SOD 活性变化

瓶插期间 SOD 活性变化见图 2,3 品种 SOD 活性总体呈先升后降的趋势。瓶插前期 SOD 活性的上升可能是花朵对逆境的一种适应能力,以维持体内各种代谢之间的平衡,但在后期这种平衡遭到破坏,SOD 活性逐渐降低,自身清除氧自由基能力下降,造成细胞内活性氧积累而产生毒害使器官快速衰老。3 个品种中,'多芽金桂'和'四季桂'达到峰值的时间较'晚籽银桂'提前 1 d,表明两者的衰老较快,即 3 品种瓶插寿命是'晚籽银桂'＞'多芽金桂'＞'四季桂'。而对照的 SOD 活性变化较处理快,也说明保鲜液对延缓桂花的衰老有一定的作用。

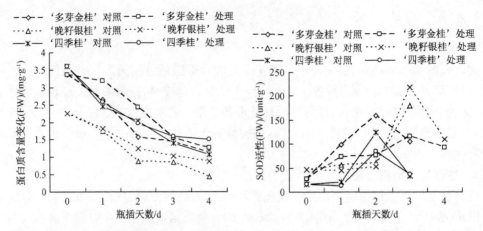

图1 3个品种切花瓶插期间蛋白质含量的变化　图2 3个品种切花瓶插期间SOD活性变化

2.3 POD活性变化

'多芽金桂'瓶插在第1天POD活性有一定的上升,第2天略有下降,随后急剧上升;'晚籽银桂'对照和处理在前3天变化较平缓,第4天急剧上升;而'四季桂'在前2天变化不是很大,第3天POD活性急剧上升(见图3)。由图3可以看出,3个品种对照的POD活性均比处理高,达到峰值的时间比处理提前,说明保鲜液对花朵的衰老有一定的作用。

2.4 MDA含量的变化

MDA为细胞膜脂氧化指标,它既是过氧化产物,又可强烈地与细胞内各种成分发生反应,使多种酶和膜系统遭受严重损伤,MDA含量的不断上升是花朵迅速衰老的标志之一。3个品种花朵中MDA含量在瓶插期间均呈上升趋势(见图4)。由图4可以看出,'多芽金桂'和'四季桂'的MDA含量较高且较为接近,而'晚籽银桂'的MDA含量较低,表明'晚籽银桂'膜脂过氧化程度比'多芽金桂'和'四季桂'低,表现出花期相对较长。3个品种的对照值均比处理要高,也即对照的膜脂氧化程度比处理高,在一定程度上反映出保鲜液可适当延迟花朵的衰老。

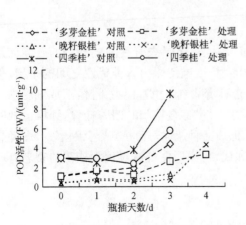

图3 3个品种切花瓶插期间POD活性的变化

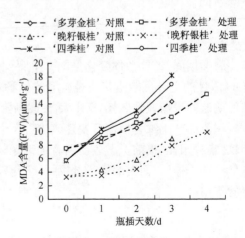

图4 3个品种切花瓶插期间MDA含量的变化

2.5 细胞质膜透性的变化

从图 5 可以看出,随着瓶插时间的延长,花朵的细胞质膜透性逐渐增大。其中'多芽金桂'和'晚籽银桂'在前两天变化缓慢,表明前期细胞膜的伤害较小,第 3 天是一个转折点,此后细胞质膜透性迅速上升,表明细胞膜受到严重伤害,电解质外渗。而'四季桂'在前 1 d 变化平缓,以后快速上升,即细胞膜的伤害比前两者要早。3 个品种的对照值均比处理值大,表明保鲜液对细胞质膜透性的伤害有一定的抑制作用。

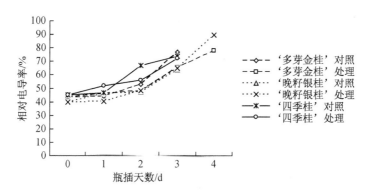

图 5　3 个品种切花瓶插期间细胞质膜相对透性变化

3　结论与讨论

(1) 瓶插寿命。3 个品种的瓶插寿命是'晚籽银桂'>'多芽金桂'>'四季桂'。这在取样的时间上也可反映出来,由于'晚籽银桂'瓶插寿命相对较长,故取样时间分为 5 次,而'多芽金桂'和'四季桂'瓶插寿命相对较短,大部分指标只取样 4 次。

(2) 保鲜液与延缓衰老的关系。本实验中,保鲜液处理后可溶性蛋白质的降解较蒸馏水处理的慢,表明营养状况(蔗糖)与切花的瓶插寿命有密切的关系。SOD、POD 的活性以及细胞质膜透性和 MDA 均比蒸馏水处理的低,说明运用保鲜液可改善桂花瓶插期间的营养状况,并可缓解细胞内由于氧自由基的产生引起的膜脂过氧化及对酶和膜的严重损伤,从而在一定程度上延长观赏期。

(3) 可溶性蛋白与桂花切花衰老的关系。Leshem 等认为细胞组成物质的水解是花朵衰老的主要代谢变化之一,其中蛋白质水解使体内蛋白质含量下降与衰老关系尤为密切[6]。宋纯鹏认为蛋白质含量下降程度可作为衰老的指标[7]。在桂花的瓶插过程中,瓶插前期蛋白质含量下降较慢,中期随着缺水情况的出现,花朵中蛋白质的降解加速,从而加速了切花的衰老。这一结果与陈洪国等[4]的研究结果一致,但与朱诚等[1]的研究结果有一定的差异。这可能是由于处理方法不同造成的,在自然条件下,由于来自植株的营养、水分等供应充足,蛋白质合成大于分解,使得其在花朵盛花期含量达到最大值,以后随着花朵逐渐衰老,蛋白质水解酶的升高致使桂花在开花和衰老期间蛋白质的分解大于合成,从而使蛋白质含量快速降低。而在瓶插过程中,由于花枝脱离母体后吸水受阻,花朵营养供应不足,蛋白质的合成受阻,使分解大于合成,蛋白质的含量变化表现为下降趋势。紫丁香[8]、月季[9]、郁金香[10]等花卉也有相同的表现。

(4) 保护酶与桂花切花衰老的关系。在植物组织中,SOD、POD 等酶促防御系统保护酶在清除活性氧中起着关键作用。SOD 的主要作用是清除超氧化物阴离子自由基,POD 的作用是清除细胞内过多的过氧化物,起到酶系统协调作用,保护膜系统免受伤害,延缓膜脂的过氧化,从而延缓衰老。从试验结果可以看出,SOD 活性呈现先升后降趋势,POD 活性也呈上升趋势,但其峰值的出现晚于 SOD,这两者酶活性的高峰出现次序不同,可能与顺序调节衰老期间自由基的产生有关[11]。

(5) 膜系统与桂花切花衰老的关系。MDA 是由膜脂中的不饱和脂肪酸发生膜脂过氧化作用而产生的,其含量是膜稳定性指标。MDA 含量增多,导致膜的渗漏,从而启动衰老,因此 MDA 与花瓣衰老密切相关。本试验结果可以看出 MDA 含量和细胞质膜透性变化呈逐步上升的趋势,这说明当 MDA 含量积累到一定程度时,就会损害细胞膜,质膜透性的增大引起细胞内代谢紊乱,酶系统遭到破坏,自由基清除功能降低,膜脂过氧化作用加强,最终导致细胞膜完整性破坏,细胞死亡。

参考文献

[1] 朱诚,曾广文.桂花花衰老过程中的某些生理生化变化[J].园艺学报,2000,27(5):356-360.

[2] 朱诚,刘非燕,郭达初,等.桂花开花和衰老过程中乙烯及膜脂过氧化水平初探[J].园艺学报,1998,25(3):27-579.

[3] 周媛,王彩云.乙烯与 $CaCl_2$ 对桂花切花衰老的影响[C]//中国观赏园艺研究进展 2005.北京:中国林业出版社,2005.

[4] 陈洪国,刘顺枝.湖北咸宁地区桂花开花和衰老过程中花瓣的某些生理生化指标变化[J].植物生理学通讯,2006,42(1):112-114.

[5] 邹琦.植物生理生化实验指导[M].北京:中国农业出版社,1997.

[6] Leshem Y Y, Halevy A H, Frenkel C.植物衰老过程和调控[M].胡文玉,译.沈阳:辽宁科学技术出版社,1990.

[7] 宋纯鹏.植物衰老生物学[M].北京:北京大学出版社,1998.

[8] 于凤鸣.紫丁香花开放与衰老中几项生化指标的研究[J].河北职业技术师范学院学报,2000,14(1):36-38.

[9] 权俊萍,闫洁,李荣,等.月季鲜切花瓶插衰老过程中保护酶活性和膜脂过氧化水平初探[J].石河子大学学报(自然科学版),2001,5(1):30-32.

[10] 丁宝莲,孙伟,陶懿伟,等.郁金香切花瓶插期间的衰老生理研究[J].上海农学院学报,1999,17(4):281-284,289.

[11] 孙守家,赵兰勇,于守超,等.平阴玫瑰鲜花花蕾采后衰老生理机制研究[J].林业科学,2004,40(5):79-83.

Studies on Senescence Physiology of *Osmanthus fragrans* Cut-Flowers during Vase Life

YANG Xiulian, JIANG Xiaozhuang, DING Yanfen, Xiang Qibai

(College of Landscape Architecture, Nanjing Forestry University, Nanjing 210037, China)

Abstract: The changes of protective enzyme activity, the cytoplasm membrane permeability, the malondialdehyde(MDA), and soluble proteins in the senescence process of three *Osmanthus fragrans* cultivars cut flowers were determined in a vase life. The results showed that the contents of soluble proteins of three *Osmanthus fragrans* cultivars were decreased, while the cytoplasm membrane permeability and MDA contents increased with the flowers growth. The activities of superoxide dismutase(SOD) increased initially and then decreased, with an apparent peak value. But the activities of peroxidase(POD) increased during the whole flower process. Comprehensively, the vase-life of cultivar 'Duoya Jin gui' was longer than cultivar 'Sijigui' but shorter than cultivar 'Wanzi Yin gui'. It was also found that the keep fresh agents had an effect on prolonging cut flowers life to some extent.

Key words: *Osmanthus fragrans*; cut flower; senescence; physiological change

第三部分
引种育种

NaCl胁迫对5个桂花品种叶片超微结构的影响

杨秀莲,王欣,高树桃,施婷婷,王良桂

(南京林业大学 风景园林学院,江苏 南京 210037)

摘 要:以5个桂花品种为材料,以Hoagland培养液为对照,设计2个NaCl含量(70、100 mmol/L),处理10 d后利用透射电镜和扫描电镜观察各处理不同品种的叶片超微结构特征,以明确桂花品种对耐NaCl胁迫的解剖结构响应机制。结果显示:(1)透射电镜观察发现:随NaCl胁迫程度的加强,5个桂花品种叶肉细胞中叶绿体结构受到不同程度的破坏;70 mmol/L NaCl处理后,5个品种的细胞核基本保持正常,而100 mmol/L NaCl处理后核内染色质发生降解;随着NaCl胁迫程度的加强,5个桂花品种的类囊体片层结构中嗜锇颗粒明显增多;在膜结构方面,'大叶银'桂的叶肉细胞被破坏程度最为严重,叶绿体膜被破坏,叶绿体形状基本不能辨认。(2)扫描电镜观察结果显示:随着NaCl浓度的增大,5个品种叶片表面的气孔密度不断增大,而张开气孔的密度却不断减小,且叶肉细胞体积均缩小;'大叶银'桂、'笑秋风'、'晚籽银桂'的栅栏组织占叶厚的比重随NaCl胁迫浓度的增大而升高,'潢川金桂'和'紫梗籽银'桂的栅栏组织占叶厚的比重则随NaCl胁迫浓度的增大而呈先升高后降低的趋势。研究表明,NaCl胁迫对桂花叶片细胞叶绿体、细胞核等的超微结构会造成损伤,且NaCl胁迫浓度越高损伤越明显。本试验可初步判断'大叶银'桂、'笑秋风'、'晚籽银桂'的耐盐性略微高于'潢川金桂'和'紫梗籽银'桂。

关键词:桂花;NaCl胁迫;细胞;超微结构

土壤盐渍化是一个世界性的资源和生态问题。盐分是影响植物生长的一个重要环境因素。盐分胁迫对植物最普遍和最显著的效应就是抑制生长,盐胁迫会造成植物发育迟缓,抑制植物组织和器官的生长和分化,使植物的发育进程提前[1]。

桂花(Osmanthus fragrans)作为中国传统十大名花之一,是集绿化、美化、香化于一体,观赏与实用兼备的优良园林树种,桂花清可绝尘,浓能远溢,香气扑鼻,含多种香料物质,可用于食用或提取香料,具有极高的观赏价值、药用价值和文化价值[2]。然而,土壤盐渍化限制了桂花在沿海地区的园林绿化应用,研究桂花在盐胁迫条件下的生理响应,有助于筛选耐盐桂花品种在盐渍化地区的绿化美化,杨秀莲[3]等通过综合考虑盐害症状初步确定3个桂花品种的耐盐性强弱排序为:'笑秋风'>'潢川金桂'>'大叶银'桂。

植物的生长与其耐盐性密切相关,研究盐渍条件下植物细胞的超微结构变化,有利于从机理上阐述植物的耐盐特征[4]。已有研究表明,经过一段时间的NaCl胁迫后,部分植物的形态结构会发生变化,如细胞膜脂过氧化程度加剧,细胞内物质大量外渗,膜的稳定性下降,叶绿体色素降低幅度较大,叶绿体结构松散变形,基粒片层模糊不清,失去完整的膜结构

---

* 原发表于《西北植物学报》,2018,38(9):1634-1645。
基金项目:江苏省农业科技自主创新资金(CX(16)1005);江苏高校品牌专业建设工程资助项目(PPZY2015A063)。

等[5]。膜结构的破坏导致植物叶片光合作用减弱,从而影响植物正常生长[6]。

桂花在逆境胁迫下的形态结构变化是桂花抗逆性研究的基础。气孔作为高等陆地植物表皮所特有的结构,是植物与环境气体交换的通道。植物叶片表皮上气孔的闭合、分布以及发育过程对外界环境的适应过程至关重要[7]。因此,本研究借助透射电镜和扫描电镜在不同 NaCl 胁迫条件下对幼苗叶片进行系统观察,比较 5 个桂花品种在 NaCl 胁迫下超微结构的差异,为进一步阐明桂花的 NaCl 胁迫反应机制提供科学依据,也有助于耐盐桂花品种的筛选。

1 材料和方法

1.1 试验材料

'大叶银'桂(O. fragrans 'Daye Yin')、'潢川金桂'(O. fragrans 'Huangchuan Jingui')、'笑秋风'(O. fragrans 'Xiao Qiufeng')、'晚籽银桂'(O. fragrans 'Wanzi Yingui')和'紫梗籽银'桂(O. fragrans 'Zigeng Zi Yin')的一年生实生苗。

1.2 试验方法

1.2.1 幼苗的培养

将不同品种的桂花种子用质量分数为 0.1% 的赤霉素浸泡 24 h[8],于 5 ℃ 的恒温箱中沙藏培养。种子发芽后移栽到小型花盆中,于温室中培养。

1.2.2 NaCl 胁迫处理

待幼苗长至 5~6 片真叶时,挑选长势一致的植株定植于基质为细沙的直径 8 cm、高度 11 cm 的花盆中,每品种 200 株,缓苗生长后进行 NaCl 胁迫处理。具体方法为:在 Hoagland 培养液中分别添加 NaCl,使培养液的最终盐浓度分别为 0、70、100 mmol/L,为避免盐冲击效应,盐浓度每天递增 30 mmol/L,直至预定浓度。处理期间,管理措施保持一致,然后每两天定时、定量按预定盐浓度浇灌一次,浇灌量为持水量的 2 倍,约 2/3 的溶液流出,从而交换以前的积余盐,以保持 NaCl 浓度的恒定。NaCl 胁迫处理 10 d 后取各品种植株顶端第二对鲜嫩小叶进行透射电镜和扫描电镜的样品制备。

1.3 样品的制备与观察

1.3.1 透射电镜观察法

将叶片沿叶脉方向用刀片快速切取大小约 2 mm×2 mm 数块,取下的样品迅速放入 2.5% 戊二醛溶液中,真空泵抽气使材料下沉,4 ℃ 过夜,磷酸缓冲液(pH=7.2)冲洗 3 次,1% 锇酸 4 ℃ 下固定 4 h,不同浓度梯度的乙醇脱水(30%、50%、70%、80%、90%、100%),每次 10~15 min,100% 的 3 次,其他浓度各一次。100% 包埋剂浸透 24 h 后,用 Epon812 环氧树脂包埋,LKB 超薄切片机切片,醋酸双氧铀染色 30 min 和柠檬酸铅染色 15 min 后,于透射电镜(HITACHI7650)下观察、拍照。

1.3.2 扫描电镜观察法

与透射电镜相同取样,用刀片将其切成约 5 mm×5 mm 的小块,快速投入 2.5% 戊二醛中,抽气使组织下沉后,于 4 ℃ 下固定 24 h,磷酸缓冲液冲洗 3 次,乙醇梯度脱水(30%、

50%、70%、80%、90%、100%),丙酮和醋酸异戊酯混合液再次进行脱水处理,然后用纯醋酸异戊酯脱水,临界点干燥后,将样品粘贴在盖玻片上,放入真空喷镀仪进行喷镀。最后用扫描电镜(日立 AMRAY 1000-B6)对桂花叶片表面和断面进行观察、拍照。

1.4 数据处理与分析

运用 SPSS 19.0 和 Excel 2010 进行数据处理和统计分析。采用 Duncan 新复极差测验法($P<0.05$)进行处理间差异显著性检验。

2 结果与分析

2.1 不同 NaCl 胁迫条件下 5 个桂花品种叶肉细胞的超微结构

经过不同浓度 NaCl 胁迫处理后,5 个桂花品种叶肉细胞的超微结构均有一定的变化(表1),正常条件(0 mmol/L NaCl)下,5 个桂花品种叶肉细胞排列疏松,各细胞器结构完整,清晰可见。叶绿体紧贴细胞壁,内部片层清晰,外形为长椭圆形,类囊体基粒清晰,并具有少量的淀粉粒。细胞核核膜清晰,内部染色质均匀分布(图1)。

NaCl 胁迫下,叶绿体超微结构的变化表现出许多类似的共同特征:在 70 mmol/L NaCl 胁迫下,叶绿体轻微肿胀,基质片层轻微扩张,其中'大叶银'桂中嗜饿颗粒增加;在 100 mmol/L NaCl 胁迫下,叶绿体表现出明显的肿胀,其中'紫梗籽银'桂叶肉细胞的细胞器最为稳定,未发生质壁分离现象;而'大叶银'桂被膜破坏,形状基本不能辨认;'笑秋风'线粒体略微肿胀(图2,图3)。

2.2 不同 NaCl 浓度胁迫对 5 个桂花品种叶片表面特征的影响

用扫描电镜观察 5 个桂花品种的幼叶,发现不同浓度 NaCl 处理下,其叶片表面的气孔器密度都有不同程度的变化,而且各品种叶片表面随着 NaCl 浓度的升高出现星芒状的蜡质纹饰,这样的特征有利于减少水分的蒸腾,以适应生境中因含有大量盐分而造成的生理干旱(图4)。

由表2可以看出,'大叶银'桂、'笑秋风'和'紫梗籽银'桂的气孔器密度在 0~70 mmol/L NaCl 时随 NaCl 浓度的升高而增大,在 100 mmol/L NaCl 时气孔器密度明显减小,呈先升高后降低的趋势;说明这 3 个品种在低盐环境下通过增大气孔器的密度减少水分的蒸腾、保持体内的水分以避免生理干旱;当盐浓度提高到 100 mmol/L 时,可能超过了植物的耐受性,气孔器密度明显减小。而'潢川金桂'和'晚籽银桂'的气孔器密度则随盐浓度的升高呈持续增大的趋势,说明这 2 个品种在高盐浓度下依然能够通过增大气孔器密度来抵抗 NaCl 胁迫。可初步判定'潢川金桂'和'晚籽银桂'耐盐性略微高于'大叶银'桂、'笑秋风'和'紫梗籽银'桂。

在盐渍环境下,植物通过减少气孔开张的数量,来减少体内水分的散失,以抵抗 NaCl 胁迫。由表2可以看出,5 个品种开张的气孔密度和开张气孔的百分比均随着盐浓度的升高而呈降低趋势。不同浓度 NaCl 胁迫处理下,气孔开张密度差异显著。说明 5 个品种均无法抵抗盐渍环境的胁迫,而使得叶片表面的气孔不得不闭合,说明这 5 个桂花品种的耐盐性均较弱。

表1 5个桂花品种叶肉细胞的超微结构对不同浓度 NaCl 胁迫的响应
Tab. 1 Influence of NaCl stress on ultrastructure of five O. fragrans varieties leaf cells

品种 Varieties	NaCl 浓度 NaCl concentration /(mmol·L^{-1})	叶绿体 Chloroplast	类囊体 Thylakoid	淀粉粒和嗜锇颗粒 Starch granules and osmiophilic particles	细胞核 Nuclear	是否质壁分离 Plasmolysis or not
'大叶银'桂 O. fragrans 'Daye Yin'	0	紧贴细胞壁排列,呈椭圆状 Chloroplasts cling to cell walls, and is oval	片层整齐排列 The layers neatly arranged	均少量 Presented a little	近一端分布 Near one end	否 No
	70	个别不贴细胞壁排列,由原来的核形肿胀成球形 Some chloroplasts not cling to cell walls, chloroplasts changed from the long oval into around ball	类囊体片层出现轻微扩张 Light expansion happened to the layers	嗜锇颗粒数量增加 Osmiophilic particles increase	无明显变化 No significant change	否 No
	100	被膜破坏,叶绿体形状基本不能辨认;破坏程度最严重 Damaged most seriously, chloroplast membranes are damaged so that the basic shape cannot be recognized	无法辨认 Cannot be recognized	嗜锇颗粒明显增多 Osmiophilic particles increase significantly	部分核膜受损 Partial nuclear membrane damaged	轻微 Slightly
'潢川金桂' O. fragrans 'Huangchuan Jingui'	0	结构完整,紧贴细胞壁排列 The structure was integrated Chloroplasts clung to cell walls	类囊体片层排列规则 The layers neatly arranged	含有少量淀粉粒 Starch granules presented a little	结构完整 The structure was integrity	否 No
	70	出现肿胀现象 Expanded was happened	类囊体片层松散,变形 The layers loosed and deformed	含少量淀粉粒,嗜锇颗粒明显增多 Starch granules presented a little and osmiophilic particles increase significantly	轻微降解 Degraded slightly	否 No
	100	肿胀;膜系统破坏程度不大 Expanded and with a little damage	类囊体片层排列不规则,有扩张现象 The layers arranged in a mess and expanded	出现大的淀粉粒,淀粉粒扩张出现空隙;嗜锇颗粒明显增多 Bigger starch granules appeared and osmiophilic particles increase significantly	染色质降解 Degraded	严重的质壁分离现象 Serious plasmolysis

第三部分 引种育种

续 表

品种 Varieties	NaCl 浓度 NaCl concentration /(mmol·L^{-1})	叶绿体 Chloroplast	类囊体 Thylakoid	淀粉粒和嗜锇颗粒 Starch granules and osmiophilic particles	细胞核 Nuclear	是否质壁分离 Plasmolysis or not
'笑秋风' O. fragrans 'Xiao Qiufeng'	0	结构完整,紧贴细胞壁排列 The structure was integrated and the chloroplasts clung to cell walls	类囊体片层排列有序无扩张现象 The layers neatly arranged and without expanded	均少量 Presented a little	完整 The structure was integrity	否 No
	70	发生肿胀 Expanded	类囊体片层模糊且轻微扩张 The layer appeared to be blurring and expended slightly	嗜锇颗粒明显增多 Osmiophilic particles increase significantly	轻微降解 Degraded slightly	否 No
	100	肿胀,被膜局部被破坏且消失 Expanded and was partially destroyed and disappeared	类囊体片层排列较规则,但有轻微扩张 The layers arranged in order and expanded slightly	嗜锇颗粒明显增多 Osmiophilic particles increase significantly	轻微降解,核内染色质分布不均匀 Degraded slightly and chromatin distribution is uneven	轻微 Slightly
'晚籽银桂' O. fragrans 'Wan zi Yingui'	0	紧贴细胞壁排列,呈狭长的带状 Chloroplasts clung to cell walls and appeared to be long and narrow bands	类囊体片层排列整齐 The layers neatly arranged	少量淀粉粒 Starch granules presented a little	近一端分布 Near one end	否 No
	70	发生肿胀,不紧贴细胞壁排列 Not clung to cell walls and expended	类囊体片层轻微扩张 The layers expanded slightly	少量淀粉粒 Starch granules presented a little	近一端分布,结构正常 Near one end and the structure was integrity	否 No
	100	肿胀成椭球形 Expanded into oval	类囊体片层结构混乱 The layers arranged messy	淀粉粒大量出现;嗜锇颗粒明显增多 Starch granules and osmiophilic particles increase significantly	变化不明显 No significant change	否 No

· 291 ·

续 表

品种 Varieties	NaCl 浓度 NaCl concentration /(mmol·L^{-1})	叶绿体 Chloroplast	类囊体 Thylakoid	淀粉粒和嗜锇颗粒 Starch granules and osmiophilic particles	细胞核 Nuclear	是否质壁分离 Plasmolysis or not
	0	紧贴细胞壁排列 Chloroplasts clung to cell walls	类囊体片层结构完整，排列有序 The structure of the layers were integrity and arranged in order	均少量 Presented a little	完整 The structure was integrity	否 No
'紫梗籽银'桂 O. fragrans 'Zigeng Zi Yin'	70	轻微肿胀 Expanded slightly	类囊体片层局部扩张，不再紧密排列 The layers expanded partially and not arranged separable	淀粉粒少量；嗜锇颗粒明显增多 Starch granules presented a little and osmiophilic particles increase significantly	近一端分布 Near one end	否 No
	100	肿胀变形 Expanded and deformed	类囊体片层结构模糊不清 The layer appeared to be blurring	淀粉粒出现；嗜锇颗粒明显增多 Starch granules appeared and osmicphilic particles increase significantly	轻微降解 Degraded slightly	否 No

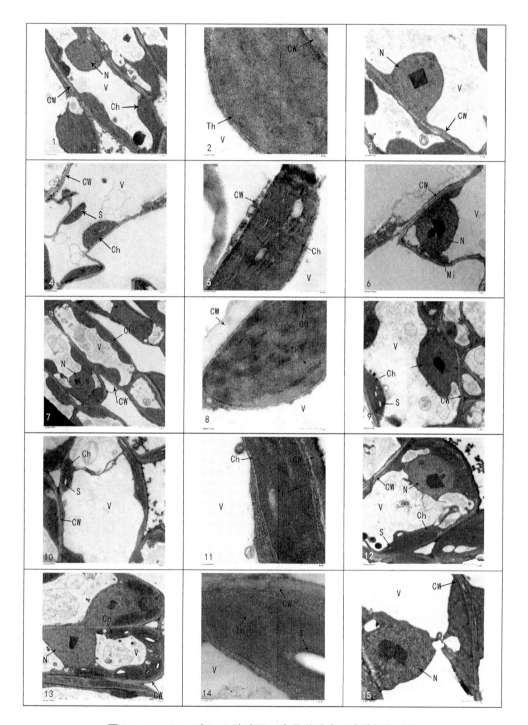

图 1　0 mmol·L⁻¹ NaCl 胁迫下 5 个品种叶肉细胞的超微结构

Ch. 叶绿体；OG. 嗜锇颗粒；Th. 类囊体片层；S. 淀粉粒；N. 细胞核；Cw. 细胞壁；V. 液泡；Mi. 线粒体。
1-3. '大叶银'桂；4-6. '潢川金桂'；7-9. '笑秋风'；10-12. '晚籽银桂'；13-15. '紫梗籽银'桂。

Fig. 1　The ultrastructure of five varieties' leaf cells under 0 mmol·L⁻¹ NaCl

Ch. Chloroplast；OG. Osmiophilic granule；Th. Thylakoid lamella；S. Starch grain；N. Cell nucleus；Cw. Cell wall；V. Vacuole；Mi. Mitochondrion.
1-3. *O. fragrans* 'Daye Yin'；4-6. *O. fragrans* 'Huangchuan Jingui'；7-9. *O. fragrans* 'Xiao Qiufeng'；10-12. *O. fragrans* 'Wanzi Yingui'；13-15. *O. fragrans* 'Zigeng Zi Yin'.

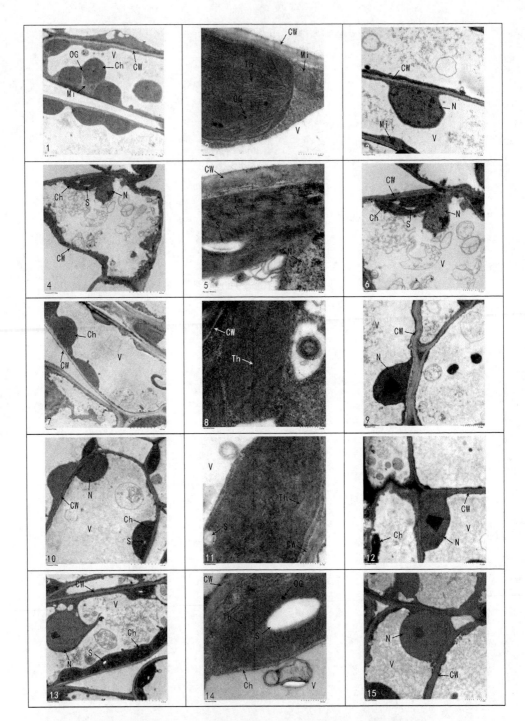

图 2 70 mmol·L^{-1} NaCl 胁迫下 5 个品种叶肉细胞的超微结构

Ch. 叶绿体；OG. 嗜锇颗粒；Th. 类囊体片层；S. 淀粉粒；N. 细胞核；Cw. 细胞壁；V. 液泡；Mi. 线粒体。
1-3. '大叶银'桂；4-6. '潢川金桂'；7-9. '笑秋风'；10-12. '晚籽银桂'；13-15. '紫梗籽银'桂。

Fig. 2 The ultrastructure of five varieties leaf cells under 70 mmol·L^{-1} NaCl

Ch. Chloroplast; OG. Osmiophilic granule; Th. Thylakoid lamella; S. Starch grain; N. Cell nucleus; Cw. Cell wall; V. Vacuole; Mi. Mitochondrion.
1-3. *O. fragrans* 'Daye Yin'; 4-6. *O. fragrans* 'Huangchuan Jingui'; 7-9. *O. fragrans* 'Xiao Qiufeng'; 10-12. *O. fragrans* 'Wanzi Yingui'; 13-15. *O. fragrans* 'Zigeng Zi Yin'.

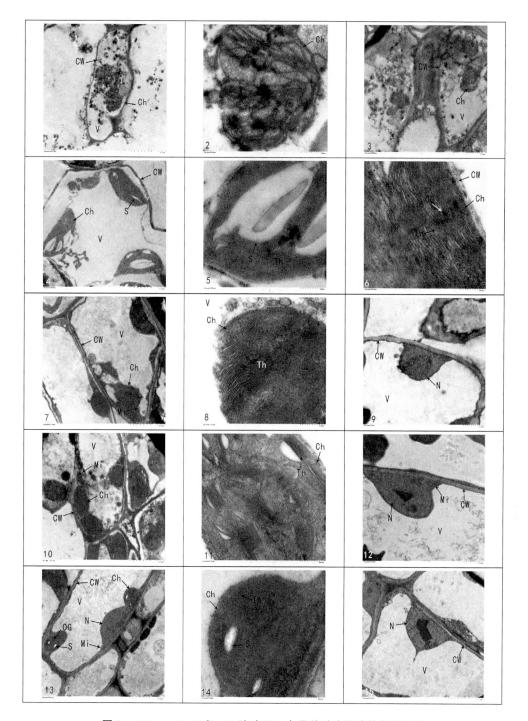

图 3　100 mmol·L^{-1} NaCl 胁迫下 5 个品种叶肉细胞的超微结构
Ch. 叶绿体；OG. 嗜锇颗粒；Th. 类囊体片层；S. 淀粉粒；N. 细胞核；Cw. 细胞壁；V. 液泡；Mi. 线粒体。
1-3.'大叶银'桂；4-6.'潢川金桂'；7-9.'笑秋风'；10-12.'晚籽银桂'；13-15.'紫梗籽银'桂。
Fig. 3　The ultrastructure of five varieties' leaf cells under 100 mmol·L^{-1} NaCl
Ch. Chloroplast; OG. Osmiophilic granule; Th. Thylakoid lamella; S. Starch grain; N. Cell nucleus; Cw. Cell wall; V. Vacuole; Mi. Mitochondrion.
1-3. *O. fragrans* 'Daye Yin'; 4-6. *O. fragrans* 'Huangchuan Jingui'; 7-9. *O. fragrans* 'Xiao Qiufeng'; 10-12. *O. fragrans* 'Wanzi Yingui'; 13-15. *O. fragrans* 'Zigeng Zi Yin'.

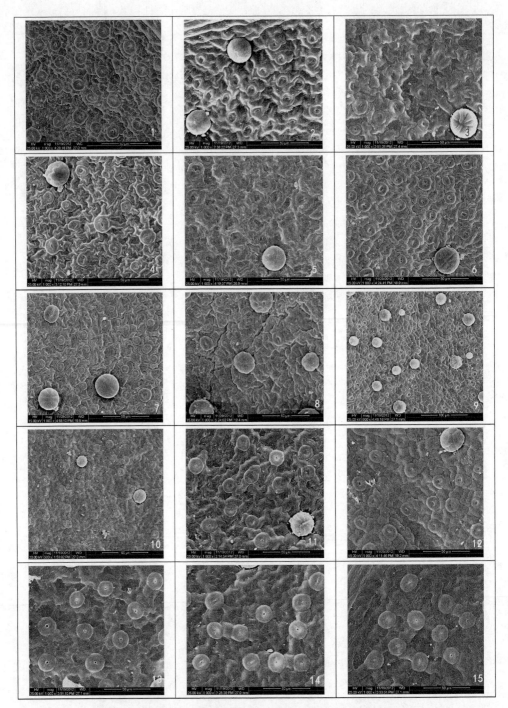

图4 不同浓度NaCl胁迫下5个品种叶片表面的扫描电镜观察

1-3. 分别示0、70、100 mmol·L^{-1} NaCl胁迫下的'大叶银'桂；4-6. 分别示0、70、100 mmol·L^{-1} NaCl胁迫下的'潢川金桂'；7-9. 分别示0、70、100 mmol·L^{-1} NaCl胁迫下的'笑秋风'；10-12. 分别示0、70、100 mmol·L^{-1} NaCl胁迫下的'晚籽银桂'；13-15. 分别示0、70、100 mmol·L^{-1} NaCl胁迫下的'紫梗籽银'桂。

Fig. 4 The scanning electron microscopy of five varieties' leaf surface in different concentration of NaCl

1-3. Shows *O. fragrans* 'Daye Yin' in 0、70、100 mmol·L^{-1} NaCl；4-6. Shows *O. fragrans* 'Huangchuan Jingui' in 0、70、100 mmol·L^{-1} NaCl；7-9. Shows *O. fragrans* 'Xiao Qiufeng' in 0、70、100 mmol·L^{-1} NaCl；10-12. Shows *O. fragrans* 'Wanzi Yingui' in 0、70、100 mmol·L^{-1} NaCl；13-15. Shows *O. fragrans* 'Zigeng Zi Yin' in 0、70、100 mmol·L^{-1} NaCl.

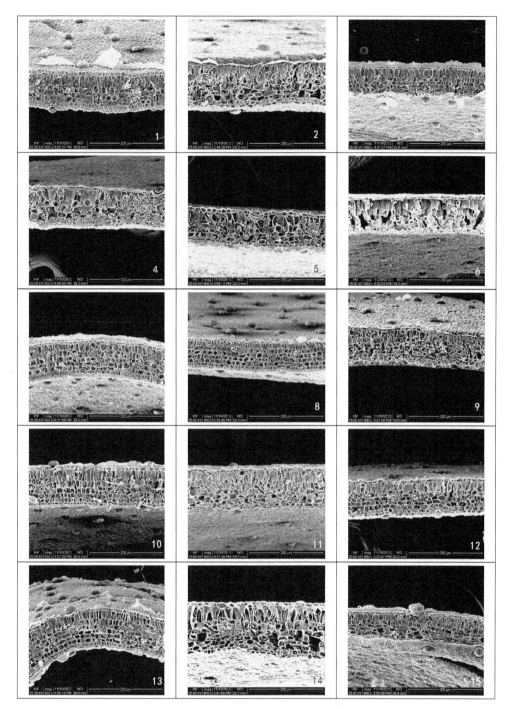

图 5 不同浓度 NaCl 胁迫下 5 个品种叶片断面的扫描电镜观察

1-3. 分别示 0、70、100 mmol·L⁻¹NaCl 胁迫下的'大叶银'桂；4-6. 分别示 0、70、100 mmol·L⁻¹NaCl 胁迫下的'潢川金桂'；7-9. 分别示 0、70、100 mmol·L⁻¹NaCl 胁迫下的'笑秋风'；10-12. 分别示 0、70、100 mmol·L⁻¹NaCl 胁迫下的'晚籽银桂'；13-15. 分别示 0、70、100 mmol·L⁻¹NaCl 胁迫下的'紫梗籽银'桂。

Fig. 5 The scanning electron microscopy of five varieties' leaf section in different concentration of NaCl

1-3. Shows *O. fragrans* 'Daye Yin' in 0、70、100 mmol·L⁻¹NaCl；4-6. Shows *O. fragrans* 'Huangchuan Jingui' in 0、70、100 mmol·L⁻¹NaCl；7-9. Shows *O. fragrans* 'Xiao Qiufeng' in 0、70、100 mmol·L⁻¹NaCl；10-12. Shows *O. fragrans* 'Wanzi Yingui' in 0、70、100 mmol·L⁻¹NaCl；13-15. Shows *O. fragrans* 'Zigeng Zi Yin' in 0、70、100 mmol·L⁻¹NaCl.

表2 不同NaCl浓度胁迫对5个品种叶片表面气孔的影响
Tab. 2 The effect on leaf surface pores of five varieties in different concentration of NaCl

品种 Varieties	浓度 Concentration /(mmol·L^{-1})	气孔器密度 Stomata density /(个·mm^{-2})	气孔开张密度 Open stomata density /(个·mm^{-2})	气孔开张百分比 Open stomata percentage/%
'大叶银'桂 O. fragrans 'Daye Yin'	0	7 200	6 400	88.89
	70	8 400	400	4.76
	100	6 000	0	0.00
'潢川金桂' O. fragrans 'Huangchuan Jingui'	0	7 200	6 000	83.33
	70	8 000	400	5.00
	100	11 600	0	0.00
'笑秋风' O. fragrans 'Xiao Qiufeng'	0	6 800	6 400	94.12
	70	14 000	1 600	11.43
	100	7 600	400	5.26
'晚籽银桂' O. fragrans 'Wanzi Yingui'	0	2 800	2 800	100.00
	70	5 600	1 600	28.57
	100	7 600	400	5.26
'紫梗籽银'桂 O. fragrans 'Zigeng Zi Yin'	0	4 800	5 200	92.86
	70	5 200	2 000	38.46
	100	4 400	800	18.18

2.3 不同NaCl浓度胁迫对5个桂花品种叶片断面特征的影响

随着盐浓度的升高,5个品种的叶片厚度都呈下降趋势,盐浓度越大,叶片厚度越小。随着盐浓度的升高,'大叶银'桂、'潢川金桂'、'紫梗籽银'桂的栅栏组织厚度呈先升高后降低的趋势;'笑秋风'和'晚籽银桂'则一直呈降低趋势。'大叶银'桂、'笑秋风'、'晚籽银桂'的栅栏组织占叶厚的比重随盐浓度的升高而增加,在70~100 mmol/L NaCl之间,增加幅度分别为32.84%、1.48%、6.39%;'潢川金桂'和'紫梗籽银'桂的栅栏组织占叶厚的比重则随盐浓度的升高而呈先增后降的趋势,在70~100 mmol/L NaCl之间,降低幅度分别为2.49%、10.73%。由图5可以看出,在正常无NaCl胁迫情况下,5个桂花品种叶片均由表皮、叶肉和叶脉三部分构成,为典型的背腹型叶。表皮细胞排列紧密,上表皮细胞较大,多数长方形。其下是叶肉组织,分化为栅栏组织和海绵组织。由2~5层柱形细胞构成的栅栏组织,排列整齐紧密,含叶绿体较多。栅栏组织下为排列疏松的海绵组织,其中含叶绿体较少,海绵组织的厚度及疏松度在不同品种间存在差异。叶肉组织下面为下表皮,细胞较小,排列不规则。在70 mmol/L NaCl胁迫下,'大叶银'桂和'紫梗籽银'桂的海绵组织细胞体积增大,细胞间隙也增大;'晚籽银桂'和'紫梗籽银'桂的叶肉细胞形态开始变得不规则;'笑秋风'的变化不明显。在100 mmol/L NaCl胁迫下,5个品种桂花的叶肉细胞体积均缩小;'大叶银'桂和'紫梗籽银'桂的栅栏组织细胞和海绵组织细胞经历了减小—增大—减小的过程,说明细胞在短时间失水收缩后体积迅速增大,以适应盐分环境,而长时间胁迫导致细胞再次失水收缩。

3 讨论

本文的结果表明,经过不同浓度NaCl胁迫处理后,5个桂花品种叶肉细胞的超微结构均有一定的变化,但叶肉细胞内各细胞器对NaCl胁迫的敏感性和耐受性不同。5个品种在正常条件(0 mmol/L NaCl)下,叶肉细胞超微结构基本不受影响,在中、高强度的NaCl胁迫下,5个品种叶肉细胞受到不同程度的伤害,尤其以叶绿体、线粒体和生物膜系统的表现较明显。

植物细胞正常的新陈代谢都是通过膜系统完成的,而生物膜是环境胁迫对植物造成伤害的原位,也是酶催化反应在整个细胞区域能有序进行的结构基础[4]。盐胁迫引起细胞器和膜结构的破坏,是植物受到盐害的共同特征,其破坏程度取决于植物耐盐性的强弱。凡是耐盐性强的品种,其细胞膜系统及细胞器的稳定性高,破坏程度小[9]。目前,盐胁迫的发生机理有2个:一是渗透胁迫,在盐胁迫下的植物细胞内渗透势大于细胞外渗透势,细胞失水导致植物缺水形成生理干旱;二是离子毒害作用,不同离子(Na^+、K^+、Cl^-等)过量渗入细胞后,使原生质凝集,蛋白质合成受到抑制且水解作用加强,造成氨基酸积累。这些氨基酸又会转化为丁二胺、戊二胺等,达到一定浓度时,细胞就会中毒死亡。而且,这些离子(Na^+、K^+、Cl^-等)的存在使得一些低浓度的矿质营养元素供应不足,降低了呼吸作用的强度,因此植物生长受抑。本试验中,NaCl浓度为70 mmol/L处理下,5个桂花品种的超微结构均出现明显变化,叶肉细胞的细胞核结构基本维持正常;而在100 mmol/L NaCl的条件下,细胞核内染色质发生凝聚,并聚集至核膜附近。推测染色质的凝聚可能是由于NaCl胁迫改变了细胞核内生物大分子(如核酸)的结构等,从而在分子水平上造成DNA损伤,对细胞核造成不可逆的损害。

嗜锇体在电镜下观察为黑色颗粒,是强烈吸收锇酸的一些脂类物质聚集在一起形成的液滴,是叶绿体脂类物质的贮存库[10]。本试验中桂花叶肉细胞内叶绿体通过淀粉粒数目增多、体积增大,嗜锇体数量增多,基粒片层结构变松散,细胞器减少等方式对NaCl胁迫进行反应,这与前人的研究结果相似[11-15]。其中淀粉粒增多主要是由于光合作用的下降,光合产物以淀粉的形式沉积,也可能与高浓度盐碱阻碍了淀粉的水解和向外运输有关[13]。一些研究发现,盐胁迫下淀粉粒数量增加、体积增大,一方面可以水解释放能量,补充因胁迫损伤造成的能量供给不足;另一方面可以升高渗透压,提高细胞液浓度,保证细胞水分正常吸收,从而缓解细胞"生理干旱"[16]。

植物生长在复杂的自然界环境中,既接受周围环境提供的生长所需条件,同时也承受着不良因素的影响。植物为了适应不良环境,会在生理及结构上做出适应性调整[17]。已有的研究结果表明,盐胁迫会减少植物叶片面积,改变叶片的组织结构,增加叶片的气孔密度,降低植物的气孔导度和净光合速率[18],这与本文的观察结果相似。光合作用是植物基本的生命活动,也是对逆境影响最敏感的生理过程之一。一般认为,盐胁迫导致光合作用下降的原因是气孔关闭导致的气孔限制[12]。气孔是植物光合作用气体进出和蒸腾作用水分散失的主要通道,气孔密度大小和气孔开张度对植物的光合作用和蒸腾作用都有重要的影响。通常为减少蒸腾水分散失,同时保证光合所需的CO_2,旱生和盐生植物的上表皮气孔密度较低,气孔多处于关闭状态,而下表皮的气孔密度较大,气孔开张度较大[19]。气孔密度越大,吸收CO_2的量就越多,从而提高植物的光合速率[20]。本研究发现随着NaCl胁迫浓度的增大,5个桂花品种叶表皮气孔密度较对照组有显著增加,而气孔开张的密度却不断减小。同时气孔的长度和宽度有所减少,说明桂花在NaCl胁迫环境中有相应的调整外部结构的适应能力,这与前人的研究结果相似[17]。

毛桂莲等[13]的研究结果表明,栅栏组织在植物适应逆境环境过程中起着重要作用,随着盐胁迫浓度的增加,植物栅栏组织结构紧密度下降,层数减少。组成栅栏组织的栅栏细胞上有大量的叶绿体,是植物进行光合同化作用的主要发生部位,因而栅栏组织越发达的植株光合效率越高。同时栅栏组织还有利于叶片内物质的横向运输[19]。又由于栅栏细胞垂直于叶面和维管组织排列,在充分利用光能的同时,有效降低了蒸腾面积,因而栅栏组织在叶肉中比例越高,越有利于防止其蒸腾作用造成的水分过度散失,保证正常代谢。本研究中,'大叶银'桂、'笑秋风'、'晚籽银桂'的栅栏组织占叶厚的比重随盐浓度的升高而增大,'潢川金桂'和'紫梗籽银'桂的栅栏组织占叶厚的比重则随盐浓度的升高而呈先升后降的趋势,由此可初步判断'大叶银'桂、'笑秋风'、'晚籽银桂'的耐盐性略微高于'潢川金桂'和'紫梗籽银'桂。

盐胁迫条件下,植物自身的形态结构变化比较复杂。明确植物结构在盐胁迫下的变化有利于认识植物的耐盐机理。本研究为后续桂花的抗逆性研究提供一定价值的参考。

参考文献

[1] 李彦,张英鹏,孙明,等. 盐分胁迫对植物的影响及植物耐盐机理研究进展[J]. 中国农学通报,2008(01):258-265.
LI Y, ZHANG Y P, SUN M, et al. Research advance in the effects of salt stress on plant and the mechanism of plant resistance[J]. Chinese Agricultural Science Bulletin, 2008(01):258-265.

[2] 梁凭. 桂花在园林绿化及园林造景中的应用[J]. 现代园艺,2015(16):115.
LIANG P. Application of *Osmanthus fragrance* in landscaping and garden landscaping[J]. Xiandai Horticulture, 2015(16):115.

[3] 杨秀莲,母洪娜,郝丽媛,等. 3个桂花品种对NaCl胁迫的光合响应[J]. 河南农业大学学报,2015(02):195-198.
YANG X L, MU H N, HAO L Y, et al. Photosynthetic response of three *Osmanthus fragrans* cultivars to NaCl stress[J]. Journal of Henan Agricultural University, 2015(02):195-198.

[4] 苏芳莉,李海福,陈曦,等. 盐胁迫对芦苇细胞超微结构的影响[J]. 西北植物学报,2011(11):2216-2221.
SU F L, LI H F, CHEN X, et al. Effect of salt on the ultrastructure of reed cell[J]. Acta Botanica Boreali-Occidentalia Sinica, 2011(11):2216-2221.

[5] 付晴晴,谭雅中,翟衡,等. NaCl胁迫对耐盐性不同葡萄株系叶片活性氧代谢及清除系统的影响[J]. 园艺学报,2018(01):30-40.
FU Q Q, TAN Y Z, QU H, et al. Effects of salt stress on the generation and scavenging of reactive oxygen species in leaves of grape strains with different salt tolerance[J]. Acta Horticulturae Sinica, 2018(01):30-40

[6] 岳健敏,张金池,尤焱煌,等. 油菜素内酯对盐胁迫刺槐苗光合作用及叶绿体超微结构的影响[J]. 西北农林科技大学学报(自然科学版),2017(10):56-66.
YUE J M, ZHANG J C, YOU Y H, et al. Effects of brassinosteroids on photosynthesis and ultrastructure of chloroplasts in *Robinia pseudoacacia* seedlings under salt stress[J]. Journal of

Northwest A&F University(Nat. Sci. Ed.), 2017(10):56-66.

[7] 史婵,杨秀清,闫海冰. 盐胁迫下唐古特白刺叶片的扫描电镜观察[J]. 山西农业大学学报(自然科学版),2017(01):35-39.
SHI C, YANG X Q, YAN H B. Microscopic structure of leaves in *Nitraria tangutorum* under salinity stress[J]. J. SHANXI AGRIC, UNIV. (Natural Science Edition) 2017(01):35-39.

[8] 杨秀莲,郝其梅. 桂花种子休眠和萌发的初步研究[J]. 浙江林学院学报,2010,27(02):272-276.
YANG X L, HAO Q M. Dormancy and germination of *Osmanthus fragrans* seeds[J]. Journal of Zhejiang Forestry, 2010,27(02):272-276.

[9] 刘爱峰,赵檀方,段友臣. 盐胁迫对大麦叶片细胞超微结构影响的研究[J]. 大麦科学,2000(03):20-22.
LIU A F, ZHAO T F, DUAN Y C. Effects of salt stress on the ultrastructure of barley leaf cells[J]. Barley Science,2000(03):20-22.

[10] 环姣姣,芦治国,韩路弯,等. 盐胁迫对'中山杉405'及其亲本叶绿体结构的影响[J]. 林业科技开发,2014(04):58-62.
HUAN J J, LU Z G, HUAN L W, et al. Effects of mixed salts solution stress on chloroplast submicroscopic structure of *Taxodium* 'Zhongshanshan 405' and its parents[J]. China Forestry Science and Technology, 2014(04):58-62.

[11] 易立,胡晓颖,韦霄,等. 遮荫对南美蟛蜞菊、蟛蜞菊及其自然杂交种叶片显微结构及叶绿体超微结构的影响[J]. 广西植物,2014(01):19-26.
YI L, HU X Y, WEI X, et al. Effects of shade on the leaf microstructure and chloroplast ultrastructure of the invasive *Wedelia trilobata*, the native *W. chinensis* and their hybrid[J]. Guihaia, 2014(01):19-26.

[12] 杨凤军,李天来,臧忠婧,等. 等渗NaCl、干旱胁迫对番茄幼苗光合特性及叶绿体超微结构的影响[J]. 应用生态学报,2017(08):2588-2596.
YANG F J, LI T L, ZANG Z J, et al. Effects of isotonic NaCl and drought stress on photosynthetic characteristics and chloroplast ultrastructure of tomato seedlings[J]. Chinese Journal of Applied Ecology, 2017(08):2588-2596.

[13] 毛桂莲,梁文裕,王盛,等. 碱性盐胁迫对宁夏枸杞生长、结构及光合参数的影响[J]. 干旱地区农业研究,2017(04):236-242.
MAO G L, LIANG W Y, WANG S, et al. Effects of alkali stress on growth, structure and photosynthetic parameters *of Lycium barbarum* L. [J]. Agricultural Research in the Arid Areas, 2017(04):236-242.

[14] Keiper F J, Chen D M, De Filippis L F. Respiratory, photosynthetic and ultrastructural changes accompanying salt adaptation in culture of *Eucalyptus microcorys*[J]. Journal of Plant Physiology, 1998,152(4):564-573

[15] Fatma B, J S J, Issam N, et al. Changes in chloroplast lipid contents and chloroplast ultrastructure in *Sulla carnosa* and *Sulla coronaria* leaves under NaCl stress.[J]. Journal of plant physiology, 2016:32-38.

[16] 解植彩,张文晋,张新慧. 不同程度盐胁迫对乌拉尔甘草叶片亚显微结构的影响[J]. 时珍国医国药,2017(05):1200-1204.
XIE Z C, ZHANG W J, ZHANG X H. Effects of different level of salt stress on ultrastructure in leaves of *Glycyrrhiza uralensis*[J]. Li Shizhen Medicine and Materia Medica Research, 2017(05):1200-1204.

[17] 罗祺圆,许涵杰,郭鹏. 短时盐胁迫对紫花苜蓿叶片气孔特征的影响[J]. 天津农业科学,2016(04):17-21.
LUO Q Y, XU H J, GUO P. Effect of short-time salt stress on stomatal feature of leaves of alfalfa

[J]. Tianjin Agricultural Sciences, 2016(04):17-21.

[18] 王碧霞,曾永海,王大勇,等. 叶片气孔分布及生理特征对环境胁迫的响应[J]. 干旱地区农业研究, 2010(02):122-126.
WANG B X, ZENG Y H, WANG D Y, et al. Responses of leaf stomata to environmental stresses indistribution and physiological characteristics[J]. Agricultural Research in the Arid Areas, 2010 (02):122-126.

[19] 左凤月. 盐胁迫对3种白刺生长、生理生化及解剖结构的影响[D]. 重庆:西南大学, 2013.
ZUO F Y. The Effect of NaCl treatment on the seed germination, the seedling growth, the leaf physiology and biochemistry and the anatomical structures of leaf and stem of 3 Nitraria species[D]. Chongqing: Southwest University, 2013.

[20] 杨国,罗洁,林雅晨,等. 盐胁迫对两种吊兰光合特性和气孔形态的影响[J]. 湖北农业科学, 2016 (19):4982-4986.
YANG G, LUO J, LIN Y C, et al. Effects of salt stress on photosynthetic characteristics and stoma structure of two kinds of bracketplants[J]. Hubei Agricultural Science, 2016(19):4982-4986.

Effect of NaCl Stress on the Leaf Ultrastructure of Five *Osmanthus fragrans* Varieties

YANG Xiulian, WANG Xin, GAO Shutao, SHI Tingting, WANG Lianggui

(College of Landscape Architecture, Nanjing Forestry University, Nanjing 210037, China)

Abstract: In this study, 5 *Osmanthus fragrans* varieties were used as materials, with Hoagland culture liquid as control, 2 NaCl content (70 and 100 mmol/L) were designed, and the ultrastructure characteristics of different varieties were observed by TEM and SEM after 10d treatment, to clarify the anatomical structure response mechanism of *Osmanthus fragrans* cultivars to NaCl stress tolerance. The results showed that (1)TEM analysis showed that the chloroplast structure in mesophyll cells of 5 varieties are destroyed in different degrees with the increase of NaCl concentration; the nucleus structure of 5 varieties maintain normal in 70 mmol/L NaCl; in 100 mmol/L NaCl, chromatin within the nucleus begins to degrade. With the NaCl stress strengthening, osmiophilic particles in the thylakoid lamellar structure of the 5 varieties increase significantly. In terms of membrane structures, the mesophyll cells of 'Daye Yin' are damaged most seriously, chloroplast membranes are damaged so that the basic shape cannot be recognized. (2)SEM photos indicated that the stomatal density on leaf surface of 5 varieties increases gradually while the openned porosity density decrease with the increase of salt concentration. The proportion of palisade tissue accounting for leaf thickness of 'Daye Yin', 'Xiao Qiufeng' and 'Wanzi Yingui' increases at first and then decreases with the increase of salt concentration of 'Huangchuan Jingui' and 'Zigeng Zi Yin'. It proves that the higher the salt concentration was, the more obvious of the damage effect was. This experiment can preliminarily judge that the salt stress tolerance of 'Daye Yin', 'Xiao Qiufeng' and 'Wanzi Yingui' is slightly higher than that of 'Huangchuan Jingui' and 'Zigeng Zi Yin'.

Key words:*Osmanthus fragrans*; NaCl stress; cell; ultrastructure

^{60}Co-γ 辐射对桂花种子萌发及幼苗生长的影响*

耿兴敏,王良桂,李娜,杨秀莲

(南京林业大学 风景园林学院,江苏 南京 210037)

摘 要:为探讨辐射对桂花种子的诱变作用,以不同剂量的^{60}Co-γ射线对7个桂花品种的种子进行辐射处理,观察其对桂花种子萌发及出苗状况的影响。结果表明,辐射处理对桂花种子萌发状况的影响存在显著的品种间差异。除'籽银桂'外,其他6个品种种子发芽率均与辐射剂量呈负相关,随着辐射剂量的增加,6个桂花种子的相对发芽率和出苗率明显降低。6个桂花种子的半致死剂量大小顺序依次为'潢川金桂'>'紫梗籽银'桂>'短柄籽银'桂>'米花籽银'桂>'金盏碧珠'>'宽叶籽银'桂,分别为267、151、146、135、98和55 Gy。低剂量(50 Gy)的辐射能显著提高'潢川金桂'、'短柄籽银'桂和'金盏碧珠'3个品种种子的萌发率及出苗率;高剂量(200 Gy)辐射后种子虽能萌发,但出苗率受到明显的抑制,尤其是250 Gy时,除了不受辐射剂量影响的'籽银桂'外,其余6个品种出苗率均为0。本研究结果可为桂花辐射育种提供依据,促进桂花育种进程。

关键词:辐射;桂花;诱变育种;种子萌发;幼苗生长

桂花(*Osmanthus fragrans* Lour.)是我国十大传统名花之一,栽培历史悠久,是一种集绿化、美化、香化于一体,观赏和实用兼备的优良园林树种。长期的演化过程中,在自然选择和人工选择的双重作用下,桂花形成了丰富的变异类型,至今已有160多个品种。目前,中国已有23个市县把桂花定为省(区)花、市花、县花[1],可见,桂花在人们心中占据重要地位。但桂花花朵小,花粉量少,柱头小,因此杂交育种困难。大多数栽培品种的桂花因子房退化而不能结实,致使桂花的种子不易获得,且桂花种子具有休眠特性,需层积很长时间才能萌发[2],这些都影响了桂花杂交育种的开展。

辐射诱变育种是利用γ射线等诱发植物基因突变和染色体畸变,获得有价值的突变体,从而育成优良品种[3]。辐射诱变育种在我国观赏植物育种中广泛应用,并取得显著成果,但在桂花新品种选育中还未见应用。辐射剂量的选择是辐射诱变育种成功的关键因素之一[4],本研究以桂花种子为供试材料,观察^{60}Co-γ射线不同辐射剂量对7个桂花品种种子的萌发状况及出苗率的影响,以期为其辐射诱变育种适宜剂量的选择提供依据,加快桂花的育种进程,促进新花色、大花、抗性强的桂花新品种的选育。

---

* 原文发表于《核农学报》,2016,30(2):0216-0223。
基金项目:江苏省高校优势学科建设工程资助项目(PAPD);"十二五"科技支撑林木种质资源发掘与创新利用(2013BA001B)。

1 材料与方法

1.1 材料

2013年5月在南京林业大学校内采集7个品种的桂花种子,其品种分类、品种名及学名见表1。

表 1 桂花种子品种分类和品种名
Tab. 1 List of *Osmanthus fragrans* variety classification and scientific name

分类 Classification	品种名 Varieties	学名 Scientific name
银桂	'籽银桂' 'Zi Yingui'	*O. fragrans* 'Zi Yin gui'
	'紫梗籽银'桂 'Zigeng Zi Yin'	*O. fragrans* 'Zigeng Zi Yin'
	'米花籽银'桂 'Mihua Zi Yin'	*O. fragrans* 'Mihua Zi Yin'
	'短柄籽银'桂 'Duanbing Zi Yin'	*O. fragrans* 'Duanbing Zi Yin'
	'宽叶籽银'桂 'Kuanye Zi Yin'	*O. fragrans* 'Kuanye Zi Yin'
	'金盏碧珠' 'Jinzhan Bizhu'	*O. fragrans* 'Jinzhan Bizhu'
金桂	'潢川金桂' 'Huangchuan Jingui'	*O. fragrans* 'Huangchuan Jingui'

1.2 辐射处理

采种后适当处理,阴干后用自封袋装好,5月14日于江苏省农业原子能农业利用研究所进行辐射处理。钴源放射剂量率为 1.3 Gy·min^{-1},辐射剂量分别为:0(CK)、50、100、150、200、250 Gy。每个处理区300粒种子(仅'紫梗籽银'桂每个处理区100粒),设置3个重复。

辐射后,立即用0.1%赤霉素浸泡24 h,浸泡后取出种子,于4 ℃冷藏室沙藏,沙子要求手握成团,松手即散开[5]。种子沙藏后每隔2周翻动1次。

1.3 沙藏期间种子萌发状况的统计

7个桂花品种种子在低温沙藏后,分别在30、60、90、120 d后统计种子的发芽率。胚根突破种皮,有露白现象即为发芽。用游标卡尺测定胚根的长度,统计胚根的长度在≤0.5 cm、0.5~1.0 cm、≥1.0 cm各个区间的种子数目。

1.4 半致死剂量的计算及桂花品种辐射敏感型的划分

参照周小梅等[6]的方法,以沙藏120 d后种子的发芽率来计算相对发芽率,即各辐射处理区发芽率占对照区发芽率的百分数。

1.4.1 半致死剂量(LD$_{50}$)的计算方法

不同品种的桂花辐射后出苗率是确定辐射损伤效应的重要指标,是计算半致死剂量的依据。半致死剂量既是确定辐射敏感性的主要指标,也是辐射育种适宜引变剂量的参考。种子的出苗率为对照50%时的辐射剂量为半致死剂量。利用以下公式计算桂花种子的半致死剂量。

$$b = \frac{\Sigma xy - \frac{\Sigma x \cdot \Sigma y}{N}}{\Sigma x^2 - \frac{(\Sigma x)^2}{N}} \quad (1)$$

$$a = \frac{\Sigma y - b \cdot \Sigma x}{N} \quad (2)$$

$$LD_{50} = \frac{50 - a}{b} \quad (3)$$

式中：x 为辐射剂量；y 为不同剂量下相对出苗率×100；LD_{50} 为所求半致死剂量。

1.4.2 相关关系的确定

相关系数反映了不同品种桂花的相对出苗率与辐照剂量的相关关系。利用相关系数公式来计算相对出苗率与辐照剂量的相关系数。

$$r = \frac{\Sigma xy - \frac{\Sigma x \cdot \Sigma y}{N}}{\sqrt{\Sigma x^2 - \frac{(\Sigma x)^2}{N}} \cdot \sqrt{\Sigma y^2 - \frac{(\Sigma y)^2}{N}}} \quad (4)$$

1.4.3 辐射敏感型的划分

半致死剂量是确定辐射敏感性的主要指标。桂花品种辐射敏感型的划分方法，是以不同辐射剂量处理的各个品种相对出苗率所求出的半致死剂量为依据，并利用以下公式计算相对发芽率和半致死剂量：

$$\bar{x} = \frac{\Sigma f(x)}{N}, \quad s = \sqrt{\frac{\Sigma f(x - \bar{x})^2}{N - 1}} \quad (5)$$

根据各品种半致死剂量的平均值和标准差，将半致死剂量的平均值加上1个标准差以上的半致死剂量划分为迟钝型；以半致死剂量平均值减1个标准差以下的半致死剂量值划分为敏感型；其余的为中间型。

1.5 辐射处理对不同桂花品种出苗率的影响

沙藏4个月后，取出已发芽的种子，小心洗净细砂以免弄伤种子的胚根，在2013年9月16日至18日播种，土壤基质按原土∶草炭∶沙子＝1∶1∶0.5体积配制。播种3个月后，统计种子的出苗率，2片子叶完全展开被视为出苗。

1.6 数据处理

试验数据用SPSS 17.0软件进行方差分析，试验数据取3次数据的平均值，在对照与各辐射剂量处理区之间进行多重比较（$P<0.01$）。

2 结果与分析

2.1 辐射处理对沙藏期间桂花种子萌发状况的影响

由图1可知，各个桂花品种在沙藏期间种子萌发速度存在显著品种间差异。'短柄籽银'桂、'潢川金桂'和'籽银桂'在沙藏30 d后就有种子萌发,其他4个品种在沙藏60 d统计时才看到种子萌发,之后各品种的种子萌发率均随着沙藏期间的延长不断提高。沙藏120 d后,'金盏碧珠'种子的萌发率最低,对照区种子萌发率仅有9.0%,'宽叶籽银'桂和'米花籽银'桂种子发芽率较高,达80%以上,其他几个种子为40%～50%。

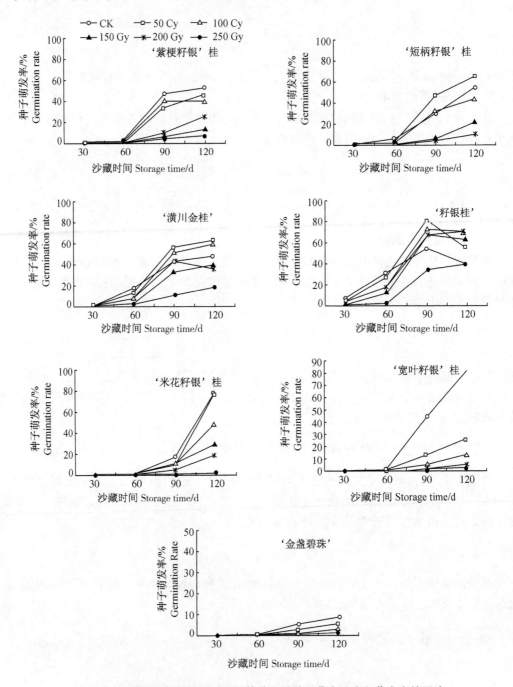

图1　不同剂量的辐射处理对不同桂花品种种子萌发速度和萌发率的影响
Fig. 1　Effects of different doses of radiation on germination speed
and rate of *Osmanthus fragrans* seeds

辐射处理对不同桂花品种种子的萌发率的影响存在品种间差异(图1、表2)。'紫梗籽银'桂、'宽叶籽银'桂、'米花籽银'桂和'金盏碧珠'4个品种随着辐射剂量的提高,种子萌发速度和萌发率显著下降;'短柄籽银'桂和'潢川金桂'在50 Gy辐射处理后,种子萌发速度和萌发率有所提高,之后随着辐射剂量的提高,种子萌发力下降;而'籽银桂'在50~200 Gy辐射剂量之间,沙藏种子的萌发速度和萌发率都比对照处理区高,虽然在100 Gy辐射区种子萌发力稍有下降。

表2 沙藏120 d后7个桂花品种种子的发芽率及不同胚根长度所占数量
Tab. 2 Seed germination and radicle length of 7 varieties of *Osmanshtus fragrans* seeds after 120 d

品种名 Varieties	辐射剂量/Gy Irradiation dose	发芽总数(发芽率)/% No. of germination (Germination rate)	胚根长度 Radicle length/cm		
			≤0.5	0.5~1.0	≥1
'紫梗籽银'桂 'Zigeng Zi Yin'	0	158(52.7)	143 Aa	15 Ba	0 Ca
	50	141(47.0)	139 Ab	2 Bb	0 Cb
	100	120(40.0)	81 Ac	35 Bc	4 Cc
	150	41(13.7)	41 Ae	0 Be	0 Ce
	200	76(25.3)	73 Ad	3 Bd	0 Cd
	250	22(7.3)	22 Af	0 Bf	0 Cf
'短柄籽银'桂 'Duanbing Zi Yin'	0	335(57.6)	308 Ab	23 Bb	4 Cb
	50	629(69.9)	539 Aa	7 Bc	83 Ca
	100	416(46.2)	271 Ac	145 Ba	0 Cc
	150	208(23.1)	198 Ad	10 Bc	0 Cd
	200	98(10.9)	95 Ae	3 Be	0 Ce
'宽叶籽银'桂 'Kuanye Zi Yin'	0	741(82.3)	481 Ac	5 Bc	255 Cc
	50	233(25.9)	226 Aa	5 Ba	2 Ca
	100	120(13.3)	115 Ab	3 Bb	2 Cb
	150	31(3.4)	31 Ad	0 Bd	0 Cd
	200	48(5.3)	47 Ae	1 Be	0 Ce
	250	4(0.4)	4 Af	0 Bf	0 Cf
'潢川金桂' 'Huangchuan Jingui'	0	475(52.8)	292 Ab	156 Bb	27 Cb
	50	625(69.4)	341 Aa	246 Ba	38 Ca
	100	583(64.8)	501 Aa	2 Ba	80 Ca
	150	396(44.0)	390 Ac	6 Bc	0 Cc
	200	357(39.7)	324 Ae	1 Be	32 Ce
	250	182(20.2)	180 Ad	2 Bd	0 Cd
'米花籽银'桂 'Mihua Zi Yin'	0	301(82.0)	301 Ac	0 Bc	0 Bc
	50	712(79.1)	706 Aa	0 Ba	6 Ba
	100	451(50.1)	441 Ab	10 Bb	0 Bb
	150	272(30.2)	272 Ac	0 Bc	0 Bc
	200	176(19.6)	175 Ad	1 Bd	0 Bd
	250	40(4.4)	40 Ae	0 Be	0 Be

续表

品种名 Varieties	辐射剂量/Gy Irradiation dose	发芽总数(发芽率)/% No. of germination (Germination rate)	胚根长度 Radicle length/cm		
			≤0.5	0.5~1.0	≥1
'籽银桂' 'Zi Yingui'	0	370(41.1)	366 Ac	4 Cc	0 Bc
	50	498(55.3)	498 Ab	0 Cb	0 Bb
	100	636(70.7)	451 Aa	5 Ca	180 Ba
	150	573(63.7)	537 Ab	1 Cb	35 Bb
	200	646(71.8)	607 Aa	30 Ca	9B a
	250	364(40.4)	358 Ac	6 Cc	0 Bc
'金盏碧珠' 'Jinzhan Bizhu'	0	81(9.0)	81 Aa	0 Ba	0 Ba
	50	51(5.7)	51 Ab	0 Bb	0 Bb
	100	32(3.6)	32 Ac	0 Bc	0 Bc
	150	18(2.0)	18 Ad	0 Bd	0 Bd
	200	11(1.2)	11 Ae	0 Be	0 Be
	250	0(0)	0 Af	0 Bf	0 Bf

注:不同大写字母代表同一品种桂花种子的胚根长度之间差异显著,不同小写字母代表同一品种桂花种子的辐射处理之间差异显著。

Note: Different capital letters indicate significant differences of the radicle length among different cultivars, and different lower-case letters indicate significant differences among different irradiation dose of the same variety.

由表2可知,沙藏120 d后,'短柄籽银'桂、'宽叶籽银'桂、'潢川金桂'和'籽银桂'4个品种的部分种子胚根长度达到1 cm以上(≥1 cm);'紫梗籽银'桂和'籽银桂'也有少部分种子胚根伸长到0.5 cm,但还未达到1.0 cm。各个品种各处理区种子胚根长度大多超过0.5 cm,其中'金盏碧珠'不仅萌发率低,所有种子胚根最长均未超过0.5 cm。

适宜剂量的辐射处理促进了部分桂花品种,如'短柄籽银'桂、'潢川金桂'和'籽银桂'种子的胚根伸长生长,这几个品种的部分辐射处理区胚根长度达到0.5 cm以上的种子数量及比率,与对照相比明显增加;各辐射剂量的辐射处理明显抑制了'宽叶籽银'桂的胚根生长,并且辐射剂量越大,抑制作用越强;对其他品种的胚根伸长生长未见明显的抑制作用。

2.2 半致死剂量的计算及桂花品种辐射敏感型的划分

由表3可知,7个桂花种子在不同辐射剂量的照射下其相对发芽率均有所差异。其中'潢川金桂'和'籽银桂'在辐射剂量为50 Gy时相对发芽率达到最大值,分别为131.6%和134.6%,随着辐射剂量增加,'潢川金桂'的相对发芽率逐渐下降,并于250 Gy时最低降至38.3%,而'籽银桂'的相对发芽率则与辐射剂量无相关性,在整个辐射剂量内保持较高的相对发芽率,最高在200 Gy时相对发芽率为174.6%,最低在250 Gy时相对发芽率为98.3%,且与对照无显著差异。其他处理材料除了'短柄籽银'桂在辐射剂量为50 Gy时相对发芽率高于对照以外,其余材料的相对发芽率均低于对照,且随着辐射剂量的增大,相对发芽率显著降低,并于250 Gy时降至最低,这说明除了'籽银桂'以外其余材料的种子萌发均受到明显的抑制作用。

表 3 不同品种桂花种子相对发芽率、半致死剂量及相关系数
Tab. 3 Relative germination rate, semi-lethal dose and their correlation coefficient of different varieties of *Osmansthus fragrans* seeds

品种名 Varieties	相对发芽率 Relative germination rate/%						半致死剂量 LD_{50}/Gy	相关系数 r Correlation coefficient
	0 Gy	50 Gy	100 Gy	150 Gy	200 Gy	250 Gy		
'紫梗籽银'桂	100 a	89.2 a	75.9 b	26.0 d	48.1 c	13.9 e	151	−0.92
'宽叶籽银'桂	100 a	31.5 b	16.2 c	4.2 d	6.5 d	0.5 d	55	−0.83
'潢川金桂'	100 b	131.6 a	122.7 a	83.4 c	75.2 c	38.3 d	267	−0.81
'籽银桂'	100 c	134.6 b	171.9 a	154.9 ab	174.6 a	98.3 c	—	—
'米花籽银'桂	100 a	96.5 a	61.1 b	36.9 c	23.9 d	5.4 e	135	−0.98
'金盏碧珠'	100 a	63.0 b	39.5 c	22.2 d	13.6 d	0.0 e	98	−0.97
'短柄籽银'桂	100 b	121.4 a	80.3 c	40.2 d	18.9 e	—	146	−0.91

注：小写字母代表同一品种不同辐射处理区之间的多重比较结果，不同字母表示不同处理间存在的差异显著；"—"表示未能计算出。下同。

Note: The lower-case letters indicate the results of multiple comparison among different irradiation dose of the same variety; The different letters indicate significant differences; "—"means not be calculated. The same as the following.

方差分析结果表明，除了'紫梗籽银'桂和'米花籽银'桂在辐射剂量为 50 Gy、'籽银桂'在辐射剂量为 250 Gy 时相对发芽率与对照无显著差异，其余处理与对照均达到差异显著水平，其中'宽叶籽银'桂在辐射剂量为 150～250 Gy、'潢川金桂'在 50～100 Gy 及 150～200 Gy、'籽银桂'在 100～200 Gy、'金盏碧珠'在 150～200 Gy 时差异不显著外，其余处理间差异均达到显著水平。且'米花籽银'桂和'短柄籽银'桂的各处理之间差异均达显著水平。由此可见，除了'籽银桂'，低剂量的辐射对种子的萌发具有一定的促进作用，而高剂量的辐射则明显抑制种子的萌发。

相关分析表明，除'籽银桂'种子发芽率与辐射剂量无相关性之外，其他 6 个品种的种子发芽率与辐射剂量负相关，相关值范围为 81%～97%，即随着辐射剂量的增加，其相对发芽率急剧下降。利用直线回归方程 $y = a + bx$ 和公式(1)、(2)、(3)计算出各桂花品种半致死剂量范围为 55～267 Gy，分别为：'紫梗籽银'桂 151 Gy、'宽叶籽银'桂 55 Gy、'潢川金桂' 267 Gy、'米花籽银'桂 135 Gy、'金盏碧珠' 98 Gy、'短柄籽银'桂 146 Gy，而'籽银桂'在目前的辐射范围内，发芽率与辐射剂量未见明显相关性，也未能计算出其半致死剂量。因此各个桂花种子的半致死剂量大小顺序依次为'潢川金桂'＞'紫梗籽银'桂＞'短柄籽银'桂＞'米花籽银'桂＞'金盏碧珠'＞'宽叶籽银'桂，半致死剂量最高的为'潢川金桂'、最低为'宽叶籽银'桂。根据公式(5)计算出 7 个品种桂花种子的半致死剂量的平均值为 142 Gy，标准差为 71 Gy，将半致死剂量大于 213 Gy 的材料划分为迟钝型，将半致死剂量小于 71 Gy 划分为敏感型，其余的为中间型。由表 4 可知，'宽叶籽银'桂属于辐射敏感型，'紫梗籽银'桂、'米花籽银'桂、'短柄籽银'桂和'金盏碧珠'属于中间型，而'潢川金桂'和'籽银桂'为迟钝型。由此可见，不同品种的桂花种子其适宜辐射剂量存在差异，可能与桂花品种、种子的种壳、种子的重量、种子的内含物、种子硬实性有关。

2.3 辐射处理对不同桂花品种出苗率的影响

播种 3 个月后，'米花籽银'桂和'金盏碧珠'的出苗率较高，分别达到 49.2%、23.5%，其他几个品种的出苗率都很低。'金盏碧珠'沙藏后种子发芽率虽然很低,但露白种子播种后的出苗率仅次于'米花籽银'桂。

表4　7个品种桂花种子的辐射敏感型
Tab. 4　7 kinds of *Osmanthus fragrans* seeds radiation sensitive

类型 Type	试验材料 Experimental materials	频数 Frequency
迟钝型 $LD_{50}>213$ Gy	'潢川金桂''籽银桂'	2
敏感型 $LD_{50}<71$ Gy	'宽叶籽银桂'	1
中间型 71 Gy$\leqslant LD_{50}\leqslant 213$ Gy	'紫梗籽银'桂、'米花籽银'桂、'短柄籽银'桂、'金盏碧珠'	4

表5　播种3个月后7个桂花品种的出苗率
Tab. 5　Seedling rate of seven varieties of *Osmanthus fragrans* seeds three months after sowing/%

品种名 Varieties	辐射剂量 Irradiation dose/Gy					
	0	50	100	150	200	250
'紫梗籽银'桂	0.6 d	5.7 b	6.7 a	2.4 c	6.6 a	0.0 d
'短柄籽银'桂	4.2 c	13.2 a	3.8 c	5.8 b	0.0 d	—
'宽叶籽银'桂	5.4 b	5.2 b	0.8 c	6.5 a	2.1 c	0.0 e
'潢川金桂'	6.9 c	15.0 a	8.7 b	1.8 e	3.4 d	0.0 f
'米花籽银'桂	49.2 a	11.0 b	2.4 c	0.0 c	0.0 c	0.0 c
'籽银桂'	4.1 c	40.4 a	17.8 b	7.0 c	14.7 b	0.3 d
'金盏碧珠'	23.5 b	45.1 a	6.3 c	0.0 d	0.0 d	0.0 d

注:"—"表示未进行播种。
Note: "—" means no seeds sown.

通过对同一品种,不同辐射剂量出苗率进行比较发现,辐射处理显著抑制了'米花籽银'桂的出苗,但适宜剂量的辐射处理不同程度地提高了其他6个品种的出苗率,其中'紫梗籽银'桂和'宽叶籽银'桂适宜的辐射剂量分别为100 Gy和150 Gy,其他4个品种的适宜剂量均为50 Gy,之后随着辐射剂量的提高出苗率也呈下降趋势。但辐射剂量达到250 Gy时,仅'籽银桂'有0.3%的出苗率,其他6个品种均未见出苗。

3　讨论

辐射诱变首先要确定适宜的辐射剂量才可以获得有效的突变。在一定范围内增加辐射剂量可以增加突变率、拓宽突变谱,但过高的辐射剂量会降低成活率、增加畸变率。一般都以半致死剂量作为辐射育种的适宜辐射剂量。本研究根据辐射后种子萌发率计算桂花辐射的半致死剂量范围为55~267 Gy;不同桂花品种间适宜的辐射剂量存在显著差异,根据对^{60}Co-γ辐射敏感性的不同,桂花种子可分为辐射敏感型、中间型和迟钝型3种类型。陈睿等[7]以2种杜鹃为材料研究辐射效应时,发现在25~65 Gy剂量范围内随着辐射剂量增大,花的变异率增加,花色和花瓣形态出现多样性。不同植物品种或类型由于存在遗传差异而对辐射的敏感性不同,即使是同种植物,处于不同生长状态时也对辐射的敏感性不同。朱宗文等[8]研究表明不同品种的番茄种子对射线敏感性差异显著,不同萌动状态番茄种子对辐射的敏感性也不同,经过30% PEG处理的较干种子对辐射敏感。种子辐射敏感性的差异与种子大小、含水量和一些小分子物质含量等各种外部和内部因素有关[9]。桂花种子具有一定的脱水敏感性和休眠特性,当年4—5月成熟的种子需层积至第二年2—3月才能萌发。根据Sarapultsev等[10-11]的分析可知,休眠的种子一般有着较好的耐辐射性。但也有研究表明烟草(Nicotiana debneyi)籽苗比其种子有着更好的抗辐射性,可能与其籽苗中较高的绿原酸含量有关[12-13]。本研究也尝试对层积处理后胚根突破种皮的种子,在25～250 Gy区间内进行辐射诱导,但出苗率几乎为零。因此,对于已露白或育苗期材料,可以尝试进一步降低辐射剂量。

^{60}Co-γ射线辐射植物种子,辐射剂量与发芽率和幼苗生长呈负相关的报道很多[14-16]。本研究结果表明,除'籽银桂'外,其他6个品种种子发芽率与辐射剂量整体呈负相关。辐射所导致的种子萌发率下降或许与辐射造成的种胚组织受损,对细胞生长与分裂的抑制以及萌发过程的生理活动产生了不同程度的影响有关[17-18]。但本研究也发现较低剂量的辐射处理可以提高'潢川金桂'、'籽银桂'和'短柄籽银'桂3个品种种子的发芽率,对其播种后幼苗的出苗也有一定的促进作用。王月华等[19]在高羊茅种子辐射育种中也发现较低剂量的辐射有利于种子萌发,较高剂量的辐射对种子萌发有抑制作用,且随着辐射剂量的提高抑制作用加强。在高羊茅种子萌发过程中,活性氧清除酶的活性均随辐射剂量的增加先升高后降低。此外,在万寿菊[20]、香福禄考[21]、毛竹[18]和多花木兰[22]等辐射育种中也有类似发现。适当剂量的辐射处理可以促进种胚组织细胞生长和分裂[18],有利于打破种子休眠。

辐射对出苗率的影响也因植物种类而异。王文恩等[23]研究日本结缕草种子辐射效应时发现,当辐射剂量低于250 Gy时,随着辐射剂量的增大,田间出苗率增加。牡丹种子辐射育种研究中也有类似报道,即当辐射剂量低于某一临界值时,辐射处理对牡丹种子的出苗率有正向的影响[24]。但在椿树种子[25]和五味子[26]种子辐射育种时,随着辐射剂量的增加,出苗率降低。熊秋芳等[27]研究发现,低剂量的辐射不会影响萝卜种子的最终发芽率,却抑制了出苗率,且随着辐射剂量增大抑制越明显,主要表现为抑制了根、茎生长。王慧娟等[20]在万寿菊辐射育种中发现,适宜剂量的辐射提高了万寿菊种子发芽率,但任何剂量的辐射都抑制了幼苗的生长及开花。本研究发现半致死剂量的辐射对不同品种桂花的出苗率有不同的影响。接近'宽叶籽银'桂半致死剂量的辐射,对其出苗几乎无影响;半致死剂量的辐射对'紫梗籽银'桂和'短柄籽银'桂的出苗有一定的促进作用,但半致死剂量的辐射对'米花籽银'桂和'金盏碧珠'等桂花品种的出苗有明显的抑制作用。辐射处理提高了'潢川金桂'的发芽率,但高于150 Gy的辐射处理明显抑制了其出苗率,尤其是在接近半致死剂量时,播种3个月后未见幼苗出土。根据陶巧静等[28]的分析,辐射能量在种子中的沉积,使其胚中分生组织细胞的分裂过程受到严重抑制,并且这种抑制作用在其后的生长过程中逐渐显现,从而抑制了种子的出苗及生长发育。因此对'米花籽银'桂、'金盏碧珠'和'潢川金桂'的适宜辐射剂量不能仅考虑发芽率,还要结合其出苗率及幼苗生长情况适当降低辐射剂量,以保证幼苗的出苗率。根据相对发芽率,'米花籽银'桂半致死剂量为135 Gy、'潢川金桂'半致死剂量为267 Gy,在辐射剂量为150 Gy时,这2个品种出苗率极低,甚至根本未能出苗,因此建议辐射剂量控制在100 Gy。

4 结论

根据对 ^{60}Co-γ 辐射敏感性的不同,桂花种子可分为辐射敏感型、中间型和迟钝型3种类型。除'籽银桂'外,其他6个品种种子发芽率与辐射剂量呈负相关,较低剂量的辐射提高了'潢川金桂'、'短柄籽银'桂和'籽银桂'等3个品种种子的萌发率。其中'潢川金桂'辐射半致死剂量为 267 Gy、'紫梗籽银'桂为 151 Gy、'短柄籽银'桂为 146 Gy、'米花籽银'桂为 135 Gy、'金盏碧珠'为 98 Gy、'宽叶籽银'桂为 55 Gy。综合萌发率、出苗率等情况,建议辐射剂量应控制在 100 Gy。

参考文献

[1] 杨秀莲,郝其梅.桂花种子休眠和萌发的初步研究[J].浙江林学院学报,2010,27(2):272-276.

[2] 向其柏,刘玉莲.中国桂花品种图志[M].中英文本.杭州:浙江科学技术出版社,2008.

[3] 杨再强,王立新.观赏植物辐射诱变育种研究进展[J].四川林业科技,2006,27(3):19-23.

[4] 杜晓华,张菊平.园林植物遗传育种学[M].北京:中国水利水电出版社,2013.

[5] 杨秀莲,郝其梅.桂花种子休眠和萌发的初步研究[J].浙江林学院学报,2010,27(2):272-276.

[6] 周小梅,赵运林,蒋建雄,易自力.几种冷季型草坪草辐射敏感性及其辐射育种半致死剂量的确定[J].湘潭师范学院学报(自然科学版),2005,27(1):75-78.

[7] 陈睿,鲜小林,万斌,等. ^{60}Co-γ 辐射对两个杜鹃品种主要性状的影响[J].北方园艺,2015(2):46-50.

[8] 朱宗文,查丁石,朱为民,等. ^{60}Co-γ 射线对番茄种子萌发及早期幼苗生长的影响[J].种子,2010,29(8):15-22.

[9] Grodzinskii A M. Radiobiology of plants [M]. Naukova Dumka, Kiev, 1989:42 [in Russian].

[10] Sarapultsev B I, Geraskin S A. Radiotaxonomy of higher plants at the dormant seed stage [J]. Radiobiologiya, 1989, 29: 94-99.

[11] Sarapultsev B I, Geraskin S A. Species radioresistance and phylogenesis of higher plants [J]. Radiobiologiya, 1989, 29: 100-107.

[12] Wada H, Koshiba T, Matsui T, et al. Involvement of peroxidase in differential sensitivity to γ-adiation in seedling of two *Nicotiana* species [J]. Plant Science, 1998, 132: 109-119.

[13] Sato M, Matsui T. Differential radiosensitivity between seeds and seedlings of *Nicotiana debneyi* in respect of chlorogenic acid content [J]. Plant Science, 1995, 109: 139-144.

[14] 李志能,刘国锋,包满珠.悬铃木种子 ^{60}Co-γ 辐照及其苗期生物学性状调查[J].核农学报,2006,20(4):299-302.

[15] 敖妍,张国盛,鲁韧强,等.扶芳藤种子与枝条的 ^{60}Co-γ 辐射效应[J].核农学报,2006,20(3):202-204.

[16] 王瑞静,王瑞文,沈宝仙. ^{60}Co-γ 射线对杨树种子的辐射效应[J].核农学报,2009,23(5):762-765.

[17] Borzouei A, Kafi M, Khazaei H, et al. Effects of gamma radiation on germination and physiological aspects of wheat (*Triticum aestivum* L.) seedlings [J]. Pakistan Journal of Botany, 2010, 42(4): 2281-2290.

[18] 蔡春菊,高健,牟少华. ^{60}Co-γ 射线对毛竹种子活力及早期幼苗生长的影响[J].核农学报,2007,21(5):436-440.

[19] 王月华,韩烈保,尹淑霞,等.γ射线辐射对高羊茅种子萌发及酶活性的影响[J].核农学报,2006,20

(3):199-201.
[20] 王慧娟,孟月娥,赵秀山,等.^{60}Co-γ射线辐射万寿菊对发芽率及生长的影响[J].中国农学通报,2009,25(19):161-163.
[21] 费金喜,强继业,张杰.^{60}Co-γ射线辐射对香福禄考发芽率及幼苗长势的影响[J].种子,2007,26(7):85-86.
[22] 罗天琼,谭金玉,莫本田,等.多花木蓝辐射的敏感性和适宜辐射剂量研究[J].安徽农业科学,2013,41(6):2483-2486,2502.
[23] 王文恩,包满珠,张俊卫.^{60}Co-γ射线对日本结缕草干种子的辐射效应研究[J].草业科学,2009,26(5):155-160.
[24] 苏美和,赵兰勇,房涛.^{60}Co-γ射线对牡丹种子发芽率及幼苗生长的影响[J].农学学报,2012,2(2):18-21.
[25] 聂莉莉,刘仲齐,张越,等.椿树辐射诱变育种初报[J].核农学报,2009,23(4):577-580.
[26] 张悦,徐海军,魏殿文.五味子辐射育种初报[J].国土与自然资源研究,2009(3):90-91.
[27] 熊秋芳,陈玉霞,张雪清,等.^{60}Co-γ辐射对萝卜种子萌发和幼苗生长的影响[J].浙江农业科学,2014(3):356-359.
[28] 陶巧静,臧丽丽,刘蓉,等.^{137}Cs-γ辐射对葡萄种子发芽和幼苗生长及叶绿素荧光特性的影响[J].核农学报,2015,29(4):761-768.

Study on the Seed Germination and Seedling Growth of *Osmanthus fragrans* under ^{60}Co-γ Irradiation

GENG Xingmin, WANG Lianggui, LI Na, YANG Xiulian

(College of Landscape Architecture, Nanjing Forestry University, Nanjing 210037, China)

Abstract: Traditional cross breeding of *Osmanthus fragrans* is very difficult due to small size of flowers, styles, and very few of pollens. To investigate the effects of irradiation on *Osmanthus fragrans* seeds, seven varieties of *Osmanthus fragrans* were irradiated with different radiation doses of ^{60}Co-γ, and the effects of radiation on seed germination and seedling growth were observed. The results indicated that the effects of radiation on seed germination varied among different varieties. Except for 'Zi Yingui', seed germination rate of the other six varieties was negatively correlated with radiation doses, The rate of germination and seedling decreased with the increase of doses. Semi-lethal dose of six varieties of *Osmanthus fragrans* was 267, 151, 146, 135, 98 and 55 Gy for 'Huangchuang Jingui', 'Zhigeng Zi Yin', 'Duanbing Zi Yin' 'Mihua Zi Yin' 'Jinzhan Bizhu' and 'Kuanye Zi Yin' respectively. Lower dose (50 Gy) of radiation increased the germination of 'Huangchuan Jingui' 'Duanbing Zi Yin' and 'Jinzhan Bizhu'. With high dose (200 Gy), the rate of germination and seedling were inhibited greatly, especially with 250 Gy, all materials except 'Zi Yingui', the others cannot germinate. The results will provide a theory basis for mutation breeding and accelerate the breeding process of *Osmanthus fragrans*.

Key words: radiation; *Osmanthus fragrans*; mutation breeding; seed germination; seedling growth

'晚籽银桂''多芽金桂'花芽的形态分化*

杨秀莲，向其柏

(南京林业大学 风景园林学院，江苏 南京 210037)

摘要：采用石蜡切片法观察了'晚籽银桂''多芽金桂'花芽形态分化过程。结果表明：'晚籽银桂'花芽分化从6月下旬苞片原基分化至9月初雌蕊原基形成历时近3个月，其过程可分为苞片分化期、花序原基分化期、小花原基分化期、顶花花萼分化期、花瓣分化期、雄蕊分化期和雌蕊分化期7个时期。其中雄蕊分化期历时长，分化较慢，其他6个时期历时较短。聚伞花序的中间顶花先分化，侧花后分化，各小花几乎同时完成花芽分化进程。'多芽金桂'的花芽分化除雌蕊发育与'晚籽银桂'不同外，其余基本一致。

关键词：'晚籽银桂'；'多芽金桂'；花芽；形态分化

桂花(*Osmanthus fragrans* Lour.)是我国传统的十大名花之一，是一种绿化、美化和香化相结合，观赏与实用兼备的优良园林树种。桂花在园林中应用十分广泛，因其花期正值中秋佳节，且花香清可绝尘、浓能溢远，花时香闻数里，有"独占三秋压群芳"之誉。了解桂花花芽分化规律是控制、保证桂花开花质量与数量及确定采收期的基础。前人观察了'厚瓣金桂'、'早籽银'桂等品种的花芽分化过程[1-4]，笔者以开花较晚但能结实的'晚籽银桂'和不结实的'多芽金桂'为对象，研究其花芽分化的规律和时间，以期了解不同开花时期桂花品种花芽分化进程的差异，以及结实和不结实品种的花芽形态差异。

1 材料和方法

1.1 供试材料

供试材料为'晚籽银桂'和'多芽金桂'，采自南京林业大学校园。采样树高约4 m，冠幅4～5 m，列植和对植，长势优良。

1.2 取样及处理

从2005年6月15日开始至9月30日，每周取样1次。取样时，从树冠中上部外围朝南方向选择生长良好、中等长度的当年生枝条，去除叶片，摘取自顶端向下的第2～3节茎节上叠生芽的上位芽，水洗后用50%的FAA固定。制作常规石蜡切片，切片厚8 μm，番红-固绿对染，加拿大树胶封固，LEICA DM LB2显微镜(徕卡股份有限公司，韦茨拉尔，德国)观察拍照。

---

* 原文发表于《南京林业大学学报(自然科学版)》，2007，31(5)：105-108。
基金项目：中美合作国家自然科学基金重大项目(39894000)。

2 结果与分析

2.1 桂花形态结构特点

桂花花芽着生在新老梢连接部至新梢顶端。每节有2～4对叠生芽,能先后长出4～8个花芽,每个花芽又各开出5～9朵小花。花序为2至多歧聚伞花序。每朵小花均有一长0.5～1.5 cm 的花柄,有绿、淡绿和淡紫等色。花萼4枚,淡绿色。花冠裂片(花瓣)分离,基部合生,有长1～1.5 mm 的花冠筒。花冠裂片常为4枚。花较小,花瓣通常长3～5 mm,宽2～3 mm,颜色因品种不同差异较大,有白、黄和橘红等色。雄蕊2枚,着生在花冠筒上,花丝很短,约长0.5 mm。雌蕊柱头两裂,子房2室,每室有2个胚珠。花期一般9—10月,常分批开放。

2.2 '晚籽银桂'花芽形态分化时期的划分及特征

根据观察结果,把'晚籽银桂'聚伞花序的花芽分化分为7个时期(图1)。

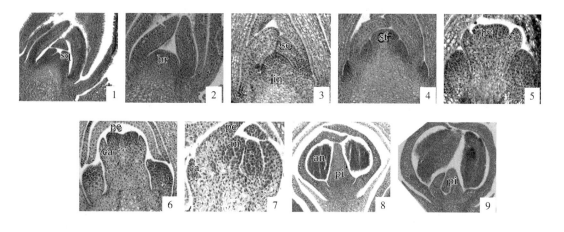

图1 '晚籽银桂'和'多芽金桂'花芽分化观察
Fig. 1 Differentiation of flower bud of *O. fragrans* 'Wanzi Yingui' and 'Duoya Jingui'

1. 营养生长期,15×;2. 苞片原基,15×;3. 花序原基分化期,20×;4. 花蕾原基分化期,15×;5. 顶花花萼分化期,20×;6. 顶花花瓣分化期,20×;7. 雄蕊分化期,40×;8. 雌蕊分化期,15×;9. '多芽金桂'的雌蕊发育,40×。sq. 鳞片;br. 苞片;ca. 花萼;in. 花序;sf. 小花原基;pe. 花瓣原基;an. 雄蕊原基;pi. 雌蕊原基。

1. nutritional growth, 15×; 2. sepal primordium, 15×; 3. inflorescence primordium differentiation, 20×; 4. flower bud primordium differentiation, 15×; 5. flowering stage, 20×; 6. flowering stage, 20×; 7. stamen differentiation, 40×; 8. pistil differentiation, 15×; 9. pistil development of multiple buds, 40×; sq. squama ca. calyx; br. bract; in. inflorescence; sf. small flower primordium; pe. petal primordium; an. androecium primordium; pi. pistil primordium.

(1) 苞片原基分化期:从6月15日至6月22日,由切片观察到初期芽内生长锥呈尖锐的圆锥状,尚处于营养生长期(图1-1)。至6月底,由营养生长转为生殖生长,生长锥顶端渐平渐宽,此期为苞片原基分化期(图1-2)。

(2) 花序原基分化期:6月底至7月初,苞片内部生长锥增大、变圆,隆起呈半球形。生长锥下部细胞大而疏松,并可见其外围的原形成层。此时苞片基部的外侧略向外凸出,由于此期历时较短,只看到分化结束的状态(图1-3),而未能看到完整的过程。

(3) 小花原基分化期:7月初,生长锥进入花序原基分化期后进一步发育,生长锥细胞继续分裂分化,花序轴延伸,在花序轴的中央皮层升高,在中央及其下部周缘隆起,并逐渐分离形成多个突起,即小花花蕾原基,中央顶端的突起逐渐变圆加宽,此为聚伞花序的顶花原基,下部及苞片基部外侧的突起形成花序中其他的侧花原基,顶部较尖,呈仙桃形。此期聚伞花序基本成形(图1-4)。

(4) 顶花花萼分化期:7月中旬,顶花原基变平后,立即转入花萼原基的分化。在基部伸长顶部膨大的顶花原基周围形成4个小突起,纵切面上是顶花原基生长点的边缘形成2个突起,即为花萼原基,花萼原基生长很快,细胞分裂使突起渐渐伸长,稍向上翘。其他小花原基变化稍有膨大变粗,顶端逐渐变圆钝(图1-5)。

(5) 顶花花瓣分化期:7月26日,在顶花花萼原基的内侧又形成两个突起,为顶花花瓣原基(图1-6)。花瓣原基生长较快。同时侧花原基明显膨大,顶端变圆变宽,开始花器官的分化。由于取样原因,未能观察到侧花花萼、花瓣的分化。

(6) 雄蕊分化期:8月上旬顶花花瓣内侧出现雄蕊原基,开始是两个小突起,逐渐发育、增大,至8月下旬,分化成花丝、花药、药隔、药室等。花丝较短,着生在花瓣内侧基部,花药大,呈豆瓣状(图1-7)。至9月中旬,花粉母细胞形成。

(7) 雌蕊分化期:自8月中旬开始,已发育成形的花药内侧生长点中央部位首先形成1个突起。突起继续发育,中心逐渐形成一个小孔,到后期小孔延伸,形成两个合在一起的雌蕊原基(图1-8)。雌蕊原基继续发育,伸长,形成子房、花柱和柱头。

2.3 '晚籽银桂'花芽分化进程及外部形态观察

在花芽分化的同时,芽体外观也在发生着相应的变化。芽体逐渐加长变宽,最后鳞片裂开,近白色珠状花蕾脱颖而出。整个过程历时约3个半月(表1)。

2.4 '多芽金桂'花芽的形态分化

由于'多芽金桂'雌性不育,从切片看出,前期的分化基本上和'晚籽银桂'相同,但花药形成的同时(8月中旬),内侧生长点中央先形成1个突起并发育,逐渐形成两个明显分离的心皮原基。随着进一步的发育,两个分离的心皮原基并不愈合,而是发育成两片叶状体(图1-9)。

3 讨论

3.1 花芽分化的时期划分

确定花芽分化的时期,在生产上具有十分重要的意义。目前对果树的花芽分化研究较多[5-7],通过花芽分化时期的确定,可以采取相应的栽培管理措施以达到果实增收的目的。对于以观花为主的花木,了解花芽分化时期,可采取相应的管理措施控制花芽分化数量,以取得较好的开花质量和数量。花芽分化时期的划分以各部分的原基出现为划分标准,大多数将花芽分化划分为未分化期、花芽分化始期、萼片分化期、花瓣分化期、雄蕊分化期和雌蕊

表 1 '晚籽银桂'花芽分化进程及外观形态

Tab. 1 The bud differentiation course and appearance configuration of 'Wan Zi Yin'

日期(年-月-日) Date (Year-Month-Day)	平均芽长/mm Average length of buds	平均芽宽/mm Average width of buds	芽长/芽宽 Length/ Width	外观形态 Appearance shape	分化进程 Differentiation course
2005-06-15	3.58	1.64	2.18	芽体较小,外鳞片紧闭,且高度角质化,用手难以剥离 The bud is smaller, the outer scales are tightly closed and highly keratinized. It is difficult to peel off with hands	未分化期 Undifferentiated stage
2005-06-22	3.61	1.68	2.15		苞片原基分化 Bract primordia differentiation stage
2005-06-27	3.65	1.73	2.11		花序原基分化期 Inflorescence primordium differentiation stage
2005-07-04	3.74	1.98	1.89		小花原基分化期 Floret primordia differentiation stage
2005-07-12	3.98	2.01	1.98	芽鳞开裂,形成"V"形狭缝 Bud scale cracks, forming a "V" slit	顶花花萼分化期 Calyx apex differentiation stage
2005-07-19	3.84	2.15	1.79		顶花花瓣分化期 Petal differentiation stage
2005-07-26	3.9	2.27	1.72	芽鳞开裂程度逐渐增大 The degree of bud scale cracking is gradually increasing	
2005-08-08	3.92	2.36	1.66	芽鳞片尖端稍向外卷,芽体膨大,芽尖已伸出鳞片之外 The bud scale tips are slightly curling up outwards, the buds are swollen, and the bud tips are extending beyond scales	雌、雄蕊分化期 Pistil and stamen differentiation stage

续 表

日期(年-月-日) Date (Year-Month-Day)	平均芽长/mm Average length of buds	平均芽宽/mm Average width of buds	芽长/芽宽 Length/Width	外观形态 Appearance shape	分化进程 Differentiation course
2005-08-17	3.97	2.41	1.65		
2005-08-24	3.98	2.43	1.64		
2005-09-03	4.12	2.55	1.62	用手可轻易剥除鳞片,并可见内部绿白色小花蕾 The scales could be easily removed by hand, and the small green buds could be seen inside	
2005-09-06	4.23	2.62	1.61		
2005-09-13	4.52	2.75	1.64		花芽形态发育完成进行性细胞的发育 Development of progressive cells after flower bud morphological development
2005-09-19	4.57	2.83	1.61	鳞片进一步展开,逐渐由绿色变为褐色,芽体膨大如珠,颜色变浓 The scalesopen further and gradually turn from green to brown, the buds expand like beads, and the color becomes lighter	
2005-09-25	4.55	2.98	1.53		
2005-09-29	4.63	3.12	1.48		

注:观察芽数均为10个。
Note: The number of buds was 10.

分化期等 6 个时期[8-11]。根据此次切片观察结果，笔者将'晚籽银桂''多芽金桂'花芽的形态分化分为 7 个时期是比较合理的，与'厚瓣金桂'的分化时期一致，但分化各时期的出现稍晚于'厚瓣金桂'[3]，这可能也是'晚籽银桂'花期较晚的原因。该结果有别于陈昳琦的研究结果[2]。

3.2 花芽分化与雌雄蕊的可育性

研究发现，'晚籽银桂'和'多芽金桂'的雄蕊发育基本一致，但是，雌蕊的发育存在明显的差异。能育的'晚籽银桂'雌蕊原基出现时是愈合在一起的 1 个突起，而后继续发育成完整的雌蕊，即正常的柱头、花柱和子房；而'多芽金桂'的雌蕊原基是分离的两个突起，继续发育成两片完全分离的叶状体，不能形成正常的柱头、花柱和子房，开花后可见发育不全的柱头。因此，笔者认为'多芽金桂'的不育是雌蕊发育不正常所导致的。

致谢：实验在南京林业大学江苏省、国家林业局林木遗传与基因工程重点实验室完成，施季森教授、席梦利博士给予了指导和帮助。

参考文献

[1] 万云先.桂花花芽分化的研究[J].华中农业大学学报,1988,7(4):364-366.
[2] 陈昳琦.桂花花芽分化的初步观察[J].中国桂花,1995(2):34-37.
[3] 王彩云.'厚瓣金桂'桂花花芽形态分化的研究[J].园艺学报,2002,29(1):52-56.
[4] 尚富德.桂花生物学研究[D].南京:南京林业大学,2003.
[5] 孙建云,王庆亚,黄清渊.李花芽形态分化的研究[J].江西农业大学学报,2005,27(3):414-416.
[6] 邱金淡,吴定尧,张海岚.石硖龙眼花芽分化的研究[J].华南农业大学学报,2001,22(1):27-30.
[7] 孔云,沈红香,姚允聪,等.玉巴达杏花芽形态分化时期和芽体特征变化[J].北京农学院学报,2006,21(1):38-40.
[8] 陈旭辉,江莎,李一帆,等.连翘花芽分化及发育的初步研究[J].园艺学报,2006,33(2):426-428.
[9] 吴昌陆,胡南珍.蜡梅花部形态和开花习性研究[J].园艺学报,1995,22(3):277-282.
[10] 韦三立,陈琰,韩碧文.大丽花的花芽分化研究[J].园艺学报,1995,22(3):272-276.
[11] 欧阳彤.阿月浑子花芽分化的初步研究[J].经济林研究,1999,17(4):8-11.

Morphological Differentiation of Flower Bud of *Osmanthus fragrans* 'Wanzi Yingui' and 'Duoya Jingui'

YANG Xiulian, XIANG Qibai

(College of Landscape Architecture, Nanjing Forestry University, Nanjing 210037, China)

Abstract: The paraffin cut method was used to observe the morphological differentiation process of the flower bud of *Osmanthus fragrans* 'Wanzi Yingui' and 'Duoya Jingui'. The result showed that the flower bud differentiation process lasted for about 3 months from bract primordial differentiation in the end of June to pistil primordial formation in the beginning of September. The morphological differentiation of osmanthus flower bud was divided into seven phases: bract differentiation phase, inflorescence primordium differentiation phase, single flower primordial differentiation phase, sepal differentiation phase of top flower, patal differentiation phase, stamen differentiation phase and pistil differentiation phase. The phase of bract differentiation and stamen differentiation were relatively long. The top flower of the inflorescence differentiation was earlier and slower than the lateral flowers. Later on all the top and lateral flowers finished differentiation almost at the same time. The prophase bud differentiation of 'Duoya Jingui' was the same as 'Wanzi Yingui' except for pistil differentiation.

Key words: 'Wanzi Yingui'; 'Duoya Jingui'; flower bud; morphological differentiation

桂花花粉活力测定与'晚籽银桂'柱头可授性分析*

杨秀莲,向其柏

(南京林业大学,江苏 南京 210037)

摘 要:采用 TTC 测定法、无机酸测定法和离体萌发法等方法测定了不同桂花品种的花粉活力,用联苯胺-过氧化氢法测定柱头可授性,用扫描电镜观察了'晚籽银桂'柱头的分泌物。结果表明 TTC 测定法、无机酸测定法不适合用于桂花花粉活力的测定,而用一定浓度的蔗糖、硼酸制成的固体培养基则能较好地测定花粉活力。桂花的不同品种和不同的开花期其花粉活力均具有显著差异。'晚籽银桂'的柱头为湿型柱头,在整个花期中均具有较强的可授性。'晚籽银桂'属雌蕊先熟型,其花粉活力与柱头可授性有 4--5 d 的相遇期。

关键词:桂花;花粉活力;柱头可授性

在植物有性繁殖过程中,两性之间的传粉是基础,传粉过程始于花药开裂和成熟花粉的散出,携带着雄配子体或其前体的花粉粒被暴露在干燥条件下,必须在具有活性时到达适宜接受的柱头。因此,花粉的活力与寿命、柱头的可授性便成为传粉生态学所必须研究的内容。近几年,花粉生活力与寿命及柱头的可授性研究已经引起不少学者的重视,并开展了大量的工作[1-5]。

桂花(*Osmanthus fragrans*)隶属于木犀科木犀属,是一种观赏兼食用、药用的常绿园林树木。多年来对桂花的研究较多地集中在品种的收集、分类、繁殖、栽培、生殖生物学以及对花的利用方面[6-8]。对于桂花的传粉生物学、生殖生态学等方面的研究目前还未见报道。本文利用生理生化研究方法,对不同桂花品种的花粉活力、柱头可授性等进行了测定。旨在揭示桂花的传粉生物学基础,为进一步研究传粉生物学特性、探讨传粉机制对桂花繁殖的影响等奠定理论基础。

1 材料与方法

1.1 桂花花粉活力测定

1.1.1 材料

选自南京林业大学校园内栽植的桂花品种:'四季桂'、'早银桂'、'紫梗籽银'桂、'银桂'、'籽银桂'、'宽叶籽银'桂、'早籽银'桂、'晚籽银桂'、'波叶金桂'、'金桂'、'多芽金桂'、

---

* 原文发表于《林业科技开发》,2007,21(3):22-25。
基金项目:北京市科技计划项目"观赏型侧柏、桧柏无性系选育研究"(Y0604017040831)。

'籽丹桂'、'早红'13个品种。剪取开花当天、盛花期(指开花第 3 天)、开花末期(开花第 5 天)的花朵,带回实验室后将花粉直接置于培养液和培养基上。

1.1.2 花粉活力测定方法

采用 TTC 测定法、无机酸测定法[9]、离体萌发法[10,11]。其中离体萌发法采用 6 种配方(100 mg/L 硼酸＋2％琼脂、5％蔗糖＋100 mg/L 硼酸＋2％琼脂、10％蔗糖＋100 mg/L 硼酸＋2％琼脂、50 mg/L 硼酸＋2％琼脂、5％蔗糖＋50 mg/L 硼酸＋2％琼脂、10％蔗糖＋50 mg/L 硼酸＋2％琼脂)培养基上做花粉萌发对比试验,确定桂花花粉萌发的适宜培养基为 10％蔗糖＋100 mg/L 硼酸＋2％琼脂,设定培养温度为 25 ℃,培养时间为 12 h(通过预试验确定)。

1.2 柱头可授性检测

选用两性花且取材容易的'晚籽银桂'进行试验。从开花当日起,每天中午采集一定数量的柱头,将其浸入凹面载玻片中含有联苯胺-过氧化氢反应液(1％联苯胺：3％过氧化氢：水＝4：11：22,体积比)的凹陷处。若柱头具可授性,则柱头周围呈现蓝色并有大量气泡出现[3]。

1.3 柱头扫描电镜样品制作

从开花当日起,每天中午采集一定数量的'晚籽银桂'柱头,经 4％戊二醛(磷酸缓冲液 pH7.2)固定,磷酸缓冲液清洗,1％锇酸(磷酸缓冲液 pH7.2)固定,磷酸缓冲液清洗,乙醇系列脱水,乙酸异戊酯置换,二氧化碳临界点干燥,粘台,喷金,菲力普 SEM-505 扫描电镜观察并拍照。

2 结果与分析

2.1 花粉活力测定

2.1.1 TTC 测定法

显微镜下,所观察的视野内均未见着色花粉粒,13 个品种的结果基本一致。因此,该方法不适合检测桂花花粉活力。

2.1.2 无机酸测定法

滴上硝酸,3 h 后观察,未见有花粉管萌发,因此此法也不宜用于检测桂花花粉活力。

2.1.3 固体培养基法

对桂花 13 个品种不同开花时期(开花当天、盛花期、开花末期)的花粉进行了活力测定,结果见表 1。由表 1 可知桂花的结实品种('紫梗籽银'桂、'籽银桂'、'宽叶籽银'桂、'早籽银'桂、'晚籽银桂'、'籽丹桂')和不结实品种('四季桂''银桂''早银桂''波叶金桂''金桂''多芽金桂''早红')的花粉均有萌发,但不同品种间及不同开花期的花粉萌发率差异较大。'籽丹桂'在 3 个不同时期的花粉萌发率均为最高,但盛花期花粉的萌发率(51.06％)明显高于其他两个时期(开花当天 11.37％和开花末期 18.5％),而'多芽金'桂花粉的萌发率在所有品种中是最低的,其盛花期花粉的萌发率仅为 3.36％。结实品种的花粉萌发率一般都在 20% 以上,不结实品种中也有高达 43.23% 的花粉萌发率('银桂'),这也证明不结实品种不能产生种子的原因不在花粉育性差,而是由于雌蕊败育造成的[8]。

对不同开花时期的花粉活力进行了方差分析,结果表明不同品种在开花当天和开花末期差异不显著,但这两个时期与开花盛期相比,花粉活力差异达到了极显著水平($S^*_{0.01}$)。(见表1)

表1 桂花品种不同开花阶段花粉的萌发率

品种	花粉萌发率/%					
	开花当天	$S^*_{0.01}$	盛花期	$S^*_{0.01}$	开花末期	$S^*_{0.01}$
'四季桂'	5.27±0.15	EF	19.20±1.39	D	7.90±0.30	D
'紫梗籽银'桂	10.13±0.40	B	30.63±8.37	C	11.87±1.21	D
'银桂'	9.67±0.71	BC	43.23±4.86	B	18.47±0.91	A
'籽银桂'	8.80±0.36	C	25.73±1.40	CD	8.67±1.21	CD
'早银桂'	2.97±0.21	G	11.83±0.95	E	3.73±0.42	F
'宽叶籽银'桂	5.27±0.35	EF	20.33±2.83	D	5.57±0.15	E
'早籽银'桂	4.80±0.20	F	25.50±1.32	CD	9.47±1.00	C
'晚籽银桂'	5.97±0.25	E	26.23±0.50	CD	7.80±0.40	D
'波叶金桂'	1.33±0.15	H	6.73±0.93	EF	1.50±0.20	G
'金桂'	9.93±0.32	B	32.76±2.15	G	11.03±0.74	B
'多芽金桂'	0.20±0.35	I	3.36±0.55	F	0.00	H
'早红'	6.97±0.21	D	26.00±0.00	CD	7.13±0.31	D
'籽丹桂'	11.37±0.90	A	51.0±63.04	A	18.50±0.89	A

注:$S^*_{0.01}$代表差异显著性。

对同一品种在不同开花期的花粉活力进行方差分析,结果表明'四季桂'、'紫梗籽银'桂、'早籽银'桂、'晚籽银桂'、'籽丹桂'在不同开花期各自的花粉活力差异均达到了极显著水平($S^*_{0.01}$)。而其他8个品种开花当天和末期差异不显著,但开花盛期达到了极显著水平($S^*_{0.01}$)。

由此可见,13个桂花品种花粉萌发率均较低,仅'籽丹桂'在盛花期达51.06%,其余品种均在50%以下。花粉萌发率在盛花期最高,其中前6位分别为:'籽丹桂'>'银桂'>'金桂'>'紫梗籽银'桂>'晚籽银桂'>'早红'>'籽银桂'。

2.2 柱头可授性检测

各结实桂花品种的柱头从开花当天直至第6天,均有可授性,在开花第4天可授性最强,到第7天时柱头变褐,过氧化物酶活性减弱至无,柱头失去活性(见表2)。

表2 柱头可授期联苯胺-过氧化氢法检测结果

品种	开花天数/d						
	1	2	3	4	5	6	7
'紫梗籽银'桂	+/−	+	++	+++	+/−	+/−	−
'籽银桂'	+/−	+	++	+++	+/−	+/−	−
'宽叶籽银'桂	+/−	++	++	+++	+/−	+/−	+/−

续表

品　种	开花天数/d						
	1	2	3	4	5	6	7
'早籽银'桂	+/−	++	++	+++	+/−	+/−	−
'晚籽银桂'	+/−	++	++	+++	+	+/−	+/−
'籽丹桂'	+/−	+	++	+++	+/−	+/−	+/−

注:"−"示柱头不具可授性;"+"示柱头具可授性;"++"示柱头具较强可授性;"+++"示柱头具强可授性;"(+/−)"示部分柱头具可授性(有个体差异),下同。

2.3 '晚籽银桂'柱头分泌物的扫描电镜观察

'晚籽银桂'两性花柱头圆锥状,有一纵沟,柱头顶面中间有一横沟(见图1)。开花前期横沟较窄,随着花朵的开放,横沟逐渐增大。柱头表面乳突细胞长圆形(见图2),在开花当天分泌物较少,未见黏附其上的花粉粒。在开放的第3、4天,柱头分泌物达最多,可见柱头上黏附有较多的花粉粒,有部分花粉已经萌发(见图3～5)。至开花第8天,柱头乳突细胞逐渐萎缩、衰退(见图6),整个柱头枯萎、皱缩。

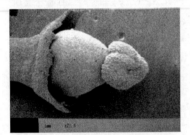

图1　'晚籽银桂'柱头
(开花当天)(71.5×)

图2　柱头乳突
(开花当天)(1 310×)

图3　柱头乳突细胞和萌发的花粉
(开花第4天)(1 500×)

图4　柱头分泌物和黏附的花粉粒
(开花第3天)(5 000×)

图5　柱头上花粉的萌发
(开花第4天)(2 000×)

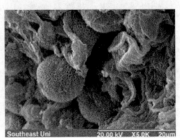

图6　柱头乳突细胞衰退
(开花第8天)(5 000×)

2.4 桂花花粉活力和可授期比较

桂花为雌雄异熟花[8],本研究结果表明柱头在花药未开裂前即已成熟,可以执行其受粉功能,故认为其雌蕊先熟。

桂花柱头在花开放当天即具可授性,随着花的展开,可授性增强,至第3、4天达最高,随后可授性逐渐降低,至花瓣脱落,部分柱头仍具活力。而桂花的花药一般在花开放后的第2或第3天才开裂,花粉活力在25℃室温下保持7天左右,桂花单花开放时期约为6~7天,至花瓣散落,花粉活力基本丧失。因此花粉活力与柱头可授性有3~4天的相遇期,表3以'晚籽银桂'为例说明此问题。由此可见,桂花虽然是雌蕊先熟,但其雌雄蕊功能在时间上是相遇的,这对延长桂花的授粉时间,提高桂花结实率是有利的。

表3 '晚籽银桂'花粉活力与柱头可授性

开花后时期	柱头可授性	花粉萌发率%
开花当天	+	5.97
开花盛期	+++	26.23
开花末期		7.80

3 讨论

花粉具有活力是完成受精的必要条件,因此准确测定花粉活力十分重要。花粉活力的检测方法较多,但不同的植物花粉活力的检测方法不尽相同[1-2,9]。

本试验结果表明,用TTC检测法和无机酸检测法,所得各品种的花粉活力均为零。可见这两种方法不适合用来检测桂花的花粉活力。笔者也曾用液体培养基培养离体花粉进行花粉活力检测,结果萌发率也很低,而固体培养基则是较好的选择,在10%蔗糖+100 mg/L硼酸+2%琼脂组成的培养基中离体培养的花粉活力最高。

从花粉活力测定结果来看,桂花品种花粉活力均较低,但不管是雄花还是两性花,其花粉均具有一定活力,其中'银桂'盛花期的花粉活力在所有不结实品种中是最高的,达43.23%,仅次于所有品种中最高的'籽丹桂'。这佐证了尚富德的结论,即不结实桂花品种不能结实的原因不是雄性不育,而是与雌蕊退化有关[8]。对花粉活力的研究可为桂花进行人工育种时父本选择提供重要的参考。

植物的柱头一般有两种不同的状态:一种是湿型柱头,在盛开时有溢出物,可作为昆虫的诱物,黏着花粉,又为花粉萌发提供必要的基质,如鸢尾、矮牵牛、百合的柱头;另一种是干型柱头,在盛开时不产生分泌物,柱头表面存在蛋白质薄膜,具有亲水性,能通过其下层的角质层中断处吸水,辅助黏着花粉和使花粉获得萌发必要的水分,如十字花科和石竹科植物的柱头[12]。本研究通过对桂花柱头的电镜观察,认为桂花的柱头应为湿型柱头。

柱头可授期是花朵成熟过程中的一个重要时期,它能在很大程度上影响自花传粉率、开花不同阶段的传粉成功率、各种传粉者的相对重要性、雄性和雌性功能之间的相互干扰、不同基因型花粉之间的竞争以及配子体选择的机会,等等。不同植物的柱头可授期所持续时间的长短不等,花朵的开放期、一天内的不同时间及柱头分泌物的有无等对其也有影响。对桂花来说,在自然条件下,柱头可授期为7天左右,花粉散发刚好在柱头可授性最强的时期,且桂花的柱头和花药处于同一平面,不存在空间隔离,花粉和柱头之间有足够的时间完成受精作用。

在开花当天的柱头扫描电镜照片中,可见乳突细胞表面有很多白色颗粒状物,而后期的柱头上却未发现。这是何物,对柱头可授性有何影响等问题有待于进一步研究。

参考文献

[1] 红雨,刘强.芍药花粉活力和柱头可授性的研究[J].广西植物,2003,23(1):90-92.
[2] 刘林德,张萍,张丽,等.锦带花花粉活力、柱头可授性及传粉者的观察[J].西北植物学报,2004,24(8):1431-1434.
[3] 刘林德,张洪军,祝宁,等.刺五加花粉活力和柱头可授性的研究[J].植物研究,2001,21(3):375-379.
[4] Yi W, Law S E, McCoy D, Wetzstein H Y. Stigma development and receptivity in almond (*Prunus dulcis*)[J]. Annals of Botany, 2006, 97(1):57-63.
[5] Ananthakalaiselvi A, Krishnasamy V, Vijaya J. Stigma receptivity and pollen viability studies in hybrid pearl millet km[J]. The Madras Agricultural Journal, 1999, 86(10-12):603-605.
[6] 向其柏.中国桂花:申报桂花品种国际登录权论文集(Ⅱ)[G].长春:吉林科学技术出版社,2002.
[7] 王彩云.'厚瓣金桂'桂花花芽形态分化的研究[J].园艺学报,2002,29(1):52-56.
[8] 尚富德.桂花生物学研究[D].南京:南京林业大学,2003.
[9] 周莉花,郝日明,赵宏波.蜡梅花粉活力检测方法筛选和保存时间观察[J].浙江林学院学报,2006,23(3):270-274.
[10] 王源秀,江聪,徐立安.柳树花粉生活力分析[J].林业科技开发,2007,21(1):28-30.
[11] 徐进,王章荣.杂种鹅掌楸及其亲本花部形态和花粉活力的遗传变异[J].植物资源与环境学报,2001,10(2):31-34.
[12] 胡适宜.被子植物胚胎学[M].北京:人民教育出版社,1982.

Pollen Viability and Stigma Receptivity of *Osmanthus fragrans*

YANG Xiulian, XIANG Qibai

(College of Landscape Architecture, Nanjing Forestry University, Nanjing 210037, China)

Abstract: The pollen viability and the stigma receptivity of the *Osmanthus fragrans* varieties were determined by methods of TTC, inorganic acid and gemination *in vitro*, and by method of benzidine-H_2O_2, and the stigma secretion of 'Wanzi Yingui' was also estimated with scanning electron microscopy (SEM) in this paper. The results showed that the methods of TTC and inorganic acid in vitro medium couldn't be used to determine the pollen viability, however, the methods of cane sugar and boric acid *in vitro* were great in determing the pollen viability. The significant differences of pollen germination among the different varieties and the different anthesis were observed. The results also indicated that the stigma was wet-typed, and its receptivity was high during the anthesis. The pistil of *Osmanthus fragrans* was mature earlier than its stamen, and there were 4～5 days to make an appointment between pollen and stigma receptivity.

Key words: *Osmanthus fragrans*; pollen viability; stigma receptivity

'晚籽银桂'胚和胚乳发育的研究

杨秀莲,向其柏

(南京林业大学 风景园林学院,江苏 南京 210037)

摘 要:用常规石蜡切片技术研究了'晚籽银桂'的胚和胚乳的发育过程。结果表明,'晚籽银桂'的胚胎发生类型为柳叶菜型。卵细胞受精后,合子经过一段时间休眠,于10月下旬进行分裂,经棒形胚、球形胚、心形胚、鱼雷胚,至果实成熟时发育为子叶胚。心形胚时期胚柄最为发达。'晚籽银桂'的胚乳发育类型为细胞型。初生胚乳核先于合子分裂,胚乳细胞分裂较快,胚周围的胚乳细胞有降解现象。

关键词:'晚籽银桂';胚;胚乳;发育

桂花(*Osmanthus fragrans* Lour.)是我国传统的十大名花之一,是一种绿化、美化和香化相结合,观赏与实用兼备的优良园林树种,是我国著名的芳香植物,深受我国人民的喜爱。据文字记载,桂花在我国的栽培历史长达2 500年以上[1]。在人们引种驯化的漫长过程中,桂花由天然生长而至人工栽培,由山野而进入宫苑、寺庙、庭院,由南方向北方推进。人们对桂花的认识逐渐深化,应用范围日益扩大,对桂花的研究也由表及里从各个角度和层次展开和深入。有关桂花的生殖生物学方面的研究也在不同的种和品种中展开,杨琴军、尚富德分别研究了结实金桂和'籽银桂'的受精和胚胎发育[2-4]。作者以结实的'晚籽银桂'为对象,对其胚胎和胚乳的发育进行了研究,以进一步了解桂花的生殖生物学知识,同时也为桂花的育种提供理论依据。

1 材料与方法

1.1 材料

供试材料为'晚籽银桂',采自南京林业大学幼儿园内。采样树高约4 m,冠幅3~5 m,列植,长势优良。2005年9月30日起,开花期间前3 d每天取自顶端向下的第二、三茎节上位花1次,每次取10朵;花后至11月中旬,每2 d取相同部位的幼小种子1次,每次10个;11月下旬至2006年3月,每周取相同部位的种子1次,每次10个。

1.2 方法

取材后直接用50%的FAA固定。较大的种子用刀片切去种子下半部分,只保留顶端带胚部分(因桂花的胚是倒生胚珠)。常规石蜡切片,切片厚度8 μm,番红-固绿对染,

---

* 原文发表于《安徽农业大学学报》,2007,34(2):232-234。
基金项目:江苏省农业三项工程项目(BK 2005133)。

加拿大树胶封固，LEICA DM LB2 显微镜（徕卡股份有限公司，韦茨拉尔，德国）观察，数码拍照。

2 结果与分析

2.1 胚的发育

桂花开花后，花粉在柱头上萌发（见图1-1），花粉管通过柱头组织到达胚囊，完成双受精[2]。受精后，合子要经过一个短暂的休眠期。大约于10月下旬，合子开始分裂。合子的第一次分裂为横分裂，形成上下两个细胞，靠近珠孔端的一个为基细胞，远离珠孔端的一个为顶细胞，这为二细胞原胚（见图1-2）。基细胞体积较大，细胞质较稀，将来发育成胚柄；顶细胞体积较小，细胞质较浓，将来发育成胚体。基细胞随后进行多次的横向分裂，形成胚柄（见图1-3）。胚柄在心形胚时期最发达，以后逐渐退化，成熟的种子中看不到胚柄的存在。

顶细胞经过两次纵向分裂形成四细胞原胚，随后又进行多次的横裂和纵裂形成棒状原胚（见图1-4）。从11月初至12月中旬或下旬，胚的本体继续进行各个方向的连续分裂，成为一团组织，称球形胚（见图1-5～1-6）。12月下旬至1月上旬，球形胚两侧细胞分裂速度加快，形成两个突起，即子叶原基，将来发育成子叶，突起间形成1个凹陷，胚体分化为心形胚（见图1-7～1-9），此时胚柄最为发达。子叶原基迅速发育伸长，成为两片子叶，胚轴也得到延伸，到1月下旬至2月上旬，整个胚体呈鱼雷形（见图1-10）。随着子叶进一步生长，子叶间的分生组织开始分裂，形成胚芽，胚体下部形成胚根。胚根和胚芽之间形成了胚轴（见图1-11）。2月下旬至3月上中旬，胚体的组织和器官分化基本完成，发育为成熟胚。果实也接近成熟，表面观外果皮由绿色逐渐变为紫黑色，中果皮解体，内果皮强烈木质化，硬度加大，形成果核。成熟种子中剥出的成熟胚呈乳白色，长5 mm左右，子叶明显。整个胚位于乳白色胚乳中间（见图1-12、1-13）。

2.2 胚乳的发育

初生胚乳核的分裂比合子分裂早一些。'晚籽银桂'的胚乳发育为细胞型，即胚乳核的每一次分裂都伴随着细胞质的分裂，形成胚乳细胞（见图1-14～1-16）。原胚时期，胚乳细胞分裂较快，形成大量的胚乳细胞包围着发育中的胚。在胚发育的整个阶段，胚体周围和胚囊壁附近有很多呈离散状态解体退化的胚乳细胞，这些胚乳细胞的降解产物为胚发育提供营养（见图1-4～1-9）。至果实成熟，胚周围的胚乳细胞被降解形成空腔，但外围的胚乳细胞不降解。在成熟的种子中，胚乳坚实，呈乳白色，占有较大的面积（见图1-13），因此桂花种子为有胚乳种子。

3 讨论

'晚籽银桂'的合子首先横向分裂为两个细胞（顶细胞和基细胞），然后顶细胞再纵向分裂，逐渐发育成胚，基细胞不参加胚体的形成。胡适宜将种子植物的胚胎发育类型分为6种，即胡椒型、柳叶菜型（又叫十字花型）、紫菀型、石竹型、茄型和藜型[5]，并提出'晚籽银桂'

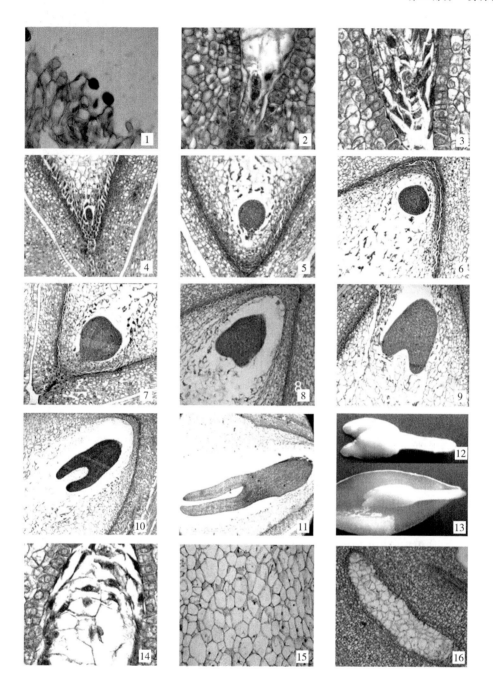

图1 '晚籽银桂'胚和胚乳发育过程

1.花粉在柱头上萌发(20×);2.二细胞原胚(20×);3.棒状胚初期(20×);4.棒状胚(15×);5-6.球形胚(15×);7-10:心形胚(15×);11.子叶胚(10×);12.未发育成熟的胚(10×);13.成熟的胚及胚乳(6×);14.胚乳的发育(200×);15.胚乳的发育(100×);16.胚乳的发育(40×)。

的胚胎发生类型应为柳叶菜型或称为十字花型。这与'金桂''籽银桂'的胚胎发育类型是一致的。由此可见,同一种不同品种之间胚的发育过程基本上是一致的,但由于花期不一致可能会导致发育进程上出现一定的差异,表现出种子的早熟与晚熟。

'晚籽银桂'的初生胚乳核的每一次分裂,都伴随着细胞质的分裂。胡适宜认为被子植物胚乳发育的类型有3种,即核型胚乳(初生胚乳核分裂初期先形成一定数目的游离细胞核,然后再形成胚乳细胞,如柿、蜡梅[6,7])、细胞型胚乳(初生胚乳核的每一次分裂都伴随细胞质的分裂,没有游离核时期,大多数双子叶合瓣花植物如番茄、烟草和芝麻等皆属此类[8])和沼生目型胚乳(介于核型胚乳和细胞型胚乳之间,多限于沼生目种类)。'晚籽银桂'的胚乳发育类型明显属于细胞型胚乳。在成熟的种子中,有丰富的乳白色胚乳存在。

桂花雌配子体通常具有两个胚囊,4个胚珠,在胚胎发育过程中绝大多数只有1个胚珠能正常受精并继续发育为种子,其余3个退化,但偶见两个胚珠在两个胚囊内同时发育成熟的。

参考文献

[1] 刘玉莲.南京地区桂花栽培品种研究[J].南京林学院学报,1985,9(1):30-37.
[2] 杨琴军,黄燕文,李和平,等.桂花受精作用的研究[J].华中农业大学学报,2001,20(4):382-390.
[3] 杨琴军,黄燕文,李和平,等.桂花胚和胚乳发育过程的研究[J].华中农业大学学报,2003,22(2):175-178.
[4] 尚富德.桂花生物学研究[D].南京:南京林业大学,2004.
[5] 胡适宜.被子植物胚胎学[M].北京:高等教育出版社,1982:80-196.
[6] 钟云,刘勇,蒋侬辉,等.柿胚胎发生发育的研究[J].园艺学报,2004,31(3):353-356.
[7] 黄坚钦,章滨森,金水虎,等.蜡梅的受精作用及胚胎发生[J].植物学通报,1999,16(6):686-690.
[8] 强胜.植物学[M].北京:高等教育出版社,2006.

Formation and Development of Embryo and Endosperm of *Osmanthus fragrans* 'Wanzi Yingui'

YANG Xiulian, XIANG Qibai

(College of Landscape Architecture, Nanjing Forestry University, Nanjing 210037)

Abstract: The development process of embryo and endosperm of *Osmanthus fragrans* 'Wanzi Yingui' was observed by means of traditional paraffin wax sections. The following results were obtained: The development of embryo was of on grad type. After undergoing a rest period, the zygote embarked on its first division, which took place at the later October, then the club-shaped, the globular embryo, the heart-shaped embryo and the fish-torpedo shape embryo, the cotyledon-shaped was developed be fore the fruit became ripe. At the heart-shaped embryo stage, the suspensor was well developed. The endosperm development of *Osmanthus fragrans* 'Wanzi Yingui' was cellular type. The primary endosperm nucleus division was often prior to zygote, and it divided quickly. The endosperm cells surrounding the embryo were disintegrated during the development of embryo.

Key words: *Osmanthus fragrans*; embryo; endosperm; development

第四部分
繁殖栽培技术

覆盖物对土壤性质及微生物碳氮的影响

倪雪[1,2]，张焕朝[3]，杨秀莲[1]，王良桂[1]

（1. 南京林业大学 风景园林学院，江苏 南京 210037；2. 安徽工业大学 建筑与工程学院，安徽 马鞍山 243000；3. 南京林业大学 林学院，江苏 南京 210037）

摘　要：本研究旨在比较在不同覆盖物处理的影响下，'日香桂'种植地土壤性质和微生物碳氮变化特点。试验基地位于中国江苏省句容市，试验为期12个月。研究涵盖五种覆盖物处理种类，其中包括空白无覆盖对照（CK）；马尼拉草皮覆盖（T）；松鳞树皮覆盖（W）；鹅卵石砾覆盖（G）；陶土颗粒覆盖（C）；每种覆盖物有三块重复样地，所有样地按照随机区组排列设计。有机覆盖为W处理，无机覆盖包括G和C处理，生物材料为T处理。在试验过程中取样两次，通过多项土壤实验检测得出结果分析比较。结果表明：在0～5 cm土壤深度，相比于无机覆盖G和C处理，覆盖T与W更能有效提高土壤养分，且显著增加了土壤有机质和土壤脲酶活性。覆盖材料对土壤容重和全氮的影响不显著，却普遍提高了土壤有效氮、有效磷、有机质含量、脲酶、磷酸酶以及微生物碳和氮等含量，但有机覆盖物处理要显著于无机覆盖物处理。在0～5 cm土壤深度，有机覆盖降低了土壤的pH值，而无机覆盖增加了pH值。深层的土壤受到覆盖物的影响要小于浅层土壤。对于土壤性质的改善方面，树皮较无机覆盖，马尼拉草皮更体现出正面效应，此外，有机覆盖物也更能够明显提高土壤微生物量。

关键词：土壤性质；土壤微生物；树皮；马尼拉草皮；鹅卵石砾；陶土粒

　　一些可循环的有机物以及无机物垃圾可以作为覆盖物用来提升农作物的土壤质量[1]。有研究指出，覆盖可以提高土壤水分温度和营养状态，对防止水土流失、防止土壤盐回流以及控制杂草有一定的作用[2]。近十年的研究中，许多种类的覆盖物材料被用来试验是否能对农作物土壤理化性质产生良好的改善作用。

　　常见覆盖物材料主要包括有机材料，如作物残留物；无机材料，如塑料薄膜、砾石或沙子、岩石碎片、火山灰、城市垃圾；生物材料，如马尼拉草，黑麦草，白车轴草等。

　　Erenstein[3]指出，对于一些亚热带国家，利用作物残留物作为覆盖是一种将资源保护与提高生产力二者相结合的充满前景的技术。例如，稻草类覆盖被指出在多种作物降低疾病方面有一定贡献，并且可以减少水土流失和增加土壤营养[4-5]。前人就不同有机覆盖物及用量对土壤物理及养分状况的影响做了研究，试图探索寻找有利于对土壤活动和作物生长的最佳有机物覆盖方式[6-9]；塑料薄膜覆盖也被认为是收集雨水的有效手段，并且是提高产量和提高水资源利用效率最有效的措施之一[10-11]。

　　在发展中国家，砾石覆盖物由于具有价格低廉和可应用范围广等优点而受到青睐，它可以有效地减少土壤水分蒸发，水土流失，增加土壤渗透，使土温变化平稳和保持土壤肥

---
* 原文发表于《南京林业大学学报（自然科学版）》，2017，41(1)：89-95。
基金项目："十二五"国家科技支撑计划（2013BA001B06）；江苏高校优势学科建设工程项目（PAPD）；江苏省农业科技自主创新资金项目（CX(14)2031）。

力[12]。砾石覆盖可以通过提高土壤含水量和稳定土温间接地增加农作物产量[13]。树皮覆盖物广泛地应用于景观绿化方面[14],它可以保持土壤湿度,减少杂草和稳定土温,同时还可促进植物生长,提高农作物产量和质量[14-16]。生物覆盖物尽管吸收土壤中水分,但它可以通过蒸腾作用释放水分从而降低土表温度[17]。此外,在合适的水分温度条件下,生物覆盖物会更快地分解腐烂释放出可供植物或者微生物吸收的养分。然而,覆盖物的影响力终究取决于覆盖物种类、土壤的化学性质和覆盖物所释放出养分的重要性[18]。

尽管覆盖物有众多优点,它们也可能对土壤的质量产生一定破坏,甚至减缓植物生长。砾石、鹅卵石和碎砖块会增加覆盖上下表面的温度引发土壤碱化,导致植物根茎的损伤[19]。树皮类覆盖物有众多局限,包括造成暂时的土壤养分缺失[20],增加火灾隐患[21],以及提高了城市景观植物被未降解树皮所带来的外来植物病菌所侵扰的风险[14]。生物覆盖物常与植物,尤其是含有较多土壤养分的景观植物争夺养分和水分。

1 材料与方法

1.1 试验地概况

所选试验地位于江苏省句容市(31°57′39″ N, 119°12′25″ E)华阳镇下甸村188号的淳盛苗木基地。基地属于亚热带湿润气候地区,年平均气温15.2 ℃,年均降水量1 012 mm,土壤为黄褐色壤质黏土。

1.2 试验设计

研究的五种处理分别为:(1) 空白无覆盖对照(CK);(2) 25 cm × 25 cm 带土带根马尼拉(Manila)草皮(T)根部向下随机平铺覆盖;(3) 大小约3.5 cm × 2 cm,12.5×10^4 kg·ha^{-1}的干燥腐熟松鳞[巧家五针松(*Pinus squamata*)树皮](W)均匀覆盖;(4) 直径约4 cm椭圆白色鹅卵石砾(G)均匀覆盖;(5) 直径约2 cm陶土颗粒(C)均匀覆盖。每块样地均为6 m×2 m,相邻样地间隔4 m。试验开始时间,即移植时间为2013年4月,所移'日香桂'(*Osmanthus fragrans* L.)[23]为一年生苗木,株高约30 cm,每块样地平均种植12株,每平方米1株。移植完成后,铺设相应种类的覆盖物。每种覆盖物每种处理分别设3块重复样地,均完全按照随机区组设计进行划分。移植后两个月内每周浇一次水,之后冬季基本不浇水,遇夏季每半月一次,如遇雨季只浇水维持土壤无干裂。因本试验不就覆盖物对杂草的抑制作用进行研究,为减少杂草因素的影响,2013年4月中旬开始对每个处理每20～30天除草一次,直到11月下旬结束。所有样地的后期养护管理方法一致。由于施肥会影响土壤的团粒结构和其他性质[13],故试验全程未施肥。取样时间分别为2013年8月、2014年2月。每个样地的土样分别在0～5 cm、5～10 cm以及10～15 cm深度进行取样,因10～15 cm深度经分析差异不明显,故本文不加以研究。

1.3 样品的预处理

每个样地的土样分别在深0～5 cm和5～10 cm的位置随机取20个点混合均匀。再把每个土样划分成两部分,一部分储存在4 ℃的冰箱中保鲜,另一部分在实验室常温(25 ℃)

风干。所有土样通过 2 mm 筛网以剔除根和石砾残留。

1.4 样品分析

全氮(N_T),有效氮(N_A),全磷(P_T),有效磷(P_A)的分析依据 Tan et al.(2005)的文章[25]。

微生物碳(MBC)和微生物氮(MBN)[26]用氯仿熏蒸提取法浸提后测定[27]。浸提液中有机碳量(TOC)使用分析仪测定(TOC-C VPN,Shimadzu,Japan)。MBC 的计算公式:
MBC =(熏蒸土样中 C_{org} 一未熏蒸土样中 C_{org})/k_{ec},k_{ec}= 0.33,此数值用来将 C 转换成 MBC[28]。

浸提液中的全氮使用 Lachat flow injection 分析仪测定(Lachat Instruments,Hach Company,Loveland,CO,USA)。MBN 的计算公式:
MBN =(熏蒸土样中的 N_T 一未熏蒸土样中的 N_T)/k_{en},k_{en}= 0.45,此数值用来将 N 转换成 MBN[29]。

1.5 数据分析

所有数据采用方差统计分析法通过 SPSS 20.0 (SPSS Inc,Chicago,IL,USA)软件进行分析,并经过 Duncan's Multiple Range Test 的 5% 有效性验证。

2 试验结果

试验地不同深度处土壤的 pH(图 1A)在不同覆盖处理下呈现出一定的变化。整体看来,空白对照位于 0~5 cm 与 5~10 cm 深度处土壤的 pH 值基本变化不大。相对于空白对照,0~5 cm 深度处土壤的 pH 值明显地受到了覆盖材料的影响。其中,G 覆盖的样地中土壤 pH 值显著高于其他处理,而有机的覆盖处理的土壤 pH 值比对照低,W 与 T 覆盖的样地之间没有表现出显著的差异。与空白对照相比,覆盖物下的样地在 5~10 cm 处土壤的 pH 值却无显著性差异。

土壤容重的变化相较之下甚微。覆盖物处理下的土壤容重在 0~5 cm 和 5~10 cm 均未观察到与对照明显的差异性。W 处理下 0~5 cm 深度的土壤容重较其他处理低,却不显著。整体看来,0~5 cm 土层的容重,有机覆盖略微低于无机覆盖。除了 C 覆盖,其他处理 5~10 cm 深度的土壤容重均低于对照值(图 1B),且个体差异较小。

图 2A 显示了不同处理不同深度处的土壤全氮(N_T)指标。覆盖处理的样地指标均高于对照,而覆盖处理的样地之间未表现出显著性差异。图中最高全氮数值(0.6 g·kg^{-1})出现在 T 覆盖处理的 0~5 cm 深度处。整体看来,0~5 cm 的全氮值略高于 5~10 cm 土壤。

各个处理的土壤全磷(P_T)在 0~5 cm 和 5~10 cm 深度处表现有较大不同(图 2B)。总体来说,0~5 cm 处土壤 P_T 普遍略高于 5~10 cm 处。P_T 在 G 覆盖处理的不同深度处均表现出最低值,分别为 0.218 9 和 0.169 5 g·kg^{-1}。与对照相比,C 和 T 覆盖处理的 P_T 均在 0~5 cm 深度处有一定的增加,却不太显著,两处数值分别为 0.300 5(C)和 0.360 5 g·kg^{-1}(T)。与对照相比,除了 G 覆盖,其他处理在 5~10 cm 深度处的 P_T 有升高趋势,但与 CK 对照之间未表现出显著性差异。

从图 2C 中可看出,不同处理的土壤有效氮(N_A)在 0~5 cm 和 5~10 cm 深度处均有明

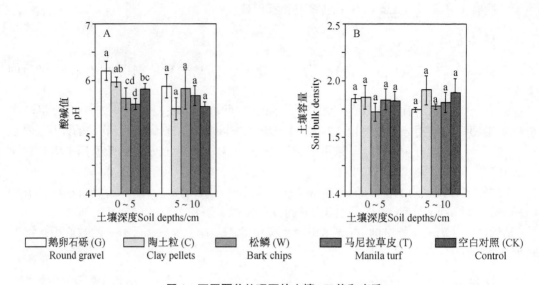

图 1 不同覆盖处理下的土壤 pH 值和容重
Fig. 1 Soil pH and bulk density under different treatments

采用 Duncan's multiple range test 方法分析,同一列不同字母表示显著性差异($P<0.05$)。
For each parameter, columns with the same letter are not significantly different ($P<0.05$) according to Duncan's multiple range test.

显的差异性。相对于其他处理,W 覆盖下的 N_A 在 0~5 cm 呈现出最明显地增加,为 52.59 mg·kg^{-1}。在 5~10 cm 土壤深度,W 处理的表现也很突出,值为 43.72 mg·kg^{-1},但低于浅层土。还可以发现,无机覆盖处理的 N_A 均低于有机覆盖处理。

有效磷(P_A)的含量在覆盖处理后,不同深度均比对照有一定的上升(图 2D)。与无机覆盖相比,有机覆盖物更显著地提高了 0~5 cm 土层的 P_A 含量。而相比对照,在 5~10 cm 深度,无论无机还是有机覆盖材料均能很大程度上增加 P_A 含量,各处理 P_A 含量差异不显著。

土壤脲酶(urease)活性在不同种类覆盖物的影响下表现不同(图 3A)。在 0~5 cm 深度,相比其他处理,两种有机覆盖 W 和 T 对脲酶活性产生了更大的刺激作用而使其呈现出较高数值,T 和 W 处理的脲酶活性无显著差异。所有处理的 5~10 cm 深度土壤中脲酶活性普遍较 0~5 cm 低。与对照相比,在 5~10 cm 处,除了 C 处理,其他处理对脲酶活性均表现出一定的促进作用,但不突出。

在 0~5 cm 土壤深度,每种覆盖处理均对磷酸酶(phosphatase)活性起到一定程度的提升作用(图 3B),但不显著。在 5~10 cm 土壤深度,有机覆盖 W 对磷酸酶活性有一定促进作用,且表现出同深度的最高值。总的来说,磷酸酶在 5~10 cm 土层的数值普遍低于 0~5 cm 土层。

从图 3C 可观察出,最多(236.26 mg·kg^{-1})的和最少(79.42 mg·kg^{-1})的微生物碳(MBC)含量分别出现在 W 覆盖和 G 覆盖的地块。相比于对照,W 覆盖和 T 覆盖的微生物碳量都有一定上升,但是 T 处理上升不大。相比之下,无机覆盖 G 和 C 处理地块的指标却有所下降,且 G 和 C 处理之间差异较小。

W 处理(323.06 mg·kg^{-1})和 T 处理(201.75 mg·kg^{-1})的微生物氮(MBN)含量也

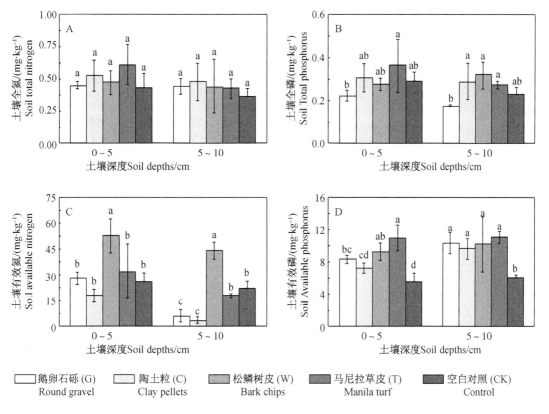

图 2 不同覆盖处理下的土壤全氮,有效氮,全磷,有效磷含量
Fig. 2 Soil total and available nitrogen (N_T and N_A) and phosphorus (P_T and P_A) concentrations under different treatments

采用 Duncan's multiple range test 方法分析,同一列不同字母表示显著性差异($P<0.05$)。
For each parameter, columns with the same letter are not significantly different ($P<0.05$) according to Duncan's multiple range test.

明显高于对照(75.09 mg·kg^{-1})(图 3D)。无机覆盖 G 和 C 对 MBN 也产生一定的促进作用,尽管程度低于有机覆盖。5~10 cm 土壤深度的 MBN 较 0~5 cm 有显著降低,不同覆盖的 MBN 按照数值递减的次序排列为 W(99.13 mg·kg^{-1}) > T(68.88 mg·kg^{-1}) > G(61.53 mg·kg^{-1}) > C(37.35 mg·kg^{-1}) > CK(34.32 mg·kg^{-1})。

3 讨论

各种不同覆盖物已被应用于改善土壤水分、养分和土壤结构。除了可被划分为无机覆盖物和有机覆盖物,覆盖物还可以被划分为化学、物理和生物类覆盖物[30]。在本研究中,鹅卵石砾和陶土颗粒可以被认为是物理类覆盖物和无机覆盖物,而树皮和草皮可以被认为是生物类覆盖物和有机覆盖物。生物类覆盖物已被证明可以保持水分、降低地表温度以及减少水分蒸发[1]。

有机覆盖物分解所产生的酸性物质,可以降低土壤的 pH 值。从土壤 pH 值(图 1A)的结果可看出,W 与 T 覆盖的土壤 pH 值比对照确有一定降低。与树皮相比,草皮是活体材

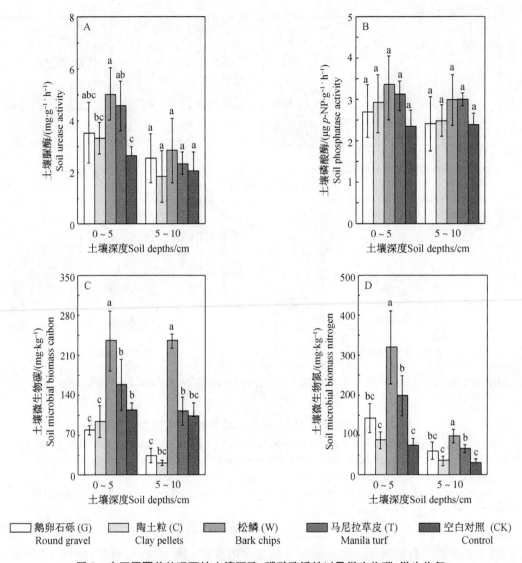

图3 在不同覆盖处理下的土壤脲酶、磷酸酶活性以及微生物碳、微生物氮
Fig. 3 Soil urease activity, phosphatase activity and soil microbial biomass C, microbial biomass N under different treatments

采用 Duncan's multiple range test 方法分析,同一列不同字母表示显著性差异($P<0.05$)。
For each parameter, columns with the same letter are not significantly different ($P<0.05$) according to Duncan's multiple range test.

料,会分泌一些酸性物质从而降低土壤的 pH 值。Hinsinger 等[31]也指出了根系可通过累积 CO_2 而降低土壤 pH 值。另外,G 覆盖的样地中 pH 值显著高于其他处理。土壤有机质分解释放出的离子(NH_4^+)可能导致 pH 值的升高[32]。Billeaud 等[33]指出四种有机覆盖物(过筛松树皮,阔叶树皮,柏树皮,矿化的装饰用松树皮)均可以显著降低沙壤土的 pH 值。Duryea 等[22]也发现松树皮覆盖可以降低土壤 pH 值。然而,Iles 等[19]发现在 Nicollet 沙壤土中无机覆盖物(如鹅卵石和岩石)和有机覆盖物(如树皮)会导致土壤 pH 值升高。结合到本研究中,这些结果均显示了覆盖物对于土壤 pH 值的影响取决于覆盖材料、土壤类型和内部构成。

由于容重与其他指标(如孔隙度、土壤水分和水力传导率)的密切关系,容重作为一项与土壤质量密切相关的数值。在本研究中,各种覆盖处理对土壤容重未产生显著的影响。这可能是由于试验田在整个试验期内未进行土壤耕作而导致[34]。

土壤养分含量是土壤质量与植物生长的重要因素。已有的研究有提及[35],养分存在于不同的土壤粒级中,不同覆盖物影响下的土壤全氮、有效氮、全磷和有效磷有一定的差异(图2)。几乎所有的土样养分指标(除了全氮),有机覆盖的土壤均高于无机覆盖。这可能由于有机覆盖物改变了土壤有机质含量(图1B),在合适的水分和温度条件下,有机覆盖物会分解并释放出有利于植物根摄取以及土壤微生物利用的养分[18],并改善了氮和磷循环的微生物环境[36]。有机覆盖物可以增加土壤的碳输入,并会显著影响微生物的生物量和活性[26]。在本研究中,图3C显示有机覆盖处理下土壤的微生物碳含量高于空白对照以及无机覆盖的土壤,结果与张桂玲[6]、梁贻仓等[8]、张林森等[9]的研究一致。有机覆盖物(尤其是W)的应用,导致土壤有效氮浓度增加,可能会导致微生物量的上升,并可能会影响氮矿化。Bonde等[37]还预测在40周的试验周期内,土壤微生物约会贡献总矿化氮量的55%～89%。尽管不显著,但是T覆盖的5～10 cm深度处N_A有略微降低,这很大程度上跟生物覆盖与植物根系的养分争夺有关[18]。G覆盖几乎不被微生物分解释放养分,N_A较空白对照显著降低。此外N_A值明显降低也会由植物根部养分摄取所导致。

土壤脲酶和磷酸同样在土壤养分循环中起到重要的作用。脲酶可促进尿素水解而产生氨来参与氮的循环,起到调节尿素的功效,相当于施加氮肥[34]。与对照和无机覆盖物相比,脲酶的活性在有机覆盖物的刺激下得到了更好地发挥。这可解释为有机材料通过增强微生物活性来影响土壤脲酶[39]。相比而言,磷酸酶活性却并未因覆盖材料影响而产生显著变化。这可能是因为有效磷量的上升抑制了土壤磷酸酶活性,或者抑制了微生物磷酸酶的合成[40]。本研究结果还表明,地面覆盖有机材料不仅增加了土壤微生物碳量,也提高了微生物氮量(图3D)。导致这种结果的原因可能是土壤碳量的增加[41]和土壤持水力的提高。

本研究中的各种覆盖处理对于土壤性质,如结构、微生物量、养分含量等所产生的积极影响,都将对农业领域中的不同方面,如资源保护、提高生产力等方面产生促进作用。今后的研究将针对覆盖物的质量控制、覆盖厚度,以及种植物的根系活性进一步深入分析。

参考文献

[1] Cook H F, Valdes G S, and Lee H C. Mulch effects on rainfall interception, soil physical characteristics and temperature under *Zea mays* L. [J]. Soil Tillage Research, 2006, 91:227-235.

[2] Bu Y, Shao H, and Wang J. Effects of different mulch materials on corn seeding growth and soil nutrients' contents and distributions[J]. Soil Water Conservation, 2002,16:40-42.

[3] Erenstein O. Crop residue mulching in tropical and semi-tropical countries: An evaluation of residue availability and other technological implications[J]. Soil Tillage Research, 2002,67:115-133.

[4] Döring T F, Brandt M, Heß J, Finckh M R, and Saucke H. Effects of straw mulch on soil nitrate dynamics, weeds, yield and soil erosion in organically grown potatoes[J]. Field Crops Research, 2005,94:238-249.

[5] Kar G, Kumar A. Effects of irrigation and straw mulch on water use and tuber yield of potato in eastern India[J]. Agricultural Water Manage,2007,94:109-116.

[6] 张桂玲.秸秆和生草覆盖对桃园土壤养分含量、微生物数量及土壤酶活性的影响[J].植物生态学,2011, 35 (12):1236-1244.

Zhang G L. Effects of straw and living grass mulching on soil nutrients, soil microbial quantities and soil enzyme activities in a peach orchard[J]. Chinese Journal of Plant Ecology,2011,35（12）：1236-1244.

[7] 贾国梅,张宝林,刘成,等.三峡库区不同植被覆盖对土壤碳的影响[J].生态环境,2008,17(5):2037-2040.

Jia G M, Zhang B L, Liu C, et al. Effects of different vegetation cover on soil carbon in Three Gorges Reservoir[J]. Ecology and Environment,2008,17(5):2037-2040.

[8] 梁贻仓,王俊,刘全全,等.地表覆盖对黄土高原土壤有机碳及其组分的影响[J].干旱地区农田研究.2014,32(5):161-167.

Liang Y C,Wang J,Liu Q Q,et al. Effects of soil surface mulching on soil organic carbon and its fractions in a wheat field in Loess Plateau,China[J]. Agricultural Research In The Arid Areas. 2014, 32(5):161-167.

[9] 张林森,刘富庭,张永旺,等.不同覆盖方式对黄土高原地区苹果园土壤有机碳组分及微生物的影响[J].中国农业科学,2013,46(15):3180-3190.

Zhang L S, Liu F T, Zhang Y W, et al. Effects of Different Mulching on Soil Organic Carbon Fractions and Soil Microbial Community of Apple Orchard in Loess Plateau[J]. Scientia Agricultura Sinica,2013,46(15):3180-3190.

[10] Chakraborty D, Nagarajan S, Aggarwal P, Gupta V, Tomar R, Garg R, et al. Effect of mulching on soil and plant water status, and the growth and yield of wheat (Triticum aestivum L.) in a semi-arid environment[J]. Agricultural Water Manage, 2008,95:1323-1334.

[11] Wang Y, Xie Z, Malhi S S, Vera C L, Zhang Y, and Wang J. Effects of rainfall harvesting and mulching technologies on water use efficiency and crop yield in the semi-arid Loess Plateau, China [J]. Agricultural Water Manage, 2009, 96:374-382.

[12] Li X Y, Liu L Y. Effect of gravel mulch on aeolian dust accumulation in the semiarid region of northwest China[J]. Soil Tillage Res. 2003, 70:73-81.

[13] Fairbourn M L. Effect of gravel mulch on crop yield[J]. Agron J. 1973, 65:925-928.

[14] Koski R, Jacobi W R. Tree pathogen survival in chipped wood mulch[J]. J Arboric. 2004;30:165-171.

[15] Watson G W. Organic mulch and grass competition influence tree root development[J]. J Arboric. 1988,14:200-203.

[16] Sinkevičienė A, Jodaugienė D, Pupalienė R, Urbonienė M. The influence of organic mulches on soil properties and crop yield[J]. Agron Res. 2009, 7:485-491.

[17] Montague T, Kjelgren R. Energy balance of six common landscape surfaces and the influence of surface properties on gas exchange of four containerized tree species[J]. Sci Hortic. 2004, 100:229-249.

[18] Chalker-Scott L. Impact of mulches on landscape plants and the environment—a review[J]. J Environ Hortic. 2007, 25:239.

[19] Iles J K, Dosmann M S. Effect of organic and mineral mulches on soil properties and growth of fairview flame red maple trees[J]. J Arboric. 1999, 25:163-167.

[20] Ashworth S, Harrison H. Evaluation of mulches for use in the home garden[J]. Hort Science 1983, 18:180-182.

[21] Hickman G W, Perry E. Using ammonium sulfate fertilizer as an organic mulch fire retardant[J]. J Arboric. 1996, 22: 279-280.

[22] Duryea M L, English R J, Hermansen L A. A comparison of landscape mulches: Chemical, allelopathic, and decomposition properties[J]. J Arboric. 1999, 25: 88-97.

[23] Wu L C, Chang L H, Chen S H, Fan N C, and Ho J A. Antioxidant activity and melanogenesis inhibitory effect of the acetonic extract of *Osmanthus fragrans*: A potential natural and functional food flavor additive [J]. LWT-Food Science and Technology, 2009, 42: 1513-1519.

[24] Martins M R, Cora J E, Jorge R F, and Marcelo A V. Crop type influences soil aggregation and organic matter under no-tillage[J]. Soil Tillage Research, 2009, 104: 22-29.

[25] Tan K H. Soil sampling, preparation, and analysis[M]. Boca Raton: CRC Press, 2005.

[26] Tu C, Ristaino J B, and Hu S. Soil microbial biomass and activity in organic tomato farming systems: Effects of organic inputs and straw mulching[J]. Soil Biology and Biochemistry, 2006, 38: 247-255.

[27] Ross D. Influence of sieve mesh size on estimates of microbial carbon and nitrogen by fumigation-extraction procedures in soils under pasture [J]. Soil Biology and Biochemistry, 1992, 24: 343-350.

[28] Sparling G, West A W. A direct extraction method to estimate soil microbial C: calibration *in situ* using microbial respiration and ^{14}C labelled cells [J]. Soil Biology and Biochemistry, 1998, 20: 337-343.

[29] Jenkinson D S. Determination of microbial biomass carbon and nitrogen in soil[A]. Wilson J R(ed.). Advances in nutrient cycling in agricultural ecosystems[M]. Wallingford: C. A. B. International, 1988.

[30] Yang Y M, Liu X J, Li W Q, and Li C Z. Effect of different mulch materials on winter wheat production in desalinized soil in Heilonggang region of North China[J]. Journal of Zhejiang University Science B, 2006, 7: 858-867.

[31] Hinsinger P, Plassard C, Tang C, and Jaillard B. Origins of root-mediated pH changes in the rhizosphere and their responses to environmental constraints: a review [J]. Plant and soil, 2003, 248: 43-59.

[32] Tisdale S L, Nelson W L, Beaton J D, Havlin J L. Soil fertility and fertilizers[M]. New York: MacMillan, 1993. p. 634.

[33] Billeaud L A, Zajicek J M. Influence of mulches on weed control, soil pH, soil nitrogen content, and growth of *Ligustrum japonicum*[J]. J Environ Hort. 1989; 7: 155-157.

[34] Ismail I, Blevins R, and Frye W. Long-term no-tillage effects on soil properties and continuous corn yields[J]. Soil Science Society of America J, 1994, 58: 193-198.

[35] Fang S Z, Xie B D, Liu J J. Soil nutrient availability, poplar growth and biomass production on degraded agricultural soil under fresh grass mulch[J]. Forest Ecology and Management, 2008, 255: 1802-1809.

[36] Ramakrishna A, Tam H M, Wani S P, Long T D. Effect of mulch on soil temperature, moisture, weed infestation and yield of groundnut in northern Vietnam[J]. Field Crops Research, 2006, 95: 115-125.

[37] Bonde T A, Schnürer J, Rosswall T. Microbial biomass as a fraction of potentially mineralizable nitrogen in soils from long-term field experiments[J]. Soil Biology and Biochemistry, 1988, 20: 447-452.

[38] Andrews R K, Blakeley R L, and Zerner B. Urease: a Ni (II) metalloenzyme[C]// Lancaster J R. The Bioinorganic Chemistry of Nickel. New York: VCH Publishers, 1989: pp. 141-166.

[39] Garcıa C, Hernandez T, Costa F, Ceccanti B. Biochemical parameters in soils regenerated by the

addition of organic wastes[J]. Waste Management and Research, 1994,12:457-466.

[40] Nannipieri P, Grego S, Ceccanti B, Bollag J, and Stotzky G. Ecological significance of the biological activity in soil[J]. Soil Biochemistry,1990,6:293-337.

[41] 李利利,王朝辉,王西娜,等. 不同地表覆盖栽培对旱地土壤有机碳、无机碳和轻质有机碳的影响[J]. 植物营养与肥料学报,2009,15(2):478-483.

Li L L, Wang Z H, Wang X N, et al. Effects of soil-surface mulching on organic carbon, inorganic carbon and light fraction organic carbon in dryland soil[J]. Plant Nutrition and Fertilizer Science, 2009,15:478-483.

Effects of Mulch on Soil Properties and Microbial Carbon and Nitrogen

NI Xue[1,2], ZHANG Huanchao[3], YANG Xiulian[1], WANG Linggui[1]*

(1. College of Landscape Architecture, Nanjing Forestry University,, Nanjing 210037, China; 2. College of Architecture and Engineering, Anhui University of Technology, Ma'anshan 243000, China; 3 College of Forestry, Nanjing Forestry University, Nanjing 210037, China)

Abstract: The purpose of this study was to compare the characteristics of soil properties and microbial carbon and nitrogen in 'Rixiang Gui' under different mulch treatments. The experimental base was located in Jurong City, Jiangsu Province, China. The experiment lasts for 12 months. Five kinds of mulches (CK, T, W, G, C) were used. Each mulch was repeated in three blocks and arranged in random groups. Organic mulching was W treatment, inorganic mulching included G and C treatment, and biomaterials were T treatment. During the experiment, two samples were taken, and the results were analyzed and compared through a number of soil tests. The results showed that at 0~5 cm soil depth, compared with inorganic mulching G and C, mulching T and W could effectively improve soil nutrients, and significantly increase soil organic matter and soil urease activity. Mulching materials had no significant effect on soil bulk density and total nitrogen, but generally increased soil available nitrogen, available phosphorus, organic matter content, urease, phosphatase and microbial carbon and nitrogen content, but organic mulching treatment was significantly better than inorganic mulching treatment. At 0~5 cm soil depth, organic mulching reduced the soil pH, while inorganic mulching increased the soil pH. Deep soil is less affected by mulch than shallow soil. For the improvement of soil properties, bark has a more positive effect than inorganic mulch. In addition, organic mulch can also significantly improve soil microbial biomass.

Key words: soil properties; soil microorganism; bark; manila turf; pebble gravel; clay particles.

Effects of Mulching on Soil Properties and Growth of Tea Olive (*Osmanthus fragrans*)[*]

Xue Ni[1,2], **Weiting Song**[3], **Huanchao Zhang**[3], **Xiulian Yang**[1], **Lianggui Wang**[1]

1. College of Landscape Architecture, Nanjing Forestry University, Nanjing 210037, China
2. College of Civil Engineering and Architecture, Anhui University of Technology, Maanshan 243000, Anhui, China

Abstract: Different mulches have variable effects on soil physical properties and plant growth. This study aimed to compare the effects of mulching with inorganic (round gravel, RG), organic (woodchips, WC), and living (manila turf grass, MG) materials on soil properties at 0~5 cm and 5~10 cm depths, as well as on the growth and physiological features of *Osmanthus fragrans* L. 'Rixiang Gui' plants. Soil samples were collected at three different time points from field plots of *O. fragrans* plants treated with the different mulching treatments. Moisture at both soil depths was significantly higher after mulching with RG and WC than that in the unmulched control (CK) treatment. Mulching did not affect soil bulk density, pH, or total nitrogen content, but consistently improved soil organic matter. The available nitrogen in the soil increased after RG and WC treatments, but decreased after MG treatment during the experimental period. Mulching improved plant growth by increasing root activity, soluble sugar, and chlorophyll a content, as well as by providing suitable moisture conditions and nutrients in the root zone. Plant height and trunk diameter were remarkably increased after mulching, especially with RG and WC. However, while MG improved plant growth at the beginning of the treatment, the 'Rixiang Gui' plants later showed no improvement in growth. This was probably because MG competed with the plants for water and available nitrogen in the soil. Thus, our findings suggest that RG and WC, but not MG, improved the soil environment and the growth of 'Rixiang Gui' plants. Considering the effect of mulching on soil properties and plant growth and physiology, round gravel and woodchips appear to be a better choice than manila turf grass in 'Rixiang Gui' nurseries. Further studies are required to determine the effects of mulch quality and mulch-layer thickness on shoot and root growths.

Key words: *Osmanthus fragrans*; mulches; soil properties; physiological feature

1 Introduction

Since the late 1930s, mulching has been used for the environmental modification of

---

[*] 原文发表于 *PLOS ONE*,2016,11(8):e0158228。
Fuding: This work was funded by the Ministry of Science and Technology of the People's Republic of China. The funders had no role in study design, data collection and analysis, decision to publish, or preparation of the manuscript.

forests, agriculture lands, and urban landscapes[1]. This process has many advantages: mulches are known to buffer soil temperature[2], prevent soil water loss by evaporation[3], inhibit weed germination, and suppress weed growth[4]. Further, they can protect soils from wind-, water-, and traffic-induced erosion and compaction[1]. Finally, mulch can improve crop production by enhancing soil quality by conserving soil moisture, enhancing soil biological activities, and improving the chemical and physical properties of soil[5-6]. Thus, mulching in urban or ornamental landscapes improves not only soil quality, but also plant growth.

Landscape mulches are generally inorganic (gravel, pebbles, or polyethylene film)[7], organic (wood, bark, or leaves, used individually or in mixtures), or living (turf grass, rye, and clover) materials. In developing countries, gravel is usually preferred because of its low cost and wide availability. Gravel effectively reduces evaporation and runoff, improves infiltration, moderates soil temperature, and maintains soil fertility[8]. It can also indirectly improve crop yield via the interaction between increased soil water and moderated soil temperature[9]. Wood-based mulches are commonly used to improve the appearance of landscapes[10]. They also conserve soil moisture; reduce weed invasion and soil temperature fluctuations; and they improve plant growth, yield, and quality[10-12]. Although living mulches require soil water, they can reduce surface temperatures by releasing water vapor via evapotranspiration[13]. Moreover, living mulches decompose faster under appropriate water and temperature conditions and release nutrients into the soil that can be used by plants and microbes. However, the effects of mulches and their extent depend on the mulch type, soil chemistry, and the importance of the released nutrients[1].

While there are many benefits to using mulches, they can also damage soil quality and decrease plant growth. Inorganic mulches made from rock, gravel, and crushed brick can increase temperatures above and below the mulch layer and cause soil alkalinization, resulting in injuries to plant stems[14]. Wood-based mulches also have several limitations, including temporary soil nitrogen deficiency[15], potential fire hazard[16], and increased risk of introducing exotic plant pathogens to urban landscapes from the uncomposted wood chips[10]. Living mulches often compete for nutrients and water, especially on landscapes with relatively high soil fertility. In addition, allelopathic effects of cool-season turf grasses on woody plants can inhibit tree growth[17]. Very few studies have compared the effects of inorganic, organic, and living mulches on soil quality and plant growth[14,18].

Osmanthus fragrans Lour. 'Rixiang Gui', a member of the family Oleaceae[19], is widely distributed and cultivated as an ornamental plant in southern and central China, where it is considered as one of the most popular traditional flowers[20]. This study aimed to compare the effects of three types of mulches (inorganic, gravel; organic, woodchips; and living, manila turf grass) on soil properties, plant growth, and the physiological performance of *O. fragrans*.

2　Materials and Methods

2.1　Site Description

Field experiments were conducted between April 2013 and June 2014 in a nursery (31°57′39″ N, 119°12′25″ E) at the Institute of Landscape Architecture, Nanjing Forestry University, China. The subtropical location of the nursery is characterized by its humid climate with an annual mean temperature and precipitation of 15.2 ℃ and 1 012 mm, respectively. The soil in the experimental field is classified as yellow brown soil (25.6% clay, 68.8% silt, and 5.6% sand) and was cleared and hoed manually before the onset of the experiments.

2.2　Experimental Design

Four treatments were established in 6 m×2 m plots: (1) unmulched control soil (CK); (2) inorganic mulch, approximately 1.5 cm layer of <4 cm diameter rounded gravel (RG; 1 250 t·ha^{-1}); (3) organic mulch (WC), an approximated 1 cm layer of wood chips that extended 3~4 cm in length from dried mature *Pinus squamata* X. W. Li (127.5 t·ha^{-1}); and (4) living mulch (MG), a 5 cm layer of 25 cm ×25 cm pieces of manila turf grass with soil and roots attached. *O. fragrans* were first grown to approximately 30 cm in height in a greenhouse and then transplanted to the plots at a density of 12 plants per plot (one plant per square meter). The mulches were applied after all *O. fragrans* had been planted. The experiment was a randomized complete block design with three replicates.

During the experimental period, no fertilizer was applied and watering and weeding practices were consistent with those used by local farmers. After the seedlings were transplanted, roots and leaves were pump-irrigated once each day during the first three days and then irrigated once a week in the first month. Plants were irrigated every 15 days at 04:00 PM during the summer, but no irrigation was performed during the winter.

Weeds were removed manually every 20~30 days between April and November.

2.3　Soil Sampling and Analysis

Soils were sampled at three different times: May 23, 2013; October 23, 2013; and May 23, 2014. Before soil core sampling, mulch was removed from the sampling area to prevent the contamination of the cores with surface organic matter. Approximately 20 soil cores were randomly collected from each plot and divided into two layers: 0~5 cm and 5~10 cm. In the field, each sample was divided into two parts and sealed in plastic bags: one part was stored at 4 ℃ for the analysis of basic soil properties; the other part was air-dried in a ventilated room, ground, and filtered through a<2-mm mesh to remove stones, root fragments, and organic debris before performing chemical analyses.

Soil samples that had been air-dried and filtered through a 2-mm mesh were used to determine the available nitrogen (N) content and measure pH. Additional samples were passed through a 0.149 mm mesh to estimate organic matter and total N contents. Soil organic matter (SOM) was measured using $H_2SO_4 - K_2Cr_2O_7$ wet oxidation, followed by titration with $FeSO_4$ according with the Walkley–Black procedure[21]; soil total nitrogen (STN) was determined using micro-Kjeldahl digestion, followed by colorimetric analysis[22-23]. Soil pH was measured in a 1 : 2.5 (m/v) soil : water ratio by using a pHS-3C pH/mV meter (Rex Ltd., Shanghai, China). Soil moisture was determined after the soil core samples were oven-dried at 105 ℃ for 8 h[24]. Soil available nitrogen (SAN) was determined using the alkali-hydrolytic diffusion method[25]. Soil bulk density was measured from samples obtained using a volumetric steel ring (100 cm³) and calculated as the mass of oven-dried soil (105 ℃), divided by the core volume for each measurement depth.

2.4 Plant Growth and Physiological Features

Trunk diameter and plant height were determined on the same days soil were sampled; diameter was measured at 15 cm above the soil surface using a caliper, and height was measured from the soil surface to the highest point in the tree crown. Simultaneously, roots and leaves were collected to determine root activity and the relative water content (RWC); relative electric conductivity (REC); and chlorophyll, soluble sugar, and free proline content of the leaves. Root activity was measured using the triphenyl tetrazolium chloride (TTC) method[26]; RWC, using Barrs and Weatherley's method[27-28]; and REC, which indicates the permeability of a leaf, was measured with a DDS-11A meter (Rex Ltd., Shanghai, China)[29]. Chlorophyll content was measured spectrophoto metrically by the method and equations proposed by Lorenzen[30]. Leaf soluble sugar and proline contents were quantified in extracts of fresh leaves (0.1 g) in potassium phosphate buffer (50 mM, pH=7.5)[31]. The extracts were filtered through four layers of cheesecloth and centrifuged at 15 500 rpm for 15 min at 4 ℃, and the resulting supernatant was collected and stored at 4 ℃. Soluble sugar was analyzed using the anthrone reagent and a Bauschand Lomb spectrophotometer[32]. Free proline was estimated by spectrophotometric analysis of a ninhydrin reaction solution[33] at 515 nm in a UV-2900 spectrophotometer(HITACHI, Japan).

2.5 Statistical Analysis

All data were subjected to analysis of variance (ANOVA) tests, and the means were compared using Student's t-tests by using JMP version 9.0 (SAS Institute Inc., Cary, NC, USA). A repeated measures ANOVA was performed for soil properties, plant height, and trunk diameter to analyze the effects of mulch type and sampling time. Differences were considered significant at $P < 0.05$.

3 Results and Discussion

3.1 Mulching Materials Have Differential Effects on Soil Properties

Soil properties showed varying effects over time to the different mulching treatments. The average soil moisture at the 0~5 cm depth layer of CK, RG, WC, and MG plots was 18.0%, 20.3%, 21.6%, and 20.3%, respectively (Tab. 1). Similarly, the average soil moisture at the 5~10 cm depth was 19.2%, 21.7%, 21.6%, and 20.6%, respectively (Tab. 2). Soil moisture values were higher in June 2014 than in May 2013 in the top and bottom layers of all treatments, indicating that mulching increases soil moisture. WC treatment had a stronger effect on soil moisture than the RG and MG treatments (Tabs. 1 and 2).

Tab. 1 Soil properties at the 0~5 cm depth at the three sampling time points with and without mulching treatments.

Sampling time	Treatment	Moisture/%	Bulk density/(g·m^{-3})	pH	STN/(g·kg^{-1})	SAN/(g·kg^{-1})	SOM/(g·kg^{-1})	C/N
2013-05-23	CK	18.6±1.7 b	1.33±0.05 a	5.8±0.2 bc	0.58±0.09 a	50.2±4.7 a	8.4±15 b	8.4±0.7 a
	RG	21.3±1.8 ab	1.37±0.06 a	6.1±0.1 a	0.56±0.03 a	54.5±2.6 a	10.4±1.5 ab	10.8±1.9 a
	WC	24.3±0.3 a	1.32±0.08 a	6.0±0.1 ab	0.56±0.05 a	56.1±1.7 a	12.9±1.8 a	13.5±2.7 a
	MG	20.5±2.4 ab	1.35±0.07 a	5.7±0.2 c	0.56±0.05 a	52.8±5.6 a	10.8±1.1 ab	11.4±2.0 a
2013-10-23	CK	14.2±0.6 b	1.30±0.05 a	5.8±0.2 a	0.55±0.07 a	34.8±7.0 ab	8.3±1.2 c	8.8±1.0 b
	RG	16.6±1.4 a	1.35±0.06 a	6.0±0.1 a	0.50±0.05 a	39.1±4.5 ab	11.5±0.9 b	13.6±2.5 a
	WC	16.0±0.5 a	1.29±0.08 a	5.9±0.2 a	0.48±0.07 a	43.9±5.5 a	13.6±0.5 a	15.1±0.5 a
	MG	15.8±0.6 a	1.29±0.10 a	5.8±0.2 a	0.51±0.03 a	28.4±5.9 b	11.1±1.2 b	12.5±1.7 a
2014-06-23	CK	21.3±1.5 b	1.32±0.08 a	5.9±0.1 a	0.55±0.13 a	23.1±1.1 b	8.5±1.4 c	9.3±2.2 b
	RG	22.9±1.3 ab	1.33±0.07 a	5.9±0.1 a	0.50±0.03 a	29.9±1.0 a	9.8±0.6 bc	10.9±1.6 b
	WC	24.6±1.9 a	1.34±0.02 a	5.8±0.1 a	0.51±0.10 a	31.7±0.7 a	13.8±1.1 a	14.2±1.3 a
	MG	24.6±1.8 a	1.32±0.02 a	5.9±0.1 a	0.53±0.07 a	21.8±1.2 b	11.3±0.9 b	12.4±1.5 ab

CK: no mulching; RG: mulching with round gravel; WC: wood chips; MG: manila turf grass; STN: soil total nitrogen; SOM: soil organic matter; SAN: soil available nitrogen; C/N: carbon to nitrogen ratio. Values with different letters in the same column indicate significant differences between treatments ($P < 0.05$, $n = 3$). Data are means ± standard deviation.

Tab. 2 Soil properties at the 5~10 cm depth at the three sampling time points with and without mulching treatments

Sampling time	Treatment	Moisture/%	Bulk density/(g·m^{-3})	pH	STN/(g·kg^{-1})	SAN/(g·kg^{-1})	SOM/(g·kg^{-1})	C/N
2013-05-23	CK	20.4±3.3 b	1.32±0.08 a	5.7±0.2 a	0.57±0.04 a	50.4±5.1 a	9.3±0.8 b	9.5±1.0 b
	RG	23.2±2.0 ab	1.38±0.07 a	5.8±0.1 a	0.56±0.06 a	54.8±2.8 a	11.3±0.9 a	11.7±0.3 ab
	WC	23.1±0.4 a	1.31±0.06 a	5.7±0.1 a	0.52±0.03 a	53.2±4.1 a	11.8±1.3 a	13.2±1.9 a
	MG	21.4±2.6 b	1.36±0.03 a	5.7±0.1 a	0.55±0.07 a	54.6±4.7 a	11.3±0.7 a	12.1±1.3 a
2013-10-23	CK	14.8±1.2 b	1.31±0.02 a	5.9±0.0 a	0.60±0.07 a	34.1±5.1 ab	8.4±1.3 b	8.2±1.5 b
	RG	17.4±0.8 a	1.33±0.10 a	6.0±0.1 a	0.52±0.04 a	39.3±2.6 a	10.8±1.0 a	12.1±0.6 a
	WC	17.3±1.4 a	1.31±0.05 a	5.9±0.2 a	0.49±0.08 a	40.8±2.5 a	11.4±0.7 a	14.6±2.3 a
	MG	17.7±1.0 a	1.35±0.05 a	5.8±0.1 a	0.54±0.07 a	31.2±5.6 b	10.5±0.8 a	11.4±2.1 ab

Cont. Tab. 2

Sampling time	Treatment	Moisture/%	Bulk density/(g·m^{-3})	pH	STN/(g·kg^{-1})	SAN/(g·kg^{-1})	SOM/(g·kg^{-1})	C/N
2014-6-23	CK	22.4±0.9 b	1.33±0.05 a	5.9±0.1 a	0.52±0.04 a	23.1±1.8 b	8.6±0.5 b	9.7±0.4 c
	RG	24.3±1.5 ab	1.36±0.09 a	5.9±0.1 a	0.47±0.08 a	29.9±1.4 a	11.8±1.2 a	14.7±1.5 ab
	WC	24.5±0.6 a	1.37±0.06 a	5.8±0.0 a	0.45±0.05 a	30.2±1.9 a	12.7±1.9 a	15.0±0.9 a
	MG	22.7±1.0 ab	1.36±0.08 a	6.0±0.1 a	0.53±0.07 a	20.3±1.8 b	10.5±1.2 ab	12.7±1.4 bc

CK: no mulching; RG: mulching with round gravel; WC: wood chips; MG: manila turf grass; STN: soil total nitrogen; SOM: soil organic matter; SAN: soil available nitrogen; C/N: carbon to nitrogen ratio. Values with different letters in the same column indicate significant differences between treatments ($P < 0.05$, $n = 3$). Data are means ± standard deviation.

All of mulch types had significant effects on soil moisture in both soil layers measured except for MG at the 5~10 cm depth (Tab. 3). These results are consistent with previous studies that suggest mulching with gravel[14, 34-35], wood chips[2-3, 11-12, 14], and grass[36] sequesters water and prevents water loss from the soil through evaporation; additionally, organic mulches conserve water more effectively than inorganic ones[14, 37]. Adequate water is essential for plant growth. However, some studies show that living mulches might compete with plants for water and hence, mulched soils can show lower moisture content than bare soils[11]. In the present study, mulching with turf grass significantly increased soil moisture at the 0~5 cm depth, but had no effect on soil moisture at the 5~10 cm depth (Tab. 3). These results maybe influenced by the high precipitation at the study site.

Tab. 3. Significance ($P < 0.05$) of soil properties, plant height, and trunk diameter at the two soil depths (0~5 cm and 5~10 cm) over time after the three mulching treatments assessed by ANOVA of three replicates

Parameters	Moisture		pH		SOC		C/N		Available N		Plant height	Trunk diameter
	0~5 cm	5~10 cm	0~5 cm	5~10 cm	0~5 cm	5~10 cm	0~5 cm	5~10 cm	0~5 cm	5~10 cm		
RG	0.006	0.015	0.003	0.142	<0.001	<0.001	0.004	0.142	0.020	0.006	<0.001	<0.001
Time	<0.001	<0.001	0.774	0.033	0.921	0.631	0.310	0.033	<0.001	<0.001	<0.001	<0.001
RG*Time	0.807	0.907	0.292	0.545	0.951	0.740	0.303	0.545	0.829	0.828	0.521	0.179
WC	<0.001	0.008	0.252	0.628	<0.001	<0.001	<0.001	0.628	0.002	0.008	<0.001	<0.001
Time	<0.001	<0.001	0.817	0.113	0.789	0.803	0.534	0.114	<0.001	<0.001	<0.001	<0.001
WC*Time	0.066	0.936	0.412	0.675	0.833	0.904	0.724	0.675	0.785	0.547	0.291	0.006
MG	<0.001	0.149	0.836	0.964	0.004	0.004	0.001	0.964	0.468	0.807	0.004	0.001
Time	<0.001	<0.001	0.368	0.021	0.460	0.302	0.555	0.021	<0.001	<0.001	<0.001	<0.001
MG*Time	0.617	0.483	0.738	0.548	0.321	0.236	0.920	0.548	0.307	0.315	0.010	0.682

RG: round gravel; WC: wood chips; MG: manila turf grass; SOC: soil organic carbon; C/N: carbon to nitrogen ratio. Significant values are italicized.

Mulching had no effect on the bulk density of either soil layer (Tabs. 1 and 2). At the 0~5 cm depth, soil pH was 5.2% higher in the RG treatment compared to the CK treatment (Tab. 1); no significant change in soil pH was noted in the other treatments. However, pH values in the RG treatment did not change in the following sampling times (Tabs. 1 and 2), indicating that the effect was ephemeral. The elevated pH values possibly resulted from the leaching of basic cations (NH_4^+) from the decomposing SOM[38]. At the 5~10 cm depth, no significant differences in soil pH were observed among treatments. Billeaud and Zajicek[39] reported that mulching with four types of organic mulches (screened pine bark, hardwood, cypress, and decorative pine bark nuggets) significantly decrease soil pH in a soil composed of fine sandy loam. Duryea et al.[18] also found that pine bark mulches decrease soil pH. However, Iles and Dosmann[14] found that mulching with inorganic (e. g. , river rock and lava rock) and organic (e. g. , wood chips and shredded bark) mulches remarkably increased soil pH in a Nicollet fine sandy loam soil. Taken together with our findings, these results suggest that the effect of mulches on soil pH depends on the mulching material as well as soil composition/type.

The three mulching treatments analyzed in this study did not affect STN content at any soil depth during the experimental period (Tabs. 1 and 2). At the first and second sampling times, no obvious differences were found in the SAN content between mulched soils and bare soil at both the sampling depths. Nonetheless, WC and MG treatments increased SAN contents by 29.4% and 37.2% at the 0~5 cm depth, and by 29.4% and 30.7% at the 5~10 cm depth, respectively, at the third sampling time (Tabs. 1 and 2). Repeated measures ANOVA showed that RG and WC treatments significantly altered SAN content, whereas the MG treatment did not (Tab. 3). Since organic mulches decompose under appropriate water and temperature levels, nutrients are released to the soil and become available for root uptake or microbial use[1]. Although not significantly, SAN contents decreased in the MG treatment over time (Tabs. 1 and 2); this may be mainly attributed to the competition for nutrients between turf grass and plants[1]. Gravel mulches always contain fewer nutrients and are difficult for microorganisms to decompose. Thus, the increase in SAN content after RG treatment maybe because the gravel provided suitable conditions for the growth of microorganisms that released more nutrients via SOM decomposition. In addition, gravel mulch can trap dirt, which contains nitrogen and organic matter. Therefore, nutrient content of gravel-mulched fields is high[8]. The significant decrease in SAN contents over time can be caused by plant nutrient uptake.

SOM is derived from the decay of dead organisms and consists of organic (carbon-based) compounds[40]. It positively contributes to tree and environmental health through its effects on soil physical, chemical, and biological properties[41]. Mulches increased SOM content and the ratio of carbon (C) to N at both soil depths tested (Tabs. 1 and 2) and repeated measures ANOVA also showed that the three mulching treatments significantly altered SOM content at both soil depths ($P < 0.05$). Although the three treatments had significant effects on the ratios of C to N at the 0~5 cm depth, they had no significant effect on these ratios at the 5~10 cm depth (Tab. 3). Previous studies have shown that organic materials increase SOM by directly improving soil properties[42], increasing photosynthesis, and by having an impact on belowground C allocation[43]. Manila turf grass remarkably influences soil properties and processes, including the increase in

SOM[44-45]. The stimulation of plant growth after the three mulching treatments can be attributed to an increase in photosynthesis, resulting in the higher sequestration of C.

3.2 Mulching Materials Varyingly Affect Plant Physiological Features

Several plant physiological features were measured to evaluate the effects of the mulches on plant health. Although root activity is an important physiological parameter for evaluating ion uptake, few studies have considered how it is affected by mulch treatment. Root activity as measured by TTC reducing capacity, was highest in WC, followed by RG, MG, and CK (Tab. 4). Moreover, root activity significantly increased after WC, RG, and MG treatments, suggesting that plants grown in mulched soils take up more nutrients than those grown in un mulched soils. Thus, the present findings are consistent with those of Chalker-Scott[1] who showed that root development and density are greater in soils treated with organic mulches than in those treated with nothing or plastic or living mulches.

Tab. 4. Physiological features of plants measured on May 23, 2014 in bare soil (CK) or soils treated with round gravel (RG), wood chips (WC), and manila turf grass (MG)

Treatment	Root activity/ ($\mu g \cdot g^{-1} \cdot h^{-1}$)	RWC/ %	REC/ %	Proline/ ($\mu g \cdot g^{-1}$)	Soluble sugar/ %	Chlorophyll a/ ($mg \cdot g^{-1}$)	Chlorophyll b/ ($mg \cdot g^{-1}$)	Chlorophyll ($mg \cdot g^{-1}$)
CK	232.1±9.0 cb	82.9±2.8 a	34.5±5.1 a	0.10±0.00 a	0.64±0.09 b	0.80±0.10 c	0.60±0.04 a	1.39±0.11 c
RG	280.1±9.7 a	83.3±2.6 a	33.8±8.9 a	0.12±0.02 a	0.88±0.04 a	1.09±0.08 ab	0.51±0.04 a	1.60±0.12 ab
WC	289.0±12.6 a	83.7±2.4 a	41.0±4.3 a	0.08±0.02 a	0.86±0.11 a	1.17±0.10 a	0.57±0.03 a	1.74±0.13 a
MT	253.4±11.3 b	83.5±3.2 a	38.5±6.8 a	0.09±0.01 a	0.79±0.06 a	0.97±0.02 b	0.51±0.06 a	1.48±0.07 bc

RWC: relative water content of leaves; REC: relative electric conductivity. chlorophyll is the sum of chlorophyll a and chlorophyll b. Different letters in the same column indicate significant differences between treatments ($P < 0.05$, $n = 3$). Data are means ± standard deviation.

The different mulches did not affect the RWC, REC, and proline content of leaves. However, the soluble sugar content increased by 37.5%, 34.4%, and 23.4% in RG, WC, and MG treatments, respectively, suggesting that mulches stimulated stress resistance of plants. Mulches did not change the chlorophyll b content, although chlorophyll a content increased by 36.3%, 46.3%, and 21.3% in the RG, WC, and MG treatments, respectively, indicating that mulches enhanced the photosynthetic rate in the leaves of plants grown under these conditions.

Some studies have shown that soil mulching decreases plant health by increasing plant stem temperature[14] and causing soil alkalinization or acidification[11,14]. In addition, uncomposted wood chips derived from wood packing materials can increase the risk of introducing exotic plant pathogens to urban landscapes[10]. Conversely, the findings of this study suggest that mulching with round gravel, woodchips, and manila turf grass positively affect plant physiological features, thus benefiting the health and growth of 'Rixiang Gui'.

3.3 Mulches Differentially Affect Plant Height and Trunk Diameter

Compared to those in CK, the plants in the RG and WC treatments showed significantly increased plant height; however, MG only increased plant height at the first sampling time and not at later time points (Fig. 1). RG, WC, and MG treatments also increased trunk diameter at the first two sampling times; at the last time point, the trunk diameters increased in RG and WC treatments but not in MG (Fig. 1). Repeated measures ANOVA analysis showed that the three mulching treatments significantly affected plant height and trunk diameter (Tab. 3), suggesting that soil mulching stimulates plant growth.

The maximum plant height and trunk diameter were 59.2 ± 1.3 cm and 0.91 ± 0.2 cm, respectively, in the WC treatment (Fig. 1C and 1F). Mulching with wood chips significantly improved plant growth, likely though improving soil properties (e.g., moisture, SAN, and SOM) and plant physiological parameters. Scharenbroch[43] also reported that organic material could be beneficial for tree establishment, weed control, and root decay. However, Iles and Dosmann[14] found that tree height and stem diameter of red maple trees (*Acer rubrum* L.) were not affected after two years of mulching with organic materials, although organic mulches are known to remarkably influence soil temperature, moisture, and pH. Ferrini et al.[46] also found that mulching with pine bark did not significantly affect the height or trunk diameter of ornamental trees. These contradictory findings might be attributed to the fact that organic matter content differs across different soil layers, with the 0～10 cm soil layer having more positive impacts on shoot growth and physiological attributes of plants, and >15 cm soil layers suffer from decreased water penetration[43], increased soil tension, reduced shoot growth, and enhanced plant stress.

Plants of the RG treatment were significantly taller (Fig. 1) and this finding is consistent with that of Fairbourn[9], who suggested that gravel mulch could increase crop yield by improving the interaction between the increased soil water content and soil temperature. Holloway[48] also found that five woody plant species grew better after stone mulch treatments than after other mulch treatments. However, Iles and Dosmann[14] found that the height and stem diameter of red maple were not affected after two years of mulching with crushed red brick, pea gravel, lava rock, carmel rock, and river rock.

In this study, mulching with MG increased soil moisture and SOM content, there by stimulating plant growth (Fig. 1, first sampling time). However, this effect decreased with time as shown by the non-significant increase in plant growth at the second and third time points. This might be attributed to the fact that manila turf grass competed with 'Rixiang Gui' for nutrients and water, leading to nutritional deficiencies in the mulched plants[1]. In fact, SAN was lower in MG than in CK, although this difference was not significant at the second and third time points (Tabs. 1 and 2). Previous studies show that tree establishment and growth are inhibited by turf grass mulches such as Bermuda grass,

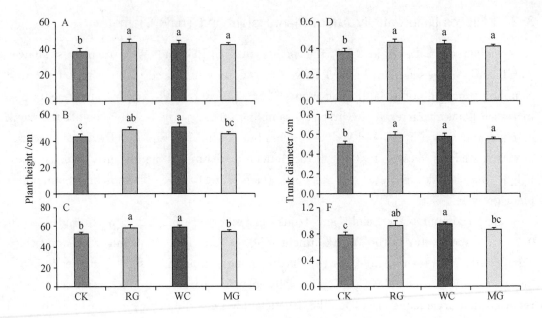

Fig. 1 Effects on plant height and trunk diameter according to mulch type

The physiological parameters of plants grown in unmulched soil (CK) and those grown in soil mulched with round gravel (RG), wood chips (WC), and manila turf grass (MG) were measured on May 23, 2013 (A, D), October 23, 2013 (B, E), and May 23, 2014 (C, F). Different letters above bars indicate significant differences between treatments ($P < 0.05$, $n = 3$). Bars indicate standard deviation.

tall fescue, and Kentucky bluegrass[17]. Watson[11] also reported that, compared to plants grown in bare soil, root density was remarkably reduced in plants mulched with grass. Hence, mulching with turf grass might not be beneficial for plant growth in the long term.

4 Conclusions

Our findings suggest that round gravel, woodchips, and manila turf grass help create a healthy soil environment and that different mulches have different effects on the soil properties at the two soil depths that were sampled. Soil moisture and SOM increased at both soil depths, whereas mulching had no effect on the bulk density, pH, or STN. SAN increased in soils mulched with round gravel and wood chips, but not with manila turf grass mulch. Root activity, soluble sugar, and chlorophyll a contents increased in all the mulched soils, especially those mulched with round gravel and wood chips. Plant height and trunk diameter were significantly higher after mulching with round grave land wood chips; however, the stimulating effect of manila turf grass decreased gradually because of the competition for SAN between turf grass and 'Rixiang Gui'. Therefore, considering the effect of mulching on soil properties and plant growth and physiology, round gravel and wood chips are a better choice than manila turf grass in 'Rixiang Gui' nurseries and plantations. Further studies are required to determine the effects of mulch quality and mulch-layer thickness on shoot and root growth.

Author Contributions

Conceived and designed the experiments: Xue Ni, Huanchao Zhang. Performed the experiments: Xue Ni, Weiting Song. Analyzed the data: Xue Ni. Contributed reagents/materials/analysis tools: Huanchao Zhang. Wrote the paper: Xue Ni. Funding support: Lianggui Wang.

References

[1] Chalker-Scott L. Impact of mulches on landscape plants and the environment—a review. J Environ Hortic. 2007; 25:239.

[2] Greenly K M, Rakow D A. The effect of wood mulch type and depth on weed and tree growth and certain soil parameters. J Arboric. 1995; 21:225-225.

[3] Gleason M L, Iles J K. Mulch matters: The proper use of organic mulch offers numerous benefits for your woody landscape plants. Am Nurseryman. 1998; 187:2431.

[4] Rathinasabapathi B, Ferguson J, Gal M. Evaluation of allelopathic potential of wood chips for weed suppression in horticultural production systems. HortScience 2005; 40:711-713.

[5] Cooper A J. Root temperature and plant growth, a review. Slough: Commonwealth Agricultural Bureaux;1973.

[6] Hanada T. The effect of mulching and row covers on vegetable production. Kyoto: Food and Fertilizer Technology Center;1991.

[7] Black R J, Gilman EF, Knox GW, Ruppert KC. Mulches for the landscape. Florida Cooperative Extension Service Fact Sheet ENH 103. Gainesville: University of Florida; 1994, p. 4.

[8] Li X Y, Liu L Y. Effect of gravel mulch on Aeolian dust accumulation in the semiarid region of northwest China. Soil Tillage Res. 2003; 70: 73-81.

[9] Fairbourn M L. Effect of gravel mulch on crop yield. Agron J. 1973; 65: 925-928.

[10] Koski R, Jacobi WR. Tree pathogen survival in chipped wood mulch. J Arboric. 2004; 30:165-171.

[11] Watson G W. Organic mulch and grass competition influence tree root development. J Arboric. 1988; 14:200-203.

[12] Sinkevičienė A, Jodaugienė D, Pupalienė R, Urbonienė M. The influence of organic mulches on soil properties and cropyield. Agron Res. 2009; 7: 485-491.

[13] Montague T, Kjelgren R. Energy balance of six common landscape surfaces and the influence of surface properties on gas exchange of four containerized tree species. Sci Hortic. 2004; 100:229-249.

[14] Iles J K, Dosmann M S. Effect of organic and mineral mulches on soil properties and growth of fairview flame red maple trees. J Arboric. 1999; 25:163-167.

[15] Ashworth S, Harrison H. Evaluation of mulches for use in the home garden. HortScience 1983; 18: 180-182.

[16] Hickman G W, Perry E. Using ammonium sulfate fertilizer as an organic mulch fire retardant. J Arboric. 1996; 22: 279-280.

[17] Griffin J J, Reid W R, Bremer D J. Turf species affect establishment and growth of redbud and

pecan. HortScience 2007; 42: 267-271.

[18] Duryea M L, English R J, Hermansen L A. A comparison of landscape mulches: Chemical, allelopathic, and decomposition properties. J Arboric. 1999; 25: 88-97.

[19] Ômura H, Honda K, Hayashi N. Floral scent of *Osmanthus fragrans* discourages foraging behavior of cabbage butterfly, *Pieris rapae*. J Chem Ecol. 2000; 26: 655-666.

[20] Wang H, Pan Y, Tang X, Huang Z. Isolation and characterization of melanin from *Osmanthus fragrans*' seeds. LWT-Food Sci Technol. 2006; 39: 496-502.

[21] Nelson D W, Sommers L E. Total carbon, organic carbon, and organic matter. In: Page AL, Miller RH, Keeney DR, editors. Methods of Soil Analysis. Part 2. Madison: American Society of Agronomy; 1982. pp. 539-579.

[22] Nelson D W, Sommers L E. Total nitrogen analysis of soil and plant tissues. J Assoc Off Anal Chem. 1980; 63: 770-778.

[23] Hook P B, Burke I C, Lauenroth W K. Heterogeneity of soil and plant N and C associated with individual plants and openings in North-American short grass steppe. Plant Soil 1991; 138: 247-256.

[24] Top G C, Ferre P A. Methods for measurement of soil water content. Thermogravimetric using convective oven-drying. In: Dane JH, Top GC, editors. Methods of Soil Analysis, Part 4: Physical Methods. Madison: Soil Science Society of America; 2002. pp. 422-424.

[25] Page A L, Miller R H, Keeney DR. Methods of soil analysis, Part 2: Chemical and microbiological properties. 2nd ed. Madison: American Society of Agronomy; 1983.

[26] Chen C L. Measurement of plant root activity(TTC). In: Li HS, editor. Principle and Technology of Plant Physiological and Biochemical Experiments. Beijing: Higher Education Press; 2003. pp. 119-120.

[27] Barrs H D, Weatherley P E. A re-examination of the relative turgidity technique for estimating water deficits in leaves. Aust J Biol Sci. 1962; 15: 413-428.

[28] Saeed R, Mirza S, Ahmad R. Electrolyte leakage and relative water content as affected by organic mulch in okra plant (*Abelmoschus esculentus* L.) moench grown under salinity. FUUAST J. Biol. 2014, 4(2): 221.

[29] Zhu G R, Zhong H W, Zhang A Q. Plant physiology experiment. Beijing: Beijing University Press; 1990. pp. 242-245.

[30] Lorenzen C J. Determination of chlorophyll and pheo-pigments: Spectrophotometric equations. Limnol Oceanogr. 1967; 12: 343-346.

[31] Goicoechea N, Aguirreolea J, Cenoz S, Garcia-Mina J M. *Verticillium dahliae* modifies the concentrations of proline, soluble sugars, starch, soluble protein, and abscisic acid in pepper plants. Eur J Plant Pathol. 2000; 106: 19-25.

[32] Yemm E W, Willis A J. The estimation of carbohydrates in plant extracts by anthrone. Biochem J. 1954; 57: 508. PMID: 13181867.

[33] Irigoyen J J, Einerich D W, Sánchez-Díaz M. Water stress induced changes in concentrations of proline and total soluble sugars in nodulated alfalfa (*Medicago sativa*) plants. Physiol Plant. 1992; 84: 55-60.

[34] Nachtergaele J, Poesen J, van Wesemael B. Gravel mulching in vineyards of southern Switzerland. Soil Tillage Res. 1998; 46: 51-59.

[35] Yamanaka T, Inoue M, Kaihotsu I. Effects of gravel mulch on water vapor transfer above and below the soil surface. Agr Water Manage. 2004; 67: 145-155.

[36] Hartman J R, Pirone T P, Sall M A. Pirone's tree maintenance. Oxford: Oxford University Press; 2000.

[37] Singh B, Gupta G N, Prasad K G, Mohan S. Use of mulches in establishment and growth of tree species on dry lands. Indian Forester. 1988; 114: 307-316.

[38] Tisdale S L, Nelson W L, Beaton J D, Havlin J L. Soil fertility and fertilizers. New York: MacMillan; 1993. p. 634.

[39] Billeaud L A, Zajicek J M. Influence of mulches on weed control, soil pH, soil nitrogen content, and growth of *Ligustrum japonicum*. J Environ Hort. 1989; 7: 155-157.

[40] Brady N C, Weil R R. Soil colloids: seat of soil chemical and physical acidity. In: Brady NC, Weil RR, editors. The Natureand Properties of Soils. New Jersey: Pearson EducationInc.; 2008. pp. 311-358.

[41] Magdoff F R. Soil organic matter fractions and implications for interpreting organic matter tests. In: Magdoff F R, Tabatabai M A, Hanlon E A Jr, editors. Soil Organic Matter: Analysis and Interpretation. Madison: Soil Science Society of America Special Publication No. 56; 1996. pp. 11-21.

[42] Scharenbroch B C, Lloyd J E. Particulate organic matter and soil nitrogen availability in urban landscapes. Arboric Urban For. 2006; 32: 180.

[43] Scharenbroch B C. A meta-analysis of studies published in Arboriculture & Urban Forestry relating to organic materials and impacts on soil, tree, and environmental properties. J Arboric. 2009; 35: 221.

[44] Skroch W A, Shribbs J M. Orchard floor management: Anoverview. HortScience 1986; 21: 390-394.

[45] Hogue E J, Neilsen G H. Orchard floor vegetation management. Hortic Rev. 1987; 9: 377-430.

[46] Ferrini F, Fini A, Frangi P, Amoroso G. Mulching of ornamental trees: Effects on growth and physiology. Arboric Urban For. 2008; 34: 157.

[47] Arnold M A, McDonald G V, Bryan D L. Planting depth and mulch thickness affect establishment of green ash (*Fraxinus pennsylvanica*) and bougainvillea golden rain tree (*Koelreuteria bipinnata*). J Arboric. 2005; 31: 163.

[48] Holloway P S. Aspen wood chip and stone mulches for landscape plantings in interior Alaska. J Environ Hortic. 1992; 10: 23-27.

桂花花梗与叶片的愈伤组织诱导研究

胡甜甜,王亚莉,杨秀莲,王良桂

(南京林业大学 风景园林学院,江苏 南京 210037)

摘 要：以桂花幼嫩花梗和叶片为外植体,通过植物组织培养技术,建立了桂花无菌体系,研究了不同品种群的桂花花梗在不同培养基配方下的出愈情况和植物生长调节剂、蔗糖浓度及暗处理时间等因素对不同桂花品种群的叶片愈伤组织诱导培养的影响。结果表明：桂花花梗最佳消毒方式为5‰ NaClO 溶液消毒 3 min,叶片最佳消毒方式为 0.1‰ $HgCl_2$ 溶液消毒 5 min。花梗愈伤组织的诱导培养结果表明,'波叶金桂'的最佳配方为 MS 2.00 mg·L^{-1} 6-BA+0.40 mg·L^{-1} 2,4-D；'紫梗籽银'桂为 B5+2.00 mg·L^{-1} 6-BA+0.40 mg·L^{-1} 2,4-D；'雨城丹'桂为 B5+1.00 mg·L^{-1} 6-BA+0.10 mg·L^{-1} 2,4-D；'日香桂'为 B5+2.00 mg·L^{-1} 6-BA+0.10 mg·L^{-1} 2,4-D。叶片愈伤组织的诱导培养结果表明,最佳蔗糖浓度为 25 g·L^{-1},最适宜的暗培养天数为 22 d；最利于'雨城丹'桂叶片愈伤组织诱导培养基的配方为 MS 3.00 mg·L^{-1} 6-BA+1.00 mg·L^{-1} 2,4-D；当 6-BA 浓度为 2.00 mg·L^{-1} 时是'天香台阁'的最佳诱导配方。

关键词：组织培养；愈伤组织；桂花；培养基

桂花(*Osmanthus fragrans* Lour.)属木犀科(Oleaceae)木犀属(*Osmanthus*)常绿阔叶观赏树种,素以花香著称。桂花寿命长、病虫害少、适应性广等特点使其在园林和庭院的绿化、美化、香化及装饰盆景上有特殊地位[1]。由于目前桂花食用、药用和观赏等功能的进一步研究与开发,市面上对桂花的需求不断增大,桂花繁育工作也得到广泛研究。

传统桂花育种方式如扦插、嫁接和播种等方式,因其繁殖系数小、周期长、生长缓慢等局限性[2],难以满足市场需要,而植物组织培养不受季节、气候等因素的制约,便于集约化管理和工厂化生产[3-4],Bonga[5]也曾将有性繁殖法与无性繁殖法遗传效益进行比较,认为无性繁殖法可获得更大的遗传效益,因此近年来桂花组织培养研究方兴未艾。目前桂花离体繁殖多以种胚、茎段、茎尖等为外植体诱导丛生芽,以胚、叶盘和茎尖等器官为外植体诱导愈伤组织,而对桂花花梗的愈伤组织诱导研究很少,同时,由于目前桂花叶片愈伤组织诱导培养均存在组织褐变、愈伤组织板结等未攻克问题,课题组探究了不同培养条件对桂花叶片愈伤组织诱导的影响。现以不同品种群的桂花花梗为外植体,以正交试验方法研究其愈伤组织诱导的最佳培养配方,旨在为今后桂花的组织培养研究提供参考依据。

1 材料与方法

1.1 试验材料

2013年3月采集南京林业大学校园内生长良好无病虫害的桂花新梢的幼嫩新叶,品种

---

*原文发表于《北方园艺》,2015(19):95-101。

分别为'天香台阁'('Tianxiang Taige')和'雨城丹'桂('Yucheng Dan');2013 年 9 月采集不同品种桂花初花期新鲜幼嫩花梗,分别为'四季桂'('Sijigui')、'波叶金桂'('Boye Jinggui')、'紫梗籽银'桂('Zigeng Zi Yin')、'雨城丹'桂('Yucheng Dan')和'日香桂'('Rixiang Gui')。采集时间在有阳光的 10:00～12:00 进行。

1.2 试验方法

1.2.1 无菌体系的建立

将采摘的'四季桂'新鲜花梗在流水下冲洗 1 h,置于超净工作台上用 75%酒精浸泡 30 s,无菌水漂洗 3 遍,然后用 5% NaClO 溶液分别消毒 1、3、5、8 min,无菌水漂洗 5 遍并在超净工作台上切成长 1.5 cm 左右小段,接种于添加 1.50 mg·L^{-1} 6-BA 和 0.10 mg·L^{-1} 2,4-D 的 MS 培养基上。将'天香台阁'叶片在流水下冲洗 2 h,置于超净工作台上用 75%的酒精消毒 30 s,无菌水冲洗 3 遍,由花梗消毒试验结果及资料查阅发现,当消毒时间小于 3 min 时,外植体的消毒不彻底,故使用 5% NaClO 溶液或 0.1% HgCl$_2$ 溶液对叶片消毒分别 3、5、8 min,无菌水漂洗 5 遍,沿主脉切成 1 cm×1 cm 的小块,接种于添加 1.00 mg·L^{-1} 6-BA 和 0.20 mg·L^{-1} 2,4-D 的 MS 培养基上。

1.2.2 花梗愈伤组织诱导的培养

将'波叶金桂'、'紫梗籽银'桂、'雨城丹'桂、'日香桂'的新梢幼嫩花梗消毒后,在超净工作台上切成长 1.5 cm 左右小段,平接于基本培养基(1/2MS、MS、B5)、6-BA(1.00、2.00、3.00 mg·L^{-1})、2,4-D(0.10、0.40、0.80 mg·L^{-1})3 因素 3 水平的正交设计 L$_9$(3^4)组合(表1)的基质中,进行愈伤组织诱导培养,观察花梗脱分化情况,30 d 后统计不同处理的花梗出愈率。

表 1 正交试验设计方案 L$_9$(3^4)
Tab. 1 Plan of orthogonal experiment design L$_9$(3^4)

水平 Level	因素 Factor			误差空列
	A 培养基 Medium	B 6-BA 浓度 Concentration of 6-BA/(mg·L^{-1})	C 2,4-D 浓度 Concentration of 2,4-D/(mg·L^{-1})	
1	1/2MS	1.00	0.10	
2	MS	2.00	0.40	
3	B5	3.00	0.80	

1.2.3 叶片愈伤组织诱导培养的单因素试验

暗处理时间对叶片愈伤组织诱导的影响:以'天香台阁'叶片为外植体,消毒处理后沿主脉切成 1 cm×1 cm 的小块接种于添加 2.00 mg·L^{-1} 6-BA 和 0.50 mg·L^{-1} 2,4-D 的 MS 培养基中进行诱导培养,并分别暗处理 1、7、15、22、30、35 d 后统计不同暗处理时间的叶片出愈率。蔗糖浓度对叶片愈伤组织诱导的影响:将'天香台阁'叶片(处理方式同上)接种于添加 2.00 mg·L^{-1} 6-BA 和 0.50 mg·L^{-1} 2,4-D 的 MS 培养基中进行诱导培养,并分别添加不同浓度蔗糖 15、25、40 g·L^{-1},30 d 后统计不同处理的叶片出愈率。植物生长调节剂对叶片愈伤组织诱导的影响:将'天香台阁'、'雨城丹'桂叶片(处理方式同上)分别接种于添加不同浓度 6-BA(1.00、2.00、3.00 mg·L^{-1})和 2,4-D(0.10、0.40 mg·L^{-1})双因素完全设计组合的 MS 培养基上进行愈伤组织的诱导培养,观察各种培养基配方对叶片愈伤组织的诱导效果,30 d 后统计不同处理的叶片出愈率。

1.2.4 培养条件与统计方法

无菌体系的建立阶段共接种 50 个外植体,其余每次处理均 10 瓶,每瓶接种 3 个外植体,重复 3 次;各处理(除研究蔗糖浓度对诱导培养的影响外)添加的蔗糖浓度均为 30 g·L^{-1},用于固化的琼脂均为 6.4 g·L^{-1};培养温度(25±1) ℃;光照强度为 18.75~25.00 $\mu mol·m^{-2}·s^{-1}$,连续光照 12 h·d^{-1};所有培养基使用前,用 1 mol·L^{-1} NaOH 或 HCl 调 pH 值至 5.5,并在 121 ℃、1.1 kg·cm^{-2}压力下高温高压灭菌 2 min。

1.3 项目评测

出愈率(%)=诱导愈伤组织的外植体数/接种的外植体总数×100;无菌存活率(%)=存活的无菌外植体总数/接种的外植体×100;污染率(%)=污染的外植体数/接种的外植体总数×100;死亡率(%)=死亡的外植体数/接种的外植体总数×100。

1.4 数据分析

试验中单因素和双因素试验设计所得的数据直接使用 SPSS 软件进行分析,正交设计所得的数据先采用极差分析,然后利用 SPSS 软件进行方差分析和多重比较。

2 结果与分析

2.1 消毒方式对无菌体系建立的影响

2.2.1 消毒时间对幼嫩花梗消毒效果的影响

由表 2 可知,桂花花梗在以 5% NaClO 溶液消毒处理时,外植体无菌存活率随着时间增长呈先上升后下降的趋势。消毒 3 min 时效果最好,污染率和死亡率均较低,分别为 8.00%和 12.00%;当消毒 1 min 时,无菌存活率最低,外植体污染率高于其死亡率,说明短时间的消毒不能很好地消除外植体污染;当消毒时间为 8 min 时外植体污染率虽然降低,但死亡率却显著升高,可能是外植体较幼嫩,长时间的消毒使其组织细胞受到毒害,死亡率增大。因此,花梗的最佳消毒方式为 5% NaClO 溶液灭菌 3 min。

表 2 不同消毒时间对花梗消毒效果的影响
Tab. 2 The effect of peduncle sterilization with disinfected time

消毒时间 Sterilizing time/min	接种数 Amount of explants	无菌存活率 Sterile survival rate/%	污染率 Pollution rate/%	死亡率 Death rate/%
1	50	58.00	24.00	18.00
3	50	80.00	8.00	12.00
5	50	76.00	10.00	14.00
8	50	70.00	4.00	26.00

2.1.2 消毒剂和消毒时间对幼嫩叶片消毒效果的影响

NaClO可分解产生具杀菌作用的氯气,灭菌处理后易于除去,不留残留[6];$HgCl_2$消毒效果虽好,但消毒后需要使用无菌水多次冲洗外植体再进行试验,以减小$HgCl_2$对外植体的毒害作用[7]。NaClO和$HgCl_2$分别对'天香台阁'叶片消毒不同时间的试验结果(表3)表明,使用5% NaClO溶液对叶片消毒时,污染率和死亡率随着处理时间的延长均逐渐降低,当消毒时间为8 min时效果最好,但此时外植体污染率仍然高于死亡率,说明消毒时间应适当延长使外植体得到充分消毒;使用0.1% $HgCl_2$溶液处理叶片的时间为5 min时效果较好,无菌存活率为66.70%,消毒时间继续增加时死亡率则明显增高。分析可知,在相同时间的消毒处理下,0.1% $HgCl_2$溶液比5% NaClO溶液效果更好,因此以'天香台阁'叶片为外植体的最佳消毒剂和消毒时间为0.1% $HgCl_2$溶液处理5 min。综合花梗与叶片的消毒试验结果发现,在相同消毒剂(NaClO)、消毒相同时间下,叶片的效果不及花梗,可能是因为叶片消毒表面积大于花梗,叶片表面附着的细菌也更多,需要适当延长消毒时间。

表3 不同消毒剂、消毒时间对幼嫩叶片消毒效果的影响
Tab. 3 The effect of young leaf sterilization with different disinfectants and disinfected time

NaClO 消毒时间 NaClO Sterilizing time/min	$HgCl_2$ 消毒时间 $HgCl_2$ Sterilizing time/min	污染率 Pollution rate/%	死亡率 Death rate/%	无菌存活率 Sterile survival rate/%
3	—	50.00	25.00	25.00
5	—	37.50	20.80	41.70
8	—	25.00	16.70	58.30
—	3	20.80	20.90	58.30
—	5	12.50	20.80	66.70
—	8	8.30	37.50	54.20

2.2 不同培养基配方对花梗愈伤组织诱导的影响

在'波叶金桂'花梗的愈伤组织诱导试验中发现,15 d左右花梗两端边缘部位开始慢慢出现浅白色愈伤组织突起,质地紧密(图1)。由表4可知,处理6的花梗愈伤组织诱导的效果较好,出愈率高达75.00%,与其他处理差异显著;出愈率状况较差的为处理1和处理3,出愈率低于20.00%,说明当MS大量元素减半时无法满足愈伤组织诱导所需营养。通过极差分析,3因素对试验影响效应:培养基>2,4-D>6-BA;由均值比较结果可知,试验的最优处理组合为$A_2B_2C_2$;因此,'波叶金桂'花梗愈伤诱导的最佳培养配方为MS+2.00 mg·L^{-1} 6-BA+0.40 mg·L^{-1} 2,4-D。

表 4 '波叶金桂'花梗愈伤组织诱导结果
Tab. 4 The 'Boye Jingui' peduncle callus inducement result

处理 Treatment	A 培养基 Medium	B 6-BA 浓度 Concentration of 6-BA/(mg·L^{-1})	C 2,4-D 浓度 Concentration of 2,4-D/(mg·L^{-1})	出愈率 Callus rate/%
1	1/2MS	1.00	0.10	19.44±1.20 e
2	1/2MS	2.00	0.40	41.67±1.17 d
3	1/2MS	3.00	0.80	12.50±4.17 f
4	MS	1.00	0.80	50.00±0.00 c
5	MS	2.00	0.10	66.67±3.17 b
6	MS	3.00	0.40	75.00±3.17 a
7	B5	1.00	0.40	62.50±0.67 b
8	B5	2.00	0.80	49.99±4.48 c
9	B5	3.00	0.10	54.17±3.17 c
k1	25.00	44.44	47.22	
k2	63.89	52.78	59.72	
k3	55.56	47.22	37.50	
R	38.89	8.34	22.22	

注：各数值均指试验的3个重复的平均值。小写字母表示在0.05显著水平下进行多重比较,有相同字母的表示两者差异不显著。下同。

Note: All values represent the average of three repeated experiments. Lowercase letters represent multiple comparisons significant at 0.05 level, the same lowercase letter indicating a difference between the same letter are not significant. The same below.

MS+2.00 mg·L^{-1} 6-BA+0.10 mg·L^{-1} 2,4-D

图 1 '波叶金桂'花梗愈伤组织诱导
Fig. 1 The 'Boye Jingui' peduncle callus inducement

'紫梗籽银'桂花梗的愈伤组织诱导试验结果(表5、图2)表明,处理8出愈率较高,达83.33%,与其他8个处理均存在显著差异;其次为处理5、6、7,出愈率均在70.00%~75.00%,3个处理间无显著差异。通过极差分析,3个因素对试验影响效应:培养基>6-BA>2,4-D,通过均值比较,各因素最优水平组合为$A_3B_2C_2$,因此,'紫梗籽银'桂花梗愈伤组织诱导的最佳培养配方为B5+2.00 mg·L^{-1} 6-BA+0.40 mg·L^{-1} 2,4-D。

表5 '紫梗籽银'桂花梗愈伤组织诱导结果
Tab. 5 The 'Zigeng Zi Yin' peduncle callus inducement result

处理 Treatment	A 培养基 Medium	B 6-BA 浓度 Concentration of 6-BA /(mg·L^{-1})	C 2,4-D 浓度 Concentration of 2,4-D/(mg·L^{-1})	出愈率 Callus rate/%
1	1/2MS	1.00	0.10	25.00±1.17 f
2	1/2MS	2.00	0.40	54.17±1.17 d
3	1/2MS	3.00	0.80	45.82±1.17 e
4	MS	1.00	0.80	66.67±1.00 c
5	MS	2.00	0.10	70.80±4.17 bc
6	MS	3.00	0.40	75.00±3.17 b
7	B5	1.00	0.40	70.80±0.17 bc
8	B5	2.00	0.80	83.33±5.17 a
9	B5	3.00	0.10	66.67±2.84 c
k1	41.67	54.17	54.17	
k2	70.83	69.43	66.67	
k3	73.60	62.50	65.27	
R	31.93	15.27	12.50	

B5+2.00 mg·L^{-1} 6-BA+0.80 mg·L^{-1} 2,4-D

图2 '紫梗籽银'桂愈伤组织诱导
Fig. 2 The 'Zigeng Zi Yin' peduncle callus inducement

'雨城丹'桂花梗的愈伤组织诱导培养结果(表6、图3)表明,处理7出愈率较高,达62.50%,与其他8个处理有显著差异;处理3出愈率最低,仅为4.17%。通过极差分析,3因素对试验影响效应:6-BA>培养基>2,4-D;通过均值比较,各因素的最优水平组合为 $A_3B_1C_1$,因此,'雨城丹'桂花梗愈伤诱导最佳培养配方为 B5+1.00 mg·L^{-1} 6-BA+0.10 mg·L^{-1} 2,4-D。

表6 '雨城丹'桂花梗愈伤组织诱导结果
Tab. 6 The 'Yucheng Dan' peduncle callus inducement result

处理 Treatment	A 培养基 Medium	B 6-BA 浓度 Concentration of 6-BA /(mg·L^{-1})	C 2,4-D 浓度 Concentration of 2,4-D /(mg·L^{-1})	出愈率 Callus rate/%
1	1/2MS	1.00	0.10	45.83±1.17 c
2	1/2MS	2.00	0.40	25.00±1.17 f
3	1/2MS	3.00	0.80	4.17±2.16 g
4	MS	1.00	0.80	58.33±4.17 b
5	MS	2.00	0.10	41.67±1.67 d
6	MS	3.00	0.40	29.17±1.84 e
7	B5	1.00	0.40	62.50±2.50 a
8	B5	2.00	0.80	45.83±1.17 c
9	B5	3.00	0.10	41.67±1.67 d
k1	25.00	55.55	43.06	
k2	43.06	37.50	38.89	
k3	50.00	25.00	36.11	
R	25.00	30.55	6.95	

B5+2.00 mg·L^{-1} 6-BA+0.10 mg·L^{-1} 2,4-D

图3 '雨城丹'桂花梗愈伤组织诱导
Fig. 3 The 'Yucheng Dan' peduncle callus inducement

'日香桂'花梗愈伤组织诱导试验结果(表7、图4)表明,处理5与处理9出愈率明显高于其他处理,2个处理无显著差异,出愈率分别为66.67%和70.83%;处理3出愈率最低,仅为16.67%。通过极差分析,各因素对试验结果影响效应大小为:培养基>2,4-D>6-BA,由均值比较可知,3因素的最优水平组合是 $A_3B_2C_1$,因此,'日香桂'花梗愈伤诱导的最佳培养配方为 B5+2.00 mg·L^{-1} 6-BA+0.10 mg·L^{-1} 2,4-D。

表 7 '日香桂'花梗愈伤组织诱导结果
Tab. 7 The 'Rixiang Gui' peduncle callus inducement result

处理 Treatment	A 培养基 Medium	B 6-BA 浓度 Concentration of 6-BA /(mg·L^{-1})	C 2,4-D 浓度 Concentration of 2,4-D /(mg·L^{-1})	出愈率 Callus rate /%
1	1/2MS	1.00	0.10	45.82±5.84 de
2	1/2MS	2.00	0.40	33.33±3.33 f
3	1/2MS	3.00	0.80	16.67±1.67 g
4	MS	1.00	0.80	41.67±4.67 e
5	MS	2.00	0.10	66.67±3.00 a
6	MS	3.00	0.40	50.00±4.00 cd
7	B5	1.00	0.40	58.33±1.67 b
8	B5	2.00	0.80	54.17±1.17 bc
9	B5	3.00	0.10	70.83±4.17 a
k1	31.93	48.61	61.10	
k2	52.79	51.38	47.21	
k3	61.11	45.84	37.52	
R	29.18	5.54	23.58	

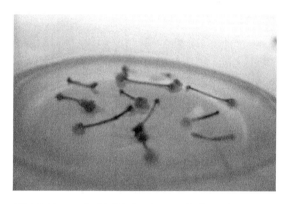

B5+2.00 mg·L^{-1} 6-BA+0.80 mg·L^{-1} 2,4-D

图 4 '日香桂'花梗愈伤组织诱导
Fig. 4 The 'Rixiang Gui' peduncle callus inducement

2.3 暗处理时间对叶片愈伤组织诱导的影响

暗处理对接种于添加 2.00 mg·L^{-1} 6-BA 和 0.50 mg·L^{-1} 2,4-D 的 MS 培养基中的'天香台阁'叶片愈伤组织诱导的试验结果(表 8)表明,暗培养 22 d 时出愈率最高,为 81.27%,暗培养 30 d 时出愈率次之,为 77.10%,2 个处理间无显著差异;暗培养 0～15 d 时出愈率无显著差异,但出愈率明显低于暗培养 22 d 和 30 d。这说明一定时间的暗培养有利于'天香台阁'叶片愈伤组织的形成,但并不是暗培养时间越长越好,超过最佳时间出愈率将会有所降低。综上,'天香台阁'叶片的愈伤组织诱导最佳的暗培养天数为 22 d。

表 8 不同暗处理时间对叶片愈伤组织诱导的影响
Tab. 8 The inducement effect of leaf callus with different dark days

处理 Treatment	暗培养时间 Dark days/d	出愈率 Callus rate/%
1	0	50.00±6.25 b
2	7	47.92±7.21 b
3	15	60.42±9.55 b
4	22	81.27±6.25 a
5	30	77.10±3.64 a

2.4 蔗糖浓度对叶片愈伤组织诱导的影响

不同蔗糖浓度对'天香台阁'叶片愈伤组织的诱导能产生显著影响(表9),蔗糖浓度为 25 g·L^{-1} 时出愈率最高达 73.59%,与蔗糖浓度为 40 g·L^{-1} 时出愈率无显著差异;当蔗糖浓度为 5 g·L^{-1} 时出愈率最低仅为 29.16%,与蔗糖浓度为 25、40 g·L^{-1} 时出愈率均存在显著性差异。因此,本着经济合理的原则,认为'天香台阁'叶片愈伤组织诱导的最佳蔗糖浓度为 25 g·L^{-1}。

表 9 不同蔗糖浓度对叶片愈伤组织诱导的影响
Tab. 9 The inducement effect of leaf callus with different concentrations of sucrose

处理 Treatment	蔗糖浓度 Concentrations of sucrose/(g·L^{-1})	出愈率 Callus rate/%
1	5	29.16±4.17 b
2	25	73.59±4.83 a
3	40	65.26±6.37 a

2.5 不同培养基配方对叶片愈伤组织诱导的影响

在 MS 培养基上添加不同浓度配比的生长调节剂的'天香台阁'叶片愈伤组织诱导试验结果(表10、图5)表明,处理3出愈率最高,可达75.00%,与其他处理差异均显著;处理6出愈率效果最差,仅为43.00%,且形成的愈伤组织较为干燥;分析发现6-BA浓度不变时,0.50 mg·L^{-1} 2,4-D诱导出愈率均优于浓度为 1.00 mg·L^{-1} 时的出愈率,说明低浓度2,4-D更利于叶片愈伤组织诱导,这可能与2,4-D趋向于抑制植物形态发生的效应有关[8]。因此,'天香台阁'叶片愈伤诱导的最佳培养配方为 MS+2.00 mg·L^{-1} 6-BA+0.50 mg·L^{-1} 2,4-D。

表 10 不同浓度 6-BA 与 2,4-D 对'天香台阁'叶片愈伤组织诱导的影响
Tab. 10 The inducement effect of 'Tianxiang Taige' leaf callus with different concentrations of 6-BA and 2,4-D

处理 Treatment	6-BA 浓度 Concentrations of 6-BA/(mg·L^{-1})	2,4-D 浓度 Concentrations of 2,4-D/(mg·L^{-1})	出愈率 Callus rate/%
1	1.00	0.50	62.67±1.82 b
2	1.00	1.00	50.50±1.50 c
3	2.00	0.50	70.00±3.60 a
4	2.00	1.00	60.23±1.76 b
5	3.00	0.50	54.53±2.25 c
6	3.00	1.00	43.00±3.22 d

MS+2.00 mg·L^{-1} 6-BA +0.50 mg·L^{-1} 2,4-D

图5 '天香台阁'叶片愈伤组织诱导

Fig. 5 'Tianxiang Taige' leaves callus inducement

以'雨城丹'桂叶片为外植体,MS为基本培养基时,6-BA与2,4-D不同浓度组合下愈伤组织诱导试验结果(表11、图6)表明,处理6出愈率最高,可达70.20%,与其他5个处理差异显著;处理1出愈率效果最差,仅为25.00%。分析发现,当6-BA浓度为1.00 mg·L^{-1}和2.00 mg·L^{-1}时的出愈率均未超过50%,且当2,4-D浓度不变时,6-BA在一定浓度范围对愈伤组织诱导的促进作用随浓度升高而升高,说明较高浓度的6-BA利于外植体的愈伤组织诱导,这与董闪等[9]、和凤美等[10]研究结果一致。因此,'雨城丹'桂叶片愈伤诱导的最佳培养配方为MS+3.00 mg·L^{-1} 6-BA+1.00 mg·L^{-1} 2,4-D。

表11 不同浓度6-BA与2,4-D对'雨城丹'桂叶片愈伤组织诱导的影响

Tab. 10 The inducement effect of 'Yucheng Dan' leaf callus with different concentrations of 6-BA and 2,4-D

处理 Treatment	6-BA浓度 Concentrations of 6-BA/(mg·L^{-1})	2,4-D浓度 Concentrations of 2,4-D/(mg·L^{-1})	出愈率 Callus rate/%
1	1.00	0.50	25.00±3.60 e
2	1.00	1.00	41.69±2.07 d
3	2.00	0.50	50.00±3.60 c
4	2.00	1.00	41.69±2.07 d
5	3.00	0.50	61.83±5.45 b
6	3.00	1.00	70.20±2.08 a

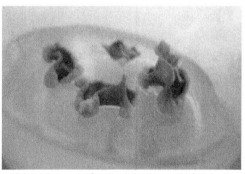

MS+3.00 mg·L^{-1} 6-BA+0.50 mg·L^{-1} 2,4-D

图6 '雨城丹'桂叶片愈伤组织诱导

Fig. 6 'Yucheng Dan' leaves callus inducement

3 结论与讨论

试验表明,外植体的取材部位、消毒方式、生长调节剂浓度配比等均能影响愈伤组织的形成。宋会访等[11]曾以无菌短枝扦插途径对桂花进行组织培养研究,发现其组培效果明显优于直接取自树体的器官进行组培试验的结果,这说明无菌体系的建立对于桂花的组织培养尤为重要,能否有效控制污染是植物组织培养成功的关键[12]。本试验中对叶片使用 0.15% $HgCl_2$ 溶液消毒 25 min 能达到较好效果,此时污染率仅 12.50%;新鲜花梗则适于使用 5% NaClO 溶液处理 3 min,污染率仅为 8.00%。

花梗愈伤组织诱导结果表明,不同桂花品种在不同配方的培养基中均能进行不同程度的脱分化,但其愈伤组织诱导效果均不同,'紫梗籽银'桂花梗的愈伤组织诱导效果明显优于其他品种,而以'雨城丹'桂花梗为外植体时,其平均出愈率仅为 39.35%,这可能与培养条件及外植体基因型相关;通过极差分析和均值比较发现,培养基与激素对不同桂花品种愈伤组织诱导试验的影响效应及其最优处理组合均不同,证明愈伤组织的诱导能表现出明显的品种特异性[13],基因型对其有较大影响,这可能与植物体内激素种类和水平不同有关[14];分析不同品种桂花花梗愈伤组织诱导试验中不同因素对试验影响效应发现,最佳基本培养基为 MS 和 B5,证明了高盐培养基和高硝态氮培养基适合于愈伤组织诱导[15];分析各桂花品种最优配方发现,当 6-BA 浓度在 1.00~2.00 mg·L^{-1},2,4-D 浓度在 0.10~0.40 mg·L^{-1} 时适于桂花梗的愈伤组织诱导,这与彭尽晖等[16]、王文房等[17]研究花梗愈伤组织诱导试验结果一致。

植物生长调节剂是培养基中的关键元素,在组织培养中用量虽然微小,但却是发挥生物学效力最强的培养因素,可以调节培养物的生长发育进程、分化方向和器官发生[18]。研究表明,在配合使用 2,4-D 和 6-BA 的情况下,愈伤组织诱导率要高于单独使用 2,4-D,但 2,4-D 浓度不宜过大,否则易引起组织褐化[19-20]。在叶片的愈伤组织诱导中,'雨城丹'桂的最佳培养配方为 MS+3.0 mg·L^{-1} 6-BA+0.5 mg·L^{-1} 2,4-D;'天香台阁'的最佳培养配方为 MS+2.0 mg·L^{-1} 6-BA+0.5 mg·L^{-1} 2,4-D。糖的使用可维持良好的渗透关系,试验发现培养基中最佳的蔗糖浓度为 25 g·L^{-1}。此外,本研究表明遮光培养有利于愈伤组织诱导[21],最适宜的暗培养天数为 22 d,这与彭尽晖等[16]研究'四季桂'暗培养结果不同,可能是由于外植体的培养环境不同和品种差异性造成的。

本试验探究了不同品种桂花花梗的组织培养条件,建立了良好的无菌体系,成功诱导其愈伤组织,并探究出'天香台阁'和'雨城丹'桂叶片组织培养的最佳配方和培养条件。但由于时间问题,本试验对可能影响愈伤组织诱导培养的其他因子,如培养温度、糖的种类等问题还未涉及,这将是桂花愈伤组织诱导培养今后的研究方向。

参考文献

[1] 聂谷华,向其柏.桂花研究现状及其存在的问题[J].九江学院学报(哲学社会科学版),2008,26(3):85-87.

[2] 张丽霞,王毅彰.桂花的离体快繁技术[J].江苏农业科学,2013,41(3):42-43.

[3] 蔡新玲,胡蕙露.桂花茎尖初代组织培养试验研究[J].安徽林业科技,2009(1):17-19.

[4] 胡海波,黄丹,许岳香.木犀科植物组织培养研究综述[J].林业科技开发,2009,23(3):5-8.

[5] Bonga J M.树木组织培养[M].阙国宁,等译.北京:中国林业出版社,1988:119-121.
[6] 潘瑞炽.植物组织培养[M].广州:广东高等教育出版社,2000:20-21.
[7] 杜燕,蒋海玉,刘其宁.贮存过期油菜种子消毒方法的研究[J].种子,2003(2):39-40,42.
[8] Wang Y P, Liu Q C, Li A X, et al. In vitro selection and identification of drought tolerant mutants in sweet potato[J]. Agricultural Sciences in China, 2003, 2(12):1314-1320.
[9] 董闪,李明,唐堃,等.白木香愈伤组织的诱导及培养[J].湖北农业科学,2014,53(15):3673-3677.
[10] 和凤美,朱永平,杨晓红,等.冬樱花愈伤组织诱导和抑制褐化初探[J].中国农学通报,2010,26(12):130-134.
[11] 宋会访,葛红,周媛,等.桂花离体培养与快速繁殖技术的初步研究[J].园艺学报,2005,32(4):738-740.
[12] 喻晓雁,刘克旺,梁文斌.穗花杉组织培养初探[J].中南林学院学报,2005,25(2):55-58.
[13] 任桂芳,王建红,冯慧,等.现代月季(Rosa hybrida)叶片植株再生体系的建立[J].园艺学报,2004,31(4):533-536.
[14] 蔡新玲.不同桂花品种群组织培养的研究[D].合肥:安徽农业大学,2007.
[15] 吕晋慧,孔冬梅.园艺植物组织培养[M].北京:中国农业科学技术出版社,2008:24.
[16] 彭尽晖,吕长平,周晨.四季桂愈伤组织诱导与继代培养[J].湖南农业大学学报(自然科学版),2003,29(2):131-133.
[17] 王文房,李修岭.樱化花柄的组织培养[J].安徽农业科学,2006,34(22):5839-5841.
[18] 王金刚,张兴.园林植物组织培养技术[M].北京:中国农业科学技术出版社,2008:27-28.
[19] 丁路明,龙瑞军,朱铁霞.2,4-D和6-BA对早熟禾愈伤组织诱导的影响[J].草原与草坪,2003(1):34-37.
[20] 周宜君,周生闰,刘玉,等.植物生长调节剂对植物愈伤组织的诱导与分化的影响[J].中央民族大学学报(自然科学版),2007,16(1):23-28
[21] 宋会访.桂花组织培养技术体系的研究[D].武汉:华中农业大学,2004.

Study on Pedicels and Leaves Callus Induction of *Osmanthus fragrans*

HU Tiantian, WANG Yali, YANG Xiulian, WANG Lianggui

(College of Landscape Architecture, Nanjing Forestry University, Nanjing 210037)

Abstract: The young pedicles and leaves of *Osmanthus fragrans* were used as the explants in the study. Through the plant tissue culture techniques, the experiment studied the establishment of *Osmanthus fragrans* sterile system, pedicels callus induction in different media formulations of varieties of *O. fragrans* groups and the influences of plant growth regulators, sucrose concentration, dark processing time *etc.* to leaves callus induction of different varieties of *O. fragrans* groups. The results showed that the best way to sterilize pedicels was using 5% NaClO for 3 min, the best way to sterilize leaves was using 0.1% $HgCl_2$ for 5 min. The best mediums for pedicels of different varieties of *O. fragrans* were as follow, 'Boye Jingui' was MS+2.00 mg·L^{-1} 6-BA+0.40 mg·L^{-1} 2,4-D, 'Zigeng Zi Yin' was B5+2.00 mg·L^{-1} 6-BA+0.40 mg·L^{-1} 2,4-D, 'Yucheng Dan' was B5+1.00 mg·L^{-1} 6-BA+0.10 mg·L^{-1} 2,4-D, 'Rixiang Gui' was B5+2.00 mg·L^{-1} 6-BA+0.10 mg·L^{-1} 2,4-D. The results of leaves callus induction showed that, the best sucrose concentration was 25 g·L^{-1}; it was best to put the mediums in dark condition for 22 days. The best medium for leaves of 'Yucheng Dan' was MS+3.00 mg·L^{-1} 6-BA+1.00 mg·L^{-1} 2,4-D. The best medium for 'Tianxiang Taige' was MS+2.00 mg·L^{-1} 6-BA+0.50 mg·L^{-1} 2,4-D.

Key words: tissue culture; callus; *Osmanthus fragrans*; mediums

控根栽培下桂花根系的动态生长与垂直分布特征

王良桂,李 霞,杨秀莲

(南京林业大学 风景园林学院,江苏 南京 210037)

摘 要:在控根栽培模式下,采用土钻法对6年生桂花嫁接苗的根系分布和季节动态进行研究。结果表明:一年中,随着时间的变化,桂花根系长度、根系体积、根长密度都呈增长趋势;须根萌发明显,细根增多,根系分布层上移,主要集中在0～10 cm土壤层,说明控根容器对桂花根系构型产生一定影响。

关键词:桂花;控根栽培;根系分布

随着经济发展和城市化进程加快,城市的园林绿化正在向高品质、精细化方向发展,对苗木质量的要求越来越高。容器育苗是提高林木质量和适应林木种苗供应方式变化的主要手段,苗木的容器化培育是一种以调控根系生长为核心的新型、快速培育方式,具有随时移植且不伤根系,移植时不截冠,移植后不需要缓苗期且成活率高的特点[1-2],因此在园林绿化中日益受到重视[3-4]。

桂花(*Osmanthus fragrans*)不仅是良好的绿化、美化环境的树种,而且有利于人类的健康,还有较好的食用价值,值得大力繁殖。笔者通过分析控根容器对根系构型的影响,探讨适合桂花生长的容器化栽培模式,为桂花苗木容器化培育和规模化发展提供理论依据。

1 材料与方法

1.1 试验地概况

试验地设在长江北岸的南京林业大学六合区苗圃(118°34′E,32°09′N),地处北亚热带湿润季风区,光能充足,热量富裕,雨热同季,年总降雨量约1 004 mm,降雨主要集中于每年的6—8月,年平均气温15.2 ℃。

1.2 供试材料

供试苗木采用6年生的桂花嫁接苗,由南京林业大学六合区苗圃提供。控根容器采用浙江台州隆基园艺材料有限公司生产的60 cm×60 cm新型K2控根容器。该容器直径20 cm,容器壁有凹凸相间的波纹,凸起点的外侧顶端留有小的修根孔。其原理是把修根孔长出的细根暴露在空气中,利用空气阻断细根的生长,从而影响容器内毛细根的生长。

1.3 控根栽培处理方法

选取生长一致的桂花嫁接苗20株分别植于控根容器内,作为试验处理(K),另取20株裸地栽植在同一块圃地上作为对照(CK),采用相同的管理措施以减少误差来源。

取处理和对照植株的东南西北4个方位,分别用内径6 cm的土钻从上而下分3层(0~10 cm、10~20 cm、20~30 cm)钻取土芯,用0.4 mm筛网流水冲洗掉泥土后,所得根系于低温(0~4 ℃)保存带回实验室。取样时间为5、7、9、10、11月的中旬,同一处理3次重复(1棵植株作为1个重复)。

将所得根系按直径分成5个等级:≥2.0 mm为Ⅰ级根,1.5~2.0 mm为Ⅱ级根,1.0~1.5 mm为Ⅲ级根,0.5~1.0 mm为Ⅳ级根,0~0.5 mm为Ⅴ级根。以Epson数字化扫描仪(Expression10000XL 1.0)将各级根系扫描后,用Win RHIZO(Pro 2004b)根系图像分析系统软件(加拿大Regent Instruments公司)对根系扫描图像进行定量分析。分析指标包括根长密度(D_{RL})、根系长度(L_R)、根系直径(d_R)、根系体积(V_R)[5]。把扫描后的各级根系在105 ℃下烘干至恒定质量,测定其生物量,并计算比根长(m/g)。

试验数据采用Excel软件分析。

2 结果与分析

2.1 桂花根系季节生长变化

2.1.1 根长的动态分布

根长(L_R)是指单位体积内植株根系的总长度,平均根长是植物根系形态的重要参数。控根栽培条件下,桂花根系的长度在5月中旬到9月中旬呈直线上升,9月中旬到11中旬增长缓慢,在7、9、10和11月中旬测得的值分别为1 222.24、2 227.51、2 319.39和2 344.06 cm,其增幅分别为40.73%、82.25%、4.12%和1.06%;而裸地栽植的桂花根系的长度在5月到7月中旬增长较快,7月中旬到11月中旬增加较缓慢,7月测得根系长度为1 166.99 cm,比5月增长了122.18%,9、10和11月中旬测得根系长度为1 170.38、1 327.21、1 341.84 cm,增幅分别为0.29%、13.39%、1.10%(图1A)。

总体看来,控根栽培的桂花根系的生长长度都比裸地生长的长度长,在控根栽培条件下整个根系的生长季根系长度平均比对照的增长了62.37%,可见控根容器促进了根系的伸长生长。

2.1.2 根长密度的动态分布

根长密度(D_{RL})指单位土体中细根的总长度(cm/cm³),是估算根系在土壤中的分布及其吸收水分、养分的重要依据。从图1B可以看出,控根栽培条件下桂花根长密度在5月中旬到9月中旬呈直线上升,9月中旬到11月中旬增长较缓慢。7月中旬测得的根长密度为2.72 cm/cm³,比5月中旬增长了112.26%,9月中旬测得的根长密度为4.45 cm/cm³,比7月份增长了63.27%。10月、11月的增幅较小,分别为10.74%和1.57%。总体来看,在根系的整个生长季,控根栽培条件下桂花苗木的根系密度平均比裸地栽培条件下的桂花苗木增长了92.19%。

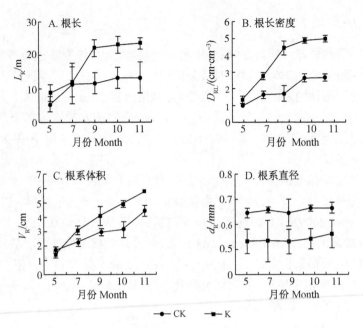

图 1　桂花根系季节生长变化

Fig. 1　The growing dynamics of *Osmanthus fragrans* seedling root system

2.1.3　根系体积的动态分布

根系体积(V_R)指标能够直观地反映根系在土层中的空间分布特征,根系体积与移栽成活率有很大的关系。从图1C中可以看出,控根栽培的桂花根系体积呈直线上升趋势,在7、9、10和11月中旬测得的值分别为3.11、4.17、4.95和5.73 cm³,增幅分别为115.63%、34.14%、18.85%和15.76%。裸地生长的根系体积也呈上升趋势,但增幅相对较小。在整个根系的生长季,控根栽培的苗木平均根系体积比裸地栽培的增长了32.35%。

2.1.4　根系直径的动态分布

根系直径是根系研究中重要的参数,从图1D中可以看出,控根栽培和裸地栽培的苗木根系平均直径在生长季内都未发生明显变化,但控根栽培的苗木平均直径明显比裸地栽培的细,二者的平均直径分别为0.543 mm和0.660 mm。

2.1.5　比根长的变化

比根长决定了根系吸收水分和养分的能力,是反映细根生理功能的重要指标之一,比根长数值的大小可以反映根系生理活性的大小,比根长越大,说明根系的生理活性越强,同时直径越小[6]。从表1可知,5—11月控根栽培的桂花苗木根系比根长在0～10 cm的土壤层达到最大值,说明在0～10 cm的土壤层根系直径最小,细根较多,生理活性较强;5—7月比根长在10～20 cm土层最小,分别为16.36和17.90 m/g,说明该土层根系直径较大,但是随着时间的变化;到9—11月时比根长在20～30 cm土层达到最小值,说明根系在该土层中直径最大。

表1 桂花根系比根长的动态变化
Tab. 1 Dynamics changes of specific root length of *O. fragrans*

土层深度/cm Soil depth	比根长 Specific root length/(m·g^{-1})				
	5月	7月	9月	10月	11月
0~10	29.25±1.27	20.10±1.17	23.03±1.73	28.81±5.07	19.78±2.04
10~20	16.36±2.62	17.90±0.62	19.03±1.21	12.84±1.58	12.11±1.69
20~30	20.66±0.63	19.99±0.70	14.02±1.56	8.67±0.79	8.86±1.00

以上数据均说明根系分布层随时间变化逐渐上移,在0~10 cm土层根系直径最细,吸收根最多,活性最强,其次是10~20 cm和20~30 cm土层。因此,土层越深,苗木根系直径越大,生理活性也相对较弱。

2.2 桂花根系垂直分布特征

桂花各级根系的根长垂直分布基本呈相同趋势(图2):根系主要集中于0~10 cm土层和10~20 cm土层,20~30 cm土层中分布较少,直径越小的根系长度越长,且随时间的变化,根系主要分布层上移。5—9月中旬Ⅴ级根在10~20 cm土层的分布均达到最大值,5、7、9月中旬在10~20 cm土层测得的根系长度分别占总长度的40%、36.3%和45.2%,11月份在0~10 cm土层的分布最多,11月中旬在0~10 cm土层测得的根系长度分别占总长度的45.9%、38.9%,根长随着分布深度的增加而逐渐减少。Ⅳ级根集中分布在0~10 cm土层和10~20 cm土层中,在20~30 cm土层中的分布较少,到9—11月份根系长度集中分布在第1层,开始往表层转移。Ⅲ级根在5—7月期间分布无明显规律,但是7—9月根系开

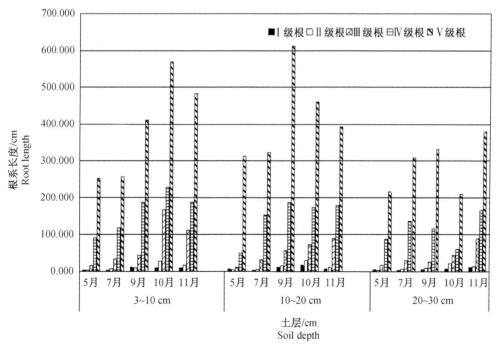

图2 桂花不同径级根系根长的垂直分布
Fig. 2 Vertical distribution of different diameter grades roots of *O. fragrans*

始慢慢集中到 10~20 cm 土层,9 月份根系长度在 0~10 cm 土层中达到最大值,说明根系向上分布转移。同样,Ⅱ级根在 9—11 月份之间根系分布层也开始往 0~10 cm 土层转移。但是Ⅰ级根主要集中在第 2、3 层中,无向上转移趋势。以上数值均说明桂花根生长的最佳环境在 10~20 cm 土层内,根系向表层集中说明 0~10 cm 土层的透气环境更好,适合细根生长。

3 讨论

(1)控根栽培条件下桂花苗木根系随着时间的变化细根明显增多,细根($d \leqslant 2$ mm)主要集中在 0~10 cm 深度的土壤中,在 20~30 cm 的土壤层分布逐渐减少,整个根系分布层上移;而裸地栽培的苗木根系并无这种生长趋势,在 20~30 cm 的土壤层也存在较多根系,说明控根效果明显。Marler 等[7]发现生长在空气断根容器中的苗木,上层土壤的新根数较多;Jones 等[8]也发现,使用 Copperblock 的杰克松容器苗出现了根系上移现象,这可能是由于控根容器中 0~10 cm 土层的透气性更好,更利于根系的生长。

(2)与空气修根影响根系构型在很多方面相似。Wenny 等[9]发现,对黑松使用 Spinout 控根试剂可以增加一级侧根长度,移栽后新根的平均长度也增加;孙盛等[10]以银杏为试材,对不同浓度的铜控根试剂进行研究,发现 0~2 mm 径级的根系体积大于裸根苗的根系体积;Whitcomb[11]研究发现在控根容器中生长的苗木根系同普通容器苗相比有更大的表面积,更多的侧根。本试验结果表明:随着时间的变化,控根栽培的桂花苗木根系长度、根系体积、根长密度都呈向上增长趋势且都比裸根苗大。

(3)本试验主要是研究控根栽培对根系构型的影响。但在试验中发现,随着栽培时间的变化,大量的新根间会出现强烈竞争,最终形成根域限制作用,从而影响根系的生长。因此今后可以从细根的生长、衰老、死亡状况,以及根系的活性和功能变化等方面进行进一步研究,以确定经济、合理的控根栽培时间,提高桂花苗木栽培的经济效益。

参考文献

[1] 秦国峰,吴天林,金国庆,等. 马尾松舒根型容器苗培育技术研究[J]. 浙江林业科技,2000,20(1):68-73.

[2] 曲良谱,喻方,张新. 苦楝容器苗育苗技术研究[J]. 林业科技开发,2008,22(6):103-106.

[3] 黄诗铿. 控根育苗技术的特点及市场分析[J]. 世界农业,2002(7):23-24.

[4] 张永青,张胜,向瑞. 容器苗培育技术研究综述[J]. 河北林果研究,2008,23(3):255-258.

[5] 鲁敏,姜凤岐,宋轩. 容器苗质量评定指标的研究[J]. 应用生态学报,2002,13(6):763-765.

[6] Fransen B, Kroon H D, Berendse F. Root morphological plasticity and nutrient acquisition of perennial grass species from habitats of different nutrient availability[J]. Oecologia,1998,115(3):351-358.

[7] Marler T E, Willis D. Chemical or air root-pruning containers improve carambola, longan, and mango seedling root morphology and initial root growth after transplanting[J]. Journal of Environmental Horticulture,1996,14(2):47-49.

[8] Jones M D, Kiiskila S, Flangan A. Field performance of pine stock types:two year results of a trial on interior lodgepole pine seedlings grown in Styroblocks, Copperblocks, or AirBlocks[J]. CAB

Abstracts,2006,30(5):40-74.
[9] Wenny D L, Woollen R L. Chemical root pruning improves the root system morphology of containerized seedlings[J]. Western Journal of Applied Forestry,1989,4(1):15-17.
[10] 孙盛,彭祚登,董凤祥,等. Cu,Zn 等制剂对银杏容器苗的控根效果[J]. 林业科学,2009,(7):156-160.
[11] Whitcomb C E. Plant Production in ContainersⅡ[M]. Stillwater: Lacebark Publications, 2006.

The Growing Dynamics and Vertical Distribution Feature of *Osmanthus fragrans* Seedling Root System under the Root-Controlling Cultivation

WANG Lianggui, LI Xia, YANG Xiulian

(College of Landscape Architecture, Nanjing Forestry University, Nanjing 210037, China)

Abstract: Under the root-controlling cultivation conditions, the distribution characteristics and seasonal dynamics of six-year-old *Osmanthus fragrans* seedling root system were studied by using soil drilling. The results showed that root length, volume and density were increased with time passed. Meanwhile, fibrous roots sprouted obviously and absorbing roots increased. Root distribution layer was up drifted and concentrated in 0～10 cm soil depth mostly which indicated that there were certain effects on root configuration under root-controlling cultivation.

Key words: *Osmanthus fragrans*; root-controlling cultivation; root distribution

桂花不同品种的种子形态比较

杨秀莲[1]，姜晓装[2]

(1.南京林业大学　风景园林学院,江苏　南京　210037；
2.江西省林业科技推广总站,江西　南昌　330038)

摘　要：在调查南京地区结实桂花品种的基础上,对不同品种种子的形态、外观纹样及内果皮结构进行了比较。结果表明：不同品种的种子长、宽和单粒重有较大差异,种皮表面纹饰以及断面细胞排列结构等均存在一定差异。因此,可以将种子的形态和外观纹饰作为桂花结实品种鉴定的辅助依据。

关键词：桂花；种子；形态

在不同的桂花品种中,存在着两类结实情况：一类是雌雄蕊发育正常,能正常结实；另一类是雄蕊正常而雌蕊退化,不能正常结实。这两种情况存在于桂花的4个品种群中,在现有的桂花品种中,目前已知结实的品种有39个,分属4个品种群。本研究在调查南京地区结实桂花品种的基础上,对不同品种的种子外观纹样及内果皮(为便于描述,以下均称种皮)结构进行了比较。

1　材料与方法

1.1　实验材料

桂花种子于2007年4—5月采自南京市玄武湖公园、灵谷寺公园、情侣园和南京林业大学校园内各结实品种,采后去除果皮,于室温风干后贮藏。品种有'宽叶籽银'桂、'早籽银'桂、'籽银桂'、'紫梗籽银'桂、'长瓣银桂'、'卷瓣银桂'、'长叶碧珠'、'潢川金桂'、'大花金'桂、'大叶籽金'桂、'墨叶金'桂、'晚金桂'、'籽丹桂'共13个。

1.2　桂花种子的长、宽测定

用游标卡尺分别测量种子(指去除外、中果皮后的种子)的纵径和横径。其中以种子的纵径作为所测种子的长度,以两次垂直横径的平均值作为所测种子的宽度。测定样品采用四分法抽取,每次测定数量为100粒。

1.3　桂花种子的单粒重和百粒重

依据林木种子检验规程(GB 2722—1999),人工随机数取100粒,使用1/1 000电子天

---

* 原文发表于《福建林业科技》,2010,37(3)：79-82。

平分别称重,而后计算其平均值,即为单粒重。

1.4 桂花种子的种皮纹饰和断面细胞排列结构

采用扫描电镜观察。取每个品种风干种子各 10 粒,在丙三醇中恒温浸泡 24 h,软化种皮,使种子容易解剖,然后将种子置于 2.5% 戊二醛中固定,经系列丙酮脱水,转入乙酸异戊酯后,进行临界点干燥、喷金,在扫描电镜下观察种子表面和横断面结构并进行记录和拍照。

2 结果与分析

2.1 桂花种子生长发育观察

本实验所涉及品种均为正常结实品种,因此这些品种的花均为两性花,柱头呈头状,雄蕊 2 枚,少数品种可见 3 枚雄蕊,雄蕊和柱头等高或稍高于柱头,子房发育正常(见图 1)。花后果实逐渐长大,果实大多长椭圆形,不同品种稍有变化。未成熟时果皮绿色,至 4 月底、5 月初,外果皮转为紫黑色时种子成熟(见图 2)。由于各品种的花期有先后,故果实成熟期也不同,前后相差 15 d 左右。每年的温度变化也会影响果实成熟的时间,在观察过程中也发现桂花结实有一定的大小年现象,这与树体的营养消耗有关。

图 1 '籽银桂'的花:示柱头和雄蕊

图 2 '籽银桂'的果实

2.2 桂花种子的大小

分别比较了 13 个结实品种种子的长度和宽度,结果见表 1。从测量结果可知,不同品种的种子长、宽和单粒重有较大差异。其中'大花金'桂的种子最长,达 1.834 cm,而'潢川金桂'种子最短,为 1.053 cm,但长宽比(1.693)却为最小,使种子外观近于圆形。由于种子的长、宽不同,不同品种单粒重也明显不同。最大的'大花金'桂种子单粒重达 0.274 g,是最小品种'籽银桂'和'大叶籽金'桂种子的 2.63 倍。

对各品种种子的长度和宽度进行了方差分析。结果表明:不同品种种子的长度($F=80.645, P=0.000$)、宽度($F=22.325, P=0.000$)以及单粒重($F=33.149, P=0.000$)均存在极显著差异。在此基础上进行 Duncan 多重对比分析,结果表明(见表 1):'大花金'桂和'晚金桂'种子最长,其他品种种子的长度有显著差异;而'早籽银'桂、'墨叶金'桂、'大花金'桂的种子宽度最大,与其他种子有显著差异;正因为长宽的不同从而使得'大花金'桂的单粒重最重,并且与其他品种的质量呈极显著差异。

表1 不同桂花结实品种种子性状比较

品种		长度/cm	宽度/cm	长/宽	单粒重/g
'宽叶籽银'桂	'Kuanye Zi Yin'	1.588 C	0.570 CD	2.786	0.145 de
'早籽银'桂	'Zao Zi Yin'	1.406 D	0.681 A	2.065	0.19 bc
'籽银桂'	'Zi Yingui'	1.388 D	0.537 DE	2.585	0.104 g
'紫梗籽银'桂	'Zigeng Zi Yin'	1.251 E	0.511 E	2.448	0.121 defg
'长瓣银'桂	'Changban Yin'	1.720 B	0.618 BC	2.783	0.178 c
'卷瓣银'桂	'Juanban Yin'	1.214 E	0.636 AB	1.909	0.121 efg
'长叶碧珠'	'Changye Bizhu'	1.384 D	0.614 BC	2.254	0.14 def
'潢川金'桂	'Huangchuan Jingui'	1.053 F	0.622 B	1.693	0.115 fg
'大花金'桂	'Dahua Jin'	1.834 A	0.656 AB	2.796	0.274 a
'大叶籽金'桂	'Daye Zi Jin'	1.438 D	0.539 DE	2.668	0.105 g
'墨叶金'桂	'Moye Jin'	1.196 E	0.674 A	1.774	0.149 d
'晚金桂'	'Wan Jingui'	1.756 AB	0.615 BC	2.856	0.213 b
'籽丹桂'	'Zi Dangui'	1.490 D	0.541 DE	2.754	0.115 fg

*:不同小写字母为0.05水平上差异显著,不同大写字母为0.01水平上差异显著。

为了解桂花种子的长度和宽度对种子重量贡献的大小,对种子的长度、宽度和种子的单粒重进行了相关分析(见表2)。从结果可知:桂花种子的长度、宽度与重量之间存在极显著正相关,其中种子的长度对种子重量的影响要比宽度大,长度越长、宽度越大,种子的单粒重就越重。

表2 桂花品种种子长、宽度与质量的相关性分析

项目	单粒重	宽度	长度
单粒重	1		
宽度	0.516**	1	
长度	0.663**	0.054	1

**表示在0.01水平下呈极显著相关。

2.3 桂花种子种皮外表纹饰比较

比较了13个桂花品种的种皮外表纹饰,不同品种种皮的纹路有差异。主要表现在种皮表面突起的棱的数量、疏密度和棱上、棱间分枝形状等(图3)。

2.3.1 '宽叶籽银'桂'Kuanye Zi Yin':种子椭圆形,两端延伸较短,头部稍尖,尾部钝。棱大多为6条,偶有7或8条,突起明显,棱间枝杈状分枝粗而短,大多有长而细的二级分叉。

2.3.2 '早籽银'桂'Zao Zi Yin':种子椭圆形,头部稍园钝,尾部平,连接两端的中心轴稍偏。多数具8~9条明显突起连接两端的棱,排列较紧密,棱间枝杈状分枝粗短。

2.3.3 '籽银桂''Zi Yingui':种子狭长椭圆形,两端较尖,基本等长,中心轴正或稍偏。棱7~8条,明显,棱间枝杈状分枝细而长。

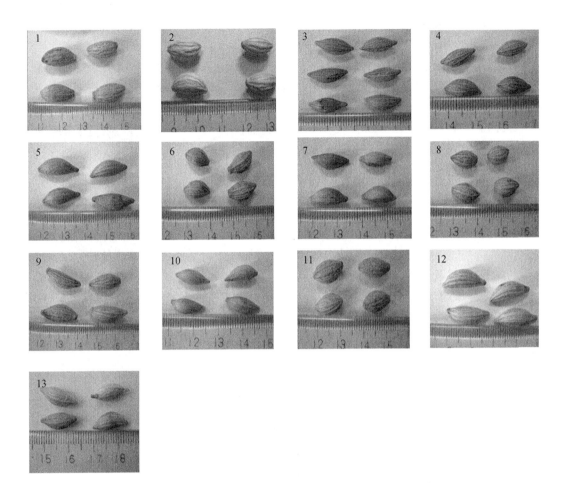

图 3　桂花不同品种的种子种皮外表纹饰比较

2.3.4 '紫梗籽银'桂 'Zigeng Zi Yin'：种子长椭圆形,两端较尖,正或稍斜。棱 5~6 条,分布较稀,棱间枝杈状分枝较细而长。

2.3.5 '长瓣银'桂 'Changban Yin'：种子狭长椭圆形,两端渐尖,尾部较先端长,中心轴稍斜。具 6~8 条稍突起的棱,棱上和棱间的分枝密且长。

2.3.6 '卷瓣银'桂 'Juanban Yin'：种子椭圆形,中心轴略歪斜,头部渐尖,尾部突出较长但先端平截。表皮上 7~8 条棱突起非常明显,每两条棱间又有小的枝杈状的小分枝。

2.3.7 '长叶碧珠' 'Changye Bizhu'：种子长椭圆形,中心轴稍偏,两端渐尖,头部先端尖,尾部先端稍钝。5~6 条棱不明显,棱上和棱间分枝较少也明显。

2.3.8 '潢川金桂' 'Huangchuan Jingui'：种子近卵圆形,头部突尖,尾部钝。5~6 条棱突起明显,每条棱及两条棱间均有枝杈状分枝,犹如一条条游龙盘缠其上,非常美观。

2.3.9 '大花金'桂 'Dahua Jin'：种子长椭圆形,中心轴稍歪斜,头部较长,先端渐尖,尾部稍长,先端平截。有 8~9 条明显突起且较宽大的棱,棱间排列紧密,棱上分枝较少且短。

2.3.10 '大叶籽金'桂 'Daye Zi Jin'：种子椭圆形,中心轴稍歪斜;头部较长,渐尖,尾部稍短,先端钝。6~7 条棱不太明显,棱上分枝较少且花纹不清晰。

2.3.11 '墨叶金'桂 'Moye Jin'：种子椭圆形，少有歪斜，两端短而钝。种子表皮上5~6条棱突起明显，每两条棱间枝杈状的分枝较长且大。

2.3.12 '晚金桂' 'Wan Jingui'：种子长椭圆形，中心轴大多歪斜，有的表现为严重弯曲畸形，两端长，尾部比头部明显加长，先端渐尖。具8~9条明显突起的棱，棱间枝杈状分枝相对较少且小。

2.3.13 '籽丹桂' 'Zi Dangui'：种子椭圆形，有的种子中心轴偏离严重，近乎拱形。两端渐尖，头部比尾部长。6~7条棱较宽，稍显。棱上的分枝较短。

2.4 种皮表面纹饰的扫描电镜观察

在肉眼判别的基础上，用扫描电镜（×30倍）比较了13个桂花品种的种皮表面纹饰，不同品种种皮表面纹饰有明显差异。有些品种的种皮表面纹饰非常清晰，种皮上突起具明显的立体感，如'长瓣银'桂种皮表面纹饰似分叉的树枝，'潢川金桂'种皮表面纹饰如展开脚爪爬行的蜥蜴，并且各级分叉纵横交错；而'籽丹桂'的种皮表面纹饰则呈现极强的立体感，多为短而大的分枝，次级侧支较少，显得干净利落。但也有一些品种的种皮表面纹饰较模糊，立体感不强。如'大花金'桂、'宽叶籽银'桂、'晚金桂'的种皮表面纹饰表现为块状结构，其纵向的突起不明显。13个品种分属3个品种群，但从供试品种的观察结果看，品种群间的种皮表面纹饰有显著的区别。

2.5 种皮断面细胞排列结构的扫描电镜观察

通过扫描电镜观察了13个桂花品种种皮的断面结构，也发现不同品种种皮断面的细胞结构有明显的差别。种子种皮一般由三种不同形状的细胞组成：最外面由3~5层横向排列的较密的长管状细胞组成；中间是十几层呈纵向规则式排列的方形细胞；最里面是4~5层棒状细胞，细胞排列较疏松，可见明显的细胞腔。细胞壁上有多数小孔，可能与透水有关。不同品种的种皮细胞层数不同，'早籽银'桂、'紫梗籽银'桂、'晚金桂'、'籽丹桂'种皮较薄，细胞层数少，其他品种相对较厚，细胞层数多。排列紧密度上也有一定的差异，'早籽银'桂、'籽银桂'、'长叶碧珠'、'潢川金桂'、'墨叶金'桂的细胞排列较紧密；'宽叶籽银'桂、'紫梗籽银'桂、'大花金'桂和'大叶籽金'桂的细胞排列较疏松。

2.6 种皮头部断面及内壁扫描电镜观察

从各品种种皮头部纵切面观察，除了种皮的厚度差别外，头部均具有一发芽孔，利于种子发芽时胚根穿透种皮。所有品种的内壁均非常光滑透亮。

3 结论与讨论

桂花不同品种种子的长度、宽度、单粒重存在极显著差异，并且长度、宽度与重量之间存在极显著正相关，其中种子的长度对种子重量的影响要比宽度大，长度越长、宽度越大，种子的单粒重就越重。一般认为种实性状表型特征的差异是基因型与环境相互作用的结果[1-4]，种子形状的大小表征了种子内营养物质的多少[5]，其变异程度会影响种子扩散和种子萌发[6]等，对幼苗定居和存活有很大影响[7-10]。同时，种子大小与果实大小有着密切的关系[11]。莫昭展等[12]对银杏种子性状进行分析认为,种子的长宽在不同品种间的差异确实存在,对种子的重量性状的影响非常大,从而影响到银杏种子的出核率,进一步影响种子的经济品质;宋丽华等[13]对不同种源的臭椿种子、曹兵等[14]对希蒙得木种子的形态特征变异分析后认为种子的千粒重影响种子的发芽率,一般千粒重越重,种子的发芽率越高,这和种子中所含营养物质有关。对于不同的桂花品种来说,种子大小也受环境因素的影响,但与种子的发芽率是否有关,有待进一步证实。通过本试验能证明桂花品种种子的形态特征具有明显的差异性,这可以为品种鉴定提供一定的辅证。

不同桂花品种的种子种皮纹饰有较大的差别,大多数品种的种皮上具有明显突起的棱,但棱的数量、清晰度以及棱上和棱间的分叉,不同品种间有较大的差别。今后应该加强这方面材料的收集,从而为品种鉴定提供一定的依据。

由不同品种的种皮断面扫描电镜图片可见,不同品种种子种皮的细胞形态和排列结构有较大差别,大部分种子细胞排列疏松,细胞腔较大,细胞壁上具有细小的孔洞,这些结构充分说明桂花种子虽然具有木质化的种皮但并不影响其吸水透气,即木质化的种皮不是造成种子休眠的原因。

参考文献

[1] 兰彦平,顾万春.北方地区皂荚种子及荚果形态特征的地理变异[J].林业科学,2006,42(7):47-51.
[2] 刘永红,杨培华,韩创举.油松不同种源种实性状的变异分析[J].浙江林学院学报,2008,25(2):163-168.
[3] Hammond D S, Brown V K. Seed size of woody plants in relation to disturbance, dispersal, soil type in wet neotropical forests[J]. Ecology, 1995, 76(8):2544-2561.
[4] Khurana E, Sagar R, Singh J S. Seed size:a key trait determining species distribution and diversity of dry tropical forest in northern India[J]. Acta Oeco-logica. 2006, 29(2):196-204.
[5] Raquel G R, Keith R P, Malcolm E R, et al. Effect of seed size and testa colour on saponin content of Spanish lentil seed [J]. Food Chem, 1997, 58(3): 223-226.
[6] Latif K M, Putul B, Uma S, et al. Seed germination and seedling fitness in Mesua ferrea L. in relation to fruit size and seed number per fruit [J]. Acta Oecol,1999,20(6):599-606.
[7] Eriksson O. Seed size variation and its effect on germination and seedling performance in the clonal herb Convallaria majalis [J]. Acta Oecol, 1999, 20(1):61-66.
[8] Khurana E, Singh J S. Influence of seed size on seedling growth of Albizia procera under different soil water levels [J]. Ann Bot, 2000, 86:1185-1192.
[9] Saverimuttu T, Westory M. Seedling longevity under deep shade in relation to seed size [J]. Journal of Ecology, 1996, 84(5):681-689.
[10] Moles A T, Westoby M. Seed size and plant strategy across the whole life cycle [J]. Oikos, 2006, 113(1):91-105.
[11] Josefa L, Juan A D, Ana O O, et al. Production and morphology of fruit and seeds in Genisteae (Fabaceae) of south west Spain [J]. Bot J Linnean Soc, 2000, 132(2):97-120.
[12] 莫昭展,曹福亮,汪贵斌,等.银杏种子性状的变异分析[J].河北林业科技,2006(4):1-5.
[13] 宋丽华,王娅丽.几个臭椿种源种子的生物学特性变异研究[J].农业科学研究,2005,26(1):18-22.
[14] 曹兵,高捍东.希蒙得木种子生物学特性研究[J].种子.2002(5):41-42.

Comparison of Seed Morphology of *Osmanthus fragrans* Cultivars

YANG Xiulian[1] JIANG Xiaozhuang[2]

(1. College of Landscape Architecture, Nanjing Forestry University, Nanjing 210037, China;
2. Jiangxi Forestry Science and Technology Popularization General Station,
Nanchang 330038, China)

Abstract: The compare was done on seed shape, appearance and inner pericarp structure based on investigation of Nanjing area. The result showed that there are distinct difference on length, width, single seed weightiness and endocarps sculpture between different cultivar seeds. So, it can be used to identify osmanthus cultivars according to the shape and appearance of seed.

Key words: sweet osmanthus; seed; shape

桂花种子休眠和萌发的初步研究

杨秀莲[1]，郝其梅[2]

（1. 南京林业大学 风景园林学院，江苏 南京 210037；2. 江苏省新沂市建设局，江苏 新沂 221400）

摘 要：为了解桂花(Osmanthus fragrans)种子的休眠原因以及掌握解除休眠的最佳方法，对'紫梗籽银'桂'Zigeng Zi Yin'种子种皮和种实的甲醇浸提液进行生物测定，并对不同质量浓度赤霉素(GA3)处理后的种子发芽率进行测定。结果表明：'紫梗籽银'桂种子的种皮和胚乳中均含有抑制白菜(Brassica campestris)籽萌发的物质，抑制作用随质量浓度增大而增加。综合前期研究结果，认为桂花种子的休眠不是外源休眠和综合休眠，而是由于由抑制物质引起的内源休眠。赤霉素处理结合低温层积可有效解除种子的休眠，其中以 1 000 mg·L^{-1} 赤霉素浸种，低温层积 75 d 发芽率最高。

关键词：植物学；桂花；种子休眠；萌发；赤霉素；低温处理

种子休眠是指具有生活力的种子处于适宜的环境条件下仍不能正常萌发的一种生理现象。种子休眠是植物在长期演化过程中为了种的生存而不断适应环境的结果[1-2]。对植物本身来说，种子的休眠对植物个体的生存、物种的延续和进化具有积极的作用，但对林木的生产带来较大的影响。通过观察，桂花(Osmanthus fragrans)种子具有一定的休眠现象。当年4—5月成熟的种子需层积至翌年2—3月才能萌发。张义等[3]利用赤霉素浸种和低温层积相结合的方法进行桂花种子的催芽试验，证明桂花种子具有休眠特性，低温层积和赤霉素(GA3)浸种可有效促进种子的萌发。袁王俊等[4]利用离体培养技术对层积3个月的桂花种子进行胚培养，打破了种子的休眠。对于桂花种子的休眠原因，目前尚未见报道。本研究通过对桂花种子的休眠和萌发的研究，旨在探讨桂花种子的休眠机制，以期寻找解除休眠的最佳方法，缩短休眠时间，提高播种繁殖率，同时也为品种间杂交和培育新品种积累经验，提供理论指导。

1 材料与方法

1.1 试验材料

'紫梗籽银'桂(Osmanthus fragrans 'Zigeng Zi Yin')种子：2007年5月初采于南京林业大学教五楼前，去除外种皮后保存于(3±1)℃冰箱中。紫梗籽银桂为银桂品种群中的一个品种，结实量较大，种子采收容易，故作为桂花种子休眠的研究对象。

白菜(Brassica campestris)种子：南京市蔬菜种子公司的'绿优1号'白菜种子，纯度95%，净度98%，发芽率85%以上。

---

* 原文发表于《浙江林学院学报》，2010，27(2)：272-276。
基金项目：江苏省农业三项工程项目(BK2005133)。

1.2 试验方法

1.2.1 种皮、种实浸提液的提取

取内种皮(以下称种皮)和种实(包含胚和胚乳两部分,下文同)各 10 g,分别置于 250 mL 容量瓶中,加入 200 mL 体积分数为 80.0%甲醇,混匀后放入冰箱内,在 2~4 ℃的恒温条件下密闭浸提,其间多次取出摇匀使其充分浸提,48 h 后过滤。所得滤液在 35 ℃下减压浓缩后定容到 0.10 g·mL^{-1}(以 1 mL 浓缩液中含有原材料的量计算)。

1.2.2 白菜种子生物测定方法

在培养皿中放置滤纸,分别加入一定量的甲醇浸提原液(0.10 g·mL^{-1}),60%原液(0.06 g·mL^{-1}),30%原液(0.03 g·mL^{-1}),每种处理设置 3 个重复,每重复 100 粒白菜种子,于 30 ℃恒温全光照条件下培养,48 h 测定发芽率,72 h 测量苗高和根长,用含体积分数为 0.2%甲醇的水溶液处理的白菜种子做对照。

1.2.3 催芽方法

采用蒸馏水(记为 G0)和质量浓度为 500 mg·L^{-1}(记为 G5)、1 000 mg·L^{-1}(记为 G10)、1 500 mg·L^{-1}(记为 G15)的赤霉素溶液浸种 48 h,后分别与湿沙按 1∶3 混合放入塑料盆中,上面覆盖保鲜膜并用剪刀剪几个小洞,利于种子呼吸,将塑料盆置于 2~4 ℃冰箱中进行低温层积处理。层积期间隔半个月取各处理种子 4 个重复,每重复 30 粒种子。在 25 ℃恒温、24 h 光照条件下做种子发芽试验,并统计种子的发芽率。

2 结果分析

2.1 甲醇浸提液对白菜籽发芽的影响

2.1.1 种皮、种实甲醇浸提液对白菜籽发芽率的影响

不同处理的种皮、种实甲醇浸提液对白菜籽发芽率的影响结果如图 1 所示。不同处理种实的甲醇浸提液处理的白菜籽发芽率分别较对照降低了 2.81%、20.83%和 29.16%;不同处理种皮浸提液处理的白菜籽发芽率分别较对照降低了 1.35%、6.98%和 11.15%。总的来说,随着浸提液质量浓度的增加,所处理的白菜籽发芽率呈逐渐降低的趋势,高质量浓度甲醇浸提液对白菜籽的抑制作用显著。

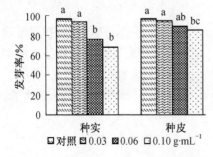

图 1 甲醇浸提液对白菜籽发芽率的影响

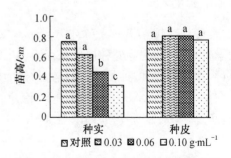

图 2 甲醇浸提液对白菜苗高的影响

2.1.2 种皮、种实甲醇浸提液对白菜苗苗高生长的影响

不同处理的种皮、种实甲醇浸提液对白菜苗的苗高生长量的影响结果见图 2。与对照相比,不同处理种实的甲醇浸提液处理的白菜苗高分别降低了 18.0%、40.0%和 58.0%;而不同处理种皮浸提液处理的白菜苗高却比对照分别增加 7.4%、7.4%和 1.6%。以上结果表明:种实浸提液处理的白菜籽苗高随着浸提液质量浓度的增加,呈现出逐渐降低的趋势,说明胚乳浸提液对白菜苗高有抑制作用,而种皮浸提液对苗高生长的抑制作用不显著。

2.1.3 种皮、种实甲醇浸提液对白菜苗的根长生长量的影响

不同处理的种皮、种实甲醇浸提液对白菜苗的根长生长量的影响结果见图 3。与对照相比,不同处理种实浸提液处理的白菜根长分别降低了 76.4%、88.76%和 91.6%;而不同处理种皮浸提液处理的白菜根长在 0.03 g·mL^{-1} 和 0.06 g·mL^{-1} 时比对照增加了 34.79%和 18.53%,质量浓度为 0.10 g·mL^{-1} 时,根长比对照降低 48.77%。以上结果表明:种实浸提液处理的白菜苗根长随着浸提液质量浓度的增加,呈现出急剧降低的趋势,说明种实浸提液对白菜苗根长生长有很强的抑制作用;而低质量浓度的种皮浸提液对白菜苗根长生长有一定的促进作用(其原因有待进一步研究),随着质量浓度的增大,逐渐转为抑制作用。

综上所述,'紫梗籽银'桂种子的种皮和种实的甲醇浸提液中含有某些抑制物质,这些物质对白菜籽的发芽率、苗高和根长生长量均有一定的抑制作用,并随着浸提液质量浓度的提高,抑制作用逐渐加强。相比而言,种实的甲醇浸提液对白菜籽发芽率的抑制作用比种皮浸提液的抑制作用更大。

2.2 不同质量浓度赤霉素处理的'紫梗籽银'桂种子发芽率的变化

未经低温层积处理的'紫梗籽银'桂种子置床 30 d 仍未见种子萌发,后将部分种子的种皮和胚乳剥除成离体胚,再置床 58 d,仅见胚体子叶部分转为绿色,胚轴仍为白色,而未剥除的种子仍无萌发迹象。这充分证明前面的研究结果:'紫梗籽银'桂种子中含有发芽抑制物质,种子存在生理休眠现象。

经不同质量浓度赤霉素处理并经低温层积的'紫梗籽银'桂种子发芽率见图 4。从图 4 中可以看出,不同处理的'紫梗籽银'桂种子在层积 60 d 后开始发芽,且以 1 000 mg·L^{-1}赤霉素处理的种子发芽率最高,随着层积时间的延续,各处理发芽率逐渐增加,至层积 105 d 时发芽率达最高,其中 G10 在 75 d 时发芽率为 71.11%,93 d 时为 67.78%,105 d 时为 72.37%,基本上保持稳定。G5 和 G15 发芽率的提高滞后于 G10,105 d 时分别为 63.82%和 70.32%。而未经赤霉素处理的种子在层积 105 d 后其发芽率仅为 26.67%。

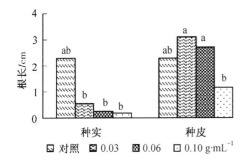

图 3 甲醇浸提液对白菜根长生长的影响

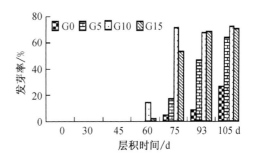

图 4 '紫梗籽银'桂种子层积过程中发芽率的变化

方差分析结果表明:不同质量浓度赤霉素、不同层积时间以及赤霉素浓度与层积时间之间的交互效应对种子发芽率的影响均达到极显著水平。由此可见,用赤霉素处理种子后再进行低温层积可显著解除休眠,其中以1 000 mg·L^{-1}赤霉素处理为最佳,只要75 d左右就可打破桂花种子的休眠。

3 结论与讨论

对'紫梗籽银'桂种子的种皮、种实的甲醇浸提液进行生物测定。结果表明:种皮、种实的甲醇浸提液对白菜籽的发芽率、白菜幼苗高生长和根生长量均有一定程度的抑制。这种抑制作用随着质量浓度的增加而增大,并且胚乳浸提液的抑制作用比种皮的大。在试验中还观察到白菜苗根系出现膨大、扭曲、向上生长等畸形现象。由此可见,'紫梗籽银'桂种子中存在一些抑制白菜籽萌发和影响生长的物质。相同的实验在南方红豆杉(*Taxus chinensis* var. *mairei*)[5]、青钱柳(*Cyclocarya paliurus*)[6]、洋白蜡(*Fraxinus pennsylvanica*)[7]等树种上均已得到证实。但'紫梗籽银'桂种子中究竟含有什么物质,多大质量浓度才会对种子休眠起作用仍需进一步研究。

种子休眠一般分为外源休眠、内源休眠和综合休眠。种子物理透性常引起外源休眠,休眠程度与种皮的不透性或硬实程度有关。如红松(*Pinus koraiensis*)种子外种皮坚硬,阻碍了种胚与外界的气体交换,进而引起种子休眠[8]。内源休眠有的因胚发育不完全所致,有的因种子萌发的生理抑制造成。如导致银杏(*Ginkgo biloba*)[9]种子深休眠的主要原因是胚发育不完全;玫瑰(*Rosa rugosa*)种子种皮中存在的抑制物脱落酸是引起其休眠的主要原因;南京椴(*Tilia mqueliana*)[10]、对节白蜡(*Fraxinus hupensis*)[11]种子的果皮、种皮及胚乳中都存在导致休眠的抑制物质。综合休眠则是内外源因素互相联系共同控制着种子的休眠,从而使种子休眠问题的研究复杂化[12]。秤锤树(*Sinojackia xylocarpa*)[13]和青钱柳[14]种子的休眠,除了种皮机械束缚和透气性差的原因外,种胚还需要一段低温的生理后熟过程,属于综合休眠类型。本研究中,'紫梗籽银'桂种子的种皮虽然较为坚硬,但种皮较薄,种子萌发时种胚可以顺利突破种皮,表明种皮不会对种胚的萌发造成机械性阻碍,对种子吸水速率的观察也表明'紫梗籽银'桂种皮具有良好的透性[15]。'紫梗籽银'桂种子的甲醇浸提液对白菜籽的发芽率、白菜苗高和根长等均有一定的抑制作用,说明种子中含有抑制物质可能是导致桂花种子休眠的主要原因,也即桂花种子的休眠属生理(内源)休眠。

种子休眠的原因各不相同,因而解除休眠的方法也各不相同。层积处理是一种在生产上应用很广泛的方法,尤其是对解除由于生理后熟引起的种子休眠很有效。张义等[3]研究发现,'紫梗籽银'桂种子胚具有生理后熟现象,低温层积能够增进种子萌发,在采收后一般需经沙藏后播种。本研究用赤霉素处理浸种后低温层积,60 d时就有部分种子萌发,并且75 d后发芽率基本达到最大值,之后种子的发芽率保持稳定,表明层积处理对解除桂花种子休眠很有效,层积75 d左右即可解除休眠。

参考文献

[1] LIU Zhen. Studies on the dormancy in *Idesia polycarpa* distributing in the subtropical zone [J]. Bull Mie Univ For, 2000, 24:107-161.

[2] NAGATA H, NAKASHIMA A, YURUGI Y. Bud dormancy in woody plants [J]. Bull Mie Univ For, 1994,18:17-42.

[3] 张义,宋春燕.赤霉素浸种与低温层积对桂花种子发芽的影响[J].中国林副特产,2005(6):9-10.
ZHANG Yi, SONG Chunyan. Impact of *Osmanthus fragrans* seed gibberellin processing and low temperature stratification on its seed germination performan [J]. Q For By-prod Spec China,2005(6): 9-10.

[4] 袁王俊,董美芳,尚富德.利用离体培养技术打破桂花种子休眠的试验[J].南京林业大学学报:自然科学版,2004,28(增刊):91-93.
YUAN Wangjun, DONG Meifang, SHANG Fude. Preliminary study on breaking the seed dormancy of *Osmanthus fragrans* by in vitro culture [J]. J Nanjing For Univ ((Nat Sci Ed), 2004,28(sup): 91-93.

[5] 张艳杰,高捍东,鲁顺保.南方红豆杉种子中发芽抑制物的研究[J].南京林业大学学报(自然科学版),2007,31(4):51-56.
ZHANG Yanjie, GAO Handong, LU Shunbao. Germination inhibitors in methanol extract from *Taxus chinensis* var. *mairei* seed [J]. J Nanjing For Univ(Nat Sci Ed),2007,31(4):51-56.

[6] 杨万霞,方升佐.青钱柳种皮甲醇浸提液的生物测定[J].植物资源与环境学报,2005,14(4):11-14.
YANG Wanxia, FANG Shengzuo. The bioassay of the methanol extract from *Cyclocarya paliurus* seed coat [J]. J Plant Resour Environ,2005,14(4):11-14.

[7] 郑彩霞,高荣孚.脱落酸和内源抑制物对洋白蜡种子休眠的影响[J].北京林业大学学报,1991,13(4):19-22.
ZHENG Caixia, GAO Rongfu. Effect of ABA and other endogenous inhibitors on *Fraxinus pennsylvanica* Marsh. seeds dormancy [J]. J Beijing For Univ, 1991,13(4):19-22.

[8] 郑彩霞,王九龄,智信.国内外红松种子休眠及催芽问题研究动态[J].世界林业研究,1997,10(5):3-9.
CHEN Caixia, WANG Jiuling, ZHI Xin. Research state of the dormancy and pregermination of *Pinus koraiensis* seed in China and aboard [J]. World For Res,1997,10(5):3-9.

[9] 蔡春菊,曹帮华,许景伟,等.银杏种子后熟生理的研究[J].山东林业科技,2002(6):7-9.
CAI Chunju, CAO Banghua, XU Jingwei, et al. Studies on delayed-ripening physiology of ginkgo seeds [J]. J Shandong For Sci Technol,2002(6):7-9.

[10] 史锋厚,沈永宝,施季森.南京椴种子发芽抑制物研究[J].福建林学院学报,2007,27(3):222-225.
SHI Fenghou, SHEN Yongbao, SHI Jisen. Study on germination inhibitor of Nanjing linden seeds [J]. J Fujian Coll For,2007,27(3):222-225.

[11] 叶要妹,王彩云,史银莲.对节白蜡种子休眠原因的探讨[J].湖北农业科学,1999,78(4):45-47.
YE Yaomei, WANG Caiyun, SHI Yinlian. Preliminary study on cause of seed dormancy of *Fraxinus hupensis*[J]. Hubei Agric Sci,1999,78(4):45-47.

[12] KHAN A A. Quantification of plant dormancy: introduction to the workshop [J]. HortScience, 1997,32(4):608-609.

[13] 史晓华,黎念林,金玲,等.秤锤树种子休眠与萌发的初步研究[J].浙江林学院学报,1999,16(3):228-233.
SHI Xiaohua, LI Nianlin, JIN Ling, et al. Seed dormancy and germination of *Sinojackia xylocarpa* [J]. J Zhejiang For Coll, 1999,16(3):228-233.

[14] 史晓华,徐本美,黎念林,等.青钱柳种子休眠与萌发的研究[J].种子,2002(5):5-8.
SHI Xiaohua, XU Benmei, LI Nianlin, et al. Study on dormancy and germination of *Cyclocarya*

paliurus (Batal.) Iljinskaja [J]. Seed, 2002(5):5-8.

[15] 杨秀莲,丁彦芬,甘习华.'紫梗籽银'桂种子休眠原因的初步探讨[J].江苏林业科技,2007,34(6):18-19,45.
YANG Xiulian, DING Yanfen, GAN Xihua. Primary research on dormancy mechanism of seeds of 'Zigengziyin' osmanthus [J]. J Jiangsu For Sci Technol,2007,34(6):18-19,45.

Dormancy and Germination of *Osmanthus fragrans* Seeds

YANG Xiulian[1], HAO Qimei[2]

(1. College of Landscape Architecture, Nanjing Forestry University, Nanjing 210037, China;
2. Construction Bureau of Xinyi City, Xinyi 221400, China)

Abstract: To detect reasons for dormancy and methods of breaking dormancy, a bioassay of a methanol extract from *Osmanthus fragrans* 'Zigeng Zi Yin' seed coats and endosperm as well as germination of seeds treated with Gibberellic acid (GA3) were studied. The cabbage seeds were germinated with methanol extract solution(0.10, 0.06, and 0.03 g · L^{-1}), by three replications each treatment. And the osmanthus seeds were soaked in different concentrations (500, 1 000, and 1 500 mg · L^{-1}) GA3 for 48 hours, then stratified with low temperature, and every 15 days, taken out 30 seeds each treatment with four replications for germination. Results showed that the restraining substances were contained in the seed coat and endosperm, and the inhibitory effects became stronger as the extraction solution concentration increased. Integrating results from the prophase, we think that seed dormancy belongs to physiological dormancy. Soaking with 1 000 mg · L^{-1} GA3 followed by cold stratification with sand for 75 d could raise the germination percentage and could be an effective method to overcome dormancy.

Key words: botany; *Osmanthus fragrans*; seed dormancy; germination; GA3; low temperature treatment

第五部分
应用研究

桂花露酒浸提及营养成分研究

杨秀莲,常兆晶,冯 洁,石 瑞,施婷婷,王良桂

(南京林业大学 风景园林学院,江苏 南京 210037)

摘 要:为研究桂花露酒的浸提条件及其营养成分变化,以白酒、新鲜桂花花瓣为试验材料,采用不同的酒精度、料液比设计两因素随机区组试验,配成不同浓度的桂花露酒,测定各处理桂花露酒的营养成分,并通过主成分分析法对各处理的综合品质进行评价,确定桂花露酒的最佳浸提条件。试验结果表明:桂花露酒最佳浸提条件为白酒酒精度为66%(V/V)、料液比为1:20(kg:L)、浸提时间为6个月,该条件下获得的桂花露酒营养成分含量相对最高,各营养成分含量为每升露酒含:总黄酮1.14 g、花青素7.49 mg、维生素C 39.9 mg、总糖2.58 g、Fe 1.14 mg、Mn 0.17 mg、Zn 0.94 mg、Cu 0.23 mg、Mg 7.69 mg、Ca 2.91 mg。实现了桂花露酒浸提条件的初步筛选和营养成分变化分析,为桂花露酒的进一步开发利用提供参考。

关键词:桂花露酒;酒精度;料液比;营养成分

根据中华人民共和国国家标准《露酒》GB/T27 588—2011对露酒的定义,露酒是以蒸馏酒、发酵酒或食用酒精为酒基,加入可食用或药食两用(或符合相关规定)的辅料或食品添加剂,进行调配、混合或再加工制成的,已经改变了其原酒基风格的饮料酒,包括植物类、动物类、动植物类露酒[1]。露酒具有营养丰富、品种繁多、风格各异的特点。桂花露酒原产中国。在中国,桂花食用历史悠久,桂花露酒最负盛名。不过在传统观念中,人们对桂花露酒的认识只停留在知道其香味甘醇、有健脾胃、助消化、活血益气之功效,至今没有人研究过桂花露酒中所含营养成分具体为何,是否还具有其他功效。桂花和白酒在我国都具有很大的市场潜力。研究开发桂花露酒,不仅为桂花,也为白酒的现代化、工业化发展提供了新的思路,具有十分广阔的市场前景。目前国内外尚没有针对桂花露酒浸提条件及营养成分进行定量定性研究,因此本试验以白酒、桂花花瓣为材料,以酒精度、料液比为试验因素设计两因素随机区组试验,研究了桂花露酒的浸提条件及营养成分含量,初步获得桂花露酒最佳浸提条件,有助于提高桂花花瓣利用率,加深人们对桂花露酒的了解,以利桂花和桂花露酒在市场上的推广。

---

* 原文发表于《食品工业科技》,2016,37(21):347-352。
基金项目:"十二五"科技支撑"林木种质资源发掘与创新利用(2013BA001B06);江苏省农业科技自主创新资金项目(CX(14)2031)。

1 材料和方法

1.1 材料与仪器

材料:白酒为江苏泗阳某酒坊提供的原酒,新鲜桂花花瓣为湖北咸宁某大型桂花基地'小金玲'盛花期的花瓣。

仪器:Lambda 950型分光光度计(PE公司),969型原子吸收光谱仪(美国热电公司)。

1.2 试验方法

采用不同的酒精度、料液比(表1),按照两因素随机区组试验设计(表2)配成不同浓度的桂花露酒,设3次重复,在干燥阴凉条件下保存,储藏1年。其间定期(2、4、6、8、10、12个月)取样,测定不同处理酒液中总黄酮、花青素、维生素C、总糖、矿质元素的含量。

表1 桂花露酒试验因素与水平
Tab. 1 *Osmanthus fragrans* liqueur's test factors and levels

组合 Combination	酒精度 Alcohol degree	料液比 Solid-liquid ratio
1	42°	1:60
2	53°	1:40
3	66°	1:20

表2 桂花露酒两因素随机区组试验
Tab. 2 *Osmanthus fragrans* liqueur's two-factor test with randomized block

组合 Combination	酒精度(A) Alcohol degree	料液比(B) Solid-liquid ratio	处理 Treatment
1	42°	1:60	A_1B_1
2	42°	1:40	A_1B_2
3	42°	1:20	A_1B_3
4	53°	1:60	A_2B_1
5	53°	1:40	A_2B_2
6	53°	1:20	A_2B_3
7	66°	1:60	A_3B_1
8	66°	1:40	A_3B_2
9	66°	1:20	A_3B_3

应用SPSS数据分析软件对每次(共6次)取样测定的各处理的桂花露酒10种营养成分指标分别进行主成分分析,提取m个主成分,提取条件为累积方差贡献率不低于85%,以各主成分对应的方差相对贡献率作为权重,对主成分得分和相应的权重进行线性加权求和构建露酒品质的评价函数。筛选出来的m个主成分是综合的、相互独立的指标,它们是原变量的正规化线性组合,主成分中各性状载荷值的大小体现了各性状在主成分中的重要程度。

1.3 指标测定及数据处理

1.3.1 指标测定 总黄酮与花青素的测定采用分光光度计方法,维生素C的测定采用2,6-二氯靛酚钠方法,总糖的测定采用直接滴定方法,矿质元素的测定采用原子吸收光谱仪方法。

1.3.2 数据处理 采用Excel 2000软件进行计算和作图,采用SPSS 19.0分析软件进行主成分分析比较。

2 结果与分析

2.1 各营养物质含量的变化情况

2.1.1 总黄酮

由图1可知,不同处理条件下,各处理的总黄酮含量均在第8个月有一个波峰,在之后4个月里逐渐下降,且均低于前半年黄酮含量的值。含量最高的为第8个月的6号处理(A_2B_3),其含量为1.94 g/L。

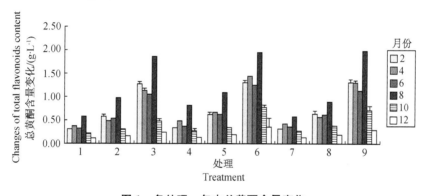

图1 各处理一年内总黄酮含量变化
Fig. 1 Flavonoids variation of each treatment in one year

表3 各因素对总黄酮含量影响的方差分析
Tab. 3 Anova analysis of the impact of various factors on total flavonoids content

源 Factor	Ⅲ型平方和 Square sum of Ⅲ	df	均方 Mean square	F	Sig.
截距 Intercept	78.835	1	78.835	40 338.787	0.000
酒精度 Alcohol degree	0.053	2	0.026	13.521	0.000
料液比 Solid-liquid ratio	14.624	2	7.312	3 741.443	0.000
酒精度×料液比 Alcohol degree × Solid-liquid ratio	0.395	4	0.099	50.490	0.000
误差 Error	0.035	18	0.002		

且由方差分析和多重对比可知,酒精度、料液比及酒精度与料液比之间的交互效应对黄酮的浸出量影响都极显著($P<0.01$)。就酒精度而言,$A_2>A_3>A_1$,且A_2与A_3、A_1之间在

$P<0.01$ 时差异显著，A_3 与 A_1 间差异不显著；就料液比而言，$B_3>B_2>B_1$，三者间在 $P<0.01$ 时差异都极显著。

由此可知，就总黄酮浸出量而言，最佳处理是 A_2B_3，即酒精度为 53%（V/V）、料液比为 1∶20（kg∶L）、浸提时间为 8 个月的条件下，总黄酮浸出量最高。

2.1.2 花青素

由图 2 可知，9 个处理中，处理 1、2、4、5、7 在全年花青素含量均变化不大，变化范围在 2~4 mg/L，处理 8 有轻微波动，在第 6 个月有一个峰值。处理 3、6、9 在全年花青素含量均明显高于其他处理，且在一年内大致均呈明显上升趋势，其中 9>6>3。这说明了，在料液比处于最高水平 1∶20 的情况下，花青素含量随浸提时间呈上升趋势。含量最高的为第 10 个月的 9 号处理（A_3B_3），其含量为 10.55 mg/L。

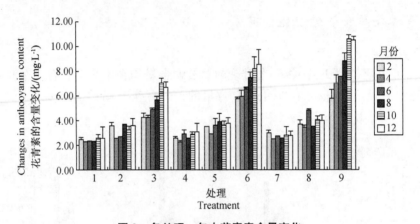

图 2 各处理一年内花青素含量变化
Fig. 2 Anthocyanins variation of each treatment in one year

表 4 各因素对花青素含量影响的方差分析表
Tab. 4 ANOVA analysis of the impact of various factors on anthocyanins content

源 Factor	Ⅲ 型平方和 Square sum of Ⅲ	df	均方 Mean square	F	Sig.
截距 Intercept	3 061.444	1	3 061.444	5 751.431	0.000
酒精度 Alcohol degree	49.585	2	24.792	46.576	0.000
料液比 Solid-liquid ratio	568.236	2	284.118	533.763	0.000
酒精度×料液比 Alcohol degree ×Solid-liquid ratio	33.779	4	8.445	15.865	0.000
误差 Error	9.581	18	0.532		

且由方差分析和多重对比可知，酒精度、料液比及酒精度与料液比之间的交互效应对花青素的浸出量影响都极显著（$P<0.01$）。就酒精度而言，$A_3>A_2>A_1$，3 个水平间差异在 $P<0.01$ 时均显著；就料液比而言，$B_3>B_2>B_1$，3 个水平间差异在 $P<0.01$ 时均显著。

这说明了，就花青素浸出量而言，最佳处理是 A_3B_3，即酒精度为 66%（V/V）、料液比为 1∶20、浸提时间为 10 个月的条件下，花青素的浸出量最高。

2.1.3 维生素 C

由图 3 可知，桂花露酒大部分处理中维生素 C 含量随时间变化为降—升—降—升，在第 6 个月达到峰值,之后逐渐下降,第 10 个月后稍有回升,这主要是由于温度的变化和维生素 C 的稳定性综合作用所致,维生素 C 本身不稳定,容易氧化,且温度的变化会同时影响维生素 C 的浸提与稳定性。整个储藏过程中维生素 C 含量较高的处理为 9、6、3,其中 9＞6＞3。含量最高的为第 6 个月的 9 号处理(A_3B_3),为 3.99 mg/100 mL。说明料液比最高的处理,其维生素 C 浸出量也最高,且料液比处于最高水平 1∶20 时,维生素 C 浸出量随酒精度上升而增加。

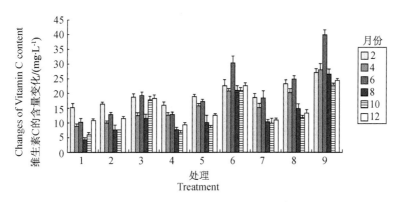

图 3　各处理一年内维生素 C 含量变化
Fig. 3　Vitamin C variation of each treatment in one year

表 5　各因素对维生素 C 浸出量影响的方差分析表
Tab. 5　ANOVA analysis of the impact of various factors on vitamin C content

源 Factor	Ⅲ 型平方和 Square sum of Ⅲ	df	均方 Mean square	F	Sig.
截距　Intercept	417.284	1	417.284	15 573.884	0.000
酒精度　Alcohol degree	16.496	2	8.248	307.841	0.000
料液比　Solid-liquid ratio	36.277	2	18.138	676.962	0.000
酒精度×料液比 Alcohol degree × Solid-liquid ratio	2.615	4	0.654	24.396	0.000
误差　Error	0.482	18	0.027		

且由方差分析和多重对比可知,酒精度、料液比及酒精度与料液比之间的交互效应对维生素 C 的浸出量影响都极显著($P<0.01$)。就酒精度而言,$A_3>A_2>A_1$,3 个水平间差异在 $P<0.01$ 时均显著;就料液比而言,$B_3>B_2>B_1$,3 个水平间差异在 $P<0.01$ 时均显著。

因此,就维生素 C 浸出量而言,最佳处理是 A_3B_3,即酒精度为 66%(V/V)、料液比为 1∶20、浸提时间为 6 个月的情况下,维生素 C 含量最高。

2.1.4　总糖

由图 4 可知,对每个处理,全年的总糖含量变化不大。整个储藏过程中总糖含量较高的处理为 9、6、3,在前半年里 6＞9＞3,后半年里 9＞6＞3。其中,处理 6 在第 4 个月达到峰值后轻微下降后趋于平稳,处理 9 一直缓慢上升在第 10 个月达到峰值后轻微下降,处理 3 表现得不稳定,先降后升再降再升。其余处理的总糖含量全过程均不高,且变化不大,大致都在第 6 到 8 个月出现波峰。含量最高的为第 10 个月的 9 号处理(A_3B_3),为 2.90 g/L。

且由方差分析和多重对比可知,酒精度、料液比及酒精度与料液比之间的交互效应对总糖的浸出量影响都极显著($P<0.01$)。就酒精度而言,$A_3>A_2>A_1$,其中 A_3 与 A_2 差异在 $P<0.01$ 时不显著,它们与 A_1 差异显著;就料液比而言,$B_3>B_2>B_1$,3 个水平间差异在 $P<0.01$ 时均显著。

这说明了,就总糖浸出量而言,最佳处理是 A_3B_3,即酒精度为 $66\%(V/V)$、料液比为 1∶20、浸提时间为 10 个月的条件下,总糖含量最高。

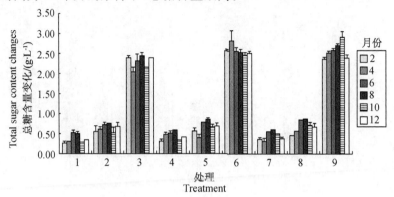

图 4　各处理一年内总糖含量变化

Fig. 4　Total sugar variation of each treatment in one year

表 6　各因素对总糖浸出量影响的方差分析表

Tab. 6　ANOVA analysis of the impact of various factors on total sugar content

源 Factor	Ⅲ 型平方和 Square sum of Ⅲ	df	均方 Mean square	F	Sig.
截距 Intercept	225.498	1	225.498	35 397.943	0.000
酒精度 Alcohol degree	0.586	2	0.293	46.013	0.000
料液比 Solid-liquid ratio	138.062	2	69.031	10 836.299	0.000
酒精度×料液比 Alcohol degree × Solid-liquid ratio	0.473	4	0.118	18.553	0.000
误差 Error	0.115	18	0.006		

2.1.5　矿质元素

本试验测定的具有代表性的矿质元素为有益微量元素 Fe、Zn、Cu、Mn,有害微量元素 Pb、Hg,大量元素 Ca、Mg。由图可以看出:

(1)有益微量元素中,就 Fe 的浸出量而言,如图 5,处理 4、5、6、7、8、9 从第 4 个月开始大幅上升,处理 1、2、3 从第 6 个月才开始大幅上升。上升之后,除个别处理之外,到第 8 个月都轻微下降,然后保持较稳定水平。可见酒精度对 Fe 的浸出量影响比较明显,酒精度超过 $53\%(V/V)$ 的白酒对桂花花瓣中 Fe 的浸出效果比 $42\%(V/V)$ 白酒好,其中 Fe 浸出量最高的为第 10 个月的 7 号处理(A_3B_1),为 2.44 mg/L。

就 Mn 的浸出量而言,如图 6,处理 3、6 的浸出量较为突出,其中,处理 3 在第 4 个月上升之后下降,之后保持较稳定水平;而处理 6 在第 4 个月下降,之后保持较稳定水平。其余的处理 Mn 的浸出量变化不是很明显,含量介于 0.08~0.15 mg/L。Mn 浸出量最高的为第 2 个月的 6 号处理(A_2B_3),为 0.62 mg/L。

就 Zn 的浸出量而言,如图 7,各处理浸出量均在第 4 个月上升,之后逐渐缓慢下降(个别处理除外),处理 4、5、6、7、8、9 从第 4 个月开始大幅上升,处理 1、2、3 从第 6 个月才开始大幅上升,可见酒精度对 Zn 的浸出量影响比较明显,酒精度超过 53%(V/V)的白酒对 Zn 浸出较快。Zn 浸出量最高的为第 4 个月的 8 号处理(A_3B_2),为 1.09 mg/L。

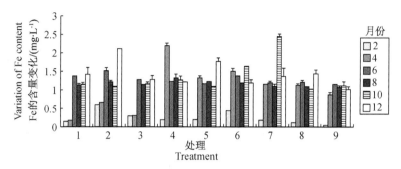

图 5　各处理在一年内 Fe 含量变化
Fig. 5　Fe content variation of each treatment in one year

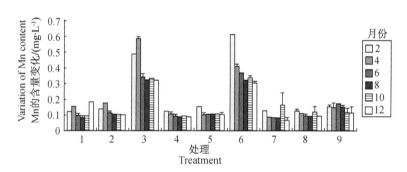

图 6　各处理在一年内 Mn 含量变化
Fig. 6　Mn content variation of each treatment in one year

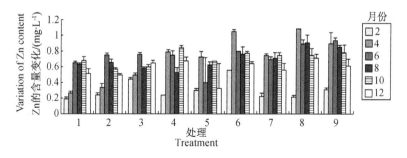

图 7　各处理在一年内 Zn 含量变化
Fig. 7　Zn content variation of each treatment in one year

就 Cu 的浸出量而言,如图 8,处理 3、6、9 变化趋势为升—降—升,在第 4 个月上升,之后下降,到第 12 个月浸出量上升,其余的处理都保持较稳定水平。可见料液比对 Cu 的浸出量影响比较明显,料液比为最大的处理均有较明显的变化趋势。Cu 浸出量最高的为第 4 个月的 6 号处理(A_2B_3),为 0.53 mg/L。

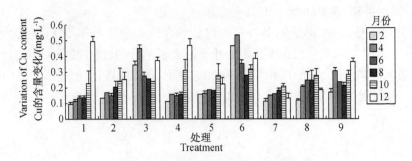

图 8　各处理在一年内 Cu 含量变化

Fig. 8　Cu content variation of each treatment in one year

(2) 有害微量元素 Pb、Hg 的含量都极低,其中 Hg 的含量为 $0.1\sim0.9~\mu g/L$,Pb 的含量为 $0.2\sim0.7~\mu g/L$。说明该批露酒健康无害。

(3) 大量元素中,对于 Mg 而言,如图 9,各处理 Mg 的浸出量在第 4 个月开始下降,之后保持较稳定水平。Mg 浸出量最高的为第 2 个月的 2 号处理(A_1B_2),为 6.50 mg/L。

对于 Ca 的浸出量而言,如图 10,处理 1、2、3 在第 6 个月达到峰值,而其他处理均在第 4 个月达到峰值。说明酒精度对 Ca 的浸出量有明显影响。3 个酒精度水平中,酒精度高的处理能促进花瓣中 Ca 的浸出。Ca 浸出量最高的为第 4 个月的 4 号处理(A_2B_2),为 4.07 mg/L。

由以上分析可知,除 Mg 元素浸出量是逐渐下降外,其余矿质元素浸出量几乎都在中间某个月有峰值,之后开始下降,这可能是由于在桂花露酒的浸泡后期,从花瓣中浸提出来的矿质元素部分与有机酸结合而损失。这与高海生等对安梨酒陈酿过程矿质元素的分析研究一致[2]。

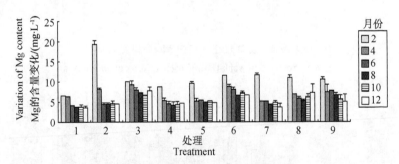

图 9　各处理在一年内 Mg 含量变化

Fig. 9　Mg content variation of each treatment in one year

对露酒的浸提而言,酒精度、料液比、浸提时间、浸提环境、酒精种类、助提手段等都会影响浸提效果。本试验研究了酒精度、料液比、浸提时间对浸提效果的影响,初步对桂花露酒进行研究。

桂花的营养成分中,有些物质比较容易溶于酒精,有些则不易溶于酒精,所以需要根据各营养成分溶于酒精的难易,以及露酒成品欲达成的功效来决定酒精的度数。李瑶等[3]对大蒜露酒最佳浸提工艺条件的研究,以大蒜素的浸出量为评定标准,结果表明对于大蒜素的浸出量而言,大蒜露酒的最佳浸提酒精度为 60°。鲍晓华等[4]对野生棠梨果露酒的研究表

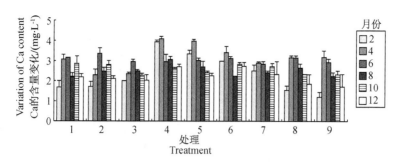

图 10 各处理在一年内 Ca 含量变化

Fig. 10 Ca content variation of each treatment in one year

明,用于野生棠梨果露酒调配用的酒最好是高粱酒,酒精度最好是 45%~47%(V/V),说明不同材料的浸提酒精度可能不同。理论上对于一定酒精度、一定体积的白酒而言,露酒中营养成分浸出量与加入的桂花花瓣含量呈正相关,即花瓣含量越高,营养成分浸出量越大。但桂花花瓣质量较轻而浮于白酒表面,失去白酒保护易引起霉变,也不利于充分浸提,所以要采取一定的助沉措施。方堃等[5]对蓝莓露酒最佳浸提条件的研究得到的最佳浸提条件中料液比为 1∶1,张翔等[6]用热水浸提法制备淫羊藿露酒,结果表明当料液比为 1∶30 时,淫羊藿总黄酮浸提率最高,说明不同材料的浸提料液比也可能不同,需增加试验次数加以确定,本试验的最佳料液比不一定为桂花露酒浸提的最佳料液比。在一定的时间内,浸提量随浸提时间的延长而增加,但过长的浸提时间不仅对浸提效果的提升起不到多大作用,往往还可能导致浸提液中营养成分含量下降甚至产生无效杂质。这可能是由于光照、温度、空气、pH 值、酶类等影响营养物质的稳定性,导致其降解,这一点在试验结果中也得到了充分的体现,在以后的浸提中可采取添加稳定剂等措施加以改善。前人[4,6-7]对多种花果露酒进行了研究,其浸提时间随着浸提条件不同而不同,浸提时间几小时到几十天不等。由于本试验是在常温下浸提,且并未用加热[8-9]、超声[10]等方式助提,所以最佳浸提时间相对较长,以后可进行桂花露酒浸提助提方法的研究,综合多种因素进行试验。

2.2 基于主成分分析的桂花露酒各处理品质综合评价

由上可知,露酒营养成分浸出评价指标共 10 个,各指标的最佳处理及处理时间并不统一,若只进行单一指标的比较,很难对各处理露酒的品质做出正确、客观的评价,也无法得出最佳浸提时间。采用主成分分析法对各处理的露酒中 10 种营养成分浸出量的 6 次观测值作更为客观的系统分析评价,结果如下:

第 2 个月时,9 个处理综合品质从高到低的排名为:6>9>3>2>8>5>7>4>1。
第 4 个月时,9 个处理综合品质从高到低的排名为:6>9>3>8>5>4>2>7>1。
第 6、10 个月时,9 个处理综合品质从高到低的排名都为:9>6>3>8>5>7>2>4>1。
第 8 个月时,9 个处理综合品质从高到低的排名为:9>6>3>8>5>2>7>4>1。
第 12 个月时,9 个处理综合品质从高到低的排名为:9>6>3>8>5>2>4>7>1。
将一年中 6 次观测值的主成分分析得分汇总如图 11 所示。

由图 11 可知,在这一年 6 次观测值各处理主成分分析得分中,排名前 5 名的处理为 9、6、3、8、5,其中除了第 2、4 个月最高分为处理 6(A_2B_3)外,其余月份中处理 9(A_3B_3)得分

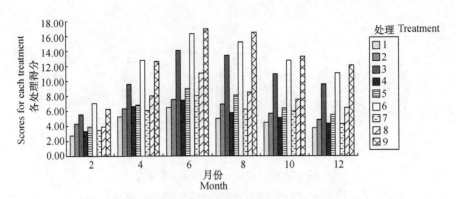

图 11　一年中各处理在主成分分析法中的得分
Fig. 11　The score of each treatment in the method of chief component analysis over a year

最高,前 5 位排名大致为:9＞6＞3＞8＞5;而处理 1 则在每次观测中都处于最低分;4 号处理在倒数第 2 名的位置出现 4 次,出现频率最高,故得分倒数第 2 名是 4 号处理;至于处理 2 和处理 7,两者均在第 6 位的位置上出现两次,而处理 2 在第 7 位上出现 3 次,在第 8 位上不出现,处理 7 在第 7 位上出现 2 次,在第 8 位上出现 2 次,故大致判断,处理 2 优于处理 7。综合分析各处理在每个月的表现,排名大致为:9＞6＞3＞8＞5＞2＞7＞4＞1。这大致可以说明,料液比对浸提的影响大于酒精度,料液比越大,得分排名越靠前;在相同的料液比条件下,酒精浓度越高,其得分排名越靠前。

对于浸提时间而言,6~8 个月是得分最高的区段。得分最高的处理为第 6 个月的 9 号处理(A_3B_3),其各营养成分的含量为:总黄酮 1.14 mg/mL、花青素 7.49 mg/L、维生素 C 3.99 mg/100 mL、总糖 2.58 g/L、Fe 1.14 mg/L、Mn 0.17 mg/L、Zn 0.94 mg/L、Cu 0.23 mg/L、Mg 7.69 mg/L、Ca 2.91 mg/L。

由此可知,桂花露酒浸提最佳处理为 9 号(A_3B_3),浸提最佳时间为 6 个月。即在酒精度为 66%(V/V),料液比为 1∶20(kg∶L)的条件下浸提半年,桂花露酒中营养成分浸出效果最佳。

主成分分析法应用范围非常广泛,被用于各行各业[11-15],郭宝林等[16]对仁用杏选种,冯娟[17]对富士苹果果实品质的评价,刘科鹏等[18]对猕猴桃果实品质的评价,濮玲等[19]对中草药中微量元素的评价均采用主成分分析法。而在酒类营养成分综合评价上,主成分分析法的应用还未见报道。本研究表明主成分分析法可作为评价酒类营养价值或品质的综合分析方法,并值得推广使用。但对于桂花露酒,感官评价是极其重要的,本试验只简单进行了口感尝试,并未系统进行安全、美味评价,以后可采取主成分分析与感官评价相结合的方式进行综合筛选。

3　结论

本试验以不同料液比和酒精度,将桂花花瓣浸泡于白酒中,制备不同处理的桂花露酒,经过一年浸泡后的桂花露酒,色泽浅黄或金黄色,酒质清亮透明,无悬浮沉淀物,具有桂花清香,且口感醇和,用传统的物质测定方法可以准确地测定酒样中各营养成分的含量。两因素随机区组试验结果表明各种营养成分的最佳浸提条件不尽相同,而检测出来的重金属含量都极低,说明了所有酒样健康无害;用主成分分析法对各处理桂花露酒品质进行综合评价,可以选出综合得分最高的浸提处理条件,此法所得的结论与传统分析方法所得结论基本一致,说明本研究中基于评价函数的评价结果与前面的试验测定结果有很好的一致性。这也是该方法在露酒研究领域中的首次应用。

总的来说,桂花露酒在桂花浸泡过程中溶入了桂花中极有营养价值的黄酮、花青素、维生素C、糖类以及多种有益矿质元素,大大增加了酒体的营养价值,为酒业发展及桂花应用提供了新的思路。

参考文献

[1] 中华人民共和国国家质量监督检验检疫总局,中国国家标准化管理委员会.露酒 GB/T27 588—2011[S].北京:中国标准出版社,2012.
[2] 高海生,柴菊华,张建才,侍朋宝.安梨酒的酿造工艺及营养成分分析[J].食品与发酵工业,2006,32(12):73-76.
[3] 李瑶,周文化,普义鑫.大蒜露酒最佳浸提工艺条件研究[J].中国酿造,2010,29(9):52-54.
[4] 鲍晓华,毕廷菊,李秀,等.野生棠梨果露酒的研制[J].安徽农业科学,2011(11):6526-6528.
[5] 方堃,陆胜民,夏其乐.蓝莓露酒主要成分变化及浸提条件优化[J].食品安全质量检测学报,2013(6):1769-1777.
[6] 张翔,张华峰,牛丽丽,等.淫羊藿保健露酒的研制及其功能评价[J].陕西师范大学学报(自然科学版),2013(5):103-108.
[7] 张玲,曾婉玲,李春海.菠萝蜜露酒加工工艺研究[J].食品研究与开发,2015(4):94-98.
[8] 靳熙茜,汪海波.桂花总黄酮提取及其体外抗氧化性能研究[J].粮食与油脂,2009(11):42-45.
[9] 陈培珍,林志銮,胡燕萍.桂花总黄酮的提取及其抗氧化活性研究[J].云南民族大学学报(自然科学版),2014(4):239-242.
[10] 陈培珍,林志銮,苏丽鳗.超声波辅助提取桂花总黄酮及其抗氧化活性[J].湖北农业科学,2013(20):5023-5025.
[11] 聂宏展,聂耸,乔怡,等.基于主成分分析法的输电网规划方案综合决策[J].电网技术,2010(6):134-138.
[12] 刘臣辉,吕信红,范海燕.主成分分析法用于环境质量评价的探讨[J].环境科学与管理,2011(3):183-186.
[13] 黄古博,李雨真.基于主成分分析法的商品住宅特征价格模型改进[J].华中农业大学学报(社会科学版),2011(4):93-97.
[14] 杨永恒,胡鞍钢,张宁.基于主成分分析法的人类发展指数替代技术[J].经济研究,2005(7):4-17.
[15] 万金保,曾海燕,朱邦辉.主成分分析法在乐安河水质评价中的应用[J].中国给水排水,2009(16):104-108.
[16] 郭宝林,杨俊霞,李永慈,等.主成分分析法在仁用杏品种主要经济性状选种上的应用研究[J].林业科学,2000(6):53-56.
[17] 冯娟.不同产地富士苹果果实品质分析与比较[D].银川:宁夏大学,2013.
[18] 刘科鹏,黄春辉,冷建华,等.'金魁'猕猴桃果实品质的主成分分析与综合评价[J].果树学报,2012(5):867-871.
[19] 濮玲,李海朝,濮御,等.十二种常用中草药中微量元素的主成分分析[J].广西师范大学学报(自然科学版),2014(4):96-100.

Study on *Osmanthus fragrans* Wine's Extraction and Nutrient Components

YANG Xiulian, CHANG Zhaojing, FENG Jie, SHI Rui, SHI Tingting, WANG Lianggui

(College of Landscape Architecture, Nanjing Forestry University, Nanjing 210037, China)

Abstract: In order to explore the different extractive conditions and nutrient components of *Osmanthus fragrans* wine, we took the white spirit and fresh petals of *Osmanthus fragrans* to produce the different concentrations *Osmanthus fragrans* wine using different alcoholic degrees and solid-liquid ratios. Then, obtained the best extractive condition of *Osmanthus fragrans* wine by principal component analysis. The results showed that the best *Osmanthus fragrans* wine extractive condition was soaking the petals into the 66 degree alcoholic under the 1 : 20 (kg : L) solid-liquid ratio for six months. In this condition, the contents of each nutrient components were the highest (total flavonoids 1.14 g, anthocyanin 7.49 mg, vitamin C 39.9 mg, total sugar 2.58 g, Fe 1.14 mg, Mn 0.17 mg, Zn 0.94 mg, Cu 0.23 mg, Mg 7.69 mg, CA 2.91 mg). Taken together, this study has achieved the preliminary screening of *Osmanthus fragrans* wine extractive condition and would be helpful for the further exploration of *Osmanthus fragrans* wine.

Key words: *Osmanthus fragrans* wine; alcoholic; solid-liquid ratio; nutrient component

桂花专类园植物景观综合评价

徐 晨,王良桂,杨秀莲

(南京林业大学 风景园林学院,江苏 南京 210037)

摘 要:选择长江流域建成多年的3个桂花专类园为研究对象,利用层次分析法,以景观质量、生态效益和服务功能3个准则层为主,建立了桂花专类园的植物景观评价体系;并对30个群落样地进行了景观评价。结果表明,准则层的权重值从高到低为:生态效益＞景观质量＞服务功能,综合评价指数的平均值从高到低为上海桂林公园(82.60％)＞南京灵谷寺公园(78.10％)＞苏州桂花公园(68.60％)。结果显示:该评价体系适合于植物专类园的植物景观评价,尤其是可利用定量分析的评价结果来更好地指导植物专类园的植物景观设计。

关键词:桂花专类园;植物景观;综合评价

随着城市园林事业的蓬勃发展,植物造景已经成为园林建设的主流,以独立性质的专类园造景形式在城市园林和风景区中也已非常普遍,出现了规模大小不等的植物专类园,有些则是在传统专类栽培的基础上加以完善而形成[1-7]。植物专类园主要有2种布局形式:其一是以植物纯林布局造景,其二是与其他园林植物配合布置[8]。但目前鲜有文献涉及植物专类园植物景观的综合评价,笔者选择桂花专类园作为研究对象,设置景观质量、生态效益、服务功能3个准则层,包含12个因子层,运用层次分析法(AHP),通过案例分析,介绍评价方法的应用过程[9-12]。

1 材料与方法

1.1 样地选择

本研究的案例在长江流域建成多年的桂花专类园:上海桂林公园、苏州桂花公园、南京灵谷寺公园,在3个桂花专类园中分别选取10个植物群落样地进行评价(表1)。选择的样方在研究地区内均匀分布,力求较为全面展现植物群落的状况,并不特定选择较优秀或者较差的植物样方,选取的样方具有普遍意义。

表1 桂花专类园公园概况及群落样地数量

公园名称	公园类型	群落样地数量/个
上海桂林公园	江南古典园林	10
苏州桂花公园	城市滨水公园	10
南京灵谷寺公园	纪念性园林	10

---

* 原文发表于《福建林业科技》,2013,40(4):123-125。
基金项目:国家科技部十二五科技支撑计划(2013BAD01B06-4);国家林业局公益性行业科研专项项目(201204607)。

1.2 评价体系的建立

1.2.1 体系的确定

应用AHP方法,结合植物专类园的特色,选择若干对专类园植物景观效果贡献较大的定性和定量指标(表2),建立桂花专类园的植物景观评价体系,对30个样地进行景观评价,并根据综合评价指数对它们进行分级。多层次评价系统分为:最高层是综合评价的目标层(A)、第2层为确定综合评价值的主要原则,即评价的准则层(B)、第3层为隶属各主要构成要素的评价因子层(C)[13-15]。

1.2.2 权重的确定

按照综合评价模型建立的层次结构关系,进行判断比较,分别构成A—B、B_1—C、B_2—C、B_3—C两两比较判断矩阵,得到层次单排序后,再对判断矩阵进行一致性检验,确保CR值小于0.1,保证数据有效。最后根据各判断矩阵得出桂花主题园景观评价准则层与因子层各指标权重值。

表2 各评价因子的含义

评价因子	含义
植物形态	植物形态的观赏性和植物的韵律美
色彩与季相	色彩变化的丰富度,季相变化的鲜明度
绿视率	绿色在人视野中的比例大小
枯落物	枯落物的多少和分布
层次丰富度	植物群落的层次多样性
群落稳定性	植物群落的异龄程度,更新的幼苗数量
物种多样性	高等植物物种的丰富程度
植物品种多样性	植物群落内植物品种的数量
群落的乡土性	植物对绿地环境的生长适应性,乡土植物所占的比例大小
群落的合理程度	与周围硬质景观协调性,具有愉悦和舒适的美感
绿地可达性	绿地周围的游步道数量,植物郁闭度对游人产生的影响
养护管理程度	群落依靠自身进行维持的程度

2 结果与分析

2.1 评价指标权重值分析

在已建立的植物景观评价体系中,评价因子的权重值各不相同,即各评价因子对植物景观的重视程度不同。从表3可以看出,准则层的权重值从高到低的排序为生态效益、景观质量、服务功能,说明在植物造景过程中,首要必须在满足生态效益基础上才能营造出具有美感的景观,服务功能的权重较低,说明人们更重视植物营造出来的景观效果。

在生态效益评价中,物种多样性和群落的合理程度对植物景观的贡献度较大,决定了景观质量的好坏;在景观质量评价中,植物形态和植物层次丰富度给人以最为直观的感受,当植物展示出美的形态,才能发挥出景观的重要功能;在服务功能评价中,绿地的可达性比养护管理更为重要,强调了人类亲近自然的需求。权重值的高低反映了专家对植物专类园的植物景观评价因子的取向,同时也体现出专类园植物景观美的构成。

表3 准则层与因子层各指标权重

目标层(A)	目标层权重	准则层(B)	准则层权重	因子层(C)	因子层权重
桂花专类园的植物景观评价	1	景观质量评价(B_1)	0.369	植物形态(C_1)	0.114
				色彩与季相(C_2)	0.080
				绿视率(C_3)	0.030
				枯落物(C_4)	0.016
				植物层次丰富度(C_5)	0.129
		生态效益评价(B_2)	0.473	群落稳定性(C_6)	0.047
				物种多样性(C_7)	0.095
				植物品种多样性(C_8)	0.142
				群落的乡土性(C_9)	0.063
				群落的合理程度(C_{10})	0.126
		服务功能评价(B_3)	0.158	绿地可达性(C_{11})	0.095
				养护管理程度(C_{12})	0.063

2.2 评价结果分析

根据各指标的权重值,利用方程:$V = \sum^{n} B\omega$,式中:V为综合得分;B为因子评分;ω为因子权重值;n为因子数。计算出各群落样地的植物景观综合评价值。然后再利用公式$CEI = S/S_0 \times 100\%$,式中:S为评价分数值;S_0为理想值。计算出综合评价指数,以差值百分比分级法分出各样地的评价等级,并求出综合评价指数的平均值(表4)。

表4 植物群落景观综合评价指数(CEI)平均值

公园名称	综合评价指数平均值/%	Ⅰ级植物景观数量	Ⅱ级植物景观数量	Ⅲ级植物景观数量
上海桂林公园	80.60	2	6	2
苏州桂花公园	68.60	2	2	6
南京灵谷寺公园	78.10	2	7	1

由表4可知,Ⅰ级植物群落景观有6个,每个专类园各占2个;Ⅱ级植物群落景观15个;Ⅲ级植物群落景观9个。Ⅰ级植物景观群落的特点主要有:①植物种类丰富,配置合理,形成乔、灌、草相搭配,层次丰富的植物景观。②植物品种较多,即单个样地内栽植了不同品种的桂花,提高了植物的观赏性,体现出植物的个体美。③植物群落稳定性高,运用了较多的乡土树种,具有良好的稳定性。上海桂林公园的综合评价指数的平均值为82.60%,苏州桂花公园的综合评价指数的平均值68.60%,南京灵谷寺公园的综合评价指数的平均值为78.10%。由此可以看出,上海桂林公园作为桂花专类园的代表,园区内的植物配置较为精致细腻,且群落样地内桂花品种较多,因此植物综合评价的分值较高。

3　小结

确定植物专类园植物景观的主要影响因子,采用 AHP 层次分析法,对桂花专类园的 30 个群落样地进行景观评价,建立了适于植物专类园的植物景观评价体系。分析结果表明,植物专类园的植物造景首先应具有较多的植物品种,起到收集种质资源的作用;通过孤植、列植、群植等布局形式展示植物专类园的形、神、韵;注重植物物种的多样性,选择与其他观花、观叶植物配置,避免单一的植物群落带来美感度上的缺失,营造出具有季相变化的植物景观;注重植物生态性,合理配置乔木、灌木、草本植物等,建立疏密有致、景观层次丰富的立体植物群落。

参考文献

[1] 苏雪痕.植物造景[M].北京:中国林业出版社,1994.
[2] 朱钧珍.中国园林植物景观艺术[M].北京:中国建筑工业出版社,2003.
[3] 王磊,汤庚国.植物造景的基本原理及应用[J].林业科技开发,2003(5):71-73.
[4] 汤珏,包志毅.植物专类园的类别和应用[J].风景园林,2005(1):61-64.
[5] 胡永红.专类园在植物园中的地位和作用及对上海辰山植物园专类园设置的启示[J].中国园林,2006(7):50-55.
[6] 王文姬.无锡市植物专类园建设[J].中国园林,2008(12):39-44.
[7] 黄仕训,周太久,骆文华,等.桂林植物园的专类园建设方法和特点[J].广西科学院学报,2007(1):49-53.
[8] 向其柏,刘玉莲.中国桂花品种图志[M].杭州:浙江科学技术出版社,2009.
[9] 宁惠娟,邵锋,孙茜茜,等.基于 AHP 法的杭州花港观鱼公园植物景观评价[J].浙江农业学报,2011,23(4):717-724.
[10] 芦建国,李舒仪.公园植物景观综合评价方法及其应用[J].南京林业大学学报:自然科学版,2009(6):139-142.
[11] 庞玉成.基于 AHP 和模糊综合评价的高校校园规划评价研究[J].青岛理工大学学报,2009(6):53-60.
[12] 邹建勤,宋丁全.定量 AHP 模型在城市居住区植物景观评价中的应用[J].金陵科技学院学报,2009(1):66-69.
[13] 徐晨.桂花主题园的植物景观研究[D].南京:南京林业大学,2010.
[14] 刘清.明孝陵景区植物景观研究[D].南京:南京林业大学,2009.
[15] 李舒仪.南京玄武湖公园植物景观评价与优化[D].南京:南京林业大学,2009.

Study on the Comprehensive Assessment of Osmanthus Specialized Park Plant Landscape

XU Chen, WANG Lianggui, YANG Xiulian

(College of Landscape Architecture, Nanjing Forestry University, Nanjing 210037, China)

Abstract: Using the AHP method and 3 criteria layers of landscape quality, ecological efficiency and service function as the main objects, taking the *Osmanthus* special garden as the research objects, the plant landscape evaluation system of specialized park was established in this paper. And, this paper evaluated the landscape of 30 community sample plots. The result indicated that criteria layer weight value order was ecological benefit > landscape quality > service function, the synthesized evaluation index average value order was Shanghai Guilin Park(82.60%) > Nanjing Linggu Temple Park(78.10%) > Suzhou Guihua Park (68.60%). The results showed the evaluation system was suitable for the plant landscape evaluation of specialized park, especially could be used to guide the plant specialized park design by utilizing the result of quantitative analysis.

Key words: osmanthus specialized park; plant landscape; comprehensive assessment

图书在版编目(CIP)数据

桂花研究/杨秀莲等编著. —南京:东南大学出版社,2018.12
 ISBN 978-7-5641-8209-0

Ⅰ.①桂… Ⅱ.①杨… Ⅲ.①木犀-研究-文集 Ⅳ.①S685.13-53

中国版本图书馆 CIP 数据核字(2018)第 293981 号

桂花研究 GUIHUA YANJIU

编　　著:	杨秀莲　王良桂等
出版发行:	东南大学出版社
社　　址:	南京市四牌楼 2 号　邮编:210096
出 版 人:	江建中
责任编辑:	姜　来　朱震霞　李　婧
网　　址:	http://www.seupress.com
电子邮箱:	press@seupress.com
经　　销:	全国各地新华书店
印　　刷:	江苏凤凰数码印务有限公司
开　　本:	787 mm×1092 mm　1/16
印　　张:	25.75
字　　数:	724 千字
版　　次:	2018 年 12 月第 1 版
印　　次:	2018 年 12 月第 1 次印刷
书　　号:	ISBN 978-7-5641-8209-0
定　　价:	79.00 元

本社图书若有印装质量问题,请直接与营销部联系。电话:025-83791830